CONVERSION FACTORS FROM ENGLISH TO SI UNITS

Length:

1 ft	=	0.3048 m
1 ft	=	30.48 cm
1 ft	=	304.8 mm
1 in.	=	0.0254 m
1 in.	=	2.54 cm
1 in.	=	25.4 mm

Area:

$1\ ft^2$	=	$929.03 \times 10^{-4}\ m^2$
$1\ ft^2$	=	$929.03\ cm^2$
$1\ ft^2$	=	$929.03 \times 10^2\ mm^2$
$1\ in.^2$	=	$6.452 \times 10^{-4}\ m^2$
$1\ in.^2$	=	$6.452\ cm^2$
$1\ in.^2$	=	$645.16\ mm^2$

Volume:

$1\ ft^3$	=	$28.317 \times 10^{-3}\ m^3$
$1\ ft^3$	=	$28.317 \times 10^3\ cm^3$
$1\ in.^3$	=	$16.387 \times 10^{-6}\ m^3$
$1\ in.^3$	=	$16.387\ cm^3$

Section modulus:

$1\ in.^3$	=	$0.16387 \times 10^5\ mm^3$
$1\ in.^3$	=	$0.16387 \times 10^{-4}\ m^3$

Hydraulic conductivity:

1 ft/min	=	0.3048 m/min
1 ft/min	=	30.48 cm/min
1 ft/min	=	304.8 mm/min
1 ft/sec	=	0.3048 m/sec
1 ft/sec	=	304.8 mm/sec
1 in./min	=	0.0254 m/min
1 in./sec	=	2.54 cm/sec
1 in./sec	=	25.4 mm/sec

Coefficient of consolidation:

$1\ in.^2/sec$	=	$6.452\ cm^2/sec$
$1\ in.^2/sec$	=	$20.346 \times 10^3\ m^2/yr$
$1\ ft^2/sec$	=	$929.03\ cm^2/sec$

Force:

1 lb	=	4.448 N
1 lb	=	$4.448 \times 10^{-3}\ kN$
1 lb	=	0.4536 kgf
1 kip	=	4.448 kN
1 U.S. ton	=	8.896 kN
1 lb	=	0.4536×10^{-3} metric ton
1 lb/ft	=	14.593 N/m

Stress:

$1\ lb/ft^2$	=	$47.88\ N/m^2$
$1\ lb/ft^2$	=	$0.04788\ kN/m^2$
$1\ U.S.\ ton/ft^2$	=	$95.76\ kN/m^2$
$1\ kip/ft^2$	=	$47.88\ kN/m^2$
$1\ lb/in.^2$	=	$6.895\ kN/m^2$

Unit weight:

$1\ lb/ft^3$	=	$0.1572\ kN/m^3$
$1\ lb/in.^3$	=	$271.43\ kN/m^3$

Moment:

1 lb-ft	=	$1.3558\ N \cdot m$
1 lb-in.	=	$0.11298\ N \cdot m$

Energy:

1 ft-lb	=	1.3558 J

Moment of inertia:

$1\ in.^4$	=	$0.4162 \times 10^6\ mm^4$
$1\ in.^4$	=	$0.4162 \times 10^{-6}\ m^4$

Principles of Geotechnical Engineering

Seventh Edition

SI Edition

BRAJA M. DAS

CENGAGE
Learning™

Australia • Brazil • Japan • Korea • Mexico • Singapore • Spain • United Kingdom • United States

CENGAGE
Learning™

Principles of Geotechnical Engineering, 7th Edition
SI Edition
Braja M. Das

SI Author
Braja M. Das

Executive Director, Global Publishing Program:
Chris Carson

Senior Developmental Editor: Hilda Gowans

Editorial Assistant: Nancy Saundercook

Associate Marketing Manager: Lauren Betsos

Director, Content and Media Production:
Barbara Fuller Jacobsen

Content Project Manager: Emily Nesheim

Production Service: RPK Editorial Services

Copyeditor: Shelly Gerger-Knechtl

Proofreader: Nancy Benziger

Indexer: Braja M. Das

Compositor: Integra Software Services

Senior Art Director: Michelle Kunkler

Internal Designer: Carmela Pereira

Cover Designer: Andrew Adams

Cover Images: H. Turan Durgunoglu, ZETAS,
Istanbul, Turkey

Permissions Account Manager, Text:
Mardell Glinski Shultz

Permissions Account Manager, Images:
Deanna Ettinger

Text and Images Permissions Researcher:
Kristiina Paul

Senior First Print Buyer: Doug Wilke

For product information and technology assistance, contact us at
Cengage Learning Customer & Sales Support, 1-800-354-9706.

For permission to use material from this text or product,
submit all requests online at **www.cengage.com/permissions.**
Further permissions questions can be emailed to
permissionrequest@cengage.com.

Library of Congress Control Number: 2009905986

ISBN-13: 978-0-495-41132-1

ISBN-10: 0-495-41132-9

Cengage Learning
200 First Stamford Place, Suite 400
Stamford, CT 06902
USA

Cengage Learning is a leading provider of customized learning
solutions with office locations around the globe, including
Singapore, the United Kingdom, Australia, Mexico, Brazil, and Japan.
Locate your local office at: **international.cengage.com/region.**

Cengage Learning products are represented in Canada by
Nelson Education Ltd.

For your course and learning solutions, visit
www.cengage.com/engineering.

Purchase any of our products at your local college store or at our
preferred online store **www.ichapters.com.**

Printed in the United States of America

2 3 4 5 6 7   13 12 11 10

To our granddaughter Elizabeth Madison

Contents

Preface

Principles of Geotechnical Engineering was originally published with a 1985 copyright and was intended for use as a text for the introductory course in geotechnical engineering taken by practically all civil engineering students, as well as for use as a reference book for practicing engineers. The book was revised in 1990, 1994, 1998, 2002, and 2006. This Seventh Edition is the twenty-fifth anniversary edition of the text. As in the previous editions of the book, this new edition offers an overview of soil properties and mechanics, together with coverage of field practices and basic engineering procedures without changing the basic philosophy in which the text was written originally.

Unlike the Sixth Edition that had 17 chapters, this edition has 18 chapters. For better understanding and more comprehensive coverage, Weight-Volume Relationships and Plasticity and Structure of Soil are now presented in two separate chapters (Chapters 3 and 4). Most of the example and homework problems have been changed and/or modified. All units in this text are in SI, including the example and homework problems. Other noteworthy changes for the Seventh Edition are

- New scanning electron micrographs for quartz, mica, limestone, sand grains, and clay minerals such as kaolinite and montmorillonite have been added to Chapter 2.
- A summary of recently published empirical relationships between liquid limit, plastic limit, plasticity index, activity, and clay-size fractions in a soil mass have been incorporated in Chapter 4.
- The USDA Textural Classification of Soil has now been added to Chapter 5 (Classification of Soil).
- Additional empirical relationships for hydraulic conductivity for granular and cohesive soils have been added, respectively, to Chapter 7 (Permeability) and Chapter 17 (Landfill Liners and Geosynthetics).
- The presentation of the filter design criteria has been improved in Chapter 8 (Seepage).
- In Chapter 11 (Compressibility of Soil), the procedure for estimating elastic settlement of foundations has been thoroughly revised with the inclusions of theories by Steinbrenner (1934) and Fox (1948). A case study related to the consolidation settlement due to a preload fill for construction of the Tampa VA Hospital is also added to this chapter.

- The presentation on estimation of active force on retaining walls with earthquake forces in Chapter 13 (Lateral Earth Pressure: At-Rest, Rankine, and Coulomb) has been improved.
- Chapter 14 (Lateral Earth Pressure: Curved Failure Surface) now includes the procedure to estimate the passive earth pressure on retaining walls with inclined backface and horizontal granular backfill using the method of triangular slices. It also includes the relationships for passive earth pressure on retaining walls with a horizontal granular backfill and vertical backface under earthquake conditions determined by using the pseudo-static method.
- Chapter 15 (Slope Stability) now includes a case history of slope failure in relation to a major improvement program of Interstate Route 95 in New Hampshire.
- A method to calculate the ultimate bearing capacity of eccentrically loaded shallow strip foundations in granular soil using the reduction factor has been added to Chapter 16 (Soil-Bearing Capacity for Shallow Foundations).

I am grateful to my wife Janice for her help in getting the manuscript ready for publication. Finally, many thanks are due to Christopher Carson, Executive Director, Global Publishing Programs; Hilda Gowans, Senior Development Editor; and the production staff of Cengage Learning (Engineering) for the final development and production of the book.

Braja M. Das
Henderson, Nevada

1 Geotechnical Engineering— A Historical Perspective

For engineering purposes, *soil* is defined as the uncemented aggregate of mineral grains and decayed organic matter (solid particles) with liquid and gas in the empty spaces between the solid particles. Soil is used as a construction material in various civil engineering projects, and it supports structural foundations. Thus, civil engineers must study the properties of soil, such as its origin, grain-size distribution, ability to drain water, compressibility, shear strength, and load-bearing capacity. *Soil mechanics* is the branch of science that deals with the study of the physical properties of soil and the behavior of soil masses subjected to various types of forces. *Soils engineering* is the application of the principles of soil mechanics to practical problems. *Geotechnical engineering* is the subdiscipline of civil engineering that involves natural materials found close to the surface of the earth. It includes the application of the principles of soil mechanics and rock mechanics to the design of foundations, retaining structures, and earth structures.

1.1 *Geotechnical Engineering Prior to the 18th Century*

The record of a person's first use of soil as a construction material is lost in antiquity. In true engineering terms, the understanding of geotechnical engineering as it is known today began early in the 18th century (Skempton, 1985). For years, the art of geotechnical engineering was based on only past experiences through a succession of experimentation without any real scientific character. Based on those experimentations, many structures were built—some of which have crumbled, while others are still standing.

Recorded history tells us that ancient civilizations flourished along the banks of rivers, such as the Nile (Egypt), the Tigris and Euphrates (Mesopotamia), the Huang Ho (Yellow River, China), and the Indus (India). Dykes dating back to about 2000 B.C. were built in the basin of the Indus to protect the town of Mohenjo Dara (in what became Pakistan after 1947). During the Chan dynasty in China (1120 B.C. to 249 B.C.) many dykes were built for irrigation purposes. There is no evidence that measures were taken to stabilize the foundations or check erosion caused by floods (Kerisel, 1985). Ancient Greek civilization used isolated pad footings and strip-and-raft foundations for building structures. Beginning around 2750 B.C., the five most important pyramids were built in Egypt in a period of less than a century (Saqqarah, Meidum, Dahshur South and North, and Cheops). This posed formidable challenges regarding foundations, stability of slopes, and construction of

1

underground chambers. With the arrival of Buddhism in China during the Eastern Han dynasty in 68 A.D., thousands of pagodas were built. Many of these structures were constructed on silt and soft clay layers. In some cases the foundation pressure exceeded the load-bearing capacity of the soil and thereby caused extensive structural damage.

One of the most famous examples of problems related to soil-bearing capacity in the construction of structures prior to the 18th century is the Leaning Tower of Pisa in Italy. (See Figure 1.1.) Construction of the tower began in 1173 A.D. when the Republic of Pisa was flourishing and continued in various stages for over 200 years. The structure weighs about 15,700 metric tons and is supported by a circular base having a diameter of 20 m. The tower has tilted in the past to the east, north, west and, finally, to the south. Recent investigations showed that a weak clay layer exists at a depth of about 11 m below the ground surface compression of which caused the tower to tilt. It became more than 5 m out of plumb with the 54 m height. The tower was closed in 1990 because it was feared that it would either fall over or collapse. It recently has been stabilized by excavating soil from under the north side of the tower. About 70 metric tons of earth were removed in 41 separate extractions that spanned the width of the tower. As the ground gradually settled

Figure 1.1 Leaning Tower of Pisa, Italy (*Courtesy of Braja M. Das, Henderson, Nevada*)

Figure 1.2 Tilting of Garisenda Tower (left) and Asinelli Tower (right) in Bologna, Italy (*Courtesy of Braja M. Das, Henderson, Nevada*)

to fill the resulting space, the tilt of the tower eased. The tower now leans 5 degrees. The half-degree change is not noticeable, but it makes the structure considerably more stable. Figure 1.2 is an example of a similar problem. The towers shown in Figure 1.2 are located in Bologna, Italy, and they were built in the 12th century. The tower on the left is usually referred to as the *Garisenda Tower*. It is 48 m in height and weighs about 4210 metric tons. It has tilted about 4 degrees. The tower on the right is the Asinelli Tower, which is 97 m high and weighs 7300 metric tons. It has tilted about 1.5 degrees.

After encountering several foundation-related problems during construction over centuries past, engineers and scientists began to address the properties and behaviors of soils in a more methodical manner starting in the early part of the 18th century. Based on the emphasis and the nature of study in the area of geotechnical engineering, the time span extending from 1700 to 1927 can be divided into four major periods (Skempton, 1985):

1. Pre-classical (1700 to 1776 A.D.)
2. Classical soil mechanics—Phase I (1776 to 1856 A.D.)
3. Classical soil mechanics—Phase II (1856 to 1910 A.D.)
4. Modern soil mechanics (1910 to 1927 A.D.)

Brief descriptions of some significant developments during each of these four periods are discussed below.

Preclassical Period of Soil Mechanics (1700–1776)

This period concentrated on studies relating to natural slope and unit weights of various types of soils, as well as the semiempirical earth pressure theories. In 1717 a French royal engineer, Henri *Gautier* (1660–1737), studied the natural slopes of soils when tipped in a heap for formulating the design procedures of retaining walls. The *natural slope* is what we now refer to as the *angle of repose*. According to this study, the natural slope of *clean dry sand* and *ordinary earth* were 31° and 45°, respectively. Also, the unit weight of clean dry sand and ordinary earth were recommended to be 18.1 kN/m³ and 13.4 kN/m³, respectively. No test results on clay were reported. In 1729, Bernard Forest de Belidor (1671–1761) published a textbook for military and civil engineers in France. In the book, he proposed a theory for lateral earth pressure on retaining walls that was a follow-up to Gautier's (1717) original study. He also specified a soil classification system in the manner shown in the following table.

Classification	Unit weight (kN/m³)
Rock	—
Firm or hard sand	16.7 to
Compressible sand	18.4
Ordinary earth (as found in dry locations)	13.4
Soft earth (primarily silt)	16.0
Clay	18.9
Peat	—

The first laboratory model test results on a 76-mm-high retaining wall built with sand backfill were reported in 1746 by a French engineer, Francois Gadroy (1705–1759), who observed the existence of slip planes in the soil at failure. Gadroy's study was later summarized by J. J. Mayniel in 1808.

Classical Soil Mechanics—Phase I (1776–1856)

During this period, most of the developments in the area of geotechnical engineering came from engineers and scientists in France. In the preclassical period, practically all theoretical considerations used in calculating lateral earth pressure on retaining walls were based on an arbitrarily based failure surface in soil. In his famous paper presented in 1776, French scientist Charles Augustin Coulomb (1736–1806) used the principles of calculus for maxima and minima to determine the true position of the sliding surface in soil behind a retaining wall. In this analysis, Coulomb used the laws of friction and cohesion for solid bodies. In 1820, special cases of Coulomb's work were studied by French engineer Jacques Frederic Francais (1775–1833) and by French applied mechanics professor Claude Louis Marie Henri Navier

(1785–1836). These special cases related to inclined backfills and backfills supporting surcharge. In 1840, Jean Victor Poncelet (1788–1867), an army engineer and professor of mechanics, extended Coulomb's theory by providing a graphical method for determining the magnitude of lateral earth pressure on vertical and inclined retaining walls with arbitrarily broken polygonal ground surfaces. Poncelet was also the first to use the symbol ϕ for soil friction angle. He also provided the first ultimate bearing-capacity theory for shallow foundations. In 1846 Alexandre Collin (1808–1890), an engineer, provided the details for deep slips in clay slopes, cutting, and embankments. Collin theorized that in all cases the failure takes place when the mobilized cohesion exceeds the existing cohesion of the soil. He also observed that the actual failure surfaces could be approximated as arcs of cycloids.

The end of Phase I of the classical soil mechanics period is generally marked by the year (1857) of the first publication by William John Macquorn Rankine (1820–1872), a professor of civil engineering at the University of Glasgow. This study provided a notable theory on earth pressure and equilibrium of earth masses. Rankine's theory is a simplification of Coulomb's theory.

1.4 Classical Soil Mechanics—Phase II (1856–1910)

Several experimental results from laboratory tests on sand appeared in the literature in this phase. One of the earliest and most important publications is one by French engineer Henri Philibert Gaspard Darcy (1803–1858). In 1856, he published a study on the permeability of sand filters. Based on those tests, Darcy defined the term *coefficient of permeability* (or hydraulic conductivity) of soil, a very useful parameter in geotechnical engineering to this day.

Sir George Howard Darwin (1845–1912), a professor of astronomy, conducted laboratory tests to determine the overturning moment on a hinged wall retaining sand in loose and dense states of compaction. Another noteworthy contribution, which was published in 1885 by Joseph Valentin Boussinesq (1842–1929), was the development of the theory of stress distribution under loaded bearing areas in a homogeneous, semiinfinite, elastic, and isotropic medium. In 1887, Osborne Reynolds (1842–1912) demonstrated the phenomenon of dilatency in sand.

1.5 Modern Soil Mechanics (1910–1927)

In this period, results of research conducted on clays were published in which the fundamental properties and parameters of clay were established. The most notable publications are described below.

Around 1908, Albert Mauritz Atterberg (1846–1916), a Swedish chemist and soil scientist, defined *clay-size fractions* as the percentage by weight of particles smaller than 2 microns in size. He realized the important role of clay particles in a soil and the plasticity thereof. In 1911, he explained the consistency of cohesive soils by defining liquid, plastic, and shrinkage limits. He also defined the plasticity index as the difference between liquid limit and plastic limit (see Atterberg, 1911).

In October 1909, the 17-m high earth dam at Charmes, France, failed. It was built between 1902–1906. A French engineer, Jean Fontard (1884–1962), carried out investigations to determine the cause of failure. In that context, he conducted undrained double-shear tests on clay specimens (0.77 m^2 in area and 200 mm thick) under constant vertical stress to determine their shear strength parameters (see Frontard, 1914). The times for failure of these specimens were between 10 to 20 minutes.

Arthur Langley Bell (1874–1956), a civil engineer from England, worked on the design and construction of the outer seawall at Rosyth Dockyard. Based on his work, he developed relationships for lateral pressure and resistance in clay as well as bearing capacity of shallow foundations in clay (see Bell, 1915). He also used shear-box tests to measure the undrained shear strength of undisturbed clay specimens.

Wolmar Fellenius (1876–1957), an engineer from Sweden, developed the stability analysis of saturated clay slopes (that is, $\phi = 0$ condition) with the assumption that the critical surface of sliding is the arc of a circle. These were elaborated upon in his papers published in 1918 and 1926. The paper published in 1926 gave correct numerical solutions for the *stability numbers* of circular slip surfaces passing through the toe of the slope.

Karl Terzaghi (1883–1963) of Austria (Figure 1.3) developed the theory of consolidation for clays as we know today. The theory was developed when Terzaghi was teaching

Figure 1.3 Karl Terzaghi (1883–1963) (*Photo courtesy of Ralph B. Peck*)

at the American Roberts College in Istanbul, Turkey. His study spanned a five-year period from 1919 to 1924. Five different clay soils were used. The liquid limit of those soils ranged between 36 to 67, and the plasticity index was in the range of 18 to 38. The consolidation theory was published in Terzaghi's celebrated book *Erdbaumechanik* in 1925.

1.6   *Geotechnical Engineering after 1927*

The publication of *Erdbaumechanik auf Bodenphysikalisher Grundlage* by Karl Terzaghi in 1925 gave birth to a new era in the development of soil mechanics. Karl Terzaghi is known as the father of modern soil mechanics, and rightfully so. Terzaghi was born on October 2, 1883 in Prague, which was then the capital of the Austrian province of Bohemia. In 1904 he graduated from the Technische Hochschule in Graz, Austria, with an undergraduate degree in mechanical engineering. After graduation he served one year in the Austrian army. Following his army service, Terzaghi studied one more year, concentrating on geological subjects. In January 1912, he received the degree of Doctor of Technical Sciences from his alma mater in Graz. In 1916, he accepted a teaching position at the Imperial School of Engineers in Istanbul. After the end of World War I, he accepted a lectureship at the American Robert College in Istanbul (1918–1925). There he began his research work on the behavior of soils and settlement of clays and on the failure due to piping in sand under dams. The publication *Erdbaumechanik* is primarily the result of this research.

In 1925, Terzaghi accepted a visiting lectureship at Massachusetts Institute of Technology, where he worked until 1929. During that time, he became recognized as the leader of the new branch of civil engineering called soil mechanics. In October 1929, he returned to Europe to accept a professorship at the Technical University of Vienna, which soon became the nucleus for civil engineers interested in soil mechanics. In 1939, he returned to the United States to become a professor at Harvard University.

The first conference of the International Society of Soil Mechanics and Foundation Engineering (ISSMFE) was held at Harvard University in 1936 with Karl Terzaghi presiding. The conference was possible due to the conviction and efforts of Professor Arthur Casagrande of Harvard University. About 200 individuals representing 21 countries attended this conference. It was through the inspiration and guidance of Terzaghi over the preceding quarter-century that papers were brought to that conference covering a wide range of topics, such as

- Effective stress
- Shear strength
- Testing with Dutch cone penetrometer
- Consolidation
- Centrifuge testing
- Elastic theory and stress distribution
- Preloading for settlement control
- Swelling clays
- Frost action
- Earthquake and soil liquefaction
- Machine vibration
- Arching theory of earth pressure

For the next quarter-century, Terzaghi was the guiding spirit in the development of soil mechanics and geotechnical engineering throughout the world. To that effect, in 1985, Ralph Peck wrote that "few people during Terzaghi's lifetime would have disagreed that he was not only the guiding spirit in soil mechanics, but that he was the clearing house for research and application throughout the world. Within the next few years he would be engaged on projects on every continent save Australia and Antarctica." Peck continued with, "Hence, even today, one can hardly improve on his contemporary assessments of the state of soil mechanics as expressed in his summary papers and presidential addresses." In 1939, Terzaghi delivered the 45th James Forrest Lecture at the Institution of Civil Engineers, London. His lecture was entitled "Soil Mechanics—A New Chapter in Engineering Science." In it, he proclaimed that most of the foundation failures that occurred were no longer "acts of God."

Following are some highlights in the development of soil mechanics and geotechnical engineering that evolved after the first conference of the ISSMFE in 1936:

- Publication of the book *Theoretical Soil Mechanics* by Karl Terzaghi in 1943 (Wiley, New York);
- Publication of the book *Soil Mechanics in Engineering Practice* by Karl Terzaghi and Ralph Peck in 1948 (Wiley, New York);
- Publication of the book *Fundamentals of Soil Mechanics* by Donald W. Taylor in 1948 (Wiley, New York);
- Start of the publication of *Geotechnique,* the international journal of soil mechanics in 1948 in England.

After a brief interruption for World War II, the second conference of ISSMFE was held in Rotterdam, The Netherlands, in 1948. There were about 600 participants, and seven volumes of proceedings were published. In this conference, A. W. Skempton presented the landmark paper on $\phi = 0$ concept for clays. Following Rotterdam, ISSMFE conferences have been organized about every four years in different parts of the world. The aftermath of the Rotterdam conference saw the growth of regional conferences on geotechnical engineering, such as

- European Regional Conference on Stability of Earth Slopes, Stockholm (1954)
- First Australia-New Zealand Conference on Shear Characteristics of Soils (1952)
- First Pan American Conference, Mexico City (1960)
- Research conference on Shear Strength of Cohesive Soils, Boulder, Colorado, USA (1960)

Two other important milestones between 1948 and 1960 are (a) the publication of A. W. Skempton's paper on *A* and *B* pore pressure parameters which made effective stress calculations more practical for various engineering works and (b) publication of the book entitled *The Measurement of Soil Properties in the Triaxial Text* by A. W. Bishop and B. J. Henkel (Arnold, London) in 1957.

By the early 1950's, computer-aided finite difference and finite element solutions were applied to various types of geotechnical engineering problems. They still remain an important and useful computation tool in our profession. Since the early days, the profession of geotechnical engineering has come a long way and has matured. It is now an established branch of civil engineering, and thousands of civil engineers declare geotechnical engineering to be their preferred area of speciality.

In 1997, the ISSMFE was changed to ISSMGE (International Society of Soil Mechanics and Geotechnical Engineering) to reflect its true scope. These international conferences have

Table 1.1 Details of ISSMFE (1936–1997) and ISSMGE (1997–present) Conferences

Conference	Location	Year
I	Harvard University, Boston, U.S.A.	1936
II	Rotterdam, the Netherlands	1948
III	Zurich, Switzerland	1953
IV	London, England	1957
V	Paris, France	1961
VI	Montreal, Canada	1965
VII	Mexico City, Mexico	1969
VIII	Moscow, U.S.S.R.	1973
IX	Tokyo, Japan	1977
X	Stockholm, Sweden	1981
XI	San Francisco, U.S.A.	1985
XII	Rio de Janeiro, Brazil	1989
XIII	New Delhi, India	1994
XIV	Hamburg, Germany	1997
XV	Istanbul, Turkey	2001
XVI	Osaka, Japan	2005
XVII	Alexandria, Egypt	2009

been instrumental for exchange of information regarding new developments and ongoing research activities in geotechnical engineering. Table 1.1 gives the location and year in which each conference of ISSMFE/ISSMGE was held, and Table 1.2 gives a list of all of the presidents of the society. In 1997, a total of 30 technical committees of ISSMGE was in place. The names of these technical committees are given in Table 1.3.

Table 1.2 Presidents of ISSMFE (1936–1997) and ISSMGE (1997–present) Conferences

Year	President
1936–1957	K. Terzaghi (U. S. A.)
1957–1961	A. W. Skempton (U. K.)
1961–1965	A. Casagrande (U. S. A.)
1965–1969	L. Bjerrum (Norway)
1969–1973	R. B. Peck (U. S. A.)
1973–1977	J. Kerisel (France)
1977–1981	M. Fukuoka (Japan)
1981–1985	V. F. B. deMello (Brazil)
1985–1989	B. B. Broms (Singapore)
1989–1994	N. R. Morgenstern (Canada)
1994–1997	M. Jamiolkowski (Italy)
1997–2001	K. Ishihara (Japan)
2001–2005	W. F. Van Impe (Belgium)
2005–2009	P. S. Sêco e Pinto (Portugal)

Table 1.3 ISSMGE Technical Committees for 1997–2001 (based on Ishihara, 1999)

Committee number	Committee name
TC-1	Instrumentation for Geotechnical Monitoring
TC-2	Centrifuge Testing
TC-3	Geotechnics of Pavements and Rail Tracks
TC-4	Earthquake Geotechnical Engineering
TC-5	Environmental Geotechnics
TC-6	Unsaturated Soils
TC-7	Tailing Dams
TC-8	Frost
TC-9	Geosynthetics and Earth Reinforcement
TC-10	Geophysical Site Characterization
TC-11	Landslides
TC-12	Validation of Computer Simulation
TC-14	Offshore Geotechnical Engineering
TC-15	Peat and Organic Soils
TC-16	Ground Property Characterization from In-situ Testing
TC-17	Ground Improvement
TC-18	Pile Foundations
TC-19	Preservation of Historic Sites
TC-20	Professional Practice
TC-22	Indurated Soils and Soft Rocks
TC-23	Limit State Design Geotechnical Engineering
TC-24	Soil Sampling, Evaluation and Interpretation
TC-25	Tropical and Residual Soils
TC-26	Calcareous Sediments
TC-28	Underground Construction in Soft Ground
TC-29	Stress-Strain Testing of Geomaterials in the Laboratory
TC-30	Coastal Geotechnical Engineering
TC-31	Education in Geotechnical Engineering
TC-32	Risk Assessment and Management
TC-33	Scour of Foundations
TC-34	Deformation of Earth Materials

1.7 *End of an Era*

In Section 1.6, a brief outline of the contributions made to modern soil mechanics by pioneers such as Karl Terzaghi, Arthur Casagrande, Donald W. Taylor, Laurits Bjerrum, and Ralph B. Peck was presented. The last of the early giants of the profession, Ralph B. Peck, passed away on February 18, 2008, at the age of 95.

Professor Ralph B. Peck (Figure 1.4) was born in Winnipeg, Canada to American parents Orwin K. and Ethel H. Peck on June 23, 1912. He received B.S. and Ph.D. degrees in 1934 and 1937, respectively, from Rensselaer Polytechnic Institute, Troy, New York. During the period from 1938 to 1939, he took courses from Arthur Casagrande at Harvard University in a new subject called "soil mechanics." From 1939 to 1943, Dr. Peck worked as an assistant to Karl Terzaghi, the "father" of modern soil mechanics, on the Chicago

Figure 1.4 Ralph B. Peck (*Photo courtesy of Ralph B. Peck*)

Subway Project. In 1943, he joined the University of Illinois at Champaign-Urban and was a professor of foundation engineering from 1948 until he retired in 1974. After retirement, he was active in consulting, which included major geotechnical projects in 44 states in the United States and 28 other countries on five continents. Some examples of his major consulting projects include

- Rapid transit systems in Chicago, San Francisco, and Washington, D.C.
- Alaskan pipeline system
- James Bay Project in Quebec, Canada
- Heathrow Express Rail Project (U.K.)
- Dead Sea dikes

His last project was the Rion-Antirion Bridge in Greece. On March 13, 2008, *The Times* of the United Kingdom wrote, "Ralph B. Peck was an American civil engineer who invented a controversial construction technique that would be used on some of the modern engineering wonders of the world, including the Channel Tunnel. Known as

Figure 1.5 Professor Ralph B. Peck (right) with the author, Braja Das, during his trip to attend the XVI ISSMGE Conference in Osaka, Japan—the last conference of its type that he would attend (*Courtesy of Braja M. Das, Henderson, Nevada*)

'the godfather of soil mechanics,' he was directly responsible for a succession of celebrated tunneling and earth dam projects that pushed the boundaries of what was believed to be possible."

Dr. Peck authored more than 250 highly distinguished technical publications. He was the president of the ISSMGE from 1969 to 1973. In 1974, he received the National Medal of Science from President Gerald R. Ford. Professor Peck was a teacher, mentor, friend, and counselor to generations of geotechnical engineers in every country in the world. The 16th ISSMGE Conference in Osaka, Japan (2005) would be the last major conference of its type that he would attend. During his trip to Osaka, even at the age of 93, he was intent on explaining to the author the importance of field testing and sound judgment in the decision-making process involved in the design and construction of geotechnical engineering projects (which he had done to numerous geotechnical engineers all over the world) (Figure 1.5).

This is truly the end of an era.

References

ATTERBERG, A. M. (1911). "Über die physikalische Bodenuntersuchung, und über die Plastizität de Tone," International Mitteilungen für Bodenkunde, *Verlag für Fachliteratur.* G.m.b.H. Berlin, Vol. 1, 10–43.

BELIDOR, B. F. (1729). *La Science des Ingenieurs dans la Conduite des Travaux de Fortification et D'Architecture Civil,* Jombert, Paris.

BELL, A. L. (1915). "The Lateral Pressure and Resistance of Clay, and Supporting Power of Clay Foundations," *Min. Proceeding of Institute of Civil Engineers,* Vol. 199, 233–272.

BISHOP, A. W. and HENKEL, B. J. (1957). *The Measurement of Soil Properties in the Triaxial Test,* Arnold, London.

BOUSSINESQ, J. V. (1885). *Application des Potentiels â L'Etude de L'Équilibre et du Mouvement des Solides Élastiques,* Gauthier-Villars, Paris.

COLLIN, A. (1846). *Recherches Expérimentales sur les Glissements Spontanés des Terrains Argileux Accompagnées de Considérations sur Quelques Principes de la Mécanique Terrestre,* Carilian-Goeury, Paris.

COULOMB, C. A. (1776). "Essai sur une Application des Règles de Maximis et Minimis à Quelques Problèmes de Statique Relatifs à L'Architecture," *Mèmoires de la Mathèmatique et de Phisique,* présentés à l'Académie Royale des Sciences, par divers savans, et lûs dans sés Assemblées, De L'Imprimerie Royale, Paris, Vol. 7, Annee 1793, 343–382.

DARCY, H. P. G. (1856). *Les Fontaines Publiques de la Ville de Dijon,* Dalmont, Paris.

DARWIN, G. H. (1883). "On the Horizontal Thrust of a Mass of Sand," *Proceedings,* Institute of Civil Engineers, London, Vol. 71, 350–378.

FELLENIUS, W. (1918). "Kaj-och Jordrasen I Göteborg," *Teknisk Tidskrift.* Vol. 48, 17–19.

FRANCAIS, J. F. (1820). "Recherches sur la Poussée de Terres sur la Forme et Dimensions des Revêtments et sur la Talus D'Excavation," *Mémorial de L'Officier du Génie,* Paris, Vol. IV, 157–206.

FRONTARD, J. (1914). "Notice sur L'Accident de la Digue de Charmes," *Anns. Ponts et Chaussées* 9th *Ser.,* Vol. 23, 173–292.

GADROY, F. (1746). *Mémoire sur la Poussée des Terres,* summarized by Mayniel, 1808.

GAUTIER, H. (1717). *Dissertation sur L'Epaisseur des Culées des Ponts . . . sur L'Effort et al Pesanteur des Arches . . . et sur les Profiles de Maconnerie qui Doivent Supporter des Chaussées, des Terrasses, et des Remparts.* Cailleau, Paris.

ISHIHARA, K. (1999). Personal communication.

KERISEL, J. (1985). "The History of Geotechnical Engineering up until 1700," *Proceedings,* XI International Conference on Soil Mechanics and Foundation Engineering, San Francisco, Golden Jubilee Volume, A. A. Balkema, 3–93.

MAYNIEL, J. J. (1808). *Traité Experimentale, Analytique et Pratique de la Poussé des Terres.* Colas, Paris.

NAVIER, C. L. M. (1839). *Leçons sur L'Application de la Mécanique à L'Establissement des Constructions et des Machines,* 2nd ed., Paris.

PECK, R. B. (1985). "The Last Sixty Years," *Proceedings,* XI International Conference on Soil Mechanics and Foundation Engineering, San Francisco, Golden Jubilee Volume, A. A. Balkema, 123–133.

PONCELET, J. V. (1840). *Mémoire sur la Stabilité des Revêtments et de seurs Fondations,* Bachelier, Paris.

RANKINE, W. J. M. (1857). "On the Stability of Loose Earth," *Philosophical Transactions,* Royal Society, Vol. 147, London.

REYNOLDS, O. (1887). "Experiments Showing Dilatency, a Property of Granular Material Possibly Connected to Gravitation," *Proceedings,* Royal Society, London, Vol. 11, 354–363.

SKEMPTON, A. W. (1948). "The $\phi = 0$ Analysis of Stability and Its Theoretical Basis," *Proceedings,* II International Conference on Soil Mechanics and Foundation Engineering, Rotterdam, Vol. 1, 72–78.

SKEMPTON, A. W. (1954). "The Pore Pressure Coefficients A and B," *Geotechnique,* Vol. 4, 143–147.

SKEMPTON, A. W. (1985). "A History of Soil Properties, 1717–1927," *Proceedings,* XI International Conference on Soil Mechanics and Foundation Engineering, San Francisco, Golden Jubilee Volume, A. A. Balkema, 95–121.

TAYLOR, D. W. (1948). *Fundamentals of Soil Mechanics,* John Wiley, New York.

TERZAGHI, K. (1925). *Erdbaumechanik auf Bodenphysikalisher Grundlage,* Deuticke, Vienna.

TERZAGHI, K. (1939). "Soil Mechanics—A New Chapter in Engineering Science," *Institute of Civil Engineers Journal,* London, Vol. 12, No. 7, 106–142.

TERZAGHI, K. (1943). *Theoretical Soil Mechanics,* John Wiley, New York.

TERZAGHI, K., and PECK, R. B. (1948). *Soil Mechanics in Engineering Practice,* John Wiley, New York.

2 Origin of Soil and Grain Size

In general, soils are formed by weathering of rocks. The physical properties of soil are dictated primarily by the minerals that constitute the soil particles and, hence, the rock from which it is derived. This chapter provides an outline of the rock cycle and the origin of soil and the grain-size distribution of particles in a soil mass.

2.1 Rock Cycle and the Origin of Soil

The mineral grains that form the solid phase of a soil aggregate are the product of rock weathering. The size of the individual grains varies over a wide range. Many of the physical properties of soil are dictated by the size, shape, and chemical composition of the grains. To better understand these factors, one must be familiar with the basic types of rock that form the earth's crust, the rock-forming minerals, and the weathering process.

On the basis of their mode of origin, rocks can be divided into three basic types: *igneous, sedimentary,* and *metamorphic.* Figure 2.1 shows a diagram of the formation cycle of different types of rock and the processes associated with them. This is called the *rock cycle.* Brief discussions of each element of the rock cycle follow.

Igneous Rock

Igneous rocks are formed by the solidification of molten *magma* ejected from deep within the earth's mantle. After ejection by either *fissure eruption* or *volcanic eruption,* some of the molten magma cools on the surface of the earth. Sometimes magma ceases its mobility below the earth's surface and cools to form intrusive igneous rocks that are called *plutons.* Intrusive rocks formed in the past may be exposed at the surface as a result of the continuous process of erosion of the materials that once covered them.

The types of igneous rock formed by the cooling of magma depend on factors such as the composition of the magma and the rate of cooling associated with it. After conducting several laboratory tests, Bowen (1922) was able to explain the relation of the rate of magma cooling to the formation of different types of rock. This explanation—known as *Bowen's reaction principle*—describes the sequence by which new minerals are formed as magma cools. The mineral crystals grow larger and some of them settle. The crystals that remain suspended in the liquid react with the remaining melt to form a new mineral

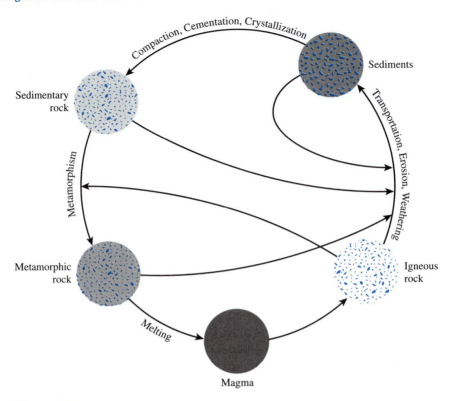

Figure 2.1 Rock cycle

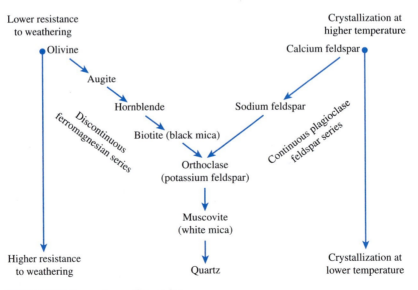

Figure 2.2 Bowen's reaction series

Table 2.1 Composition of Minerals Shown in Bowen's Reaction Series

Mineral	Composition
Olivine	$(Mg, Fe)_2SiO_4$
Augite	$Ca, Na(Mg, Fe, Al)(Al, Si_2O_6)$
Hornblende	Complex ferromagnesian silicate of Ca, Na, Mg, Ti, and Al
Biotite (black mica)	$K(Mg, Fe)_3AlSi_3O_{10}(OH)_2$
Plagioclase { calcium feldspar	$Ca(Al_2Si_2O_8)$
Plagioclase { sodium feldspar	$Na(AlSi_3O_8)$
Orthoclase (potassium feldspar)	$K(AlSi_3O_8)$
Muscovite (white mica)	$KAl_3Si_3O_{10}(OH)_2$
Quartz	SiO_2

at a lower temperature. This process continues until the entire body of melt is solidified. Bowen classified these reactions into two groups: (1) *discontinuous ferromagnesian reaction series,* in which the minerals formed are different in their chemical composition and crystalline structure, and (2) *continuous plagioclase feldspar reaction series,* in which the minerals formed have different chemical compositions with similar crystalline structures. Figure 2.2 shows Bowen's reaction series. The chemical compositions of the minerals are given in Table 2.1. Figure 2.3 is a scanning electron micrograph of a fractured surface of quartz showing glass-like fractures with no discrete planar cleavage. Figure 2.4 is a scanning electron micrograph that shows basal cleavage of individual mica grains.

Thus, depending on the proportions of minerals available, different types of igneous rock are formed. Granite, gabbro, and basalt are some of the common types of igneous rock generally encountered in the field. Table 2.2 shows the general composition of some igneous rocks.

Table 2.2 Composition of Some Igneous Rocks

Name of rock	Mode of occurrence	Texture	Abundant minerals	Less abundant minerals
Granite	Intrusive	Coarse	Quartz, sodium feldspar, potassium feldspar	Biotite, muscovite, hornblende
Rhyolite	Extrusive	Fine		
Gabbro	Intrusive	Coarse	Plagioclase, pyroxines, olivine	Hornblende, biotite, magnetite
Basalt	Extrusive	Fine		
Diorite	Intrusive	Coarse	Plagioclase, hornblende	Biotite, pyroxenes (quartz usually absent)
Andesite	Extrusive	Fine		
Syenite	Intrusive	Coarse	Potassium feldspar	Sodium feldspar, biotite, hornblende
Trachyte	Extrusive	Fine		
Peridotite	Intrusive	Coarse	Olivine, pyroxenes	Oxides of iron

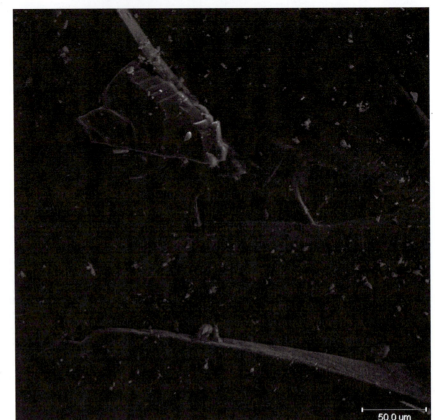

Figure 2.3
Scanning electron micrograph of fractured surface of quartz showing glass-like fractures with no discrete planar surface (*Courtesy of David J. White, Iowa State University, Ames, Iowa*)

50.0 um

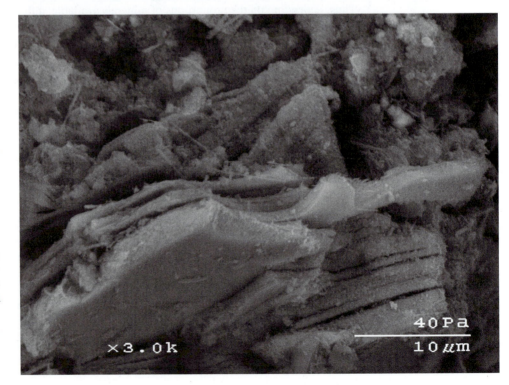

Figure 2.4
Scanning electron micrograph showing basal cleavage of individual mica grains (*Courtesy of David J. White, Iowa State University, Ames, Iowa*)

×3.0k

40Pa

10 μm

Weathering

Weathering is the process of breaking down rocks by *mechanical* and *chemical processes* into smaller pieces. Mechanical weathering may be caused by the expansion and contraction of rocks from the continuous gain and loss of heat, which results in ultimate disintegration. Frequently, water seeps into the pores and existing cracks in rocks. As the temperature drops, the water freezes and expands. The pressure exerted by ice because of volume expansion is strong enough to break down even large rocks. Other physical agents that help disintegrate rocks are glacier ice, wind, the running water of streams and rivers, and ocean waves. It is important to realize that in mechanical weathering, large rocks are broken down into smaller pieces without any change in the chemical composition. Figure 2.5 shows several examples of mechanical erosion due to ocean waves and wind at Yehliu in Taiwan. This area is located at a long and narrow sea cape at the northwest side of Keelung, about 15 kilometers between the north coast of Chin Shan and Wanli.

In chemical weathering, the original rock minerals are transformed into new minerals by chemical reaction. Water and carbon dioxide from the atmosphere form carbonic acid, which reacts with the existing rock minerals to form new minerals and soluble salts. Soluble salts present in the groundwater and organic acids formed from decayed organic matter also cause chemical weathering. An example of the chemical weathering of orthoclase to form clay minerals, silica, and soluble potassium carbonate follows:

$$H_2O + CO_2 \rightarrow H_2CO_3 \rightarrow H^+ + (HCO_3)^-$$
Carbonic acid

$$2K(AlSi_3O_8) + 2H^+ + H_2O \rightarrow 2K^+ + 4SiO_2 + Al_2Si_2O_5(OH)_4$$

Orthoclase   Silica   Kaolinite
                                                           (clay mineral)

Most of the potassium ions released are carried away in solution as potassium carbonate is taken up by plants.

The chemical weathering of plagioclase feldspars is similar to that of orthoclase in that it produces clay minerals, silica, and different soluble salts. Ferromagnesian minerals also form the decomposition products of clay minerals, silica, and soluble salts. Additionally, the iron and magnesium in ferromagnesian minerals result in other products such as hematite and limonite. Quartz is highly resistant to weathering and only slightly soluble in water. Figure 2.2 shows the susceptibility of rock-forming minerals to weathering. The minerals formed at higher temperatures in Bowen's reaction series are less resistant to weathering than those formed at lower temperatures.

The weathering process is not limited to igneous rocks. As shown in the rock cycle (Figure 2.1), sedimentary and metamorphic rocks also weather in a similar manner.

Thus, from the preceding brief discussion, we can see how the weathering process changes solid rock masses into smaller fragments of various sizes that can range from large boulders to very small clay particles. Uncemented aggregates of these small grains in various proportions form different types of soil. The clay minerals, which are a product of chemical weathering of feldspars, ferromagnesians, and micas, give the plastic property to soils. There are three important clay minerals: (1) *kaolinite,* (2) *illite,* and (3) *montmorillonite.* (We discuss these clay minerals later in this chapter.)

Figure 2.5 Mechanical erosion due to ocean waves and wind at Yehliu, Taiwan (*Courtesy of Braja M. Das, Henderson, Nevada*)

Figure 2.5 (Continued)

Transportation of Weathering Products

The products of weathering may stay in the same place or may be moved to other places by ice, water, wind, and gravity.

The soils formed by the weathered products at their place of origin are called *residual soils.* An important characteristic of residual soil is the gradation of particle size. Fine-grained soil is found at the surface, and the grain size increases with depth. At greater depths, angular rock fragments may also be found.

The transported soils may be classified into several groups, depending on their mode of transportation and deposition:

1. *Glacial soils*—formed by transportation and deposition of glaciers
2. *Alluvial soils*—transported by running water and deposited along streams
3. *Lacustrine soils*—formed by deposition in quiet lakes
4. *Marine soils*—formed by deposition in the seas
5. *Aeolian soils*—transported and deposited by wind
6. *Colluvial soils*—formed by movement of soil from its original place by gravity, such as during landslides

Sedimentary Rock

The deposits of gravel, sand, silt, and clay formed by weathering may become compacted by overburden pressure and cemented by agents like iron oxide, calcite, dolomite, and quartz. Cementing agents are generally carried in solution by ground-water. They fill the spaces between particles and form sedimentary rock. Rocks formed in this way are called *detrital sedimentary rocks.*

All detrital rocks have a *clastic* texture. The following are some examples of detrital rocks with clastic texture.

Particle size	Sedimentary rock
Granular or larger (grain size 2 mm–4 mm or larger)	Conglomerate
Sand	Sandstone
Silt and clay	Mudstone and shale

In the case of conglomerates, if the particles are more angular, the rock is called *breccia.* In sandstone, the particle sizes may vary between $\frac{1}{16}$ mm and 2 mm. When the grains in sandstone are practically all quartz, the rock is referred to as *orthoquartzite.* In mudstone and shale, the size of the particles are generally less than $\frac{1}{16}$ mm. Mudstone has a blocky aspect; whereas, in the case of shale, the rock is split into platy slabs.

Sedimentary rock also can be formed by chemical processes. Rocks of this type are classified as *chemical sedimentary rock.* These rocks can have *clastic* or *nonclastic* texture. The following are some examples of chemical sedimentary rock.

Composition	Rock
Calcite ($CaCO_3$)	Limestone
Halite (NaCl)	Rock salt
Dolomite [$CaMg(CO_3)$]	Dolomite
Gypsum ($CaSO_4 \cdot 2H_2O$)	Gypsum

Limestone is formed mostly of calcium carbonate deposited either by organisms or by an inorganic process. Most limestones have a clastic texture; however, nonclastic textures also are found commonly. Figure 2.6 shows the scanning electron micrograph of a fractured surface of Limestone. Individual grains of calcite show rhombohedral cleavage. Chalk is a sedimentary rock made in part from biochemically derived calcite, which are skeletal fragments of microscopic plants and animals. Dolomite is formed either by chemical deposition of mixed carbonates or by the reaction of magnesium in water with limestone. Gypsum and anhydrite result from the precipitation of soluble $CaSO_4$ due to evaporation of ocean water. They belong to a class of rocks generally referred to as *evaporites*. Rock salt (NaCl) is another example of an evaporite that originates from the salt deposits of seawater.

Sedimentary rock may undergo weathering to form sediments or may be subjected to the process of *metamorphism* to become metamorphic rock.

Metamorphic Rock

Metamorphism is the process of changing the composition and texture of rocks (without melting) by heat and pressure. During metamorphism, new minerals are formed, and mineral grains are sheared to give a foliated-texture to metamorphic rock. Gneiss is a metamorphic rock derived from high-grade regional metamorphism of igneous rocks, such as granite, gabbro, and diorite. Low-grade metamorphism of shales and mudstones results in slate. The clay minerals in the shale become chlorite and mica by heat; hence, slate is composed primarily of mica flakes and chlorite. Phyllite is a metamorphic rock, which is derived from slate with

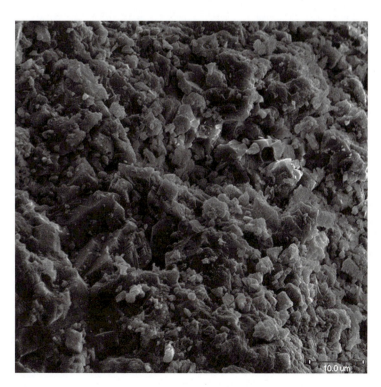

Figure 2.6 Scanning electron micrograph of the fractured surface of limestone (*Courtesy of David J. White, Iowa State University, Ames, Iowa*)

further metamorphism being subjected to heat greater than 250 to 300°C. Schist is a type of metamorphic rock derived from several igneous, sedimentary, and low-grade metamorphic rocks with a well-foliated texture and visible flakes of platy and micaceous minerals. Metamorphic rock generally contains large quantities of quartz and feldspar as well.

Marble is formed from calcite and dolomite by recrystallization. The mineral grains in marble are larger than those present in the original rock. Green marbles are colored by hornblends, serpentine, or talc. Black marbles contain bituminous material, and brown marbles contain iron oxide and limonite. Quartzite is a metamorphic rock formed from quartz-rich sandstones. Silica enters into the void spaces between the quartz and sand grains and acts as a cementing agent. Quartzite is one of the hardest rocks. Under extreme heat and pressure, metamorphic rocks may melt to form magma, and the cycle is repeated.

2.2 Soil-Particle Size

As discussed in the preceding section, the sizes of particles that make up soil vary over a wide range. Soils generally are called *gravel, sand, silt,* or *clay,* depending on the predominant size of particles within the soil. To describe soils by their particle size, several organizations have developed particle-size classifications. Table 2.3 shows the particle-size classifications developed by the Massachusetts Institute of Technology, the U.S. Department of Agriculture, the American Association of State Highway and Transportation Officials, and the U.S. Army Corps of Engineers and U.S. Bureau of Reclamation. In this table, the MIT system is presented for illustration purposes only. This system is important in the history of the development of the size limits of particles present in soils; however, the Unified Soil Classification System is now almost universally accepted and has been adopted by the American Society for Testing and Materials (ASTM). Figure 2.7 shows the size limits in a graphic form.

Gravels are pieces of rocks with occasional particles of quartz, feldspar, and other minerals. *Sand* particles are made of mostly quartz and feldspar. Other mineral grains also

Table 2.3 Particle-Size Classifications

	Grain size (mm)			
Name of organization	**Gravel**	**Sand**	**Silt**	**Clay**
Massachusetts Institute of Technology (MIT)	>2	2 to 0.06	0.06 to 0.002	<0.002
U.S. Department of Agriculture (USDA)	>2	2 to 0.05	0.05 to 0.002	<0.002
American Association of State Highway and Transportation Officials (AASHTO)	76.2 to 2	2 to 0.075	0.075 to 0.002	<0.002
Unified Soil Classification System (U.S. Army Corps of Engineers, U.S. Bureau of Reclamation, and American Society for Testing and Materials)	76.2 to 4.75	4.75 to 0.075	Fines (i.e., silts and clays) <0.075	

Note: Sieve openings of 4.75 mm are found on a U.S. No. 4 sieve; 2-mm openings on a U.S. No. 10 sieve; 0.075-mm openings on a U.S. No. 200 sieve. See Table 2.5.

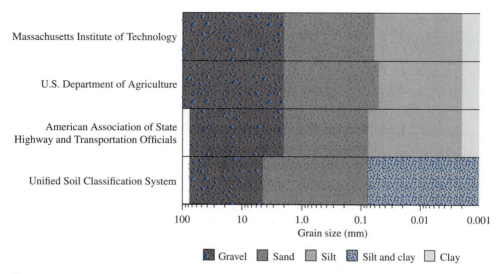

Gravel Sand Silt Silt and clay Clay

Figure 2.7 Soil-separate-size limits by various systems

may be present at times. Figure 2.8 shows the scanning electron micrograph of some sand grains. Note that the larger grains show rounding that can occur as a result of wear during intermittent transportation by wind and/or water. Figure 2.9 is a higher magnification of the grains highlighted in Figure 2.8, and it reveals a few small clay particles adhering to larger sand grains. *Silts* are the microscopic soil fractions that consist of very fine quartz

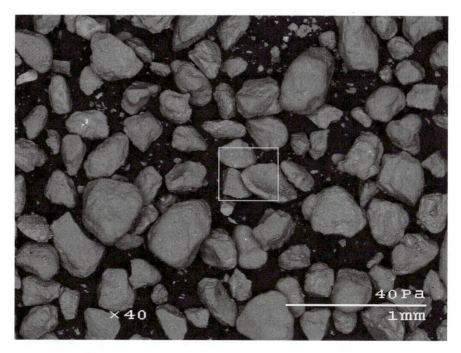

Figure 2.8 Scanning electron micrograph of some sand grains (*Courtesy of David J. White, Iowa State University, Ames, Iowa*)

Figure 2.9 Higher magnification of the sand grains highlighted in Figure 2.8 (*Courtesy of David J. White, Iowa State University, Ames, Iowa*)

grains and some flake-shaped particles that are fragments of micaceous minerals. *Clays* are mostly flake-shaped microscopic and submicroscopic particles of mica, clay minerals, and other minerals.

As shown in Table 2.3 and Figure 2.7, clays generally are defined as particles smaller than 0.002 mm. However, in some cases, particles between 0.002 and 0.005 mm in size also are referred to as clay. Particles classified as clay on the basis of their size may not necessarily contain clay minerals. Clays have been defined as those particles "which develop plasticity when mixed with a limited amount of water" (Grim, 1953). (Plasticity is the putty-like property of clays that contain a certain amount of water.) Nonclay soils can contain particles of quartz, feldspar, or mica that are small enough to be within the clay classification. Hence, it is appropriate for soil particles smaller than 2 microns (2 μm), or 5 microns (5 μm) as defined under different systems, to be called clay-sized particles rather than clay. Clay particles are mostly in the colloidal size range (<1 μm), and 2 μm appears to be the upper limit.

2.3 Clay Minerals

Clay minerals are complex aluminum silicates composed of two basic units: (1) *silica tetrahedron* and (2) *alumina octahedron.* Each tetrahedron unit consists of four oxygen atoms surrounding a silicon atom (Figure 2.10a). The combination of tetrahedral silica units gives

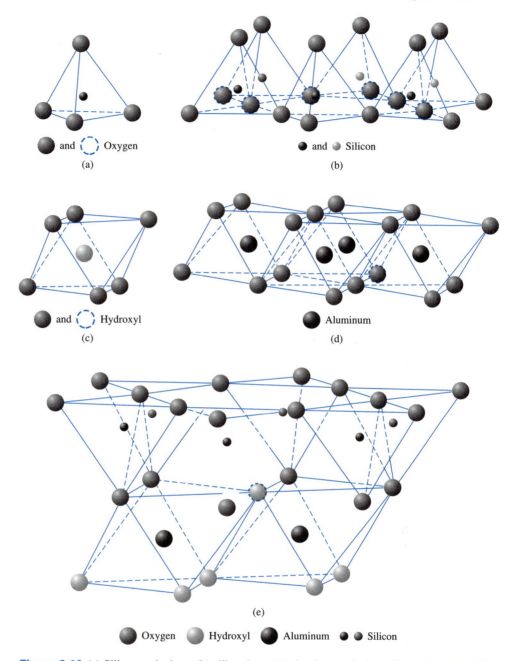

Figure 2.10 (a) Silica tetrahedron; (b) silica sheet; (c) alumina octahedron; (d) octahedral (gibbsite) sheet; (e) elemental silica-gibbsite sheet (*After Grim, 1959. With permission from ASCE.*)

a *silica sheet* (Figure 2.10b). Three oxygen atoms at the base of each tetrahedron are shared by neighboring tetrahedra. The octahedral units consist of six hydroxyls surrounding an aluminum atom (Figure 2.10c), and the combination of the octahedral aluminum hydroxyl units gives an *octahedral sheet*. (This also is called a *gibbsite sheet*—Figure 2.10d.)

Sometimes magnesium replaces the aluminum atoms in the octahedral units; in this case, the octahedral sheet is called a *brucite sheet*.

In a silica sheet, each silicon atom with a positive charge of four is linked to four oxygen atoms with a total negative charge of eight. But each oxygen atom at the base of the tetrahedron is linked to two silicon atoms. This means that the top oxygen atom of each tetrahedral unit has a negative charge of one to be counterbalanced. When the silica sheet is stacked over the octahedral sheet, as shown in Figure 2.10e, these oxygen atoms replace the hydroxyls to balance their charges.

Of the three important clay minerals, *kaolinite* consists of repeating layers of elemental silica-gibbsite sheets in a 1:1 lattice, as shown in Figures 2.11 and 2.12a. Each layer is about 7.2 Å thick. The layers are held together by hydrogen bonding. Kaolinite occurs as platelets, each with a lateral dimension of 1000 to 20,000 Å and a thickness of 100 to 1000 Å. The surface area of the kaolinite particles per unit mass is about 15 m²/g. The surface area per unit mass is defined as *specific surface*. Figure 2.13 shows a scanning electron micrograph of a kaolinite specimen.

Illite consists of a gibbsite sheet bonded to two silica sheets—one at the top and another at the bottom (Figures 2.14 and 2.12b). It is sometimes called *clay mica*. The illite layers are bonded by potassium ions. The negative charge to balance the potassium ions comes from the substitution of aluminum for some silicon in the tetrahedral sheets. Substitution of one element for another with no change in the crystalline form is known as

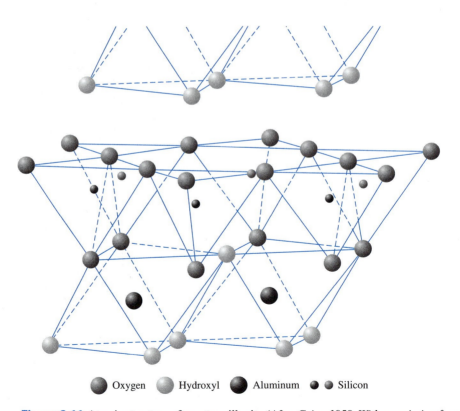

Oxygen   Hydroxyl   Aluminum   Silicon

Figure 2.11 Atomic structure of montmorillonite (*After Grim, 1959. With permission from ASCE.*)

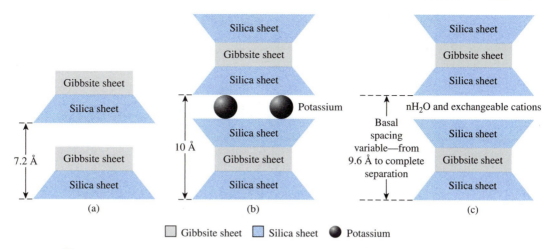

Figure 2.12 Diagram of the structures of (a) kaolinite; (b) illite; (c) montmorillonite

Figure 2.13
Scanning electron micrograph of a kaolinite specimen (*Courtesy of David J. White, Iowa State University, Ames, Iowa*)

isomorphous substitution. Illite particles generally have lateral dimensions ranging from 1000 to 5000 Å and thicknesses from 50 to 500 Å. The specific surface of the particles is about 80 m^2/g.

   Montmorillonite has a structure similar to that of illite—that is, one gibbsite sheet sandwiched between two silica sheets. (See Figures 2.15 and 2.12c.) In montmorillonite there is isomorphous substitution of magnesium and iron for aluminum in the octahedral

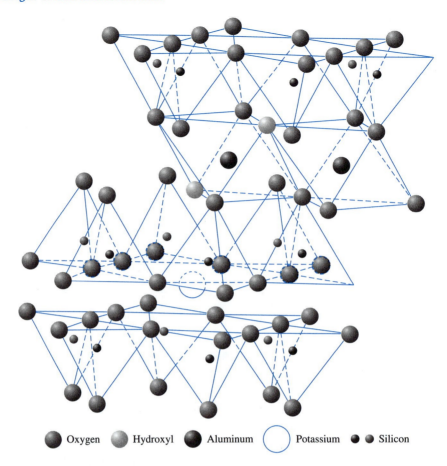

Figure 2.14
Atomic structure
of illite

⬤ Oxygen   ⚪ Hydroxyl   ⬤ Aluminum   ◯ Potassium   ● • Silicon

sheets. Potassium ions are not present as in illite, and a large amount of water is attracted into the space between the layers. Particles of montmorillonite have lateral dimensions of 1000 to 5000 Å and thicknesses of 10 to 50 Å. The specific surface is about 800 m^2/g. Figure 2.16 is a scanning electron micrograph showing the fabric of montmorillonite.

Besides kaolinite, illite, and montmorillonite, other common clay minerals generally found are chlorite, halloysite, vermiculite, and attapulgite.

The clay particles carry a net negative charge on their surfaces. This is the result both of isomorphous substitution and of a break in continuity of the structure at its edges. Larger negative charges are derived from larger specific surfaces. Some positively charged sites also occur at the edges of the particles. A list of the reciprocal of the average surface densities of the negative charges on the surfaces of some clay minerals follows (Yong and Warkentin, 1966):

Clay mineral	Reciprocal of average surface density of charge (Å^2/electronic charge)
Kaolinite	25
Clay mica and chlorite	50
Montmorillonite	100
Vermiculite	75

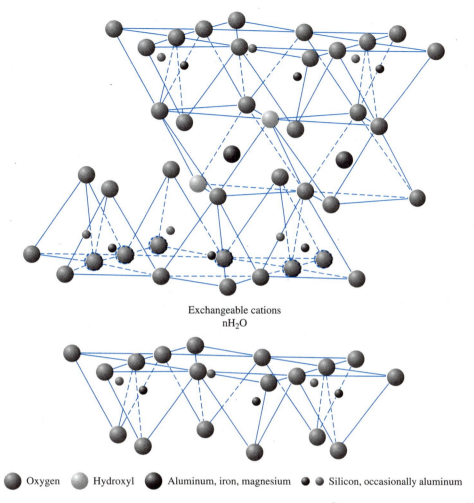

Exchangeable cations
nH₂O

● Oxygen ○ Hydroxyl ● Aluminum, iron, magnesium ● ● Silicon, occasionally aluminum

Figure 2.15 Atomic structure of montmorillonite (*After Grim, 1959. With permission from ASCE.*)

In dry clay, the negative charge is balanced by exchangeable cations like Ca^{2+}, Mg^{2+}, Na^+, and K^+ surrounding the particles being held by electrostatic attraction. When water is added to clay, these cations and a few anions float around the clay particles. This configuration is referred to as a *diffuse double layer* (Figure 2.17a). The cation concentration decreases with the distance from the surface of the particle (Figure 2.17b).

Water molecules are polar. Hydrogen atoms are not axisymmetric around an oxygen atom; instead, they occur at a bonded angle of 105° (Figure 2.18). As a result, a water molecule has a positive charge at one side and a negative charge at the other side. It is known as a *dipole.*

Dipolar water is attracted both by the negatively charged surface of the clay particles and by the cations in the double layer. The cations, in turn, are attracted to the soil particles. A third mechanism by which water is attracted to clay particles is *hydrogen bonding,* where hydrogen atoms in the water molecules are shared with oxygen atoms on the

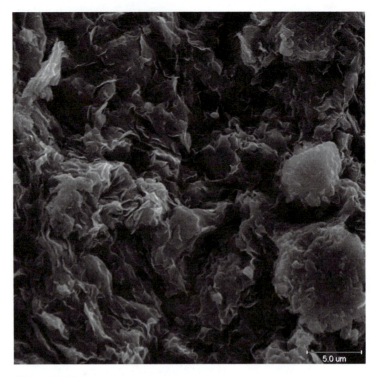

Figure 2.16 Scanning electron micrograph showing the fabric of montmorillonite (*Courtesy of David J. White, Iowa State University, Ames, Iowa*)

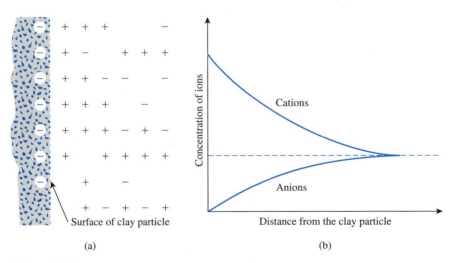

Figure 2.17 Diffuse double layer

surface of the clay. Some partially hydrated cations in the pore water are also attracted to the surface of clay particles. These cations attract dipolar water molecules. All these possible mechanics of attraction of water to clay are shown in Figure 2.19. The force of attraction between water and clay decreases with distance from the surface of the particles. All

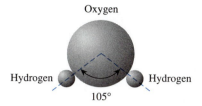

Figure 2.18 Dipolar character of water

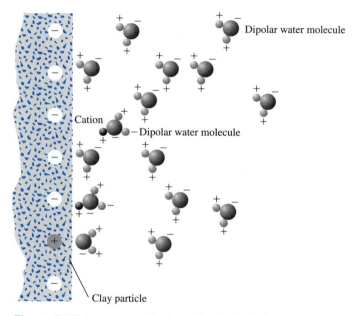

Figure 2.19 Attraction of dipolar molecules in diffuse double layer

the water held to clay particles by force of attraction is known as *double-layer water.* The innermost layer of double-layer water, which is held very strongly by clay, is known as *adsorbed water.* This water is more viscous than free water is.

Figure 2.20 shows the absorbed and double-layer water for typical montmorillonite and kaolinite particles. This orientation of water around the clay particles gives clay soils their plastic properties.

It needs to be well recognized that the presence of clay minerals in a soil aggregate has a great influence on the engineering properties of the soil as a whole. When moisture is present, the engineering behavior of a soil will change greatly as the percentage of clay mineral content increases. For all practical purposes, when the clay content is about 50% or more, the sand and silt particles float in a clay matrix, and the clay minerals primarily dictate the engineering properties of the soil.

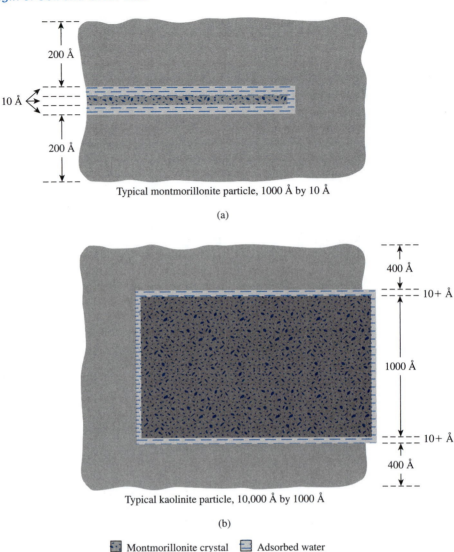

Typical montmorillonite particle, 1000 Å by 10 Å

(a)

Typical kaolinite particle, 10,000 Å by 1000 Å

(b)

Montmorillonite crystal   Adsorbed water
Kaolinite crystal   Double-layer water

Figure 2.20 Clay water (*Redrawn after Lambe, 1958. With permission from ASCE.*)

2.4 *Specific Gravity (G_s)*

Specific gravity is defined as the ratio of the unit weight of a given material to the unit weight of water. The specific gravity of soil solids is often needed for various calculations in soil mechanics. It can be determined accurately in the laboratory. Table 2.4 shows the specific gravity of some common minerals found in soils. Most of the values fall within a range of 2.6 to 2.9. The specific gravity of solids of light-colored sand, which is mostly made of quartz, may be estimated to be about 2.65; for clayey and silty soils, it may vary from 2.6 to 2.9.

Table 2.4 Specific Gravity of Common Minerals

Mineral	Specific gravity, G_s
Quartz	2.65
Kaolinite	2.6
Illite	2.8
Montmorillonite	2.65–2.80
Halloysite	2.0–2.55
Potassium feldspar	2.57
Sodium and calcium feldspar	2.62–2.76
Chlorite	2.6–2.9
Biotite	2.8–3.2
Muscovite	2.76–3.1
Hornblende	3.0–3.47
Limonite	3.6–4.0
Olivine	3.27–3.7

2.5   *Mechanical Analysis of Soil*

Mechanical analysis is the determination of the size range of particles present in a soil, expressed as a percentage of the total dry weight. Two methods generally are used to find the particle-size distribution of soil: (1) *sieve analysis*—for particle sizes larger than 0.075 mm in diameter, and (2) *hydrometer analysis*—for particle sizes smaller than 0.075 mm in diameter. The basic principles of sieve analysis and hydrometer analysis are described briefly in the following two sections.

Sieve Analysis

Sieve analysis consists of shaking the soil sample through a set of sieves that have progressively smaller openings. U.S. standard sieve numbers and the sizes of openings are given in Table 2.5.

Table 2.5 U.S. Standard Sieve Sizes

Sieve no.	Opening (mm)	Sieve no.	Opening (mm)
4	4.75	35	0.500
5	4.00	40	0.425
6	3.35	50	0.355
7	2.80	60	0.250
8	2.36	70	0.212
10	2.00	80	0.180
12	1.70	100	0.150
14	1.40	120	0.125
16	1.18	140	0.106
18	1.00	170	0.090
20	0.850	200	0.075
25	0.710	270	0.053
30	0.600		

The sieves used for soil analysis are generally 203 mm in diameter. To conduct a sieve analysis, one must first oven-dry the soil and then break all lumps into small particles. The soil then is shaken through a stack of sieves with openings of decreasing size from top to bottom (a pan is placed below the stack). Figure 2.21 shows a set of sieves in a shaker used for conducting the test in the laboratory. The smallest-sized sieve that should be used for this type of test is the U.S. No. 200 sieve. After the soil is shaken, the mass of soil retained on each sieve is determined. When cohesive soils are analyzed, breaking the lumps into individual particles may be difficult. In this case, the soil may be mixed with water to make a slurry and then washed through the sieves. Portions retained on each sieve are collected separately and oven-dried before the mass retained on each sieve is measured.

1. Determine the mass of soil retained on each sieve (i.e., $M_1, M_2, \cdots M_n$) and in the pan (i.e., M_p)
2. Determine the total mass of the soil: $M_1 + M_2 + \cdots + M_i + \cdots + M_n + M_p = \Sigma M$
3. Determine the cumulative mass of soil retained above each sieve. For the *i*th sieve, it is $M_1 + M_2 + \cdots + M_i$

Figure 2.21 A set of sieves for a test in the laboratory (*Courtesy of Braja M. Das, Henderson, Nevada*)

4. The mass of soil passing the *i*th sieve is $\Sigma M - (M_1 + M_2 + \cdots + M_i)$
5. The percent of soil passing the *i*th sieve (or *percent finer*) is

$$F = \frac{\Sigma M - (M_1 + M_2 + \cdots + M_i)}{\Sigma M} \times 100$$

Once the percent finer for each sieve is calculated (step 5), the calculations are plotted on semilogarithmic graph paper (Figure 2.22) with percent finer as the ordinate (arithmetic scale) and sieve opening size as the abscissa (logarithmic scale). This plot is referred to as the *particle-size distribution curve.*

Hydrometer Analysis

Hydrometer analysis is based on the principle of sedimentation of soil grains in water. When a soil specimen is dispersed in water, the particles settle at different velocities, depending on their shape, size, weight, and the viscosity of the water. For simplicity, it is assumed that all the soil particles are spheres and that the velocity of soil particles can be expressed by *Stokes' law,* according to which

$$v = \frac{\rho_s - \rho_w}{18\eta}D^2 \tag{2.1}$$

where v = velocity
ρ_s = density of soil particles
ρ_w = density of water
η = viscosity of water
D = diameter of soil particles

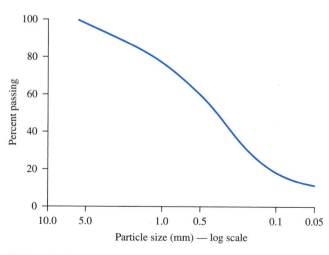

Figure 2.22 Particle-size distribution curve

Thus, from Eq. (2.1),

$$D = \sqrt{\frac{18\eta v}{\rho_s - \rho_w}} = \sqrt{\frac{18\eta}{\rho_s - \rho_w}} \sqrt{\frac{L}{t}} \qquad (2.2)$$

where $v = \dfrac{\text{Distance}}{\text{Time}} = \dfrac{L}{t}$.

Note that

$$\rho_s = G_s \rho_w \qquad (2.3)$$

Thus, combining Eqs. (2.2) and (2.3) gives

$$D = \sqrt{\frac{18\eta}{(G_s - 1)\rho_w}} \sqrt{\frac{L}{t}} \qquad (2.4)$$

If the units of η are $(\text{g} \cdot \text{sec})/\text{cm}^2$, ρ_w is in g/cm^3, L is in cm, t is in min, and D is in mm, then

$$\frac{D(\text{mm})}{10} = \sqrt{\frac{18\eta\,[(\text{g} \cdot \text{sec})/\text{cm}^2]}{(G_s - 1)\rho_w\,(\text{g/cm}^3)}} \sqrt{\frac{L\,(\text{cm})}{t\,(\text{min}) \times 60}}$$

or

$$D = \sqrt{\frac{30\eta}{(G_s - 1)\rho_w}} \sqrt{\frac{L}{t}}$$

Assume ρ_w to be approximately equal to 1 g/cm^3, so that

$$D\,(\text{mm}) = K \sqrt{\frac{L\,(\text{cm})}{t\,(\text{min})}} \qquad (2.5)$$

where

$$K = \sqrt{\frac{30\eta}{(G_s - 1)}} \qquad (2.6)$$

Note that the value of K is a function of G_s and η, which are dependent on the temperature of the test. Table 2.6 gives the variation of K with the test temperature and the specific gravity of soil solids.

Table 2.6 Values of K from Eq. (2.6)[a]

Temperature (°C)	G_s							
	2.45	2.50	2.55	2.60	2.65	2.70	2.75	2.80
16	0.01510	0.01505	0.01481	0.01457	0.01435	0.01414	0.01394	0.01374
17	0.01511	0.01486	0.01462	0.01439	0.01417	0.01396	0.01376	0.01356
18	0.01492	0.01467	0.01443	0.01421	0.01399	0.01378	0.01359	0.01339
19	0.01474	0.01449	0.01425	0.01403	0.01382	0.01361	0.01342	0.01323
20	0.01456	0.01431	0.01408	0.01386	0.01365	0.01344	0.01325	0.01307
21	0.01438	0.01414	0.01391	0.01369	0.01348	0.01328	0.01309	0.01291
22	0.01421	0.01397	0.01374	0.01353	0.01332	0.01312	0.01294	0.01276
23	0.01404	0.01381	0.01358	0.01337	0.01317	0.01297	0.01279	0.01261
24	0.01388	0.01365	0.01342	0.01321	0.01301	0.01282	0.01264	0.01246
25	0.01372	0.01349	0.01327	0.01306	0.01286	0.01267	0.01249	0.01232
26	0.01357	0.01334	0.01312	0.01291	0.01272	0.01253	0.01235	0.01218
27	0.01342	0.01319	0.01297	0.01277	0.01258	0.01239	0.01221	0.01204
28	0.01327	0.01304	0.01283	0.01264	0.01244	0.01225	0.01208	0.01191
29	0.01312	0.01290	0.01269	0.01249	0.01230	0.01212	0.01195	0.01178
30	0.01298	0.01276	0.01256	0.01236	0.01217	0.01199	0.01182	0.01169

[a]After ASTM (2007)

In the laboratory, the hydrometer test is conducted in a sedimentation cylinder usually with 50 g of oven-dried sample. Sometimes 100-g samples also can be used. The sedimentation cylinder is 457 mm high and 63.5 mm in diameter. It is marked for a volume of 1000 ml. Sodium hexametaphosphate generally is used as the *dispersing agent*. The volume of the dispersed soil suspension is increased to 1000 ml by adding distilled water. Figure 2.23 shows an ASTM 152H type of hydrometer.

When a hydrometer is placed in the soil suspension at a time t, measured from the start of sedimentation it measures the specific gravity in the vicinity of its bulb at a depth L (Figure 2.24). The specific gravity is a function of the amount of soil particles present per unit volume of suspension at that depth. Also, at a time t, the soil particles in suspension at a depth L will have a diameter smaller than D as calculated in Eq. (2.5). The larger particles would have settled beyond the zone of measurement. Hydrometers are designed to give the amount of soil, in grams, that is still in suspension. They are calibrated for soils that have a specific gravity, G_s, of 2.65; for soils of other specific gravity, a correction must be made.

By knowing the amount of soil in suspension, L, and t, we can calculate the percentage of soil by weight finer than a given diameter. Note that L is the depth measured from the surface of the water to the center of gravity of the hydrometer bulb at which the density of the suspension is measured. The value of L will change with time t. Hydrometer analysis is effective for separating soil fractions down to a size of about

Figure 2.23
ASTM 152H hydrometer
(*Courtesy of ELE International*)

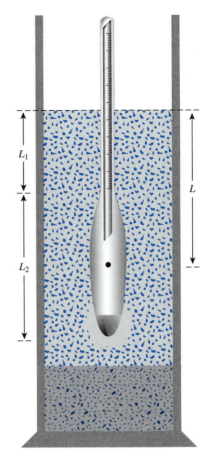

Figure 2.24 Definition of L in hydrometer test

0.5 μm. The value of L (cm) for the ASTM 152H hydrometer can be given by the expression (see Figure 2.24)

$$L = L_1 + \frac{1}{2}\left(L_2 - \frac{V_B}{A}\right) \qquad (2.7)$$

where L_1 = distance along the stem of the hydrometer from the top of the bulb to the mark for a hydrometer reading (cm)

L_2 = length of the hydrometer bulb = 14 cm

V_B = volume of the hydrometer bulb = 67 cm^3

A = cross-sectional area of the sedimentation cylinder = 27.8 cm^2

The value of L_1 is 10.5 cm for a reading of $R = 0$ and 2.3 cm for a reading of $R = 50$. Hence, for any reading R,

$$L_1 = 10.5 - \frac{(10.5 - 2.3)}{50}R = 10.5 - 0.164R \text{ (cm)}$$

Table 2.7 Variation of L with Hydrometer Reading— ASTM 152H Hydrometer

Hydrometer reading, R	L (cm)	Hydrometer reading, R	L (cm)
0	16.3	31	11.2
1	16.1	32	11.1
2	16.0	33	10.9
3	15.8	34	10.7
4	15.6	35	10.6
5	15.5	36	10.4
6	15.3	37	10.2
7	15.2	38	10.1
8	15.0	39	9.9
9	14.8	40	9.7
10	14.7	41	9.6
11	14.5	42	9.4
12	14.3	43	9.2
13	14.2	44	9.1
14	14.0	45	8.9
15	13.8	46	8.8
16	13.7	47	8.6
17	13.5	48	8.4
18	13.3	49	8.3
19	13.2	50	8.1
20	13.0	51	7.9
21	12.9	52	7.8
22	12.7	53	7.6
23	12.5	54	7.4
24	12.4	55	7.3
25	12.2	56	7.1
26	12.0	57	7.0
27	11.9	58	6.8
28	11.7	59	6.6
29	11.5	60	6.5
30	11.4		

Thus, from Eq. (2.7),

$$L = 10.5 - 0.164R + \frac{1}{2}\left(14 - \frac{67}{27.8}\right) = 16.29 - 0.164R \qquad (2.8)$$

where R = hydrometer reading corrected for the meniscus.

On the basis of Eq. (2.8), the variations of L with the hydrometer readings R are given in Table 2.7.

In many instances, the results of sieve analysis and hydrometer analysis for finer fractions for a given soil are combined on one graph, such as the one shown in Figure 2.25. When these results are combined, a discontinuity generally occurs in the range where they overlap. This discontinuity occurs because soil particles are generally irregular in shape.

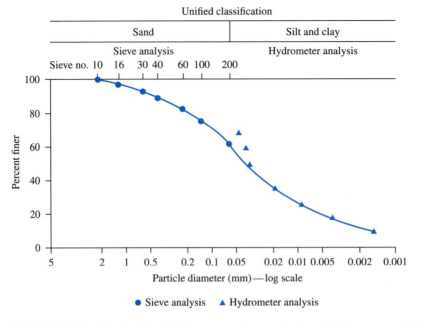

Figure 2.25 Particle-size distribution curve—sieve analysis and hydrometer analysis

Sieve analysis gives the intermediate dimensions of a particle; hydrometer analysis gives the diameter of an equivalent sphere that would settle at the same rate as the soil particle.

2.6 *Particle-Size Distribution Curve*

A particle-size distribution curve can be used to determine the following four parameters for a given soil (Figure 2.26):

1. *Effective size* (D_{10}): This parameter is the diameter in the particle-size distribution curve corresponding to 10% finer. The effective size of a granular soil is a good measure to estimate the hydraulic conductivity and drainage through soil.

2. *Uniformity coefficient* (C_u): This parameter is defined as

$$C_u = \frac{D_{60}}{D_{10}}$$ (2.9)

where D_{60} = diameter corresponding to 60% finer.

3. *Coefficient of gradation* (C_c): This parameter is defined as

$$C_c = \frac{D_{30}^2}{D_{60} \times D_{10}}$$ (2.10)

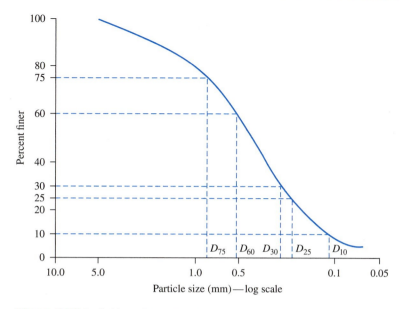

Figure 2.26 Definition of D_{75}, D_{60}, D_{30}, D_{25}, and D_{10}

4. *Sorting coefficient* (S_0): This parameter is another measure of uniformity and is generally encountered in geologic works and expressed as

$$S_0 = \sqrt{\frac{D_{75}}{D_{25}}} \tag{2.11}$$

The sorting coefficient is not frequently used as a parameter by geotechnical engineers.

The percentages of gravel, sand, silt, and clay-size particles present in a soil can be obtained from the particle-size distribution curve. As an example, we will use the particle-size distribution curve shown in Figure 2.25 to determine the gravel, sand, silt, and clay size particles as follows (according to the Unified Soil Classification System—see Table 2.3):

Size (mm)	Percent finer	
76.2	100	100 − 100 = 0% gravel
4.75	100	100 − 62 = 38% sand
0.075	62	62 − 0 = 62% silt and clay
—	0	

The particle-size distribution curve shows not only the range of particle sizes present in a soil, but also the type of distribution of various-size particles. Such types of distributions are demonstrated in Figure 2.27. Curve I represents a type of soil in which most of the soil grains are the same size. This is called *poorly graded* soil. Curve II represents a soil in which the particle sizes are distributed over a wide range, termed *well graded*. A well-graded soil has a uniformity coefficient greater than about 4 for gravels and 6 for sands, and a coefficient of gradation between 1 and 3 (for gravels and sands). A soil might have a combination of two or more uniformly graded fractions. Curve III represents such a soil. This type of soil is termed *gap graded*.

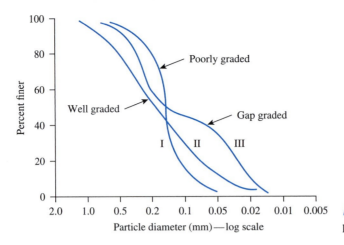

Figure 2.27 Different types of particle-size distribution curves

---

Example 2.1

Following are the results of a sieve analysis. Make the necessary calculations and draw a particle-size distribution curve.

U.S. sieve no.	Mass of soil retained on each sieve (g)
4	0
10	40
20	60
40	89
60	140
80	122
100	210
200	56
Pan	12

Solution
The following table can now be prepared.

U.S. sieve (1)	Opening (mm) (2)	Mass retained on each sieve (g) (3)	Cumulative mass retained above each sieve (g) (4)	Percent finer[a] (5)
4	4.75	0	0	100
10	2.00	40	$0 + 40 = 40$	94.5
20	0.850	60	$40 + 60 = 100$	86.3
40	0.425	89	$100 + 89 = 189$	74.1
60	0.250	140	$189 + 140 = 329$	54.9
80	0.180	122	$329 + 122 = 451$	38.1
100	0.150	210	$451 + 210 = 661$	9.3
200	0.075	56	$661 + 56 = 717$	1.7
Pan	–	12	$717 + 12 = 729 = \Sigma M$	0

$$^a \frac{\Sigma M - \text{col. 4}}{\Sigma M} \times 100 = \frac{729 - \text{col. 4}}{729} \times 100$$

The particle-size distribution curve is shown in Figure 2.28.

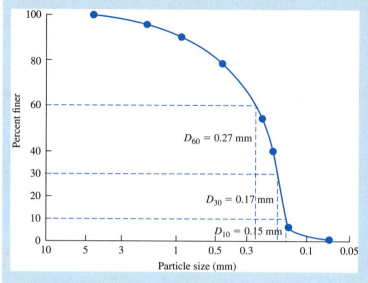

Figure 2.28 Particle-size distribution curve ∎

Example 2.2

For the particle-size distribution curve shown in Figure 2.28 determine

    a. D_{10}, D_{30}, and D_{60}
    b. Uniformity coefficient, C_u
    c. Coefficient of gradation, C_z

Solution
Part a
From Figure 2.28,

$$D_{10} = \textbf{0.15 mm}$$
$$D_{30} = \textbf{0.17 mm}$$
$$D_{60} = \textbf{0.27 mm}$$

Part b

$$C_u = \frac{D_{60}}{D_{10}} = \frac{0.27}{0.15} = \textbf{1.8}$$

$$C_z = \frac{D_{30}^2}{D_{60} \times D_{10}} = \frac{(0.17)^2}{(0.27)(0.15)} = \textbf{0.71}$$ ∎

Example 2.3

For the particle-size distribution curve shown in Figure 2.28, determine the percentages of gravel, sand, silt, and clay-size particles present. Use the Unified Soil Classification System.

Solution
From Figure 2.28, we can prepare the following table.

Size (mm)		Percent finer
76.2	100	100 − 100 = **0% gravel**
4.75	100	100 − 1.7 = **98.3% sand**
0.075	1.7	1.7 − 0 = **1.7% silt and clay**
−	0	

2.7   *Particle Shape*

The shape of particles present in a soil mass is equally as important as the particle-size distribution because it has significant influence on the physical properties of a given soil. However, not much attention is paid to particle shape because it is more difficult to measure. The particle shape generally can be divided into three major categories:

1. Bulky
2. Flaky
3. Needle shaped

Bulky particles are formed mostly by mechanical weathering of rock and minerals. Geologists use such terms as *angular, subangular, subrounded,* and *rounded* to describe the shapes of bulky particles. These shapes are shown qualitatively in Figure 2.29. Small sand particles located close to their origin are generally very angular. Sand particles carried by

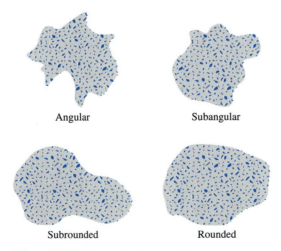

Angular   Subangular

Subrounded   Rounded

Figure 2.29 Shape of bulky particles

wind and water for a long distance can be subangular to rounded in shape. The shape of granular particles in a soil mass has a great influence on the physical properties of the soil, such as maximum and minimum void ratios, shear strength parameters, compressibility, etc.

The *angularity, A,* is defined as

$$A = \frac{\text{Average radius of corners and edges}}{\text{Radius of the maximum inscribed sphere}} \qquad (2.12)$$

The *sphericity* of bulky particles is defined as

$$S = \frac{D_e}{L_p} \qquad (2.13)$$

where D_e = equivalent diameter of the partilce = $\sqrt[3]{\frac{6V}{\pi}}$

V = volume of particle
L_p = length of particle

Flaky particles have very low sphericity—usually 0.01 or less. These particles are predominantly clay minerals.

Needle-shaped particles are much less common than the other two particle types. Examples of soils containing needle-shaped particles are some coral deposits and attapulgite clays.

2.8 Summary

In this chapter, we discussed the rock cycle, the origin of soil by weathering, the particle-size distribution in a soil mass, the shape of particles, and clay minerals. Some important points include the following:

1. Rocks can be classified into three basic categories: (a) igneous, (b) sedimentary, and (c) metamorphic.
2. Soils are formed by chemical and mechanical weathering of rocks.
3. Based on the size of the soil particles, soil can be classified as gravel, sand, silt, or clay.
4. Clays are mostly flake-shaped microscopic and submicroscopic particles of mica, clay minerals, and other minerals.
5. Clay minerals are complex aluminum silicates that develop plasticity when mixed with a limited amount of water.
6. Mechanical analysis is a process for determining the size range of particles present in a soil mass. Sieve analysis and hydrometer analysis are two tests used in the mechanical analysis of soil.

Problems

2.1 For a soil, suppose that $D_{10} = 0.08$ mm, $D_{30} = 0.22$ mm, and $D_{60} = 0.41$ mm. Calculate the uniformity coefficient and the coefficient of gradation.

2.2 Repeat Problem 2.1 with the following: $D_{10} = 0.24$ mm, $D_{30} = 0.82$ mm, and $D_{60} = 1.81$ mm.

2.3   Repeat Problem 2.1 with the following: $D_{10} = 0.18$ mm, $D_{30} = 0.32$ mm, and $D_{60} = 0.78$ mm.

2.4   The following are the results of a sieve analysis:

U.S. sieve no.	Mass of soil retained (g)
4	0
10	18.5
20	53.2
40	90.5
60	81.8
100	92.2
200	58.5
Pan	26.5

   a. Determine the percent finer than each sieve and plot a grain-size distribution curve.

   b. Determine D_{10}, D_{30}, and D_{60} from the grain-size distribution curve.

   c. Calculate the uniformity coefficient, C_u.

   d. Calculate the coefficient of gradation, C_c.

2.5   Repeat Problem 2.4 with the following:

U.S. sieve no.	Mass of soil retained (g)
4	0
10	44
20	56
40	82
60	51
80	106
100	92
200	85
Pan	35

2.6   Repeat Problem 2.4 with the following:

U.S. sieve no.	Mass of soil retained (g)
4	0
10	41.2
20	55.1
40	80.0
60	91.6
100	60.5
200	35.6
Pan	21.5

2.7 Repeat Problem 2.4 with the following results of a sieve analysis:

U.S. sieve no.	Mass of soil retained on each sieve (g)
4	0
6	0
10	0
20	9.1
40	249.4
60	179.8
100	22.7
200	15.5
Pan	23.5

2.8 The following are the results of a sieve and hydrometer analysis.

Analysis	Sieve number/ grain size	Percent finer than
Sieve	40	100
	80	97
	170	92
	200	90
Hydrometer	0.04 mm	74
	0.015 mm	42
	0.008 mm	27
	0.004 mm	17
	0.002 mm	11

a. Draw the grain-size distribution curve.
b. Determine the percentages of gravel, sand, silt, and clay according to the MIT system.
c. Repeat part b according to the USDA system.
d. Repeat part b according to the AASHTO system.

2.9 The particle-size characteristics of a soil are given in this table. Draw the particle-size distribution curve.

Size (mm)	Percent finer
0.425	100
0.033	90
0.018	80
0.01	70
0.0062	60
0.0035	50
0.0018	40
0.001	35

Determine the percentages of gravel, sand, silt, and clay:

   a. According to the USDA system.

   b. According to the AASHTO system.

2.10 Repeat Problem 2.9 with the following data:

Size (mm)	Percent finer
0.425	100
0.1	92
0.052	84
0.02	62
0.01	46
0.004	32
0.001	22

2.11 In a hydrometer test, the results are as follows: $G_s = 2.60$, temperature of water $= 24\ °C$, and $R = 43$ at 60 min after the start of sedimentation (see Figure 2.24). What is the diameter, D, of the smallest-size particles that have settled beyond the zone of measurement at that time (that is, $t = 60$ min)?

2.12 Repeat Problem 2.11 with the following values: $G_s = 2.70$, temperature $= 23\ °C$, $t = 120$ min, and $R = 25$.

References

AMERICAN SOCIETY FOR TESTING AND MATERIALS (2007). *ASTM Book of Standards*, Sec. 4, Vol. 04.08, West Conshohocken, Pa.

BOWEN, N. L. (1922). "The Reaction Principles in Petrogenesis," *Journal of Geology*, Vol. 30, 177–198.

GRIM, R. E. (1953). *Clay Mineralogy*, McGraw-Hill, New York.

GRIM, R. E. (1959). "Physico-Chemical Properties of Soils: Clay Minerals," *Journal of the Soil Mechanics and Foundations Division*, ASCE, Vol. 85, No. SM2, 1–17.

LAMBE, T. W. (1958). "The Structure of Compacted Clay," *Journal of the Soil Mechanics and Foundations Division*, ASCE, Vol. 84, No. SM2, 1655–1 to 1655–35.

YONG, R. N., and WARKENTIN, B. P. (1966). *Introduction of Soil Behavior*, Macmillan, New York.

3 Weight–Volume Relationships

Chapter 2 presented the geologic processes by which soils are formed, the description of limits on the sizes of soil particles, and the mechanical analysis of soils. In natural occurrence, soils are three-phase systems consisting of soil solids, water, and air. This chapter discusses the weight–volume relationships of soil aggregates.

3.1 Weight–Volume Relationships

Figure 3.1a shows an element of soil of volume V and weight W as it would exist in a natural state. To develop the weight–volume relationships, we must separate the three phases (that is, solid, water, and air) as shown in Figure 3.1b. Thus, the total volume of a given soil sample can be expressed as

$$V = V_s + V_v = V_s + V_w + V_a \tag{3.1}$$

where V_s = volume of soil solids
V_v = volume of voids
V_w = volume of water in the voids
V_a = volume of air in the voids

Assuming that the weight of the air is negligible, we can give the total weight of the sample as

$$W = W_s + W_w \tag{3.2}$$

where W_s = weight of soil solids
W_w = weight of water

The *volume relationships* commonly used for the three phases in a soil element are *void ratio, porosity,* and *degree of saturation. Void ratio* (e) is defined as the ratio of the volume of voids to the volume of solids. Thus,

$$e = \frac{V_v}{V_s} \tag{3.3}$$

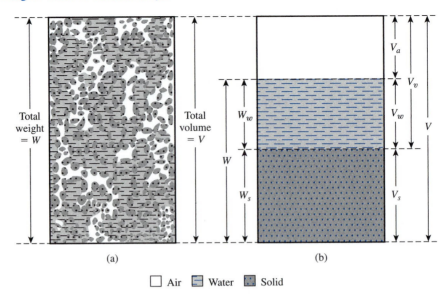

Air Water Solid

Figure 3.1 (a) Soil element in natural state; (b) three phases of the soil element

Porosity (n) is defined as the ratio of the volume of voids to the total volume, or

$$n = \frac{V_v}{V} \tag{3.4}$$

The *degree of saturation (S)* is defined as the ratio of the volume of water to the volume of voids, or

$$S = \frac{V_w}{V_v} \tag{3.5}$$

It is commonly expressed as a percentage.

The relationship between void ratio and porosity can be derived from Eqs. (3.1), (3.3), and (3.4) as follows:

$$e = \frac{V_v}{V_s} = \frac{V_v}{V - V_v} = \frac{\left(\dfrac{V_v}{V}\right)}{1 - \left(\dfrac{V_v}{V}\right)} = \frac{n}{1 - n} \tag{3.6}$$

Also, from Eq. (3.6),

$$n = \frac{e}{1 + e} \tag{3.7}$$

The common terms used for *weight relationships* are *moisture content* and *unit weight. Moisture content (w)* is also referred to as *water content* and is defined as the ratio of the weight of water to the weight of solids in a given volume of soil:

$$w = \frac{W_w}{W_s} \tag{3.8}$$

Unit weight (γ) is the weight of soil per unit volume. Thus,

$$\gamma = \frac{W}{V} \tag{3.9}$$

The unit weight can also be expressed in terms of the weight of soil solids, the moisture content, and the total volume. From Eqs. (3.2), (3.8), and (3.9),

$$\gamma = \frac{W}{V} = \frac{W_s + W_w}{V} = \frac{W_s \left[1 + \left(\dfrac{W_w}{W_s}\right)\right]}{V} = \frac{W_s(1 + w)}{V} \tag{3.10}$$

Soils engineers sometimes refer to the unit weight defined by Eq. (3.9) as the *moist unit weight*.

Often, to solve earthwork problems, one must know the weight per unit volume of soil, excluding water. This weight is referred to as the *dry unit weight, γ_d*. Thus,

$$\gamma_d = \frac{W_s}{V} \tag{3.11}$$

From Eqs. (3.10) and (3.11), the relationship of unit weight, dry unit weight, and moisture content can be given as

$$\gamma_d = \frac{\gamma}{1 + w} \tag{3.12}$$

In SI (Système International), the unit used is kilo Newtons per cubic meter (kN/m^3). Because the Newton is a derived unit, working with mass densities (ρ) of soil may sometimes be convenient. The SI unit of mass density is kilograms per cubic meter (kg/m^3). We can write the density equations [similar to Eqs. (3.9) and (3.11)] as

$$\rho = \frac{M}{V} \tag{3.13}$$

and

$$\rho_d = \frac{M_s}{V} \tag{3.14}$$

where ρ = density of soil (kg/m^3)
ρ_d = dry density of soil (kg/m^3)
M = total mass of the soil sample (kg)
M_s = mass of soil solids in the sample (kg)

The unit of total volume, V, is m^3.

The unit weight in kN/m³ can be obtained from densities in kg/m³ as

$$\gamma \, (\text{kN/m}^3) = \frac{g\rho(\text{kg/m}^3)}{1000}$$

and

$$\gamma_d \, (\text{kN/m}^3) = \frac{g\rho_d(\text{kg/m}^3)}{1000}$$

where g = acceleration due to gravity = 9.81 m/sec².

Note that unit weight of water (γ_w) is equal to 9.81 kN/m³ or 1000 kgf/m³.

3.2 Relationships among Unit Weight, Void Ratio, Moisture Content, and Specific Gravity

To obtain a relationship among unit weight (or density), void ratio, and moisture content, let us consider a volume of soil in which the volume of the soil solids is one, as shown in Figure 3.2. If the volume of the soil solids is one, then the volume of voids is numerically equal to the void ratio, e [from Eq. (3.3)]. The weights of soil solids and water can be given as

$$W_s = G_s \gamma_w$$

$$W_w = wW_s = wG_s \gamma_w$$

where G_s = specific gravity of soil solids
w = moisture content
γ_w = unit weight of water

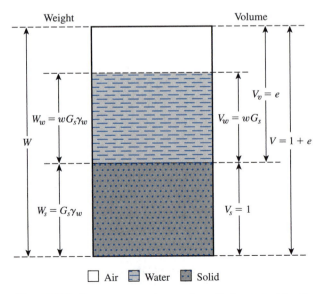

Figure 3.2 Three separate phases of a soil element with volume of soil solids equal to one

Now, using the definitions of unit weight and dry unit weight [Eqs. (3.9) and (3.11)], we can write

$$\gamma = \frac{W}{V} = \frac{W_s + W_w}{V} = \frac{G_s\gamma_w + wG_s\gamma_w}{1 + e} = \frac{(1 + w)G_s\gamma_w}{1 + e} \tag{3.15}$$

and

$$\gamma_d = \frac{W_s}{V} = \frac{G_s\gamma_w}{1 + e} \tag{3.16}$$

or

$$e = \frac{G_s\gamma_w}{\gamma_d} - 1 \tag{3.17}$$

Because the weight of water for the soil element under consideration is $wG_s\gamma_w$, the volume occupied by water is

$$V_w = \frac{W_w}{\gamma_w} = \frac{wG_s\gamma_w}{\gamma_w} = wG_s$$

Hence, from the definition of degree of saturation [Eq. (3.5)],

$$S = \frac{V_w}{V_v} = \frac{wG_s}{e}$$

or

$$Se = wG_s \tag{3.18}$$

This equation is useful for solving problems involving three-phase relationships.

If the soil sample is *saturated*—that is, the void spaces are completely filled with water (Figure 3.3)—the relationship for saturated unit weight (γ_{sat}) can be derived in a similar manner:

$$\gamma_{sat} = \frac{W}{V} = \frac{W_s + W_w}{V} = \frac{G_s\gamma_w + e\gamma_w}{1 + e} = \frac{(G_s + e)\gamma_w}{1 + e} \tag{3.19}$$

Also, from Eq. (3.18) with $S = 1$,

$$e = wG_s \tag{3.20}$$

As mentioned before, due to the convenience of working with densities in the SI system, the following equations, similar to unit–weight relationships given in Eqs. (3.15), (3.16), and (3.19), will be useful:

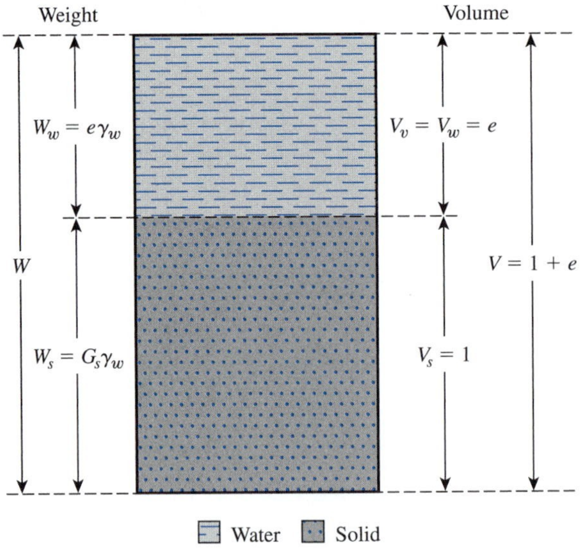

Weight   Volume

$W_w = e\gamma_w$

$V_v = V_w = e$

W

$V = 1 + e$

$W_s = G_s\gamma_w$

$V_s = 1$

☐ Water   ☐ Solid

Figure 3.3
Saturated soil element with volume
of soil solids equal to one

$$\text{Density} = \rho = \frac{(1 + w)G_s\rho_w}{1 + e} \qquad (3.21)$$

$$\text{Dry density} = \rho_d = \frac{G_s\rho_w}{1 + e} \qquad (3.22)$$

$$\text{Saturated density} = \rho_{\text{sat}} = \frac{(G_s + e)\rho_w}{1 + e} \qquad (3.23)$$

where ρ_w = density of water = 1000 kg/m³.

    Equation (3.21) may be derived by referring to the soil element shown in Figure 3.4, in which the volume of soil solids is equal to 1 and the volume of voids is equal to e.

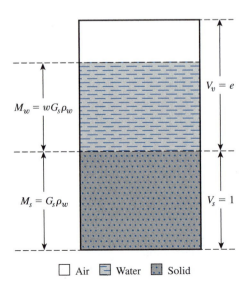

$V_v = e$

$M_w = wG_s\rho_w$

$M_s = G_s\rho_w$

$V_s = 1$

☐ Air   ☐ Water   ☐ Solid

Figure 3.4
Three separate phases of a soil element showing
mass–volume relationship

Hence, the mass of soil solids, M_s, is equal to $G_s\rho_w$. The moisture content has been defined in Eq. (3.8) as

$$w = \frac{W_w}{W_s} = \frac{(\text{mass of water}) \cdot g}{(\text{mass of solid}) \cdot g}$$

$$= \frac{M_w}{M_s}$$

where M_w = mass of water.

Since the mass of soil in the element is equal to $G_s\rho_w$, the mass of water

$$M_w = wM_s = wG_s\rho_w$$

From Eq. (3.13), density

$$\rho = \frac{M}{V} = \frac{M_s + M_w}{V_s + V_v} = \frac{G_s\rho_w + wG_s\rho_w}{1 + e}$$

$$= \frac{(1 + w)G_s\rho_w}{1 + e}$$

Equations (3.22) and (3.23) can be derived similarly.

3.3 Relationships among Unit Weight, Porosity, and Moisture Content

The relationship among *unit weight, porosity,* and *moisture content* can be developed in a manner similar to that presented in the preceding section. Consider a soil that has a total volume equal to one, as shown in Figure 3.5. From Eq. (3.4),

$$n = \frac{V_v}{V}$$

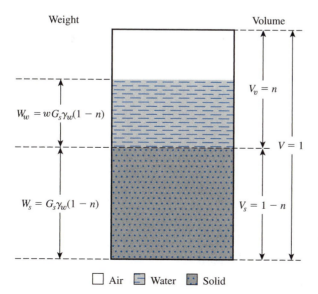

Weight

$$W_w = wG_s\gamma_w(1 - n)$$

$$W_s = G_s\gamma_w(1 - n)$$

Volume

$$V_v = n$$

$$V = 1$$

$$V_s = 1 - n$$

☐ Air   ▤ Water   ▨ Solid

Figure 3.5
Soil element with total volume equal to one

If V is equal to 1, then V_v is equal to n, so $V_s = 1 - n$. The weight of soil solids (W_s) and the weight of water (W_w) can then be expressed as follows:

$$W_s = G_s\gamma_w(1 - n) \tag{3.24}$$

$$W_w = wW_s = wG_s\gamma_w(1 - n) \tag{3.25}$$

So, the dry unit weight equals

$$\gamma_d = \frac{W_s}{V} = \frac{G_s\gamma_w(1 - n)}{1} = G_s\gamma_w(1 - n) \tag{3.26}$$

The moist unit weight equals

$$\gamma = \frac{W_s + W_w}{V} = G_s\gamma_w(1 - n)(1 + w) \tag{3.27}$$

Figure 3.6 shows a soil sample that is saturated and has $V = 1$. According to this figure,

$$\gamma_{sat} = \frac{W_s + W_w}{V} = \frac{(1 - n)G_s\gamma_w + n\gamma_w}{1} = [(1 - n)G_s + n]\gamma_w \tag{3.28}$$

The moisture content of a saturated soil sample can be expressed as

$$w = \frac{W_w}{W_s} = \frac{n\gamma_w}{(1 - n)\gamma_wG_s} = \frac{n}{(1 - n)G_s} \tag{3.29}$$

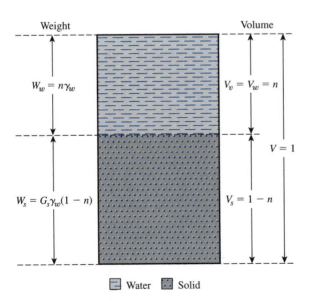

Weight   Volume

$W_w = n\gamma_w$ $V_v = V_w = n$

$V = 1$

$W_s = G_s\gamma_w(1 - n)$ $V_s = 1 - n$

☐ Water   ☐ Solid

Figure 3.6 Saturated soil element with total volume equal to one

3.4 Various Unit-Weight Relationships

In Sections 3.2 and 3.3, we derived the fundamental relationships for the moist unit weight, dry unit weight, and saturated unit weight of soil. Several other forms of relationships that can be obtained for γ, γ_d, and γ_{sat} are given in Table 3.1. Some typical values of void ratio, moisture content in a saturated condition, and dry unit weight for soils in a natural state are given in Table 3.2.

Table 3.1 Various Forms of Relationships for γ, γ_d, and γ_{sat}

Moist unit weight (γ)		Dry unit weight (γ_d)		Saturated unit weight (γ_{sat})	
Given	**Relationship**	**Given**	**Relationship**	**Given**	**Relationship**
w, G_s, e	$\dfrac{(1+w)G_s\gamma_w}{1+e}$	γ, w	$\dfrac{\gamma}{1+w}$	G_s, e	$\dfrac{(G_s+e)\gamma_w}{1+e}$
S, G_s, e	$\dfrac{(G_s+Se)\gamma_w}{1+e}$	G_s, e	$\dfrac{G_s\gamma_w}{1+e}$	G_s, n	$[(1-n)G_s+n]\gamma_w$
w, G_s, S	$\dfrac{(1+w)G_s\gamma_w}{1+\dfrac{wG_s}{S}}$	G_s, n	$G_s\gamma_w(1-n)$	G_s, w_{sat}	$\left(\dfrac{1+w_{sat}}{1+w_{sat}G_s}\right)G_s\gamma_w$
		G_s, w, S	$\dfrac{G_s\gamma_w}{1+\left(\dfrac{wG_s}{S}\right)}$	e, w_{sat}	$\left(\dfrac{e}{w_{sat}}\right)\left(\dfrac{1+w_{sat}}{1+e}\right)\gamma_w$
w, G_s, n	$G_s\gamma_w(1-n)(1+w)$	e, w, S	$\dfrac{eS\gamma_w}{(1+e)w}$	n, w_{sat}	$n\left(\dfrac{1+w_{sat}}{w_{sat}}\right)\gamma_w$
S, G_s, n	$G_s\gamma_w(1-n)+nS\gamma_w$	γ_{sat}, e	$\gamma_{sat}-\dfrac{e\gamma_w}{1+e}$	γ_d, e	$\gamma_d+\left(\dfrac{e}{1+e}\right)\gamma_w$
		γ_{sat}, n	$\gamma_{sat}-n\gamma_w$	γ_d, n	$\gamma_d+n\gamma_w$
		γ_{sat}, G_s	$\dfrac{(\gamma_{sat}-\gamma_w)G_s}{(G_s-1)}$	γ_d, S	$\left(1-\dfrac{1}{G_s}\right)\gamma_d+\gamma_w$
				γ_d, w_{sat}	$\gamma_d(1+w_{sat})$

Table 3.2 Void Ratio, Moisture Content, and Dry Unit Weight for Some Typical Soils in a Natural State

Type of soil	Void ratio, e	Natural moisture content in a saturated state (%)	Dry unit weight, γ_d (kN/m³)
Loose uniform sand	0.8	30	14.5
Dense uniform sand	0.45	16	18
Loose angular-grained silty sand	0.65	25	16
Dense angular-grained silty sand	0.4	15	19
Stiff clay	0.6	21	17
Soft clay	0.9–1.4	30–50	11.5–14.5
Loess	0.9	25	13.5
Soft organic clay	2.5–3.2	90–120	6–8
Glacial till	0.3	10	21

Example 3.1

For a saturated soil, show that

$$\gamma_{sat} = \left(\frac{e}{w} \right) \left(\frac{1 + w}{1 + e} \right) \gamma w$$

Solution

From Eqs. (3.19) and (3.20),

$$\gamma_{sat} = \frac{(G_s + e)\gamma w}{1 + e} \qquad \text{(a)}$$

and

$$e = wG_s$$

or

$$G_s = \frac{e}{w} \qquad \text{(b)}$$

Combining Eqs. (a) and (b) gives

$$\gamma_{sat} = \frac{\left(\dfrac{e}{w} + e \right)\gamma w}{1 + e} = \left(\frac{e}{w} \right)\left(\frac{1 + w}{1 + e} \right)\gamma w \qquad \blacksquare$$

Example 3.2

For a moist soil sample, the following are given.

- Total volume: $V = 1.2 \text{ m}^3$
- Total mass: $M = 2350 \text{ kg}$
- Moisture content: $w = 8.6\%$
- Specific gravity of soil solids: $G_s = 2.71$

Determine the following.

- **a.** Moist density
- **b.** Dry density
- **c.** Void ratio
- **d.** Porosity
- **e.** Degree of saturation
- **f.** Volume of water in the soil sample

Solution

Part a

From Eq. (3.13),

$$\rho = \frac{M}{V} = \frac{2350}{1.2} = \textbf{1958.3 kg/m}^3$$

Part b
From Eq. (3.14),

$$\rho_d = \frac{M_s}{V} = \frac{M}{(1 + w)V} = \frac{2350}{\left(1 + \dfrac{8.6}{100}\right)(1.2)} = \mathbf{1803.3 \ kg/m^3}$$

Part c
From Eq. (3.22),

$$\rho_d = \frac{G_s\rho_w}{1 + e}$$

$$e = \frac{G_s\rho_w}{\rho_d} - 1 = \frac{(2.71)(1000)}{1803.3} - 1 = \mathbf{0.503}$$

Part d
From Eq. (3.7),

$$n = \frac{e}{1 + e} = \frac{0.503}{1 + 0.503} = \mathbf{0.335}$$

Part e
From Eq. (3.18),

$$S = \frac{wG_s}{e} = \frac{\left(\dfrac{8.6}{100}\right)(2.71)}{0.503} = 0.463 = \mathbf{46.3\%}$$

Part f

Volume of water:

$$\frac{M_w}{\rho_w} = \frac{M - M_s}{\rho_w} = \frac{M - \dfrac{M}{1 + w}}{\rho_w} = \frac{2350 - \left(\dfrac{2350}{1 + \dfrac{8.6}{100}}\right)}{1000} = \mathbf{0.186 \ m^3}$$

Alternate Solution
Refer to Figure 3.7.

Part a

$$\rho = \frac{M}{V} = \frac{2350}{1.2} = \mathbf{1958.3 \ kg/m^3}$$

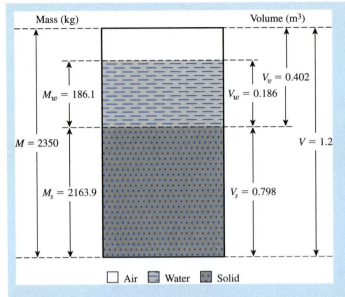

Figure 3.7

Part b

$$M_s = \frac{M}{1 + w} = \frac{2350}{1 + \dfrac{8.6}{100}} = 2163.9 \text{ kg}$$

$$\rho_d = \frac{M_s}{V} = \frac{M}{(1 + w)V} = \frac{2350}{\left(1 + \dfrac{8.6}{100}\right)(1.2)} = \mathbf{1803.3 \text{ kg/m}^3}$$

Part c

The volume of solids: $\dfrac{M_s}{G_s\rho_w} = \dfrac{2163.9}{(2.71)(1000)} = 0.798 \text{ m}^3$

The volume of voids: $V_v = V - V_s = 1.2 - 0.798 = 0.402 \text{ m}^3$

Void ratio: $e = \dfrac{V_v}{V_s} = \dfrac{0.402}{0.798} = \mathbf{0.503}$

Part d

Porosity: $n = \dfrac{V_v}{V} = \dfrac{0.402}{1.2} = \mathbf{0.335}$

Part e

$$S = \frac{V_w}{V_v}$$

Volume of water: $V_w = \dfrac{M_w}{\rho_w} = \dfrac{186.1}{1000} = 0.186 \text{ m}^3$

Hence,

$$S = \frac{0.186}{0.402} = 0.463 = \mathbf{46.3\%}$$

Part f
From Part e,

$$V_w = \mathbf{0.186 \ m^3}$$

Example 3.3

The following data are given for a soil:

- Porosity: $= 0.4$
- Specific gravity of the soil solids: $= 2.68$
- Moisture content: $= 12\%$

Determine the mass of water to be added to 10 m^3 of soil for full saturation.

Solution

Equation (3.27) can be rewritten in terms of density as

$$\rho = G_s \rho_w (1 - n)(1 + W)$$

Similarly, from Eq. (3.28)

$$\rho_{sat} = [(1 - n)G_s + n]\rho_w$$

Thus,

$$\rho = (2.68)(1000)(1 - 0.4)(1 + 0.12) = 1800.96 \ \text{kg/m}^3$$

$$\rho_{sat} = [(1 - 0.4)(2.68) + 0.4] = 2008 \ \text{kg/m}^3$$

Mass of water needed per cubic meter equals

$$\rho_{sat} - \rho = 2008 - 1800.96 = 207.04 \ \text{kg}$$

So, total mass of water to be added equals

$$207.04 \times 10 = \mathbf{2070.4 \ kg}$$

Example 3.4

A saturated soil has a dry unit weight of 16.19 kN/m^3. Its moisture content is 23%.

Determine:

- **a.** Saturated unit weight, γ_{sat}
- **b.** Specific gravity, G_s
- **c.** Void ratio, e

Solution

Part a: Saturated Unit Weight

From Eq. (3.12),

$$\gamma_{sat} = \gamma_d(1 + w) = (16.9)\left(1 + \frac{23}{100}\right) = 19.91 \text{ kN/m}^3$$

Part b: Specific Gravity, G_s

From Eq. (3.16),

$$\gamma_d = \frac{G_s \gamma_w}{1 + e}$$

Also from Eq. (3.20) for saturated soils, $e = wG_s$. Thus,

$$\gamma_d = \frac{G_s \gamma_w}{1 + wG_s}$$

So,

$$16.19 = \frac{G_s(9.81)}{1 + (0.23)(G_s)}$$

or

$$16.19 + 3.72 = 9.81$$
$$G_s = \mathbf{2.66}$$

Part c: Void Ratio, e

For saturated soils,

$$e = wG_s = (0.23)(2.66) = \mathbf{0.61}$$ ∎

3.5   *Relative Density*

The term *relative density* is commonly used to indicate the *in situ* denseness or looseness of granular soil. It is defined as

$$D_r = \frac{e_{max} - e}{e_{max} - e_{min}} \tag{3.30}$$

where D_r = relative density, usually given as a percentage
       e = *in situ* void ratio of the soil
    e_{max} = void ratio of the soil in the loosest state
    e_{min} = void ratio of the soil in the densest state

The values of D_r may vary from a minimum of 0% for very loose soil to a maximum of 100% for very dense soils. Soils engineers qualitatively describe the granular soil deposits according to their relative densities, as shown in Table 3.3. In-place soils seldom

Table 3.3 Qualitative Description of Granular Soil Deposits

Relative density (%)	Description of soil deposit
0–15	Very loose
15–50	Loose
50–70	Medium
70–85	Dense
85–100	Very dense

have relative densities less than 20 to 30%. Compacting a granular soil to a relative density greater than about 85% is difficult.

The relationships for relative density can also be defined in terms of porosity, or

$$e_{max} = \frac{n_{max}}{1 - n_{max}} \tag{3.31}$$

$$e_{min} = \frac{n_{min}}{1 - n_{min}} \tag{3.32}$$

$$e = \frac{n}{1 - n} \tag{3.33}$$

where n_{max} and n_{min} = porosity of the soil in the loosest and densest conditions, respectively. Substituting Eqs. (3.31), (3.32), and (3.33) into Eq. (3.30), we obtain

$$D_r = \frac{(1 - n_{min})(n_{max} - n)}{(n_{max} - n_{min})(1 - n)} \tag{3.34}$$

By using the definition of dry unit weight given in Eq. (3.16), we can express relative density in terms of maximum and minimum possible dry unit weights. Thus,

$$D_r = \frac{\left[\dfrac{1}{\gamma_{d(min)}}\right] - \left[\dfrac{1}{\gamma_d}\right]}{\left[\dfrac{1}{\gamma_{d(min)}}\right] - \left[\dfrac{1}{\gamma_{d(max)}}\right]} = \left[\frac{\gamma_d - \gamma_{d(min)}}{\gamma_{d(max)} - \gamma_{d(min)}}\right]\left[\frac{\gamma_{d(max)}}{\gamma_d}\right] \tag{3.35}$$

where $\gamma_{d(min)}$ = dry unit weight in the loosest condition (at a void ratio of e_{max})
γ_d = *in situ* dry unit weight (at a void ratio of e)
$\gamma_{d(max)}$ = dry unit weight in the densest condition (at a void ratio of e_{min})

In terms of density, Eq. (3.35) can be expressed as

$$D_r = \left[\frac{\rho_d - \rho_{d(min)}}{\rho_{d(max)} - \rho_{d(min)}}\right]\frac{\rho_{d(max)}}{\rho_d} \tag{3.36}$$

ASTM Test Designation D-4253 (2007) provides a procedure for determining the minimum and maximum dry unit weights of granular soils so that they can be used in Eq. (3.35) to measure the relative density of compaction in the field. For sands, this procedure involves using a mold with a volume of 2830 cm³. For a determination of the *minimum dry unit weight,* sand is poured loosely into the mold from a funnel with a 12.7 mm diameter spout. The average height of the fall of sand into the mold is maintained at about 25.4 mm. The value of $\gamma_{d(min)}$ then can be calculated by using the following equation

$$\gamma_{d(min)} = \frac{W_s}{V_m} \tag{3.37}$$

where W_s = weight of sand required to fill the mold
  V_m = volume of the mold

The *maximum dry unit weight* is determined by vibrating sand in the mold for 8 min. A surcharge of 14 kN/m² is added to the top of the sand in the mold. The mold is placed on a table that vibrates at a frequency of 3600 cycles/min and that has an amplitude of vibration of 0.635 mm. The value of $\gamma_{d(max)}$ can be determined at the end of the vibrating period with knowledge of the weight and volume of the sand. Several factors control the magnitude of $\gamma_{d(max)}$: the magnitude of acceleration, the surcharge load, and

Example 3.5

For a given sandy soil, $e_{max} = 0.75$ and $e_{min} = 0.4$. Let $G_s = 2.68$. In the field, the soil is compacted to a moist density of 1797.4 kg/m³ at a moisture content of 12%. Determine the relative density of compaction.

Solution
From Eq. (3.21),

$$\rho = \frac{(1 + w)G_s\rho_w}{1 + e}$$

or

$$e = \frac{G_s\rho_w(1 + w)}{\rho} - 1 = \frac{(2.68)(1000)(1 + 0.12)}{1797.4} - 1 = 0.67$$

From Eq. (3.30),

$$D_r = \frac{e_{max} - e}{e_{max} - e_{min}} = \frac{0.75 - 0.67}{0.75 - 0.4} = \mathbf{0.229 = 22.9\%}$$ ∎

the geometry of acceleration. Hence, one can obtain a larger-value $\gamma_{d(max)}$ than that obtained by using the ASTM standard method described earlier.

Comments on e_{max} *and* e_{min}

The maximum and minimum void ratios for granular soils described in Section 3.5 depend on several factors, such as

- Grain size
- Grain shape
- Nature of the grain-size distribution curve
- Fine contents, F_c (that is, fraction smaller than 0.075 mm)

The amount of *non-plastic fines* present in a given granular soil has a great influence on e_{max} and e_{min}. Figure 3.8 shows a plot of the variation of e_{max} and e_{min} with the percentage of nonplastic fines (by volume) for Nevada 50/80 sand (Lade, *et al.*, 1998). The ratio of D_{50} (size through which 50% of soil will pass) for the sand to that for nonplastic fines used

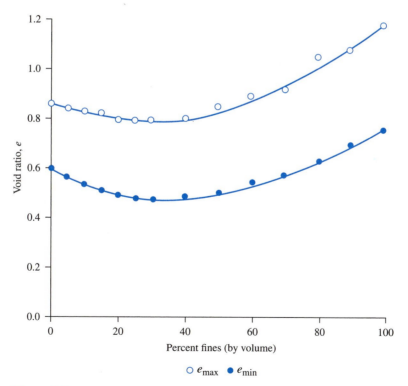

Figure 3.8 Variation of e_{max} and e_{min} (for Nevada 50/80 sand) with percentage of nonplastic fines (*Redrawn from Lade et al, 1998. Copyright ASTM INTERNATIONAL. Reprinted with permission.*)

for the tests shown in Figure 3.8 (that is, $D_{50\text{-sand}}/D_{50\text{-fine}}$) was 4.2. From Figure 3.8, it can be seen that as the percentage of fines by volume increased from zero to about 30% the magnitudes of e_{max} and e_{min} decreased. This is the filling-of-void phase where the fines tend to fill the void spaces between the larger sand particles. There is a transition zone when the percentage of fines is between 30 to 40%. However, when the percentage of fines becomes more than about 40%, the magnitudes of e_{max} and e_{min} start increasing. This is the replacement-of-solids phase where the large-sized particles are pushed out and gradually are replaced by the fines.

Cubrinovski and Ishihara (2002) studied the variation of e_{max} and e_{min} for a very larger number of soils. Based on the best-fit linear-regression lines, they provided the following relationships.

- Clean sand ($F_c = 0$ to 5%)

$$e_{max} = 0.072 + 1.53\, e_{min} \tag{3.38}$$

- Sand with fines ($5 < F_c \le 15\%$)

$$e_{max} = 0.25 + 1.37\, e_{min} \tag{3.39}$$

- Sand with fines and clay ($15 < F_c \le 30\%;\ P_c = 5$ to 20%)

$$e_{max} = 0.44 + 1.21\, e_{min} \tag{3.40}$$

- Silty soils ($30 < F_c \le 70\%;\ P_c = 5$ to 20%)

$$e_{max} = 0.44 + 1.32\, e_{min} \tag{3.41}$$

where F_c = fine fraction for which grain size is smaller than 0.075 mm
P_c = clay-size fraction (< 0.005 mm)

Figure 3.9 shows a plot of $e_{max} - e_{min}$ versus the mean grain size (D_{50}) for a number of soils (Cubrinovski and Ishihara, 1999 and 2002). From this figure, the average plot for sandy and gravelly soils can be given by the relationship

$$e_{max} - e_{min} = 0.23 + \frac{0.06}{D_{50}\,(\text{mm})} \tag{3.42}$$

3.7   *Summary*

In this chapter, we discussed weight–volume relations and the concept of relative density. The volume relationships are those for void ratio, porosity, and degree of saturation. The weight relationships include those for moisture content and dry, moist, and saturated unit weight.

Relative density is a term to describe the denseness of a granular soil. Relative density can be expressed in terms of the maximum, minimum, and *in situ* unit weights/densities of a soil and is generally expressed as a percentage.

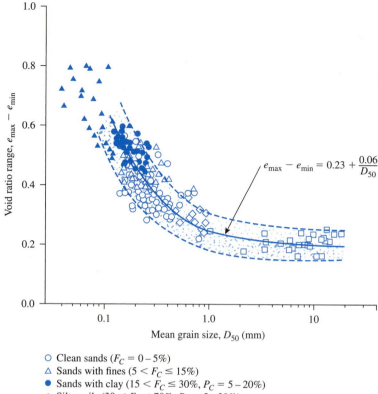

O Clean sands ($F_C = 0 - 5\%$)
△ Sands with fines ($5 < F_C \le 15\%$)
● Sands with clay ($15 < F_C \le 30\%$, $P_C = 5 - 20\%$)
▲ Silty soils ($30 < F_C \le 70\%$, $P_C = 5 - 20\%$)
◇ Gravelly sands ($F_C < 6\%$, $P_C = 17 - 36\%$)
□ Gravels

Figure 3.9 Plot of $e_{max} - e_{min}$ versus the mean grain size (*Cubrinovski and Ishihara, 2002*)

Problems

3.1 For a given soil, show that
$$\gamma_{sat} = \gamma_d + n\gamma_w$$

3.2 For a given soil, show that
$$\gamma_{sat} = \gamma_d + \left(\frac{e}{1 + e} \right)\gamma_w$$

3.3 For a given soil, show that
$$\gamma_d = \frac{eS\gamma_w}{(1 + e)w}$$

3.4 A 0.4-m³ moist soil sample has the following:
- Moist mass = 711.2 kg
- Dry mass = 623.9 kg
- Specific gravity of soil solids = 2.68

Estimate:
a. Moisture content
b. Moist density

      c. Dry density

      d. Void ratio

      e. Porosity

3.5 In its natural state, a moist soil has a volume of 9.35×10^{-3} m^3 and weighs 177.6×10^{-3} kN. The oven-dry weight of the soil is 153.6×10^{-3} kN. If $G_s = 2.67$, calculate the moisture content, moist unit weight, dry unit weight, void ratio, porosity, and degree of saturation.

3.6 The moist weight of 5.66×10^{-3} m^3 of a soil is 102.3×10^{-3} kN. The moisture content and the specific gravity of the soil solids are determined in the laboratory to be 11% and 2.7, respectively. Calculate the following:

      a. Moist unit weight (kN/m^3)

      b. Dry unit weight (kN/m^3)

      c. Void ratio

      d. Porosity

      e. Degree of saturation (%)

      f. Volume occupied by water (m^3)

3.7 The saturated unit weight of a soil is 19.8 kN/m^3. The moisture content of the soil is 17.1%. Determine the following:

      a. Dry unit weight

      b. Specific gravity of soil solids

      c. Void ratio

3.8 The unit weight of a soil is 14.94 kN/m^3. The moisture content of this soil is 19.2% when the degree of saturation is 60%. Determine:

      a. Void ratio

      b. Specific gravity of soil solids

      c. Saturated unit weight

3.9 For a given soil, the following are given: $G_s = 2.67$; moist unit weight, $\gamma = 17.61$ kN/m^3, and moisture content, $w = 10.8\%$. Determine:

      a. Dry unit weight

      b. Void ratio

      c. Porosity

      d. Degree of saturation

3.10 Refer to Problem 3.9. Determine the weight of water, in kN, to be added per cubic meter of soil for:

      a. 80% degree of saturation

      b. 100% degree of saturation

3.11 The moist density of a soil is 1680 kg/m^3. Given $w = 18\%$ and $G_s = 2.73$, determine:

      a. Dry density

      b. Porosity

      c. Degree of saturation

      d. Mass of water, in kg/m^3, to be added to reach full saturation

3.12 The dry density of a soil is 1780 kg/m^3. Given $G_s = 2.68$, what would be the moisture content of the soil when saturated?

3.13 The porosity of a soil is 0.35. Given $G_s = 2.69$, calculate:

      a. saturated unit weight (kN/m^3)

      b. Moisture content when moist unit weight = 17.5 kN/m^3

3.14 A saturated soil has $w = 23\%$ and $G_s = 2.62$. Determine its saturated and dry densities in kg/m^3.

3.15 A soil has $e = 0.75$, $w = 21.5\%$, and $G_s = 2.71$. Determine:
  a. Moist unit weight (kN/m^3)
  b. Dry unit weight (kN/m^3)
  c. Degree of saturation (%)

3.16 A soil has $w = 18.2\%$, $G_s = 2.67$, and $S = 80\%$. Determine the moist and dry unit weights of the soil in kN/m^3.

3.17 The moist unit weight of a soil is 17.66 kN/m^3 at a moisture content of 10%. Given $G_s = 2.7$, determine:
  a. e
  b. Saturated unit weight

3.18 The moist unit weights and degrees of saturation of a soil are given in the table.

γ(kN/m^3)	S (%)
16.62	50
17.71	75

  Determine:
  a. e
  b. G_s

3.19 Refer to Problem 3.18. Determine the weight of water, in kN, that will be in 0.0708 m^3 of the soil when it is saturated.

3.20 For a given sand, the maximum and minimum void ratios are 0.78 and 0.43, respectively. Given $G_s = 2.67$, determine the dry unit weight of the soil in kN/m^3 when the relative density is 65%.

3.21 For a given sandy soil, $e_{max} = 0.75$, $e_{min} = 0.46$, and $G_s = 2.68$. What will be the moist unit weight of compaction (kN/m^3) in the field if $D_r = 78\%$ and $w = 9\%$?

3.22 For a given sandy soil, the maximum and minimum dry unit weights are 16.98 kN/m^3 and 14.46 kN/m^3, respectively. Given $G_s = 2.65$, determine the moist unit weight of this soil when the relative density is 60% and the moisture content is 8%.

3.23 The moisture content of a soil sample is 18.4%, and its dry unit weight is 15.72 kN/m^3. Assuming that the specific gravity of solids is 2.65,
  a. Calculate the degree of saturation.
  b. What is the maximum dry unit weight to which this soil can be compacted without change in its moisture content?

3.24 A loose, uncompacted sand fill 1.83 m in depth has a relative density of 40%. Laboratory tests indicated that the minimum and maximum void ratios of the sand are 0.46 and 0.90, respectively. The specific gravity of solids of the sand is 2.65.
  a. What is the dry unit weight of the sand?
  b. If the sand is compacted to a relative density of 75%, what is the decrease in thickness of the 1.83 m fill?

References

AMERICAN SOCIETY FOR TESTING AND MATERIALS (2007). *Annual Book of ASTM Standards*, Sec. 4, Vol. 04.08. West Conshohocken, Pa.

CUBRINOVSKI, M., and ISHIHARA, K. (1999). "Empirical Correlation Between SPT N-Value and Relative Density for Sandy Soils," *Soils and Foundations*. Vol. 39, No. 5, 61–71.

CUBRINOVSKI, M., and ISHIHARA, K. (2002). "Maximum and Minimum Void Ratio Characteristics of Sands," *Soils and Foundations*. Vol. 42, No. 6, 65–78.

LADE, P. V., LIGGIO, C. D., and YAMAMURO, J. A. (1998). "Effects of Non-Plastic Fines on Minimum and Maximum Void Ratios of Sand," *Geotechnical Testing Journal*, ASTM. Vol. 21, No. 4, 336–347.

4 Plasticity and Structure of Soil

4.1 Introduction

When clay minerals are present in fine-grained soil, the soil can be remolded in the presence of some moisture without crumbling. This cohesive nature is caused by the adsorbed water surrounding the clay particles. In the early 1900s, a Swedish scientist named Atterberg developed a method to describe the consistency of fine-grained soils with varying moisture contents. At a very low moisture content, soil behaves more like a solid. When the moisture content is very high, the soil and water may flow like a liquid. Hence, on an arbitrary basis, depending on the moisture content, the behavior of soil can be divided into four basic states—*solid, semisolid, plastic,* and *liquid*—as shown in Figure 4.1.

The moisture content, in percent, at which the transition from solid to semisolid state takes place is defined as the *shrinkage limit.* The moisture content at the point of transition from semisolid to plastic state is the *plastic limit,* and from plastic to liquid state is the *liquid limit.* These parameters are also known as *Atterberg limits.* This chapter describes the procedures to determine the Atterberg limits. Also discussed in this chapter are soil structure and geotechnical parameters, such as activity and liquidity index, which are related to Atterberg limits, and also soil structure.

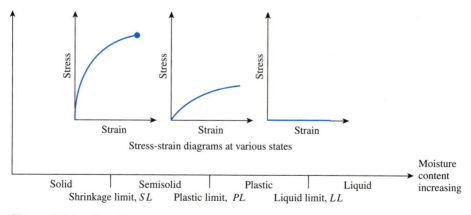

Figure 4.1 Atterberg Limits

Liquid Limit (LL)

A schematic diagram (side view) of a liquid limit device is shown in Figure 4.2a. This device consists of a brass cup and a hard rubber base. The brass cup can be dropped onto the base by a cam operated by a crank. To perform the liquid limit test, one must place a soil

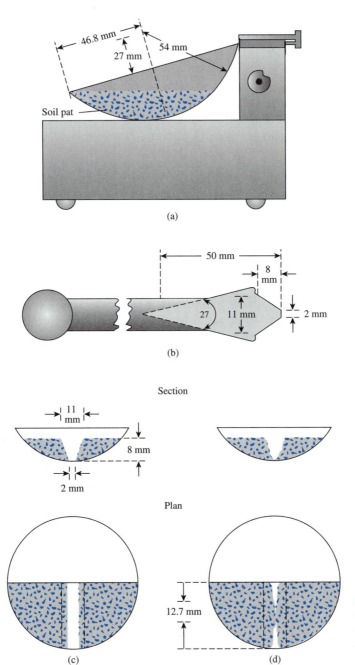

Figure 4.2 Liquid limit test: (a) liquid limit device; (b) grooving tool; (c) soil pat before test; (d) soil pat after test

paste in the cup. A groove is then cut at the center of the soil pat with the standard groov-ing tool (Figure 4.2b). By the use of the crank-operated cam, the cup is lifted and dropped from a height of 10 mm. The moisture content, in percent, required to close a distance of 12.7 mm along the bottom of the groove (see Figures 4.2c and 4.2d) after 25 blows is defined as the *liquid limit.*

It is difficult to adjust the moisture content in the soil to meet the required 12.7 mm closure of the groove in the soil pat at 25 blows. Hence, at least three tests for the same soil are conducted at varying moisture contents, with the number of blows, *N,* required to achieve closure varying between 15 and 35. Figure 4.3 shows a photograph of a liquid limit test device and grooving tools. Figure 4.4 shows photographs of the soil pat in the liquid limit device before and after the test. The moisture content of the soil, in percent, and the corresponding number of blows are plotted on semilogarithmic graph paper (Figure 4.5). The relationship between moisture content and log *N* is approximated as a straight line. This line is referred to as the *flow curve.* The moisture content corresponding to *N* = 25, determined from the flow curve, gives the liquid limit of the soil. The slope of the flow line is defined as the *flow index* and may be written as

$$I_F = \frac{w_1 - w_2}{\log\left(\dfrac{N_2}{N_1}\right)} \qquad (4.1)$$

Figure 4.3 Liquid limit test device and grooving tools (*Courtesy of ELE International*)

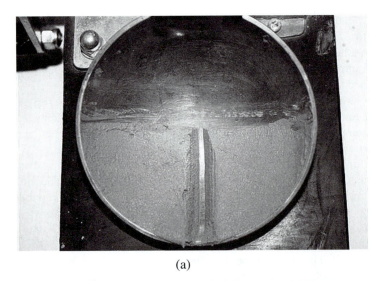

(a)

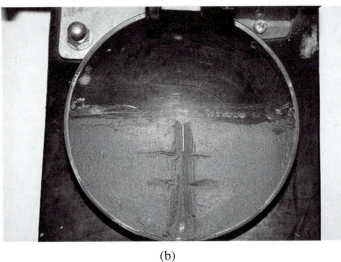

(b)

Figure 4.4
Photographs showing
the soil pat in the liquid
limit device: (a) before
test; (b) after test [*Note*:
The 12.7 mm groove
closure in (b) is marked
for clarification]
(*Courtesy of Braja M.
Das, Henderson,
Nevada*)

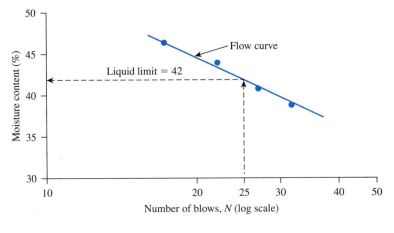

Figure 4.5
Flow curve for
liquid limit
determination of
a clayey silt

where I_F = flow index

$\quad w_1$ = moisture content of soil, in percent, corresponding to N_1 blows

$\quad w_2$ = moisture content corresponding to N_2 blows

Note that w_2 and w_1 are exchanged to yield a positive value even though the slope of the flow line is negative. Thus, the equation of the flow line can be written in a general form as

$$w = -I_F \log N + C \tag{4.2}$$

where C = a constant.

From the analysis of hundreds of liquid limit tests, the U.S. Army Corps of Engineers (1949) at the Waterways Experiment Station in Vicksburg, Mississippi, proposed an empirical equation of the form

$$LL = w_N \left(\frac{N}{25}\right)^{\tan \beta} \tag{4.3}$$

where N = number of blows in the liquid limit device for a 12.7 mm groove closure

$\quad w_N$ = corresponding moisture content

$\quad \tan \beta$ = 0.121 (but note that $\tan \beta$ is not equal to 0.121 for all soils)

Equation (4.3) generally yields good results for the number of blows between 20 and 30. For routine laboratory tests, it may be used to determine the liquid limit when only one test is run for a soil. This procedure is generally referred to as the *one-point method* and was also adopted by ASTM under designation D-4318. The reason that the one-point method yields fairly good results is that a small range of moisture content is involved when $N = 20$ to $N = 30$.

Another method of determining liquid limit that is popular in Europe and Asia is the *fall cone method* (British Standard—BS1377). In this test the liquid limit is defined as the moisture content at which a standard cone of apex angle 30° and weight of 0.78 N (80 gf) will penetrate a distance $d = 20$ mm in 5 seconds when allowed to drop from a position of point contact with the soil surface (Figure 4.6a). Due to the difficulty in achieving the liquid limit from a single test, four or more tests can be conducted at various moisture contents to determine the fall cone penetration, d. A semilogarithmic graph can then be plotted with moisture content (w) versus cone penetration d. The plot results in a straight line. The moisture content corresponding to $d = 20$ mm is the liquid limit (Figure 4.6b). From Figure 4.6(b), the *flow index* can be defined as

$$I_{FC} = \frac{w_2(\%) - w_1(\%)}{\log d_2 - \log d_1} \tag{4.4}$$

where w_1, w_2 = moisture contents at cone penetrations of d_1 and d_2, respectively.

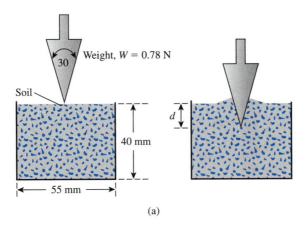

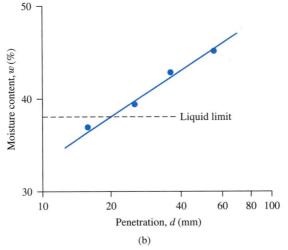

(b)

Figure 4.6
(a) Fall cone test (b) plot of moisture content vs. cone penetration for determination of liquid limit

| 4.3 | *Plastic Limit (PL)* |

The *plastic limit* is defined as the moisture content in percent, at which the soil crumbles, when rolled into threads of 3.2 in diameter. The plastic limit is the lower limit of the plastic stage of soil. The plastic limit test is simple and is performed by repeated rollings of an ellipsoidal-sized soil mass by hand on a ground glass plate (Figure 4.7). The procedure for the plastic limit test is given by ASTM in Test Designation D-4318.

As in the case of liquid limit determination, the fall cone method can be used to obtain the plastic limit. This can be achieved by using a cone of similar geometry but with a mass of 2.35 N (240 gf). Three to four tests at varying moisture contents of soil are conducted, and the corresponding cone penetrations (d) are determined. The moisture content corresponding to a cone penetration of $d = 20$ mm is the plastic limit. Figure 4.8 shows the liquid and plastic limit determination of Cambridge Gault clay reported by Wroth and Wood (1978).

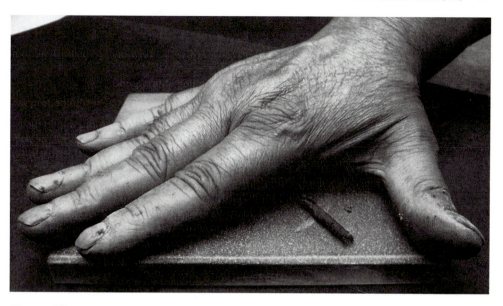

Figure 4.7 Rolling of soil mass on ground glass plate to determine plastic limit (*Courtesy of Braja M. Das, Henderson, Nevada*)

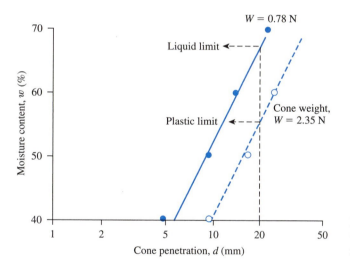

Figure 4.8
Liquid and plastic limits for Cambridge Gault clay determined by fall cone test

The *plasticity index* (*PI*) is the difference between the liquid limit and the plastic limit of a soil, or

$$PI = LL - PL \qquad (4.5)$$

Table 4.1 gives the ranges of liquid limit, plastic limit, and activity (Section 4.6) of some clay minerals (Mitchell, 1976; Skempton, 1953).

Table 4.1 Typical Values of Liquid Limit, Plastic Limit, and Activity of Some Clay Minerals

Mineral	Liquid limit, *LL*	Plastic limit, *PL*	Activity, *A*
Kaolinite	35–100	20–40	0.3–0.5
Illite	60–120	35–60	0.5–1.2
Montmorillonite	100–900	50–100	1.5–7.0
Halloysite (hydrated)	50–70	40–60	0.1–0.2
Halloysite (dehydrated)	40–55	30–45	0.4–0.6
Attapulgite	150–250	100–125	0.4–1.3
Allophane	200–250	120–150	0.4–1.3

Burmister (1949) classified the plasticity index in a qualitative manner as follows:

PI	Description
0	Nonplastic
1–5	Slightly plastic
5–10	Low plasticity
10–20	Medium plasticity
20–40	High plasticity
>40	Very high plasticity

The plasticity index is important in classifying fine-grained soils. It is fundamental to the Casagrande plasticity chart (presented in Section 4.7), which is currently the basis for the Unified Soil Classification System. (See Chapter 5.)

Sridharan, *et al.* (1999) showed that the plasticity index can be correlated to the flow index as obtained from the liquid limit tests (Section 4.2). According to their study,

$$PI\,(\%) = 4.12I_F\,(\%) \tag{4.6}$$

and

$$PI\,(\%) = 0.74I_{FC}\,(\%) \tag{4.7}$$

In a recent study by Polidori (2007) that involved six inorganic soils and their respective mixtures with fine silica sand, it was shown that

$$PL = 0.04(LL) + 0.26(CF) + 10 \tag{4.8}$$

and

$$PI = 0.96(LL) - 0.26(CF) - 10 \tag{4.9}$$

where CF = clay fraction (<2 μm) in %. The experimental results of Polidori (2007) show that the preceding relationships hold good for CF approximately equal to or greater than 30%.

4.4 *Shrinkage Limit (SL)*

Soil shrinks as moisture is gradually lost from it. With continuing loss of moisture, a stage of equilibrium is reached at which more loss of moisture will result in no further volume change (Figure 4.9). The moisture content, in percent, at which the volume of the soil mass ceases to change is defined as the *shrinkage limit.*

Shrinkage limit tests (ASTM Test Designation D-427) are performed in the laboratory with a porcelain dish about 44 mm in diameter and about 12.7 mm high. The inside of the dish is coated with petroleum jelly and is then filled completely with wet soil. Excess soil standing above the edge of the dish is struck off with a straightedge. The mass of the wet soil inside the dish is recorded. The soil pat in the dish is then oven-dried. The volume of the oven-dried soil pat is determined by the displacement of mercury. Because handling mercury may be hazardous, ASTM D-4943 describes a method of dipping the oven-dried soil pat in a melted pot of wax. The wax-coated soil pat is then cooled. Its volume is determined by submerging it in water.

By reference to Figure 4.9, the shrinkage limit can be determined as

$$SL = w_i(\%) - \Delta w\,(\%) \tag{4.10}$$

where w_i = initial moisture content when the soil is placed in the shrinkage limit dish
Δw = change in moisture content (that is, between the initial moisture content and the moisture content at the shrinkage limit)

However,

$$w_i(\%) = \frac{M_1 - M_2}{M_2} \times 100 \tag{4.11}$$

where M_1 = mass of the wet soil pat in the dish at the beginning of the test (g)
M_2 = mass of the dry soil pat (g) (see Figure 4.10)

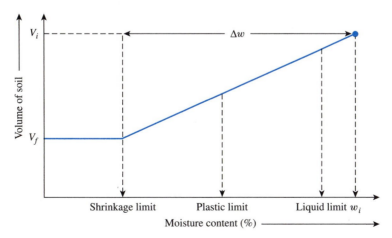

Figure 4.9 Definition of shrinkage limit

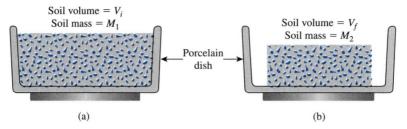

Figure 4.10 Shrinkage limit test: (a) soil pat before drying; (b) soil pat after drying

Also,

$$\Delta w \, (\%) = \frac{(V_i - V_f)\rho_w}{M_2} \times 100 \tag{4.12}$$

where V_i = initial volume of the wet soil pat (that is, inside volume of the dish, cm^3)
V_f = volume of the oven-dried soil pat (cm^3)
ρ_w = density of water (g/cm^3)

Finally, combining Eqs. (4.10), (4.11), and (4.12) gives

$$SL = \left(\frac{M_1 - M_2}{M_2} \right)(100) - \left(\frac{V_i - V_f}{M_2} \right)(\rho_w)(100) \tag{4.13}$$

Another parameter that can be determined from a shrinkage limit test is the *shrinkage ratio,* which is the ratio of the volume change of soil as a percentage of the dry volume to the corresponding change in moisture content, or

$$SR = \frac{\left(\dfrac{\Delta V}{V_f} \right)}{\left(\dfrac{\Delta M}{M_2} \right)} = \frac{\left(\dfrac{\Delta V}{V_f} \right)}{\left(\dfrac{\Delta V \rho_w}{M_2} \right)} = \frac{M_2}{V_f \rho_w} \tag{4.14}$$

where ΔV = change in volume
ΔM = corresponding change in the mass of moisture

It can also be shown that

$$G_s = \frac{1}{\dfrac{1}{SR} - \left(\dfrac{SL}{100} \right)} \tag{4.15}$$

where G_s = specific gravity of soil solids.

Example 4.1

Following are the results of a shrinkage limit test:

- Initial volume of soil in a saturated state = 24.6 cm^3
- Final volume of soil in a dry state = 15.9 cm^3
- Initial mass in a saturated state = 44.0 g
- Final mass in a dry state = 30.1 g

Determine the shrinkage limit of the soil.

Solution
From Eq. (4.13),

$$SL = \left(\frac{M_1 - M_2}{M_2}\right)(100) - \left(\frac{V_i - V_f}{M_2}\right)(\rho_w)(100)$$

$$M_1 = 44.0\text{g} \qquad V_i = 24.6 \text{ cm}^3 \qquad \rho_w = 1 \text{ g/cm}^3$$

$$M_2 = 30.1\text{g} \qquad V_f = 15.9 \text{ cm}^3$$

$$SL = \left(\frac{44.0 - 30.1}{30.1}\right)(100) - \left(\frac{24.6 - 15.9}{30.1}\right)(1)(100)$$

$$= 46.18 - 28.9 = \textbf{17.28\%}$$ ∎

Typical values of shrinkage limit for some clay minerals are as follows (Mitchell, 1976).

Mineral	Shrinkage limit
Montmorillonite	8.5–15
Illite	15–17
Kaolinite	25–29

4.5 *Liquidity Index and Consistency Index*

The relative consistency of a cohesive soil in the natural state can be defined by a ratio called the *liquidity index*, which is given by

$$LI = \frac{w - PL}{LL - PL} \tag{4.16}$$

where w = *in situ* moisture content of soil.

The *in situ* moisture content for a sensitive clay may be greater than the liquid limit. In this case (Figure 4.11),

$$LI > 1$$

These soils, when remolded, can be transformed into a viscous form to flow like a liquid.

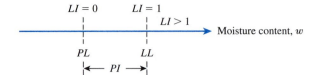

Figure 4.11 Liquidity index

Soil deposits that are heavily overconsolidated may have a natural moisture content less than the plastic limit. In this case (Figure 4.11),

$$LI < 0$$

Another index that is commonly used for engineering purposes is the *consistency index (CI)*, which may be defined as

$$CI = \frac{LL - w}{LL - PI} \tag{4.17}$$

where $w = $ *in situ* moisture content. If w is equal to the liquid limit, the consistency index is zero. Again, if $w = PI$, then $CI = 1$.

4.6   *Activity*

Because the plasticity of soil is caused by the adsorbed water that surrounds the clay particles, we can expect that the type of clay minerals and their proportional amounts in a soil will affect the liquid and plastic limits. Skempton (1953) observed that the plasticity index of a soil increases linearly with the percentage of clay-size fraction (% finer than 2 μm by weight) present (Figure 4.12). The correlations of *PI* with the clay-size fractions for different clays plot separate lines. This difference is due to the diverse plasticity characteristics of the

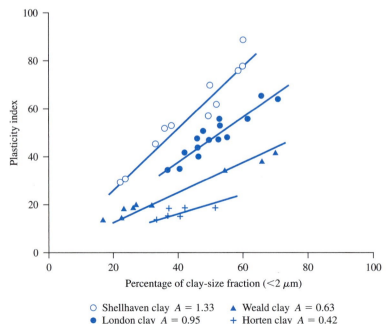

Percentage of clay-size fraction ($<2\ \mu$m)

○ Shellhaven clay $A = 1.33$ ▲ Weald clay $A = 0.63$
● London clay $A = 0.95$ + Horten clay $A = 0.42$

Figure 4.12
Activity (*Based on Skempton, 1953*)

various types of clay minerals. On the basis of these results, Skempton defined a quantity called *activity,* which is the slope of the line correlating *PI* and % finer than 2 μm. This activity may be expressed as

$$A = \frac{PI}{(\%\,\text{of clay-size fraction, by weight})} \tag{4.18}$$

where A = activity. Activity is used as an index for identifying the swelling potential of clay soils. Typical values of activities for various clay minerals are given in Table 4.1.

Seed, Woodward, and Lundgren (1964a) studied the plastic property of several artificially prepared mixtures of sand and clay. They concluded that, although the relationship of the plasticity index to the percentage of clay-size fraction is linear (as observed by Skempton), it may not always pass through the origin. This is shown in Figures 4.13 and 4.14. Thus, the activity can be redefined as

$$A = \frac{PI}{\%\,\text{of clay-size fraction} \ - \ C'} \tag{4.19}$$

where C' is a constant for a given soil.

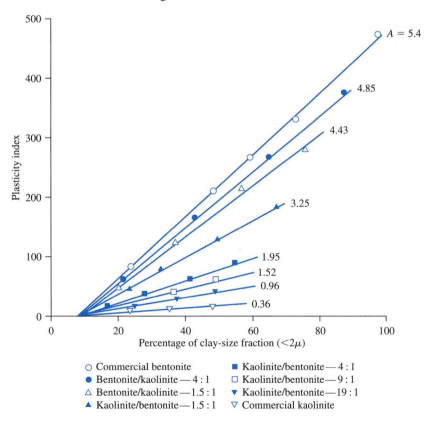

Figure 4.13 Relationship between plasticity index and clay-size fraction by weight for kaolinite/bentonite clay mixtures (*After Seed, Woodward, and Lundgren, 1964a. With permission from ASCE.*)

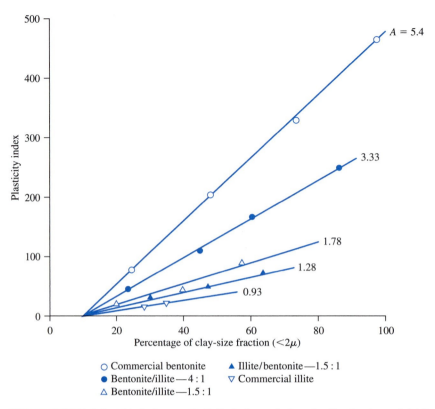

Figure 4.14 Relationship between plasticity index and clay-size fraction by weight for illite/bentonite clay mixtures (*After Seed, Woodward, and Lundgren, 1964a. With permission from ASCE.*)

For the experimental results shown in Figures 4.13 and 4.14, $C' = 9$.

Further works of Seed, Woodward, and Lundgren (1964b) have shown that the relationship of the plasticity index to the percentage of clay-size fractions present in a soil can be represented by two straight lines. This is shown qualitatively in Figure 4.15. For clay-size fractions greater than 40%, the straight line passes through the origin when it is projected back.

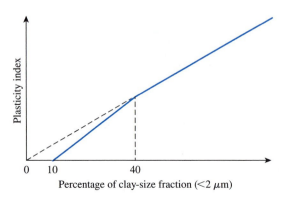

Figure 4.15 Simplified relationship between plasticity index and percentage of clay-size fraction by weight (*After Seed, Woodward, and Lundgren, 1964b. With permission from ASCE.*)

Based on Eqs. (4.8) and (4.9), Polidori (2007) provided an empirical relationship for activity as (for *CF* equal to or greater than 30%)

$$A = \frac{0.96(LL) - 0.26(CF) - 10}{CF} \tag{4.20}$$

where *CF* is the clay fraction ($<2 \ \mu$m)

4.7 *Plasticity Chart*

Liquid and plastic limits are determined by relatively simple laboratory tests that provide information about the nature of cohesive soils. Engineers have used the tests extensively for the correlation of several physical soil parameters as well as for soil identification. Casagrande (1932) studied the relationship of the plasticity index to the liquid limit of a wide variety of natural soils. On the basis of the test results, he proposed a plasticity chart as shown in Figure 4.16. The important feature of this chart is the empirical *A*-line that is given by the equation $PI = 0.73(LL - 20)$. An *A*-line separates the inorganic clays from the inorganic silts. Inorganic clay values lie above the *A*-line, and values for inorganic silts lie below the *A*-line. Organic silts plot in the same region (below the *A*-line and with *LL* ranging from 30 to 50) as the inorganic silts of medium compressibility. Organic clays plot in the same region as inorganic silts of high compressibility (below the *A*-line and *LL* greater than 50). The information provided in the plasticity chart is of great value and is the basis for the classification of fine-grained soils in the Unified Soil Classification System. (See Chapter 5.)

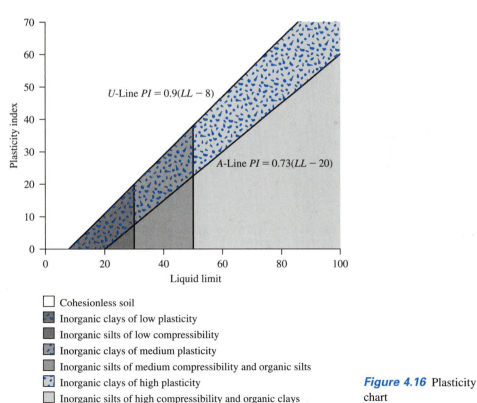

Figure 4.16 Plasticity chart

Note that a line called the *U*-line lies above the *A*-line. The *U*-line is approximately the upper limit of the relationship of the plasticity index to the liquid limit for any currently known soil. The equation for the *U*-line can be given as

$$PI = 0.9(LL - 8) \tag{4.21}$$

There is another use for the *A*-line and the *U*-line. Casagrande has suggested that the shrinkage limit of a soil can be approximately determined if its plasticity index and liquid limit are known (see Holtz and Kovacs, 1981). This can be done in the following manner with reference to Figure 4.17.

a. Plot the plasticity index against the liquid limit of a given soil such as point *A* in Figure 4.17.
b. Project the *A*-line and the *U*-line downward to meet at point *B*. Point *B* will have the coordinates of $LL = -43.5$ and $PI = -46.4$.
c. Join points *B* and *A* with a straight line. This will intersect the liquid limit axis at point *C*. The abscissa of point *C* is the estimated shrinkage limit.

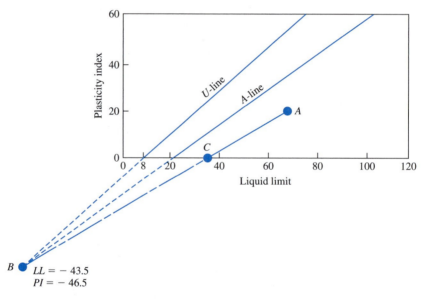

Figure 4.17 Estimation of shrinkage from plasticity chart (*Adapted from Holtz and Kovacs, 1981*)

4.8 *Soil Structure*

Soil structure is defined as the geometric arrangement of soil particles with respect to one another. Among the many factors that affect the structure are the shape, size, and mineralogical composition of soil particles, and the nature and composition of soil water. In general, soils can be placed into two groups: cohesionless and cohesive. The structures found in soils in each group are described next.

Structures in Cohesionless Soil

The structures generally encountered in cohesionless soils can be divided into two major categories: *single grained* and *honeycombed.* In single-grained structures, soil particles are in stable positions, with each particle in contact with the surrounding ones. The shape and size distribution of the soil particles and their relative positions influence the denseness of packing (Figure 4.18); thus, a wide range of void ratios is possible. To get an idea of the variation of void ratios caused by the relative positions of the particles, let us consider the mode of packing of equal spheres shown in Figure 4.19.

Figure 4.19a shows the case of a very loose state of packing. If we isolate a cube with each side measuring *d,* which is equal to the diameter of each sphere as shown in the figure, the void ratio can be calculated as

$$e = \frac{V_v}{V_s} = \frac{V - V_s}{V_s}$$

where V = volume of the cube = d^3
V_s = volume of sphere (i.e., solid) inside the cube

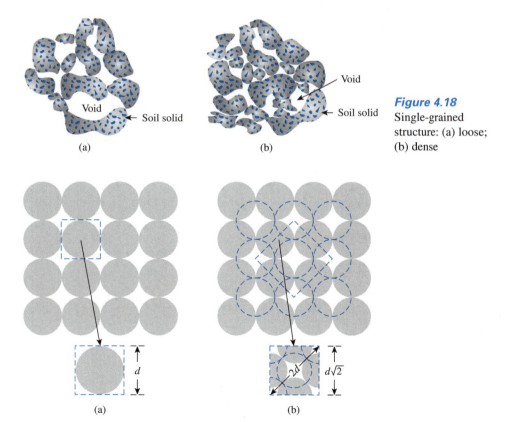

(a)   (b)

Void ← Soil solid   Void   Soil solid

Figure 4.18
Single-grained structure: (a) loose; (b) dense

(a)   (b)

d $d\sqrt{2}$

Figure 4.19 Mode of packing of equal spheres (plan views): (a) very loose packing ($e = 0.91$); (b) very dense packing ($e = 0.35$)

Noting that $V = d^3$ and $V_s = \pi d^3/6$ yields

$$e = \frac{d^3 - \left(\dfrac{\pi d^3}{6}\right)}{\left(\dfrac{\pi d^3}{6}\right)} = 0.91$$

Similarly, Figure 4.19b shows the case of a very dense state of packing. Figure 4.19b also shows an isolated cube, for which each side measures $d\sqrt{2}$. It can be shown that, for this case, $e = 0.35$.

Real soil differs from the equal-spheres model in that soil particles are neither equal in size nor spherical. The smaller-size particles may occupy the void spaces between the larger particles, thus the void ratio of soils is decreased compared with that for equal spheres. However, the irregularity in the particle shapes generally yields an increase in the void ratio of soils. As a result of these two factors, the void ratios encountered in real soils have approximately the same range as those obtained in equal spheres.

In the honeycombed structure (Figure 4.20), relatively fine sand and silt form small arches with chains of particles. Soils that exhibit a honeycombed structure have large void ratios, and they can carry an ordinary static load. However, under a heavy load or when subjected to shock loading, the structure breaks down, which results in a large amount of settlement.

Structures in Cohesive Soils

To understand the basic structures in cohesive soils, we need to know the types of forces that act between clay particles suspended in water. In Chapter 2, we discussed the negative charge on the surface of the clay particles and the diffuse double layer surrounding each particle. When two clay particles in suspension come close to each other, the tendency for interpenetration of the diffuse double layers results in repulsion between the particles. At the same time, an attractive force exists between the clay particles that is caused by van der Waals forces and is independent of the characteristics of water. Both repulsive and attractive forces increase with decreasing distance between the particles, but at different rates. When the spacing between the particles is very small, the force of attraction is greater than the force of repulsion. These are the forces treated by colloidal theories.

Soil solid

Void

Figure 4.20 Honeycombed structure

The fact that local concentrations of positive charges occur at the edges of clay particles was discussed in Chapter 2. If the clay particles are very close to each other, the positively charged edges can be attracted to the negatively charged faces of the particles.

Let us consider the behavior of clay in the form of a dilute suspension. When the clay is initially dispersed in water, the particles repel one another. This repulsion occurs because with larger interparticle spacing, the forces of repulsion between the particles are greater than the forces of attraction (van der Waals forces). The force of gravity on each particle is negligible. Thus, the individual particles may settle very slowly or remain in suspension, undergoing *Brownian motion* (a random zigzag motion of colloidal particles in suspension). The sediment formed by the settling of the individual particles has a dispersed structure, and all particles are oriented more or less parallel to one another (Figure 4.21a).

If the clay particles initially dispersed in water come close to one another during random motion in suspension, they might aggregate into visible flocs with edge-to-face contact. In this instance, the particles are held together by electrostatic attraction of positively charged edges to negatively charged faces. This aggregation is known as *flocculation.* When the flocs become large, they settle under the force of gravity. The sediment formed in this manner has a flocculent structure (Figure 4.21b).

When salt is added to a clay-water suspension that has been initially dispersed, the ions tend to depress the double layer around the particles. This depression reduces the interparticle repulsion. The clay particles are attracted to one another to form flocs and settle. The flocculent structure of the sediments formed is shown in Figure 4.21c. In flocculent sediment structures of the salt type, the particle orientation approaches a large degree of parallelism, which is due to van der Waals forces.

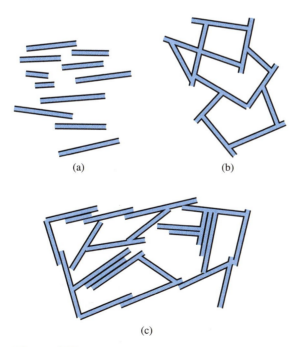

(a)

(b)

(c)

Figure 4.21 Sediment structures: (a) dispersion; (b) nonsalt flocculation; (c) salt flocculation (*Adapted from Lambe, 1958*)

Clays that have flocculent structures are lightweight and possess high void ratios. Clay deposits formed in the sea are highly flocculent. Most of the sediment deposits formed from freshwater possess an intermediate structure between dispersed and flocculent.

A deposit of pure clay minerals is rare in nature. When a soil has 50% or more particles with sizes of 0.002 mm or less, it is generally termed *clay.* Studies with scanning electron microscopes (Collins and McGown, 1974; Pusch, 1978; Yong and Sheeran, 1973) have shown that individual clay particles tend to be aggregated or flocculated in submicroscopic units. These units are referred to as *domains.* The domains then group together, and these groups are called *clusters.* Clusters can be seen under a light microscope. This grouping to form clusters is caused primarily by interparticle forces. The clusters, in turn, group to form *peds.* Peds can be seen without a microscope. Groups of peds are macrostructural features along with joints and fissures. Figure 4.22a shows the arrangement of the peds and macropore spaces. The arrangement of domains and clusters with silt-size particles is shown in Figure 4.22b.

From the preceding discussion, we can see that the structure of cohesive soils is highly complex. Macrostructures have an important influence on the behavior of soils from an engineering viewpoint. The microstructure is more important from a fundamental viewpoint. Table 4.2 summarizes the macrostructures of clay soils.

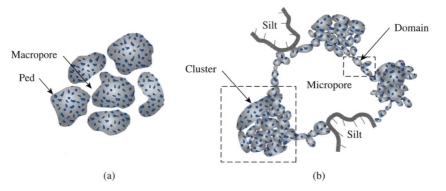

(a)   (b)

Figure 4.22 Soil structure: (a) arrangement of peds and macropore spaces; (b) arrangement of domains and clusters with silt-sized particles

Table 4.2 Structure of Clay Soils

Item	Remarks
Dispersed structures	Formed by settlement of individual clay particles; more or less parallel orientation (see Figure 4.21a)
Flocculent structures	Formed by settlement of flocs of clay particles (see Figures 4.21b and 4.21c)
Domains	Aggregated or flocculated submicroscopic units of clay particles
Clusters	Domains group to form clusters; can be seen under light microscope
Peds	Clusters group to form peds; can be seen without microscope

4.9 *Summary*

This chapter discussed two major components in the study of soil mechanics. They are (a) plasticity of soil and related topics (Sections 4.2 to 4.7), and (b) structure of soil (Section 4.8).

Liquid limit, plastic limit, and shrinkage limit tests of fine-grained soil are indicators of the nature of its plasticity. The difference between the liquid limit and plastic limit is called the plasticity index. Liquid limit and plasticity index are required parameters for classification of fine-grained soils.

The structure of cohesionless soils can be single grained or honeycombed. Honeycombed structures are encountered in relatively fine sands and silts. The macrostructure of clay soils can be broadly divided into categories such as dispersed structures, flocculent structures, domains, clusters, and peds.

Problems

4.1   Following are the results from the liquid and plastic limit tests for a soil. *Liquid limit test:*

Number of blows, N	Moisture content (%)
16	36.5
20	34.1
28	27.0

Plastic limit test: PL = 12.2%
a. Draw the flow curve and obtain the liquid limit.
b. What is the plasticity index of the soil?

4.2   Determine the liquidity index of the soil described in Problem 4.1 if $w_{\text{in situ}} = 31\%$.

4.3   Following are the results from the liquid and plastic limit tests for a soil. *Liquid limit test:*

Number of blows, N	Moisture content (%)
15	42
20	40.8
28	39.1

Plastic limit test: PL = 18.7%
a. Draw the flow curve and obtain the liquid limit.
b. What is the plasticity index of the soil?

4.4   Refer to Problem 4.3. Determine the liquidity index of the soil when the *in situ* moisture content is 26%.

4.5   A saturated soil has the following characteristics: initial volume (V_i) = 19.65 cm³, final volume (V_f) = 13.5 cm³, mass of wet soil (M_1) = 36 g, and mass of dry soil (M_2) = 25 g. Determine the shrinkage limit and the shrinkage ratio.

4.6   Repeat Problem 4.5 with the following: V_i = 24.6 cm³, V_f = 15.9 cm³, M_1 = 44 g, and M_2 = 30.1 g.

References

AMERICAN SOCIETY FOR TESTING AND MATERIALS (2004). *Annual Book of ASTM Standards,* Sec. 4, Vol. 04.08, West Conshohocken, Pa.

BS:1377 (1990). *British Standard Methods of Tests for Soil for Engineering Purposes,* Part 2, BSI, London.

BURMISTER, D. M. (1949). "Principles and Techniques of Soil Identification," *Proceedings,* Annual Highway Research Board Meeting, National Research Council, Washington, D.C., Vol. 29, 402–434.

CASAGRANDE, A. (1932). "Research of Atterberg Limits of Soils," *Public Roads,* Vol. 13, No. 8, 121–136.

COLLINS, K., and McGOWN, A. (1974). "The Form and Function of Microfabric Features in a Variety of Natural Soils," *Geotechnique,* Vol. 24, No. 2, 223–254.

HOLTZ, R. D., and KOVACS, W. D. (1981). *An Introduction to Geotechnical Engineering,* Prentice-Hall, Englewood Cliffs, NJ.

LAMBE, T. W. (1958). "The Structure of Compacted Clay," *Journal of the Soil Mechanics and Foundations Division,* ASCE, Vol. 85, No. SM2, 1654-1 to 1654-35.

MITCHELL, J. K. (1976). *Fundamentals of Soil Behavior,* Wiley, New York.

POLIDORI, E. (2007). "Relationship between Atterberg Limits and Clay Contents," *Soils and Foundations,* Vol. 47, No. 5, 887–896.

PUSCH, R. (1978). "General Report on Physico-Chemical Processes Which Affect Soil Structure and Vice Versa," *Proceedings,* International Symposium on Soil Structure, Gothenburg, Sweden, Appendix, 34.

SEED, H. B., WOODWARD, R. J., and LUNDGREN, R. (1964a). "Clay Mineralogical Aspects of Atterberg Limits," *Journal of the Soil Mechanics and Foundations Division,* ASCE, Vol. 90, No. SM4, 107–131.

SEED, H. B., WOODWARD, R. J., and LUNDGREN, R. (1964b). "Fundamental Aspects of the Atterberg Limits," *Journal of the Soil Mechanics and Foundations Division,* ASCE, Vol. 90, No. SM6, 75–105.

SKEMPTON, A. W. (1953). "The Colloidal Activity of Clays," *Proceedings,* 3rd International Conference on Soil Mechanics and Foundation Engineering, London, Vol. 1, 57–61.

SRIDHARAN, A., NAGARAJ, H. B., and PRAKASH, K. (1999). "Determination of the Plasticity Index from Flow Index," *Geotechnical Testing Journal,* ASTM, Vol. 22, No. 2, 175–181.

U.S. ARMY CORPS OF ENGINEERS (1949). *Technical Memo 3-286,* U.S. Waterways Experiment Station, Vicksburg, Miss.

WROTH, C. P., and WOOD, D. M. (1978). "The Correlation of Index Properties with Some Basic Engineering Properties of Soils," *Canadian Geotechnical Journal,* Vol. 15, No. 2, 137–145.

YONG, R. N., and SHEERAN, D. E. (1973). "Fabric Unit Interaction and Soil Behavior," *Proceedings,* International Symposium on Soil Structure, Gothenburg, Sweden, 176–184.

5 Classification of Soil

Different soils with similar properties may be classified into groups and sub-groups according to their engineering behavior. Classification systems provide a common language to concisely express the general characteristics of soils, which are infinitely varied, without detailed descriptions. Most of the soil classification systems that have been developed for engineering purposes are based on simple index properties such as particle-size distribution and plasticity. Although several classification systems are now in use, none is totally definitive of any soil for all possible applications because of the wide diversity of soil properties.

5.1 Textural Classification

In a general sense, *texture* of soil refers to its surface appearance. Soil texture is influenced by the size of the individual particles present in it. Table 2.3 divided soils into gravel, sand, silt, and clay categories on the basis of particle size. In most cases, natural soils are mixtures of particles from several size groups. In the textural classification system, the soils are named after their principal components, such as sandy clay, silty clay, and so forth.

A number of textural classification systems were developed in the past by different organizations to serve their needs, and several of those are in use today. Figure 5.1 shows the textural classification systems developed by the U.S. Department of Agriculture (USDA). This classification method is based on the particle-size limits as described under the USDA system in Table 2.3; that is

- *Sand size:* 2.0 to 0.05 mm in diameter
- *Silt size:* 0.05 to 0.002 mm in diameter
- *Clay size:* smaller than 0.002 mm in diameter

The use of this chart can best be demonstrated by an example. If the particle-size distribution of soil *A* shows 30% sand, 40% silt, and 30% clay-size particles, its textural classification can be determined by proceeding in the manner indicated by the arrows in Figure 5.1. This soil falls into the zone of *clay loam*. Note that this chart is based on only

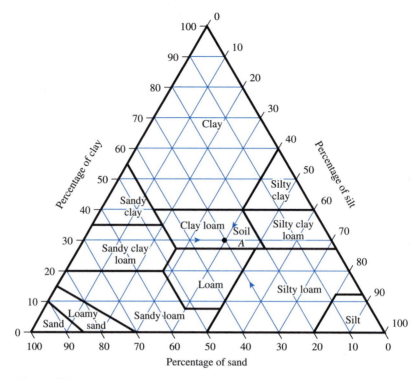

Figure 5.1 U.S. Department of Agriculture textural classification

the fraction of soil that passes through the No. 10 sieve. Hence, if the particle-size distribution of a soil is such that a certain percentage of the soil particles is larger than 2 mm in diameter, a correction will be necessary. For example, if soil *B* has a particle-size distribution of 20% gravel, 10% sand, 30% silt, and 40% clay, the modified textural compositions are

$$Sand\ size: \frac{10 \times 100}{100 - 20} = 12.5\%$$

$$Silt\ size: \frac{30 \times 100}{100 - 20} = 37.5\%$$

$$Clay\ size: \frac{40 \times 100}{100 - 20} = 50.0\%$$

On the basis of the preceding modified percentages, the USDA textural classification is *clay*. However, because of the large percentage of gravel, it may be called *gravelly clay*.

Several other textural classification systems are also used, but they are no longer useful for civil engineering purposes.

Example 5.1

Classify the following soils according to the USDA textural classification system.

Particle-size distribution (%)	A	B	C	D
		Soil		
Gravel	10	21	0	12
Sand	20	12	18	22
Silt	41	35	24	26
Clay	29	32	58	40

Solution

Step 1. Calculate the modified percentages of sand, gravel, and silt as follows:

$$\text{Modified \% sand} = \frac{\% \, \text{sand}}{100 \, - \, \% \, \text{gravel}} \times 100$$

$$\text{Modified \% silt} = \frac{\% \, \text{silt}}{100 \, - \, \% \, \text{gravel}} \times 100$$

$$\text{Modified \% clay} = \frac{\% \, \text{clay}}{100 \, - \, \% \, \text{gravel}} \times 100$$

Thus, the following table results:

Particle-size distribution (%)	A	B	C	D
		Soil		
Sand	22.2	15.2	18	25
Silt	45.6	44.3	24	29.5
Clay	32.2	40.5	58	45.5

Step 2. With the modified composition calculated, refer to Figure 5.1 to determine the zone into which each soil falls. The results are as follows:

Classification of soil			
A	B	C	D
Gravelly clay loam	Gravelly silty clay	Clay	Gravelly clay

Note: The word *gravelly* was added to the classification of soils A, B, and D because of the large percentage of gravel present in each.

5.2 Classification by Engineering Behavior

Although the textural classification of soil is relatively simple, it is based entirely on the particle-size distribution. The amount and type of clay minerals present in fine-grained soils dictate to a great extent their physical properties. Hence, the soils engineer must consider *plasticity*, which results from the presence of clay minerals, to interpret soil characteristics properly. Because textural classification systems do not take plasticity into account and are not totally indicative of many important soil properties, they are inadequate for most engineering purposes. Currently, two more elaborate classification systems are commonly used by soils engineers. Both systems take into consideration the particle-size distribution and Atterberg limits. They are the American Association of State Highway and Transportation Officials (AASHTO) classification system and the Unified Soil Classification System. The AASHTO classification system is used mostly by state and county highway departments. Geotechnical engineers generally prefer the Unified system.

5.3 AASHTO Classification System

The AASHTO system of soil classification was developed in 1929 as the Public Road Administration classification system. It has undergone several revisions, with the present version proposed by the Committee on Classification of Materials for Subgrades and Granular Type Roads of the Highway Research Board in 1945 (ASTM designation D-3282; AASHTO method M145).

The AASHTO classification in present use is given in Table 5.1. According to this system, soil is classified into seven major groups: A-1 through A-7. Soils classified under groups A-1, A-2, and A-3 are granular materials of which 35% or less of the particles pass through the No. 200 sieve. Soils of which more than 35% pass through the No. 200 sieve are classified under groups A-4, A-5, A-6, and A-7. These soils are mostly silt and clay-type materials. This classification system is based on the following criteria:

1. *Grain size*
   a. *Gravel:* fraction passing the 75-mm sieve and retained on the No. 10 (2-mm) U.S. sieve
   b. *Sand:* fraction passing the No. 10 (2-mm) U.S. sieve and retained on the No. 200 (0.075-mm) U.S. sieve
   c. *Silt and clay:* fraction passing the No. 200 U.S. sieve
2. *Plasticity:* The term *silty* is applied when the fine fractions of the soil have a plasticity index of 10 or less. The term *clayey* is applied when the fine fractions have a plasticity index of 11 or more.
3. If cobbles and *boulders* (size larger than 75 mm) are encountered, they are excluded from the portion of the soil sample from which classification is made. However, the percentage of such material is recorded.

To classify a soil according to Table 5.1, one must apply the test data from left to right. By process of elimination, the first group from the left into which the test data fit is the correct

Table 5.1 Classification of Highway Subgrade Materials

General classification	Granular materials (35% or less of total sample passing No. 200)						
	A-1		A-3	A-2			
Group classification	A-1-a	A-1-b		A-2-4	A-2-5	A-2-6	A-2-7
Sieve analysis (percentage passing)							
No. 10	50 max.						
No. 40	30 max.	50 max.	51 min.				
No. 200	15 max.	25 max.	10 max.	35 max.	35 max.	35 max.	35 max.
Characteristics of fraction passing No. 40							
Liquid limit				40 max.	41 min.	40 max.	41 min.
Plasticity index	6 max.		NP	10 max.	10 max.	11 min.	11 min.
Usual types of significant constituent materials	Stone fragments, gravel, and sand		Fine sand	Silty or clayey gravel and sand			
General subgrade rating	Excellent to good						

General classification	Silt-clay materials (more than 35% of total sample passing No. 200)			
Group classification	A-4	A-5	A-6	A-7 A-7-5[a] A-7-6[b]
Sieve analysis (percentage passing)				
No. 10				
No. 40				
No. 200	36 min.	36 min.	36 min.	36 min.
Characteristics of fraction passing No. 40				
Liquid limit	40 max.	41 min.	40 max.	41 min.
Plasticity index	10 max.	10 max.	11 min.	11 min.
Usual types of significant constituent materials	Silty soils		Clayey soils	
General subgrade rating	Fair to poor			

[a] For A-7-5, $PI \leq LL - 30$
[b] For A-7-6, $PI > LL - 30$

classification. Figure 5.2 shows a plot of the range of the liquid limit and the plasticity index for soils that fall into groups A-2, A-4, A-5, A-6, and A-7.

To evaluate the quality of a soil as a highway subgrade material, one must also incorporate a number called the *group index* (*GI*) with the groups and subgroups of the soil. This index is written in parentheses after the group or subgroup designation. The group index is given by the equation

$$GI = (F_{200} - 35)[0.2 + 0.005(LL - 40)] + 0.01(F_{200} - 15)(PI - 10) \quad (5.1)$$

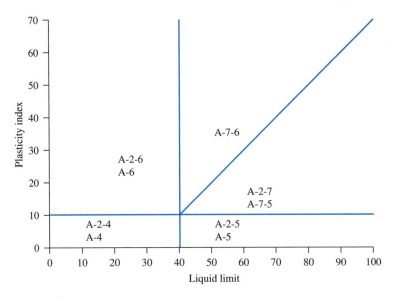

Figure 5.2 Range of liquid limit and plasticity index for soils in groups A-2, A-4, A-5, A-6, and A-7

where F_{200} = percentage passing through the No. 200 sieve
LL = liquid limit
PI = plasticity index

The first term of Eq. (5.1)—that is, $(F_{200} - 35)[0.2 + 0.005(LL - 40)]$—is the partial group index determined from the liquid limit. The second term—that is, $0.01(F_{200} - 15)(PI - 10)$—is the partial group index determined from the plasticity index. Following are some rules for determining the group index:

1. If Eq. (5.1) yields a negative value for *GI,* it is taken as 0.
2. The group index calculated from Eq. (5.1) is rounded off to the nearest whole number (for example, $GI = 3.4$ is rounded off to 3; $GI = 3.5$ is rounded off to 4).
3. There is no upper limit for the group index.
4. The group index of soils belonging to groups A-1-a, A-1-b, A-2-4, A-2-5, and A-3 is always 0.
5. When calculating the group index for soils that belong to groups A-2-6 and A-2-7, use the partial group index for *PI,* or

$$GI = 0.01(F_{200} - 15)(PI - 10) \qquad (5.2)$$

In general, the quality of performance of a soil as a subgrade material is inversely proportional to the group index.

Example 5.2

The results of the particle-size analysis of a soil are as follows:

- Percent passing the No. 10 sieve = 100
- Percent passing the No. 40 sieve = 80
- Percent passing the No. 200 sieve = 58

The liquid limit and plasticity index of the minus No. 40 fraction of the soil are 30 and 10, respectively. Classify the soil by the AASHTO system.

Solution

Using Table 5.1, since 58% of the soil is passing through the No. 200 sieve, it falls under silt-clay classifications—that is, it falls under group A-4, A-5, A-6, or A-7. Proceeding from left to right, it falls under group A-4.
From Eq. (5.1)

$$GI = (F_{200} - 35)[0.2 + 0.005(LL - 40)] + 0.01(F_{200} - 15)(PI - 10)$$
$$= (58 - 35)[0.2 + 0.005(30 - 40)] + (0.01)(58 - 15)(10 - 10)$$
$$= 3.45 \approx 3$$

So, the soil will be classified as A-4(3).   ■

Example 5.3

Ninety-five percent of a soil passes through the No. 200 sieve and has a liquid limit of 60 and plasticity index of 40. Classify the soil by the AASHTO system.

Solution

According to Table 5.1, this soil falls under group A-7 (proceed in a manner similar to Example 5.1). Since

$$40 > 60 - 30$$
$$\uparrow \quad \uparrow$$
$$PI \quad LL$$

this is an A-7-6 soil.

$$GI = (F_{200} - 35)[0.2 + 0.005(LL - 40)] + 0.01(F_{200} - 15)(PI - 10)$$
$$= (95 - 35)[0.2 + 0.005(60 - 40)] + (0.01)(95 - 15)(40 - 10)$$
$$= 42$$

So, the classification is **A-7-6(42)**.   ■

Example 5.4

Classify the following soil by the AASHTO Classification System:

- Percentage passing No. 10 sieve = 90
- Percentage passing No. 40 sieve = 76
- Percentage passing No. 200 sieve = 34
- Liquid limit (–No. 40 fraction) = 37
- Plasticity index (–No. 40 fraction) = 12

Solution

The percentage passing through the No. 200 sieve is less than 35, so the soil is a granular material. From Table 5.1, we see that it is type A-2-6. From Eq. (5.2),

$$GI = 0.01(F_{200} - 15)(PI - 10)$$

For this soil, $F_{200} = 34$ and $PI = 12$, so

$$GI = 0.01(34 - 15)(12 - 10) = 0.38 \approx 0$$

Thus, the soil is type **A-2-6(0)**. ∎

5.4 *Unified Soil Classification System*

The original form of this system was proposed by Casagrande in 1942 for use in the airfield construction works undertaken by the Army Corps of Engineers during World War II. In cooperation with the U.S. Bureau of Reclamation, this system was revised in 1952. At present, it is used widely by engineers (ASTM Test Designation D-2487). The Unified classification system is presented in Table 5.2.

This system classifies soils into two broad categories:

1. Coarse-grained soils that are gravelly and sandy in nature with less than 50% passing through the No. 200 sieve. The group symbols start with a prefix of G or S. G stands for gravel or gravelly soil, and S for sand or sandy soil.
2. Fine-grained soils are with 50% or more passing through the No. 200 sieve. The group symbols start with prefixes of M, which stands for inorganic silt, C for inorganic clay, or O for organic silts and clays. The symbol Pt is used for peat, muck, and other highly organic soils.

Other symbols used for the classification are:

- W—well graded
- P—poorly graded
- L—low plasticity (liquid limit less than 50)
- H—high plasticity (liquid limit more than 50)

Table 5.2 Unified Soil Classification System (Based on Material Passing 76.2-mm Sieve)

Criteria for assigning group symbols				Group symbol
Coarse-grained soils More than 50% retained on No. 200 sieve	**Gravels** More than 50% of coarse fraction retained on No. 4 sieve	Clean Gravels Less than 5% fines[a]	$C_u \geq 4$ and $1 \leq C_c \leq 3^c$	GW
			$C_u < 4$ and/or $1 > C_c > 3^c$	GP
		Gravels with Fines More than 12% fines[a,d]	PI < 4 or plots below "A" line (Figure 5.3)	GM
			PI > 7 and plots on or above "A" line (Figure 5.3)	GC
	Sands 50% or more of coarse fraction passes No. 4 sieve	Clean Sands Less than 5% fines[b]	$C_u \geq 6$ and $1 \leq C_c \leq 3^c$	SW
			$C_u < 6$ and/or $1 > C_c > 3^c$	SP
		Sands with Fines More than 12% fines[b,d]	PI < 4 or plots below "A" line (Figure 5.3)	SM
			PI > 7 and plots on or above "A" line (Figure 5.3)	SC
Fine-grained soils 50% or more passes No. 200 sieve	**Silts and clays** Liquid limit less than 50	Inorganic	PI > 7 and plots on or above "A" line (Figure 5.3)[e]	CL
			PI < 4 or plots below "A" line (Figure 5.3)[e]	ML
		Organic	$\dfrac{\text{Liquid limit — oven dried}}{\text{Liquid limit — not dried}} < 0.75$; see Figure 5.3; OL zone	OL
	Silts and clays Liquid limit 50 or more	Inorganic	PI plots on or above "A" line (Figure 5.3)	CH
			PI plots below "A" line (Figure 5.3)	MH
		Organic	$\dfrac{\text{Liquid limit — oven dried}}{\text{Liquid limit — not dried}} < 0.75$; see Figure 5.3; OH zone	OH
Highly Organic Soils	Primarily organic matter, dark in color, and organic odor			Pt

[a] Gravels with 5 to 12% fine require dual symbols: GW-GM, GW-GC, GP-GM, GP-GC.

[b] Sands with 5 to 12% fines require dual symbols: SW-SM, SW-SC, SP-SM, SP-SC.

[c] $C_u = \dfrac{D_{60}}{D_{10}}$; $C_c = \dfrac{(D_{30})^2}{D_{60} \times D_{10}}$

[d] If $4 \leq PI \leq 7$ and plots in the hatched area in Figure 5.3, use dual symbol GC-GM or SC-SM.

[e] If $4 \leq PI \leq 7$ and plots in the hatched area in Figure 5.3, use dual symbol CL-ML.

103

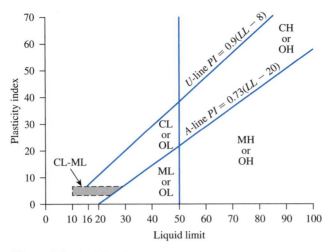

Figure 5.3 Plasticity chart

For proper classification according to this system, some or all of the following information must be known:

1. Percent of gravel—that is, the fraction passing the 76.2-mm sieve and retained on the No. 4 sieve (4.75-mm opening)
2. Percent of sand—that is, the fraction passing the No. 4 sieve (4.75-mm opening) and retained on the No. 200 sieve (0.075-mm opening)
3. Percent of silt and clay—that is, the fraction finer than the No. 200 sieve (0.075-mm opening)
4. Uniformity coefficient (C_u) and the coefficient of gradation (C_c)
5. Liquid limit and plasticity index of the portion of soil passing the No. 40 sieve

The group symbols for coarse-grained gravelly soils are GW, GP, GM, GC, GC-GM, GW-GM, GW-GC, GP-GM, and GP-GC. Similarly, the group symbols for fine-grained soils are CL, ML, OL, CH, MH, OH, CL-ML, and Pt.

More recently, ASTM designation D-2487 created an elaborate system to assign *group names* to soils. These names are summarized in Figures 5.4, 5.5, and 5.6. In using these figures, one needs to remember that, in a given soil,

- Fine fraction = percent passing No. 200 sieve
- Coarse fraction = percent retained on No. 200 sieve
- Gravel fraction = percent retained on No. 4 sieve
- Sand fraction = (percent retained on No. 200 sieve) − (percent retained on No. 4 sieve)

5.5   *Summary and Comparison Between the AASHTO and Unified Systems*

Both soil classification systems, AASHTO and Unified, are based on the texture and plasticity of soil. Also, both systems divide the soils into two major categories, coarse grained and fine grained, as separated by the No. 200 sieve. According to the

Figure 5.4 Flowchart group names for gravelly and sandy soil (*Source:* From "Annual Book of ASTM Standards, 04.08." Copyright © 2007, American Society for Testing and Materials. Reprinted with permission.)

AASHTO system, a soil is considered fine grained when more than 35% passes through the No. 200 sieve. According to the Unified system, a soil is considered fine grained when more than 50% passes through the No. 200 sieve. A coarse-grained soil that has about 35% fine grains will behave like a fine-grained material. This is because enough fine grains exist to fill the voids between the coarse grains and hold them apart. In this respect, the AASHTO system appears to be more appropriate. In the AASHTO system,

Group symbol

Group name

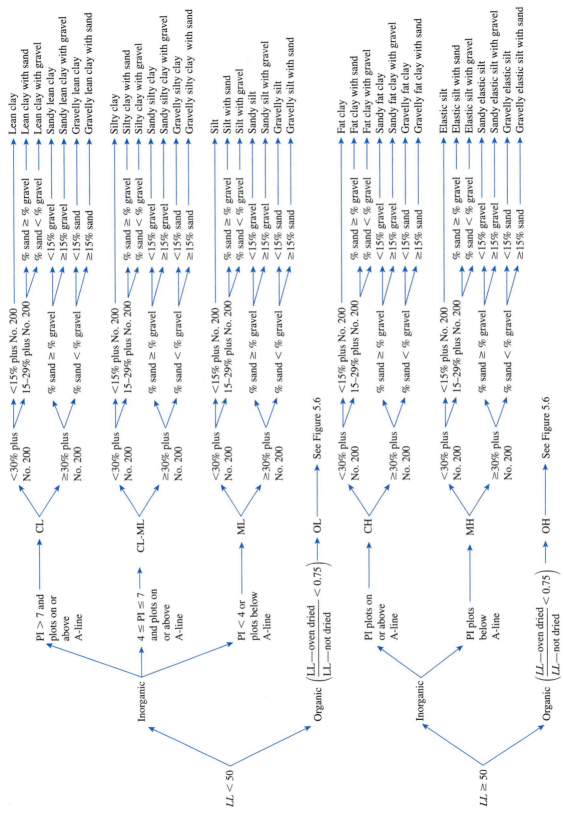

Figure 5.5 Flowchart group names for inorganic silty and clayey soils (*Source:* From "Annual Book of ASTM Standards, 04.08." Copyright © 2007, American Society for Testing and Materials. Reprinted with permission.)

Group symbol

Group name

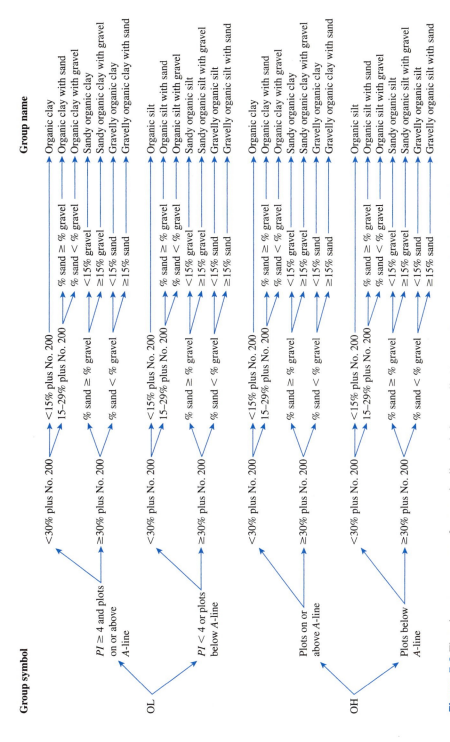

Figure 5.6 Flowchart group names for organic silty and clayey soils (*Source:* From "Annual Book of ASTM Standards, 04.08." Copyright © 2007, American Society for Testing and Materials. Reprinted with permission.)

Example 5.5

Figure 5.7 gives the grain-size distribution of two soils. The liquid and plastic limits of minus No. 40 sieve fraction of the soil are as follows:

	Soil A	Soil B
Liquid limit	30	26
Plastic limit	22	20

Determine the group symbols and group names according to the Unified Soil Classification System.

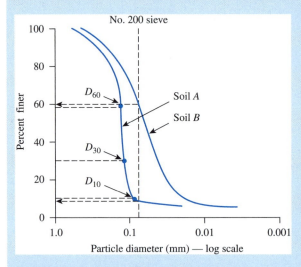

Figure 5.7 Particle-size distribution of two soils

Solution
Soil A
The grain-size distribution curve (Figure 5.7) indicates that percent passing No. 200 sieve is 8. According to Table 5.2, it is a coarse-grained soil. Also, from Figure 5.7, the percent retained on No. 4 sieve is zero. Hence, it is a sandy soil.

From Figure 5.7, $D_{10} = 0.085$ mm, $D_{30} = 0.12$ m, and $D_{60} = 0.135$ mm. Thus,

$$C_u = \frac{D_{60}}{D_{10}} = \frac{0.135}{0.085} = 1.59 < 6$$

$$C_c = \frac{D_{30}^2}{D_{60} \times D_{10}} = \frac{(0.12)^2}{(0.135)(0.085)} = 1.25 > 1$$

With $LL = 30$ and $PI = 30 - 22 = 8$ (which is greater than 7), it plots above the A-line in Figure 5.3. Hence, the group symbol is **SP-SC**.

In order to determine the group name, we refer to Figure 5.4.

$$\text{Percentage of gravel} = 0 \, (\text{which is} < 15\%)$$

So, the group name is **poorly graded sand with clay**.

Soil *B*

The grain-size distribution curve in Figure 5.7 shows that percent passing No. 200 sieve is 61 ($>$50%); hence, it is a fine-grained soil. Given: $LL = 26$ and $PI = 26 - 20 = 6$. In Figure 5.3, the PI plots in the hatched area. So, from Table 5.2, the group symbol is **CL-ML**.

For group name (assuming that the soil is inorganic), we go to Figure 5.5 and obtain Plus No. 200 sieve $= 100 - 61 = 39$ (which is greater than 30).

$$\text{Percentage of gravel} = 0; \text{percentage of sand} = 100 - 61 = 39$$

Thus, because the percentage of sand is greater than the percentage of gravel, the soil is **sandy silty clay**. ∎

Example 5.6

For a given soil, the following are known:

- Percentage passing No. 4 sieve 70
- Percentage passing No. 200 sieve 30
- Liquid limit 33
- Plastic limit 12

Classify the soil using the Unified Soil Classification System. Give the group symbol and the group name.

Solution

Refer to Table 5.2. The percentage passing No. 200 sieve is 30%, which is less than 50%. So it is a coarse-grained soil. Thus,

$$\text{Coarse fraction} = 100 - 30 = 70\%$$

Gravel fraction $=$ percent retained on No. 4 sieve $= 100 - 70 = 30\%$ Hence, more than 50% of the coarse fraction is passing No. 4 sieve. Thus, it is a sandy soil. Since more than 12% is passing No. 200 sieve, it is SM or SC. For this soil, $PI = 33 - 12 = 21$ (which is greater than 7). With $LL = 33$ and $PI = 21$, it plots above the A-line in Figure 5.3. Thus, the group symbol is **SC**.

For the group name, refer to Figure 5.4. Since the percentage of gravel is more than 15%, it is **clayey sand with gravel**. ∎

the No. 10 sieve is used to separate gravel from sand; in the Unified system, the No. 4 sieve is used. From the viewpoint of soil-separated size limits, the No. 10 sieve is the more accepted upper limit for sand. This limit is used in concrete and highway base-course technology.

In the Unified system, the gravelly and sandy soils clearly are separated; in the AASHTO system, they are not. The A-2 group, in particular, contains a large variety of soils. Symbols like GW, SM, CH, and others that are used in the Unified system are more descriptive of the soil properties than the A symbols used in the AASHTO system.

The classification of organic soils, such as OL, OH, and Pt, is provided in the Unified system. Under the AASHTO system, there is no place for organic soils. Peats usually have a high moisture content, low specific gravity of soil solids, and low unit weight. Figure 5.8 shows the scanning electron micrographs of four peat samples collected in Wisconsin. Some of the properties of the peats are given in Table 5.3.

Liu (1967) compared the AASHTO and Unified systems. The results of his study are presented in Tables 5.4 and 5.5.

Figure 5.8 Scanning electron micrographs for four peat samples (*After Dhowian and Edil, 1980. Copyright ASTM INTERNATIONAL. Reprinted with permission.*)

Table 5.3 Properties of the Peats Shown in Figure 5.8

Source of peat	Moisture content (%)	Unit weight (kN/m^3)	Specific gravity, G_s	Ash content (%)
Middleton	510	9.1	1.41	12.0
Waupaca County	460	9.6	1.68	15.0
Portage	600	9.6	1.72	19.5
Fond du Lac County	240	10.2	1.94	39.8

Table 5.4 Comparison of the AASHTO System with the Unified System*

Soil group in AASHTO system	Comparable soil groups in Unified system		
	Most probable	Possible	Possible but improbable
A-1-a	GW, GP	SW, SP	GM, SM
A-1-b	SW, SP, GM, SM	GP	—
A-3	SP	—	SW, GP
A-2-4	GM, SM	GC, SC	GW, GP, SW, SP
A-2-5	GM, SM	—	GW, GP, SW, SP
A-2-6	GC, SC	GM, SM	GW, GP, SW, SP
A-2-7	GM, GC, SM, SC	—	GW, GP, SW, SP
A-4	ML, OL	CL, SM, SC	GM, GC
A-5	OH, MH, ML, OL	—	SM, GM
A-6	CL	ML, OL, SC	GC, GM, SM
A-7-5	OH, MH	ML, OL, CH	GM, SM, GC, SC
A-7-6	CH, CL	ML, OL, SC	OH, MH, GC, GM, SM

*After Liu (1967)

Table 5.5 Comparison of the Unified System with the AASHTO System*

Soil group in Unified system	Comparable soil groups in AASHTO system		
	Most probable	Possible	Possible but improbable
GW	A-1-a	—	A-2-4, A-2-5, A-2-6, A-2-7
GP	A-1-a	A-1-b	A-3, A-2-4, A-2-5, A-2-6, A-2-7
GM	A-1-b, A-2-4, A-2-5, A-2-7	A-2-6	A-4, A-5, A-6, A-7-5, A-7-6, A-1-a
GC	A-2-6, A-2-7	A-2-4	A-4, A-6, A-7-6, A-7-5
SW	A-1-b	A-1-a	A-3, A-2-4, A-2-5, A-2-6, A-2-7
SP	A-3, A-1-b	A-1-a	A-2-4, A-2-5, A-2-6, A-2-7
SM	A-1-b, A-2-4, A-2-5, A-2-7	A-2-6, A-4	A-5, A-6, A-7-5, A-7-6, A-1-a
SC	A-2-6, A-2-7	A-2-4, A-6, A-4, A-7-6	A-7-5
ML	A-4, A-5	A-6, A-7-5, A-7-6	—
CL	A-6, A-7-6	A-4	—
OL	A-4, A-5	A-6, A-7-5, A-7-6	—
MH	A-7-5, A-5	—	A-7-6
CH	A-7-6	A-7-5	—
OH	A-7-5, A-5	—	A-7-6
Pt	—	—	—

*After Liu (1967)

Problems

5.1 Classify the following soil using the U.S. Department of Agriculture textural classification chart.

Soil	Particle-size distribution (%)		
	Sand	Silt	Clay
A	20	20	60
B	55	5	40
C	45	35	20
D	50	15	35
E	70	15	15

5.2 The sieve analysis of ten soils and the liquid and plastic limits of the fraction passing through the No. 40 sieve are given below. Classify the soils using the AASHTO classification system and give the group indexes.

Soil No.	Sieve analysis (percent finer)			Liquid limit	Plastic limit
	No. 10	No. 40	No. 200		
1	98	80	50	38	29
2	100	92	80	56	23
3	100	88	65	37	22
4	85	55	45	28	20
5	92	75	62	43	28
6	48	28	6	—	NP
7	87	62	30	32	24
8	90	76	34	37	25
9	100	78	8	—	NP
10	92	74	32	44	35

5.3 Classify the following soils using the Unified soil classification system. Give group symbols and group names.

Soil No.	Sieve analysis (percent finer)		Liquid limit	Plasticity limit	Comments
	No. 4	No. 200			
1	94	3	—	NP	$C_u = 4.48$ and $C_c = 1.22$
2	100	77	63	25	
3	100	86	55	28	
4	100	45	36	22	
5	92	48	30	8	
6	60	40	26	4	
7	99	76	60	32	

5.4 For an inorganic soil, the following grain-size analysis is given.

U.S. Sieve No.	Percent passing
4	100
10	90
20	64
40	38
80	18
200	13

For this soil, $LL = 23$ and $PL = 19$. Classify the soil by using
a. AASHTO soil classification system
b. Unified soil classification system
Give group names and group symbols.

References

AMERICAN ASSOCIATION OF STATE HIGHWAY AND TRANSPORTATION OFFICIALS (1982). *AASHTO Materials, Part I, Specifications,* Washington, D.C.

AMERICAN SOCIETY FOR TESTING AND MATERIALS (2007). *Annual Book of ASTM Standards,* Sec. 4, Vol. 04.08, West Conshohoken, Pa.

CASAGRANDE, A. (1948). "Classification and Identification of Soils," *Transactions,* ASCE, Vol. 113, 901–930.

DHOWIAN, A. W., and EDIL, T. B. (1980). "Consolidation Behavior of Peats," *Geotechnical Testing Journal,* ASTM, Vol. 3, No. 3, 105–114.

LIU, T. K. (1967). "A Review of Engineering Soil Classification Systems," *Highway Research Record No. 156,* National Academy of Sciences, Washington, D.C., 1–22.

6 Soil Compaction

In the construction of highway embankments, earth dams, and many other engineering structures, loose soils must be compacted to increase their unit weights. Compaction increases the strength characteristics of soils, which increase the bearing capacity of foundations constructed over them. Compaction also decreases the amount of undesirable settlement of structures and increases the stability of slopes of embankments. Smooth-wheel rollers, sheepsfoot rollers, rubber-tired rollers, and vibratory rollers are generally used in the field for soil compaction. Vibratory rollers are used mostly for the densification of granular soils. Vibroflot devices are also used for compacting granular soil deposits to a considerable depth. Compaction of soil in this manner is known as *vibroflotation.* This chapter discusses in some detail the principles of soil compaction in the laboratory and in the field.

6.1 *Compaction—General Principles*

Compaction, in general, is the densification of soil by removal of air, which requires mechanical energy. The degree of compaction of a soil is measured in terms of its dry unit weight. When water is added to the soil during compaction, it acts as a softening agent on the soil particles. The soil particles slip over each other and move into a densely packed position. The dry unit weight after compaction first increases as the moisture content increases. (See Figure 6.1.) Note that at a moisture content $w = 0$, the moist unit weight (γ) is equal to the dry unit weight (γ_d), or

$$\gamma = \gamma_{d(w=0)} = \gamma_1$$

When the moisture content is gradually increased and the same compactive effort is used for compaction, the weight of the soil solids in a unit volume gradually increases. For example, at $w = w_1$,

$$\gamma = \gamma_2$$

However, the dry unit weight at this moisture content is given by

$$\gamma_{d(w=w_1)} = \gamma_{d(w=0)} + \Delta\gamma_d$$

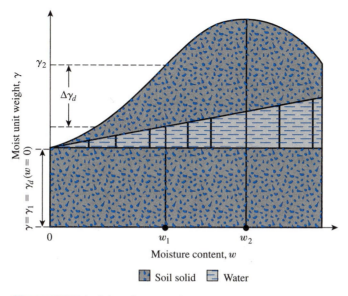

Figure 6.1 Principles of compaction

Beyond a certain moisture content $w = w_2$ (Figure 6.1), any increase in the moisture content tends to reduce the dry unit weight. This phenomenon occurs because the water takes up the spaces that would have been occupied by the solid particles. The moisture content at which the maximum dry unit weight is attained is generally referred to as the *optimum moisture content*.

The laboratory test generally used to obtain the maximum dry unit weight of compaction and the optimum moisture content is called the *Proctor compaction test* (Proctor, 1933). The procedure for conducting this type of test is described in the following section.

6.2   *Standard Proctor Test*

In the Proctor test, the soil is compacted in a mold that has a volume of 944 cm³. The diameter of the mold is 101.6 mm. During the laboratory test, the mold is attached to a baseplate at the bottom and to an extension at the top (Figure 6.2a). The soil is mixed with varying amounts of water and then compacted in three equal layers by a hammer (Figure 6.2b) that delivers 25 blows to each layer. The hammer has a mass of 2.5 kg and has a drop of 30.5 mm. Figure 6.2c is a photograph of the laboratory equipment required for conducting a standard Proctor test.

For each test, the moist unit weight of compaction, γ, can be calculated as

$$\gamma = \frac{W}{V_{(m)}} \tag{6.1}$$

where W = weight of the compacted soil in the mold
$V_{(m)}$ = volume of the mold (944 cm³)

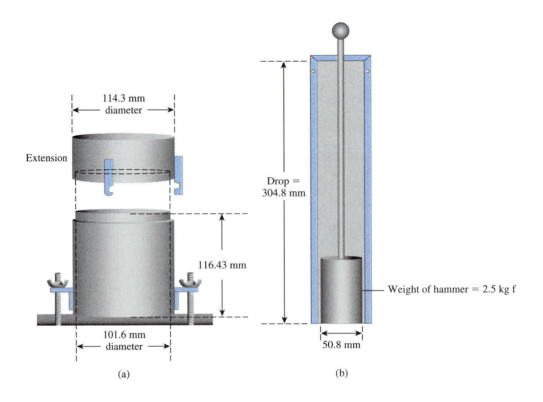

(a)   (b)

(c)

Figure 6.2 Standard Proctor test equipment: (a) mold; (b) hammer; (c) photograph of laboratory equipment used for test (*Courtesy of Braja M. Das, Henderson, Nevada*)

For each test, the moisture content of the compacted soil is determined in the laboratory. With the known moisture content, the dry unit weight can be calculated as

$$\gamma_d = \frac{\gamma}{1 + \dfrac{w\,(\%)}{100}} \tag{6.2}$$

where $w\,(\%)$ = percentage of moisture content.

The values of γ_d determined from Eq. (6.2) can be plotted against the corresponding moisture contents to obtain the maximum dry unit weight and the optimum moisture content for the soil. Figure 6.3 shows such a plot for a silty-clay soil.

The procedure for the standard Proctor test is elaborated in ASTM Test Designation D-698 (ASTM, 2007) and AASHTO Test Designation T-99 (AASHTO, 1982).

For a given *moisture content w* and *degree of saturation S*, the dry unit weight of compaction can be calculated as follows. From Chapter 3 [Eq. (3.16)], for any soil,

$$\gamma_d = \frac{G_s \gamma_w}{1 + e}$$

where G_s = specific gravity of soil solids
γ_w = unit weight of water
e = void ratio

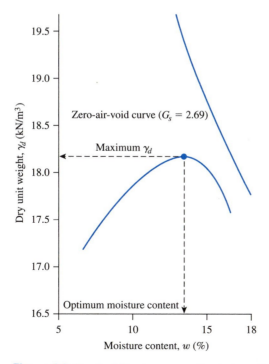

Figure 6.3 Standard Proctor compaction test results for a silty clay

and, from Eq. (3.18),

$$Se = G_s w$$

or

$$e = \frac{G_s w}{S}$$

Thus,

$$\gamma_d = \frac{G_s \gamma_w}{1 + \dfrac{G_s w}{S}} \tag{6.3}$$

For a given moisture content, the theoretical maximum dry unit weight is obtained when no air is in the void spaces—that is, when the degree of saturation equals 100%. Hence, the maximum dry unit weight at a given moisture content with zero air voids can be obtained by substituting $S = 1$ into Eq. (6.3), or

$$\gamma_{zav} = \frac{G_s \gamma_w}{1 + w G_s} = \frac{\gamma_w}{w + \dfrac{1}{G_s}} \tag{6.4}$$

where γ_{zav} = zero-air-void unit weight.
   To obtain the variation of γ_{zav} with moisture content, use the following procedure:

1. Determine the specific gravity of soil solids.
2. Know the unit weight of water (γ_w).
3. Assume several values of w, such as 5%, 10%, 15%, and so on.
4. Use Eq. (6.4) to calculate γ_{zav} for various values of w.

   Figure 6.3 also shows the variation of γ_{zav} with moisture content and its relative location with respect to the compaction curve. Under no circumstances should any part of the compaction curve lie to the right of the zero-air-void curve.

6.3   *Factors Affecting Compaction*

The preceding section showed that moisture content has a strong influence on the degree of compaction achieved by a given soil. Besides moisture content, other important factors that affect compaction are soil type and compaction effort (energy per unit volume). The importance of each of these two factors is described in more detail in the following two sections.

Effect of Soil Type

The soil type—that is, grain-size distribution, shape of the soil grains, specific gravity of soil solids, and amount and type of clay minerals present—has a great influence on the maximum dry unit weight and optimum moisture content. Figure 6.4 shows typical

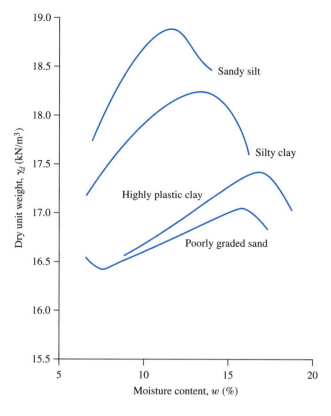

Figure 6.4 Typical compaction curves for four soils (*ASTM D-698*)

compaction curves obtained from four soils. The laboratory tests were conducted in accordance with ASTM Test Designation D-698.

Note also that the bell-shaped compaction curve shown in Figure 6.3 is typical of most clayey soils. Figure 6.4 shows that for sands, the dry unit weight has a general tendency first to decrease as moisture content increases and then to increase to a maximum value with further increase of moisture. The initial decrease of dry unit weight with increase of moisture content can be attributed to the capillary tension effect. At lower moisture contents, the capillary tension in the pore water inhibits the tendency of the soil particles to move around and be compacted densely.

Lee and Suedkamp (1972) studied compaction curves for 35 soil samples. They observed that four types of compaction curves can be found. These curves are shown in Figure 6.5. The following table is a summary of the type of compaction curves encountered in various soils with reference to Figure 6.5.

Type of compaction curve (Figure 6.5)	Description of curve	Liquid limit
A	Bell shaped	Between 30 to 70
B	1-1/2 peak	Less than 30
C	Double peak	Less that 30 and those greater than 70
D	Odd shaped	Greater than 70

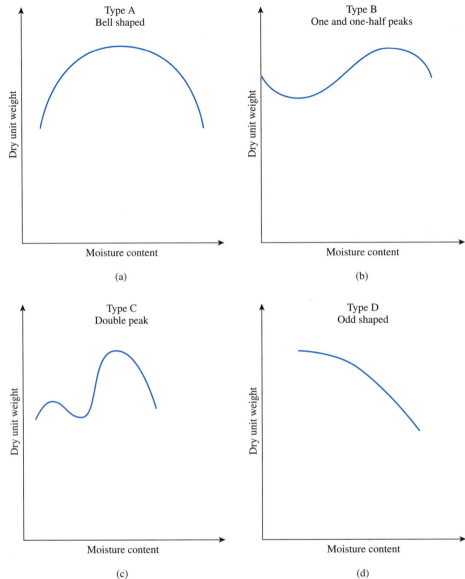

Figure 6.5
Various types
of compaction
curves encountered
in soils

Effect of Compaction Effort

The compaction energy per unit volume used for the standard Proctor test described in Section 6.2 can be given as

$$E = \frac{\left(\begin{array}{c}\text{Number}\\ \text{of blows}\\ \text{per layer}\end{array}\right) \times \left(\begin{array}{c}\text{Number}\\ \text{of}\\ \text{layers}\end{array}\right) \times \left(\begin{array}{c}\text{Weight}\\ \text{of}\\ \text{hammer}\end{array}\right) \times \left(\begin{array}{c}\text{Height of}\\ \text{drop of}\\ \text{hammer}\end{array}\right)}{\text{Volume of mold}} \quad (6.5)$$

or, in SI units,

$$E = \frac{(25)(3)\left(\dfrac{2.5 \times 9.81}{1000} \text{ kN}\right)(0.305 \text{ m})}{944 \times 10^{-6} \text{ m}^3} = 594 \text{ kN-m/m}^3 \approx 600 \text{ kN-m/m}^3$$

If the compaction effort per unit volume of soil is changed, the moisture–unit weight curve also changes. This fact can be demonstrated with the aid of Figure 6.6, which shows four compaction curves for a sandy clay. The standard Proctor mold and hammer were used to obtain these compaction curves. The number of layers of soil used for compaction was three for all cases. However, the number of hammer blows per each layer varied from 20 to 50, which varied the energy per unit volume.

From the preceding observation and Figure 6.6, we can see that

1. As the compaction effort is increased, the maximum dry unit weight of compaction is also increased.

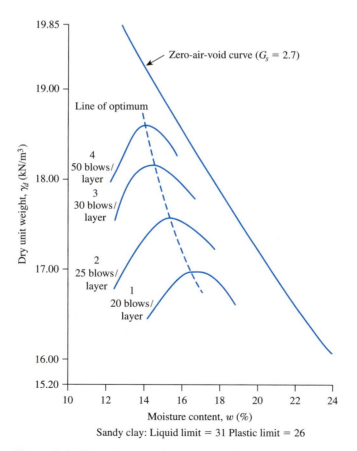

Figure 6.6 Effect of compaction energy on the compaction of a sandy clay

2. As the compaction effort is increased, the optimum moisture content is decreased to some extent.

The preceding statements are true for all soils. Note, however, that the degree of compaction is not directly proportional to the compaction effort.

6.4 Modified Proctor Test

With the development of heavy rollers and their use in field compaction, the standard Proctor test was modified to better represent field conditions. This revised version sometimes is referred to as the *modified Proctor test* (ASTM Test Designation D-1557 and AASHTO Test Designation T-180). For conducting the modified Proctor test, the same mold is used with a volume of 944 cm³, as in the case of the standard Proctor test. However, the soil is compacted in five layers by a hammer that has a mass of 4.54 kg. The drop of the hammer is 457 mm. The number of hammer blows for each layer is kept at 25 as in the case of the standard Proctor test. Figure 6.7 shows a comparison between the hammers used in standard and modified Proctor tests.

The compaction energy for this type of compaction test can be calculated as 2700 kN-m/m³.

Because it increases the compactive effort, the modified Proctor test results in an increase in the maximum dry unit weight of the soil. The increase in the maximum dry unit weight is accompanied by a decrease in the optimum moisture content.

In the preceding discussions, the specifications given for Proctor tests adopted by ASTM and AASHTO regarding the volume of the mold and the number of blows are generally those adopted for fine-grained soils that pass through the U.S. No. 4 sieve. However, under each test designation, there are three suggested methods that reflect the mold size, the number of blows per layer, and the maximum particle size in a soil aggregate used for testing. A summary of the test methods is given in Table 6.1.

Omar, *et al.* (2003) recently presented the results of modified Proctor compaction tests on 311 soil samples. Of these samples, 45 were gravelly soil (GP, GP-GM, GW, GW-GM, and GM), 264 were sandy soil (SP, SP-SM, SW-SM, SW, SC-SM, SC, and SM), and two were clay with low plasticity (CL). All compaction tests were conducted using ASTM 1557 method C to avoid over-size correction. Based on the tests, the following correlations were developed.

$$\rho_{d(\max)} \, (\text{kg/m}^3) = [4,804,574 G_s - 195.55(LL)^2 + 156,971(R\#4)^{0.5} - 9,527,830]^{0.5} \tag{6.6}$$

$$\ln(w_{\text{opt}}) = 1.195 \times 10^{-4}(LL)^2 - 1.964 G_s - 6.617 \times 10^{-5}(R\#4) + 7.651 \tag{6.7}$$

where $\rho_{d(\max)}$ = maximum dry density (kg/m³)
$\quad\quad w_{\text{opt}}$ = optimum moisture content (%)
$\quad\quad G_s$ = specific gravity of soil solids
$\quad\quad LL$ = liquid limit, in percent
$\quad\quad R\#4$ = percent retained on No. 4 sieve

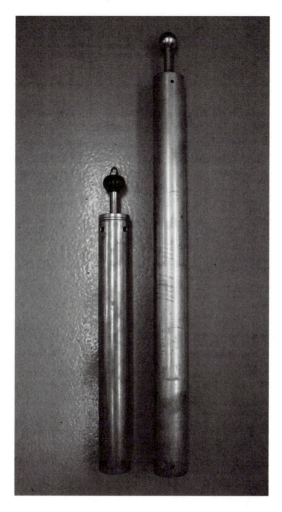

Figure 6.7 Comparison between standard Proctor hammer (left) and modified Proctor hammer (right) (*Courtesy of Braja M. Das, Henderson, Nevada*)

Gurtug and Sridharan (2004) proposed correlations for optimum moisture content and maximum dry unit weight with the plastic limit (*PL*) of cohesive soils. These correlations can be expressed as:

$$w_{\text{opt}}(\%) = [1.95 - 0.38(\log CE)](PL) \tag{6.8}$$

$$\gamma_{d(\text{max})}\,(\text{kN/m}^3) = 22.68e^{-0.0183w_{\text{opt}}(\%)} \tag{6.9}$$

where *PL* = plastic limit (%)
  CE = compaction energy (kN-m/m^3)

For modified Proctor test, *CE* = 2700 kN/m^3. Hence,

$$w_{\text{opt}}(\%) \approx 0.65(PL)$$

and

$$\gamma_{d(\text{max})}\,(\text{kN/m}^3) \approx 22.68e^{-0.012(PL)}$$

Table 6.1 Summary of Standard and Modified Proctor Compaction Test Specifications (ASTM D-698 and D-1557)

	Description	Method A	Method B	Method C
Physical data for the tests	Material	Passing No. 4 sieve	Passing 9.5 mm sieve	Passing 19 mm sieve
	Use	Used if 20% or less by weight of material is retained on No. 4 (4.75 mm) sieve	Used if more than 20% by weight of material is retained on No. 4 (4.75 mm) sieve and 20% or less by weight of material is retained on 9.5 mm sieve	Used if more than 20% by weight of material is retained on 9.5 mm sieve and less than 30% by weight of material is retained on 19 mm sieve
	Mold volume	944 cm^3	944 cm^3	2124 cm^3
	Mold diameter	101.6 mm	101.6 mm	152.4 mm
	Mold height	116.4 mm	116.4 mm	116.4 mm
Standard Proctor test	Weight of hammer	24.4 N	24.4 N	24.4 N
	Height of drop	305 mm	305 mm	305 mm
	Number of soil layers	3	3	3
	Number of blows/layer	25	25	56
Modified Proctor test	Weight of hammer	44.5 N	44.5 N	44.5 N
	Height of drop	457 mm	457 mm	457 mm
	Number of soil layers	5	5	5
	Number of blows/layer	25	25	56

Example 6.1

The laboratory test results of a standard Proctor test are given in the following table.

Volume of mold (cm^3)	Weight of moist soil in mold (N)	Moisture content, w (%)
944	16.81	10
944	17.84	12
944	18.41	14
944	18.33	16
944	17.84	18
944	17.35	20

a. Determine the maximum dry unit weight of compaction and the optimum moisture content.
b. Calculate and plot γ_d versus the moisture content for degree of saturation, $S = 80$, 90 and 100% (i.e., γ_{zav}). Given: $G_s = 2.72$.

Solution
Part a
The following table can be prepared:

Volume of mold V (cm³)	Weight of soil, W (N)	Moist unit weight, γ (kN/m³)[a]	Moisture content, w (%)	Dry unit weight, γ_d (kN/m³)[b]
944	16.81	17.81	10	16.19
944	17.84	18.90	12	16.79
944	18.41	19.50	14	17.11
944	18.33	19.42	16	17.04
944	17.84	18.90	18	16.02
944	17.35	18.38	20	15.32

[a] $\gamma = \dfrac{W}{V}$

[b] $\gamma_d = \dfrac{\gamma}{\left\{ 1 + \dfrac{w\%}{100} \right\}}$

The plot of γ_d versus w is shown at the bottom of Figure 6.8. From the plot, we see that the maximum dry unit weight, $\gamma_{d\,(max)} = \mathbf{17.15\ kN/m^3}$ and the optimum moisture content is **14.4%**.

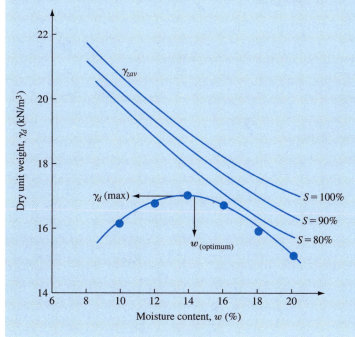

Figure 6.8
Moisture–unit weight curves

Part b

From Eq. (6.3),

$$\gamma_d = \frac{G_s\gamma_w}{1 + \dfrac{G_s w}{S}}$$

The following table can be prepared.

		γ_d (kN/m³)		
G_s	w (%)	$S = 80\%$	$S = 90\%$	$S = 100\%$
2.7	8	20.86	21.37	21.79
2.7	10	19.81	20.37	20.86
2.7	12	18.85	19.48	20.01
2.7	14	17.99	18.65	19.23
2.7	16	17.20	17.89	18.50
2.7	18	16.48	17.20	17.83
2.7	20	15.82	16.55	17.20

The plot of γ_d versus w for the various degrees of saturation is also shown in Figure 6.8. ■

Example 6.2

For a granular soil, the following are given:

- $G_s = 2.6$
- Liquid limit on the fraction passing No. 40 sieve = 20
- Percent retained on No. 4 sieve = 20

Using Eqs. (6.6) and (6.7) estimate the maximum dry density of compaction and the optimum moisture content based on the modified Proctor test.

Solution

From Eq. (6.6)

$$\rho_{d(max)} \text{ (kg/m}^3) = [4{,}804{,}574 G_s - 195.55(LL)^2 + 156{,}971(R\#4)^{0.5} - 9{,}527{,}830]^{0.5}$$
$$= [4{,}804{,}574(2.6) - 195.55(20)^2 + 156{,}971(20)^{0.5} - 9{,}527{,}830]^{0.5}$$
$$= \mathbf{1894 \ kg/m^3}$$

From Eq. (6.7)

$$\ln(w_{opt}) = 1.195 \times 10^{-4}(LL)^2 - 1.964 G_s - 6.617 \times 10^{-5}(R\#4) + 7.651$$
$$= 1.195 \times 10^{-4}(20)^2 - 1.964(2.6) - 6.617 \times 10^{-5}(20) + 7.651$$
$$= 2.591$$
$$w_{opt} = \mathbf{13.35\%}$$
■

6.5   *Structure of Compacted Clay Soil*

Lambe (1958a) studied the effect of compaction on the structure of clay soils, and the results of his study are illustrated in Figure 6.9. If clay is compacted with a moisture content on the dry side of the optimum, as represented by point *A,* it will possess a flocculent structure. This type of structure results because, at low moisture content, the diffuse double layers of ions surrounding the clay particles cannot be fully developed; hence, the interparticle repulsion is reduced. This reduced repulsion results in a more random particle orientation and a lower dry unit weight. When the moisture content of compaction is increased, as shown by point *B,* the diffuse double layers around the particles expand, which increases the repulsion between the clay particles and gives a lower degree of flocculation and a higher dry unit weight. A continued increase in moisture content from *B* to *C* expands the double layers more. This expansion results in a continued increase of repulsion between the particles and thus a still greater degree of particle orientation and a more or less dispersed structure. However, the dry unit weight decreases because the added water dilutes the concentration of soil solids per unit volume.

At a given moisture content, higher compactive effort yields a more parallel orientation to the clay particles, which gives a more dispersed structure. The particles are closer and the soil has a higher unit weight of compaction. This phenomenon can be seen by comparing point *A* with point *E* in Figure 6.9.

Figure 6.10 shows the variation in the degree of particle orientation with molding water content for compacted Boston blue clay. Works of Seed and Chan (1959) have shown similar results for compacted kaolin clay.

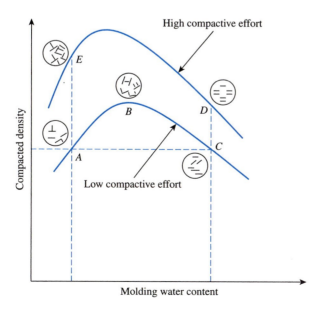

Figure 6.9

Effect of compaction on structure of clay soils (*Redrawn after Lambe, 1958a. With permission from ASCE.*)

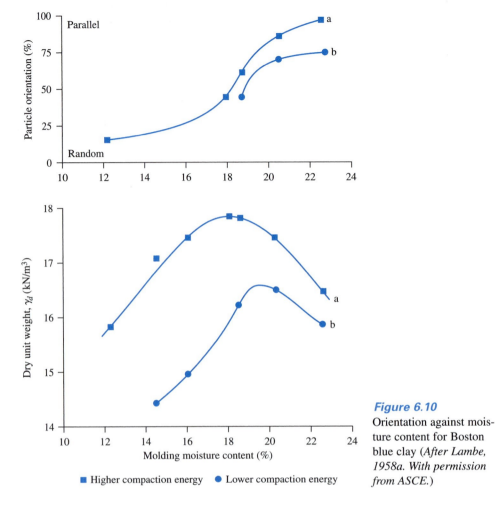

Figure 6.10
Orientation against moisture content for Boston blue clay (*After Lambe, 1958a. With permission from ASCE.*)

6.6   *Effect of Compaction on Cohesive Soil Properties*

Compaction induces variations in the structure of cohesive soils. Results of these structural variations include changes in hydraulic conductivity, compressibility, and strength. Figure 6.11 shows the results of permeability tests (Chapter 7) on Jamaica sandy clay. The samples used for the tests were compacted at various moisture contents by the same compactive effort. The hydraulic conductivity, which is a measure of how easily water flows through soil, decreases with the increase of moisture content. It reaches a minimum value at approximately the optimum moisture content. Beyond the optimum moisture content, the hydraulic conductivity increases slightly. The high value of the hydraulic conductivity on the dry side of the optimum moisture content is due to the random orientation of clay particles that results in larger pore spaces.

One-dimensional compressibility characteristics (Chapter 11) of clay soils compacted on the dry side of the optimum and compacted on the wet side of the optimum are

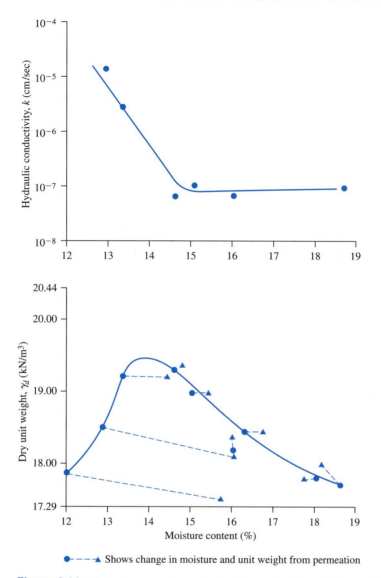

Shows change in moisture and unit weight from permeation

Figure 6.11 Effect of compaction on hydraulic conductivity of clayey soil (*Redrawn after Lambe, 1958b. With permission from ASCE.*)

shown in Figure 6.12. Under lower pressure, a soil that is compacted on the wet side of the optimum is more compressible than a soil that is compacted on the dry side of the optimum. This is shown in Figure 6.12a. Under high pressure, the trend is exactly the opposite, and this is shown in Figure 6.12b. For samples compacted on the dry side of the optimum, the pressure tends to orient the particles normal to its direction of application. The space between the clay particles is also reduced at the same time. However, for samples compacted on the wet side of the optimum, pressure merely reduces the space between the clay particles. At very high pressure, it is possible to have identical structures for samples compacted on the dry and wet sides of optimum.

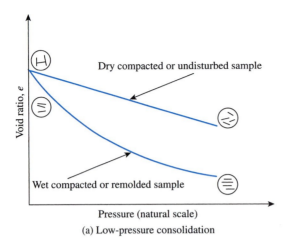

(a) Low-pressure consolidation

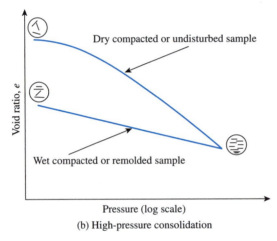

(b) High-pressure consolidation

Figure 6.12

Effect of compaction on one-dimensional compressibility of clayey soil (*Redrawn after Lambe, 1958b. With permission from ASCE.*)

The strength of compacted clayey soils (Chapter 12) generally decreases with the molding moisture content. This is shown in Figure 6.13, which is the result of several unconfined compression-strength tests on compacted specimens of a silty clay soil. The test specimens were prepared by kneading compaction. The insert in Figure 6.13 shows the relationship between dry unit weight and moisture content for the soil. Note that specimens *A, B,* and *C* have been compacted, respectively, on the dry side of the optimum moisture content, near optimum moisture content, and on the wet side of the optimum moisture content. The unconfined compression strength, q_u, is greatly reduced for the specimen compacted on the wet side of the optimum moisture content.

Some expansive clays in the field do not stay compacted, but expand upon entry of water and shrink with loss of moisture. This shrinkage and swelling of soil can cause serious distress to the foundations of structures. The nature of variation of expansion and shrinkage of expansive clay is shown in Figure 6.14. Laboratory observations such as this will help soils engineers to adopt a moisture content for compaction to minimize swelling and shrinkage.

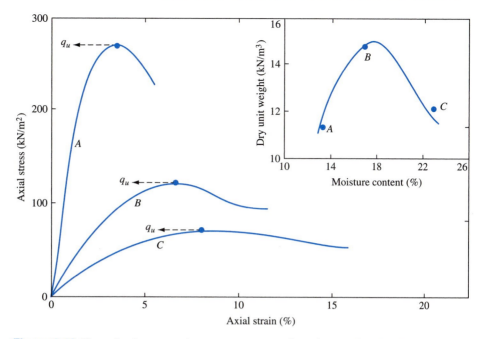

Figure 6.13 Unconfined compression test on compacted specimens of a silty clay

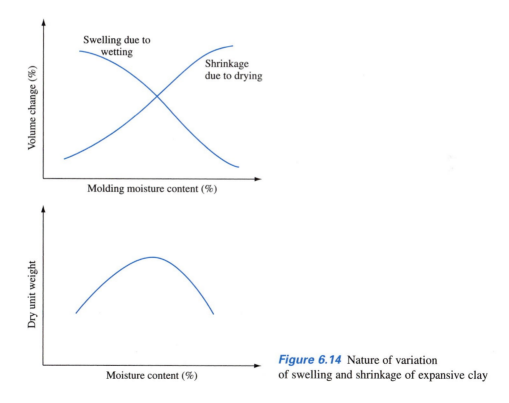

Figure 6.14 Nature of variation of swelling and shrinkage of expansive clay

6.7   *Field Compaction*

Compaction Equipment

Most of the compaction in the field is done with rollers. The four most common types of rollers are

1. Smooth-wheel rollers (or smooth-drum rollers)
2. Pneumatic rubber-tired rollers
3. Sheepsfoot rollers
4. Vibratory rollers

Smooth-wheel rollers (Figure 6.15) are suitable for proof rolling subgrades and for finishing operation of fills with sandy and clayey soils. These rollers provide 100% coverage under the wheels, with ground contact pressures as high as 310 to 380 kN/m^2. They are not suitable for producing high unit weights of compaction when used on thicker layers.

Pneumatic rubber-tired rollers (Figure 6.16) are better in many respects than the smooth-wheel rollers. The former are heavily loaded with several rows of tires. These tires are closely spaced—four to six in a row. The contact pressure under the tires can range from 600 to 700 kN/m^2, and they produce about 70 to 80% coverage. Pneumatic rollers can be used for sandy and clayey soil compaction. Compaction is achieved by a combination of pressure and kneading action.

Sheepsfoot rollers (Figure 6.17) are drums with a large number of projections. The area of each projection may range from 25 to 85 cm^2. These rollers are most effective in compacting clayey soils. The contact pressure under the projections can range from 1400 to 7000 kN/m^2. During compaction in the field, the initial passes compact the lower portion of a lift. Compaction at the top and middle of a lift is done at a later stage.

Figure 6.15 Smooth-wheel roller (*Ingram Compaction LLC*)

Figure 6.16 Pneumatic rubber-tired roller (*Ingram Compaction LLC*)

Figure 6.17 Sheepsfoot roller (*SuperStock/Alamy*)

Vibratory rollers are extremely efficient in compacting granular soils. Vibrators can be attached to smooth-wheel, pneumatic rubber-tired, or sheepsfoot rollers to provide vibratory effects to the soil. Figure 6.18 demonstrates the principles of vibratory rollers. The vibration is produced by rotating off-center weights.

Handheld vibrating plates can be used for effective compaction of granular soils over a limited area. Vibrating plates are also gang-mounted on machines. These plates can be used in less restricted areas.

Factors Affecting Field Compaction

In addition to soil type and moisture content, other factors must be considered to achieve the desired unit weight of compaction in the field. These factors include the thickness of lift, the intensity of pressure applied by the compacting equipment, and the area over which the pressure is applied. These factors are important because the pressure applied at the surface decreases with depth, which results in a decrease in the degree of soil compaction. During compaction, the dry unit weight of soil also is affected by the number of roller passes. Figure 6.19 shows the growth curves for a silty clay soil. The dry unit weight of a soil at a given moisture content increases to a certain point with the number of roller passes. Beyond this point, it remains approximately constant. In most cases, about 10 to 15 roller passes yield the maximum dry unit weight economically attainable.

Figure 6.20a shows the variation in the unit weight of compaction with depth for a poorly graded dune sand for which compaction was achieved by a vibratory drum roller. Vibration was produced by mounting an eccentric weight on a single rotating shaft within the drum cylinder. The weight of the roller used for this compaction was 55.6 kN, and the drum diameter was 1.19 m. The lifts were kept at 2.44 m. Note that, at any given depth, the dry unit weight of compaction increases with the number of roller passes. However, the rate of increase in unit weight gradually decreases after about 15 passes. Another fact to note from Figure 6.20a is the variation of dry unit weight with depth for any given number of roller passes. The dry unit weight and hence the relative density, D_r, reach maximum values at a depth of about 0.5 m and gradually decrease at lesser depths. This decrease occurs because of the lack of confining pressure toward the surface. Once the relationship between depth and relative density (or dry unit weight) for a given soil with a given number of roller

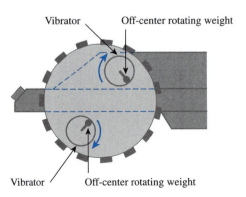

Vibrator   Off-center rotating weight

Vibrator   Off-center rotating weight   *Figure 6.18* Principles of vibratory rollers

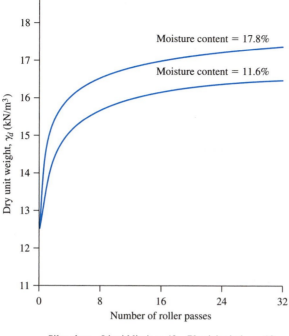

Figure 6.19
Growth curves for a silty clay—
relationship between dry unit
weight and number of passes of
84.5 kN three-wheel roller when
the soil is compacted in 229 mm
loose layers at different moisture
contents (*Adapted from Full-Scale
Field Tests on 3-Wheel Power
Rollers. In Highway Research
Bulletin 272, Highway Research
Board, National Research Council,
Washington, D.C., 1960,
Figure 15, p. 23. Reproduced with
permission of the Transportation
Research Board.*)

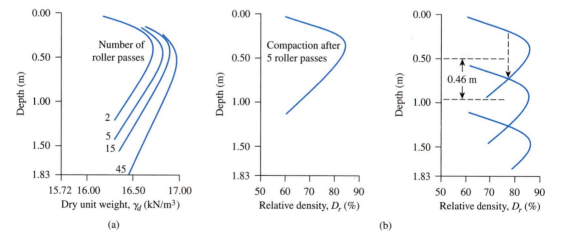

Figure 6.20 (a) Vibratory compaction of a sand—variation of dry unit weight with number of
roller passes; thickness of lift = 2.45 m; (b) estimation of compaction lift thickness for minimum
required relative density of 75% with five roller passes (*After D' Appolonia, Whitman, and
D' Appolonia, 1969. With permission from ASCE.*)

passes is determined, estimating the approximate thickness of each lift is easy. This procedure is shown in Figure 6.20b (D'Appolonia, Whitman, and D'Appolonia, 1969).

6.8   *Specifications for Field Compaction*

In most specifications for earthwork, the contractor is instructed to achieve a compacted field dry unit weight of 90 to 95% of the maximum dry unit weight determined in the laboratory by either the standard or modified Proctor test. This is a specification for relative compaction, which can be expressed as

$$R(\%) = \frac{\gamma_{d(field)}}{\gamma_{d(max-lab)}} \times 100 \tag{6.10}$$

where R = relative compaction.

For the compaction of granular soils, specifications sometimes are written in terms of the required relative density D_r or the required relative compaction. Relative density should not be confused with relative compaction. From Chapter 3, we can write

$$D_r = \left[\frac{\gamma_{d(field)} - \gamma_{d(min)}}{\gamma_{d(max)} - \gamma_{d(min)}} \right] \left[\frac{\gamma_{d(max)}}{\gamma_{d(field)}} \right] \tag{6.11}$$

Comparing Eqs. (6.10) and (6.11), we see that

$$R = \frac{R_0}{1 - D_r(1 - R_0)} \tag{6.12}$$

where

$$R_0 = \frac{\gamma_{d(min)}}{\gamma_{d(max)}} \tag{6.13}$$

On the basis of observation of 47 soil samples, Lee and Singh (1971) devised a correlation between R and D_r for granular soils:

$$R = 80 + 0.2D_r \tag{6.14}$$

The specification for field compaction based on relative compaction or on relative density is an end-product specification. The contractor is expected to achieve a minimum dry unit weight regardless of the field procedure adopted. The most economical compaction condition can be explained with the aid of Figure 6.21. The compaction curves $A, B,$ and C are for the same soil with varying compactive effort. Let curve A represent the conditions of maximum compactive effort that can be obtained from the existing equipment. Let the contractor be required to achieve a minimum dry unit weight of $\gamma_{d(field)} = R\gamma_{d(max)}$.

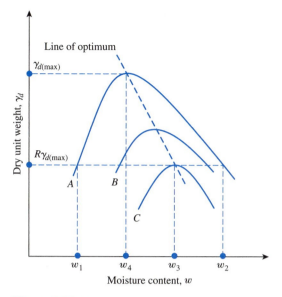

Figure 6.21 Most economical compaction condition

To achieve this, the contractor must ensure that the moisture content w falls between w_1 and w_2. As can be seen from compaction curve C, the required $\gamma_{d(\text{field})}$ can be achieved with a lower compactive effort at a moisture content $w = w_3$. However, for most practical conditions, a compacted field unit weight of $\gamma_{d(\text{field})} = R\gamma_{d(\text{max})}$ cannot be achieved by the minimum compactive effort. Hence, equipment with slightly more than the minimum compactive effort should be used. The compaction curve B represents this condition. Now we can see from Figure 6.21 that the most economical moisture content is between w_3 and w_4. Note that $w = w_4$ is the optimum moisture content for curve A, which is for the maximum compactive effort.

The concept described in the preceding paragraph, along with Figure 6.21, is attributed historically to Seed (1964) and is elaborated on in more detail in Holtz and Kovacs (1981).

Table 6.2 gives some of the requirements to achieve 95-to-100% relative compaction (based on standard Proctor maximum dry unit weight) by various field compaction equipment (U.S. Department of Navy, 1971).

6.9 *Determination of Field Unit Weight of Compaction*

When the compaction work is progressing in the field, knowing whether the specified unit weight has been achieved is useful. The standard procedures for determining the field unit weight of compaction include

1. Sand cone method
2. Rubber balloon method
3. Nuclear method

Following is a brief description of each of these methods.

Table 6.2 Requirements to Achieve R = 95 to 100% (based on standard Proctor maximum dry unit-weight)*

Equipment type	Applicability	Compacted lift thickness	Passes or coverages	Dimensions and weight of equipment		Possible variations in equipment
Sheepsfoot rollers	For fine-grained soils or dirty coarse-grained soils with more than 20% passing the No. 200 sieve. Not suitable for clean, coarse-grained soils. Particularly appropriate for compaction of impervious zone for earth dam or linings where bonding of lifts is important.	150 mm	4 to 6 passes for fine-grained soil 6 to 8 passes for coarse-grained soil	*Soil type* Fine-grained soil, PI > 30 Fine-grained soil, PI < 30 Coarse-grained soil Efficient compaction of soils on the wet side of the optimum requires less contact pressure than that required by the same soils at lower moisture contents.	*Foot contact area* 30 to 80 cm² 45 to 90 cm² 65 to 90 cm²	*Foot contact pressures* 1700 to 3400 kN/m² 1400 to 2800 kN/m² 1000 to 1800 kN/m²	For earth dam, highway, and airfield work, a drum of 1.5 m diameter, loaded to 45 to 90 kN per linear meter of drum is generally utilized. For smaller projects, a 1 m diameter drum, loaded to 22 to 50 kN per linear meter of drum, is used. Foot contact pressure should be regulated to avoid shearing the soil on the third or fourth pass.
Rubber-tired rollers	For clean, coarse-grained soils with 4 to 8% passing the No. 200 sieve. For fine-grained soils or well-graded, dirty, coarse-grained soils with more than 8% passing the No. 200 sieve.	250 mm 150 to 200 mm	3 to 5 coverages 4 to 6 coverages	Tire inflation pressures of 400 to 550 kN/m² for clean granular material or base course and subgrade compaction. Wheel load, 80 to 110 kN. Tire inflation pressures in excess of 450 kN/m² for fine-grained soils of high plasticity. For uniform clean sands or silty fine sands, use larger-size tires with pressures of 280 to 350 kN/m²		A wide variety of rubber-tired compaction equipment is available. For cohesive soils, light-wheel loads, such as provided by wobble-wheel equipment, may be substituted for heavy-wheel loads if lift thickness is decreased. For cohesionless soils, large-size tires are desirable to avoid shear and rutting.

Equipment type	Applicability	Lift thickness	Coverages	Requirements for compaction	Possible variations in equipment
Smooth-wheel rollers	Appropriate for subgrade or base course compaction of well-graded sand-gravel mixtures.	200 to 300 mm	4 coverages	Tandem-type rollers for base-course or sub-grade compaction, 90 to 135 kN weight, 53 to 88 kN per linear meter of width of rear roller.	Three-wheel rollers are obtainable in a wide range of sizes. Two-wheel tandem rollers are available in the 9 to 180 kN weight range. Three-axle tandem rollers are generally used in the 90 to 180 kN weight range. Very heavy rollers are used for proof rolling of subgrade or base course.
	May be used for fine-grained soils other than in earth dams. Not suitable for clean, well-graded sands or silty, uniform sands.	150 to 200 mm	6 coverages	Three-wheel roller for compaction of fine-grained soil; weights from 45 to 55 kN for materials of low plasticity to 90 kN for materials of high plasticity.	
Vibrating baseplate compactors	For coarse-grained soils with less than about 12% passing the No. 200 sieve. Best suited for materials with 4 to 8% passing the No. 200 sieve, placed thoroughly wet.	200 to 250 mm	3 coverages	Single pads or plates should weigh no less than 0.9 kN. May be used in tandem where working space is available. For clean, coarse-grained soil, vibration frequency should be no less than 1600 cycles per minute.	Vibrating pads or plates are available, hand-propelled or self-propelled, single or in gangs, with width of coverage from 0.45 to 4.5 m. Various types of vibrating-drum equipment should be considered for compaction in large areas.
Crawler tractor	Best suited for coarse-grained soils with less than 4 to 8% passing the No. 200 sieve, placed thoroughly wet.	250 to 300 mm	3 to 4 coverages	No smaller than D8 tractor with blade, 153 kN weight, for high compaction.	Tractor weights up to 265 kN
Power tamper or rammer	For difficult access, trench backfill. Suitable for all inorganic soils.	100 to 150 mm for silt or clay; 150 mm for coarse-grained soils	2 coverages	130 N minimum weight. Considerable range is tolerable, depending on materials and conditions.	Weights up to 1.1 kN, foot diameter, 100 to 250 mm

* After U.S. Navy (1971). Published by U.S. Government Printing Office

Sand Cone Method (ASTM Designation D-1556)

The sand cone device consists of a glass or plastic jar with a metal cone attached at its top (Figure 6.22). The jar is filled with uniform dry Ottawa sand. The combined weight of the jar, the cone, and the sand filling the jar is determined (W_1). In the field, a small hole is excavated in the area where the soil has been compacted. If the weight of the moist soil excavated from the hole (W_2) is determined and the moisture content of the excavated soil is known, the dry weight of the soil can be obtained as

$$W_3 = \frac{W_2}{1 + \dfrac{w\,(\%)}{100}} \tag{6.15}$$

where w = moisture content.

After excavation of the hole, the cone with the sand-filled jar attached to it is inverted and placed over the hole (Figure 6.23). Sand is allowed to flow out of the jar to fill the hole and the cone. After that, the combined weight of the jar, the cone, and the remaining sand in the jar is determined (W_4), so

$$W_5 = W_1 - W_4 \tag{6.16}$$

where W_5 = weight of sand to fill the hole and cone.

The volume of the excavated hole can then be determined as

$$V = \frac{W_5 - W_c}{\gamma_{d(\text{sand})}} \tag{6.17}$$

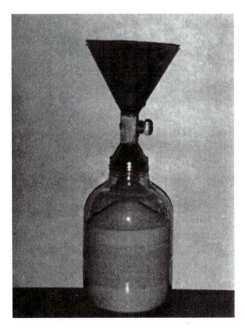

Figure 6.22
Glass jar filled with Ottawa sand with sand cone attached (*Courtesy of Braja M. Das, Henderson, Nevada*)

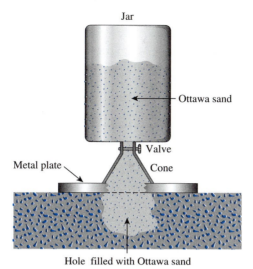

Jar

Ottawa sand

Valve

Metal plate

Cone

Hole filled with Ottawa sand

Figure 6.23 Field unit weight determined by sand cone method

where W_c = weight of sand to fill the cone only

$\gamma_{d(sand)}$ = dry unit weight of Ottawa sand used

The values of W_c and $\gamma_{d(sand)}$ are determined from the calibration done in the laboratory. The dry unit weight of compaction made in the field then can be determined as follows:

$$\gamma_d = \frac{\text{Dry weight of the soil excavated from the hole}}{\text{Volume of the hole}} = \frac{W_3}{V} \qquad (6.18)$$

Rubber Balloon Method (ASTM Designation D-2167)

The procedure for the rubber balloon method is similar to that for the sand cone method; a test hole is made and the moist weight of soil removed from the hole and its moisture content are determined. However, the volume of the hole is determined by introducing into it a rubber balloon filled with water from a calibrated vessel, from which the volume can be read directly. The dry unit weight of the compacted soil can be determined by using Eq. (6.18). Figure 6.24 shows a calibrated vessel that would be used with a rubber balloon.

Nuclear Method

Nuclear density meters are often used for determining the compacted dry unit weight of soil. The density meters operate either in drilled holes or from the ground surface. It uses a radioactive isotope source. The isotope gives off Gamma rays that radiate back to the meter's detector. Dense soil absorbs more radiation than loose soil. The instrument measures the weight of wet soil per unit volume and the weight of water present in a unit volume of soil. The dry unit weight of compacted soil can be determined by subtracting the weight of water from the moist unit weight of soil. Figure 6.25 shows a photograph of a nuclear density meter.

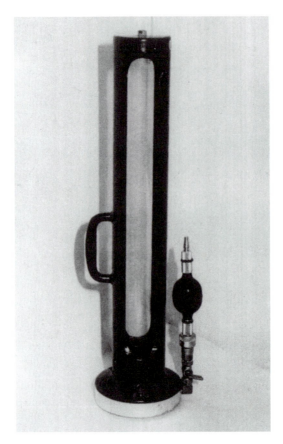

Figure 6.24 Calibrated vessel used with rubber balloon (not shown) (*Courtesy of John Hester, Carterville, Illinois*)

Figure 6.25 Nuclear density meter (*Courtesy of Braja M. Das, Henderson, Nevada*)

Example 6.3

Laboratory compaction test results for a clayey silt are given in the following table.

Moisture content (%)	Dry unit weight (kN/m³)
6	14.80
8	17.45
9	18.52
11	18.9
12	18.5
14	16.9

Following are the results of a field unit-weight determination test performed on the same soil by means of the sand cone method:

- Calibrated dry density of Ottawa sand = 1570 kg/m³
- Calibrated mass of Ottawa sand to fill the cone = 0.545 kg
- Mass of jar + cone + sand (before use) = 7.59 kg
- Mass of jar + cone + sand (after use) = 4.78 kg
- Mass of moist soil from hole = 3.007 kg
- Moisture content of moist soil = 10.2%

Determine:

a. Dry unit weight of compaction in the field
b. Relative compaction in the field

Solution
Part a

In the field,

Mass of sand used to fill the hole and cone = 7.59 kg − 4.78 kg = 2.81 kg

Mass of sand used to fill the hole = 2.81 kg − 0.545 kg = 2.265 kg

$$\text{Volume of the hole } (V) = \frac{2.265 \text{ kg}}{\text{Dry density of Ottawa sand}}$$

$$= \frac{2.265 \text{ kg}}{1570 \text{ kg/m}^3} = 0.0014426 \text{ m}^3$$

$$\text{Moist density of compacted soil} = \frac{\text{Mass of moist soil}}{\text{Volume of hole}}$$

$$= \frac{3.007}{0.0014426} = 2084.4 \text{ kg/m}^3$$

$$\text{Moist unit weight of compacted soil} = \frac{(2084.4)(9.81)}{1000} = 20.45 \text{ kN/m}^3$$

Hence,

$$\gamma_d = \frac{\gamma}{1 + \dfrac{w(\%)}{100}} = \frac{20.45}{1 + \dfrac{10.2}{100}} = \textbf{18.56 kN/m}^3$$

Part b

The results of the laboratory compaction test are plotted in Figure 6.26. From the plot, we see that $\gamma_d(\text{max}) = 19$ kN/m³. Thus, from Eq. (6.10).

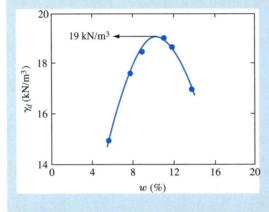

Figure 6.26 Plot of laboratory-compaction test results

$$R = \frac{\gamma_{d(\text{field})}}{\gamma_{d(\text{field})}} = \frac{18.56}{19.0} = \textbf{97.7\%}$$

■

6.10   *Compaction of Organic Soil and Waste Materials*

The presence of organic materials in a soil reduces its strength. In many cases, soils with a high organic content are generally discarded as fill material; however, in certain economic circumstances, slightly organic soils are used for compaction. In fact, organic soils are desirable in many circumstances (e.g., for agriculture, decertification, mitigation, and urban planning). The high costs of waste disposal have sparked an interest in the possible use of waste materials (e.g., bottom ash obtained from coal burning, copper slag, paper mill sludge, shredded waste tires mixed with inorganic soil, and so forth) in various landfill operations. Such use of waste materials is one of the major thrusts of present-day environmental geotechnology. Following is a discussion of the compaction characteristics of some of these materials.

Organic Soil

Franklin, Orozco, and Semrau (1973) conducted several laboratory tests to observe the effect of organic content on the compaction characteristics of soil. In the test program, various natural soils and soil mixtures were tested. Figure 6.27 shows the effect of organic content on the maximum dry unit weight. When the organic content exceeds 8 to 10%,

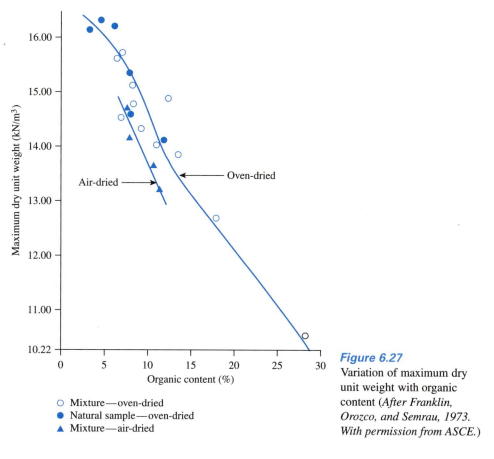

Figure 6.27
Variation of maximum dry unit weight with organic content (*After Franklin, Orozco, and Semrau, 1973. With permission from ASCE.*)

the maximum dry unit weight of compaction decreases rapidly. Conversely, the optimum moisture content for a given compactive effort increases with an increase in organic content. This trend is shown in Figure 6.28. Likewise, the maximum unconfined compression strength (see Chapter 11) obtained from a compacted soil (with a given compactive effort) decreases with increasing organic content of a soil. From these facts, we can see that soils with organic contents higher than about 10% are undesirable for compaction work.

Soil and Organic Material Mixtures

Lancaster, *et al.* (1996) conducted several modified Proctor tests to determine the effect of organic content on the maximum dry unit weight and optimum moisture content of soil and organic material mixtures. The soils tested consisted of a poorly graded sandy soil (SP-SM) mixed with either shredded redwood bark, shredded rice hulls, or municipal sewage sludge. Figures 6.29 and 6.30 show the variations of maximum dry unit weight of compaction and optimum moisture content, respectively, with organic content. As in Figure 6.27, the maximum dry unit weight decreased with organic content in all cases (see Figure 6.29). Conversely, the optimum moisture content increased with organic content for soil mixed with shredded red-wood or rice hulls (see Figure 6.30), similar to the pattern shown in Figure 6.28. However, for soil and municipal sewage sludge mixtures, the optimum moisture content remained practically constant (see Figure 6.30).

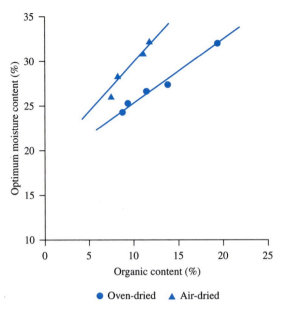

Figure 6.28 Variation of optimum moisture content with organic content (*After Franklin, Orozco, and Semrau, 1973. With permission from ASCE.*)

Bottom Ash from Coal Burning and Copper Slag

Laboratory standard Proctor test results for bottom ash from coal-burning power plants and for copper slag are also available in the literature. These waste products have been shown to be environmentally safe for use as landfill. A summary of some of these test results is given in Table 6.3.

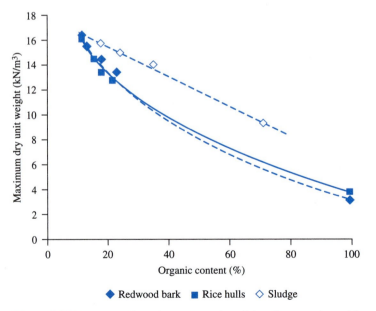

Figure 6.29 Variation of maximum dry unit weight of compaction with organic content—soil and organic material mixture. (*Source:* After "The Effect of Organic Content on Soil Compaction," by J. Lancaster, R. Waco, J. Towle, and R. Chaney, 1996. In *Proceedings, Third International Symposium on Environmental Geotechnology,* p. 159. Used with permission of the author)

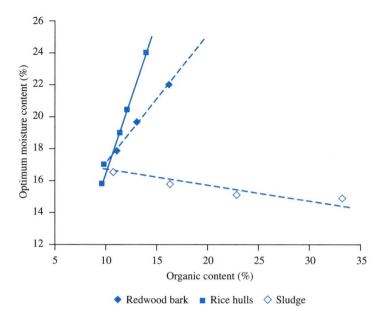

Figure 6.30 Variation of optimum moisture content with organic content — soil and organic material mixtures. (*Source:* After "The Effect of Organic Content on Soil Compaction," by J. Lancaster, R. Waco, J. Towle, and R. Chaney, 1996. In *Proceedings, Third International Symposium on Environmental Geotechnology,* p. 159. Used with permission of the author)

Table 6.3 Standard Proctor Test Results of Bottom Ash and Copper Slag

Type	Location	Maximum dry unit weight (kN/m³)	Optimum moisture content (%)	Source
Bottom ash—bituminous coal (West Virginia)	Fort Martin	13.4	24.8	Seals, Moulton, and Ruth (1972)
	Kammer	16.0	13.8	
	Kanawha River	11.4	26.2	
	Mitchell	18.3	14.6	
	Muskingham	14.3	22.0	
	Willow Island	14.5	21.2	
Bottom ash—lignite coal	Big Stone Power Plant, South Dakota	16.4	20.5	Das, Selim, and Pfeifle (1978)
Copper slag	American Smelter and Refinery Company, El Paso, Texas	19.8	18.8	Das, Tarquin, and Jones (1983)

6.11 *Special Compaction Techniques*

Several special types of compaction techniques have been developed for deep compaction of in-place soils, and these techniques are used in the field for large-scale compaction works. Among these, the popular methods are vibroflotation, dynamic compaction, and blasting. Details of these methods are provided in the following sections.

Vibroflotation

Vibroflotation is a technique for *in situ* densification of thick layers of loose granular soil deposits. It was developed in Germany in the 1930s. The first vibroflotation device was used in the United States about 10 years later. The process involves the use of a *Vibroflot* unit (also called the *vibrating unit*), which is about 2.1 m long. (As shown in Figure 6.31.) This vibrating unit has an eccentric weight inside it and can develop a centrifugal force, which enables the vibrating unit to vibrate horizontally. There are openings at the bottom and top of the vibrating unit for water jets. The vibrating unit is attached to a follow-up pipe. Figure 6.31 shows the entire assembly of equipment necessary for conducting the field compaction.

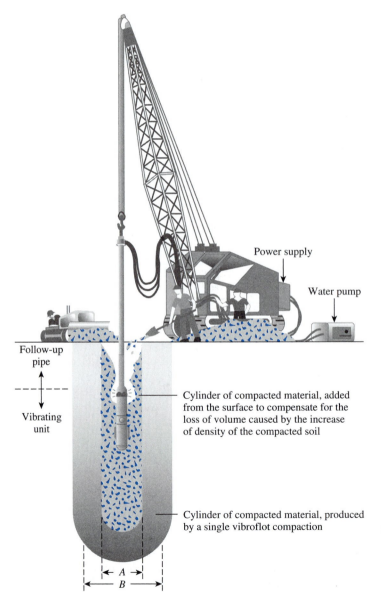

Power supply

Water pump

Follow-up pipe

Vibrating unit

Cylinder of compacted material, added from the surface to compensate for the loss of volume caused by the increase of density of the compacted soil

Cylinder of compacted material, produced by a single vibroflot compaction

A

B

Figure 6.31
Vibroflotation unit
(*After Brown, 1977.
With permission from
ASCE.*)

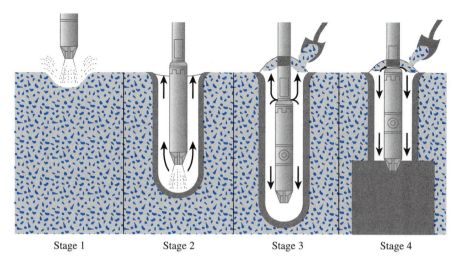

| Stage 1 | Stage 2 | Stage 3 | Stage 4 |

Figure 6.32 Compaction by vibroflotation process (*After Brown, 1977. With permission from ASCE.*)

The entire vibroflotation compaction process in the field can be divided into four stages (Figure 6.32):

Stage 1: The jet at the bottom of the Vibroflot is turned on and lowered into the ground.

Stage 2: The water jet creates a quick condition in the soil and it allows the vibrating unit to sink into the ground.

Stage 3: Granular material is poured from the top of the hole. The water from the lower jet is transferred to the jet at the top of the vibrating unit. This water carries the granular material down the hole.

Stage 4: The vibrating unit is gradually raised in about 0.3 m lifts and held vibrating for about 30 seconds at each lift. This process compacts the soil to the desired unit weight.

The details of various types of Vibroflot units used in the United States are given in Table 6.4. Note that 23 kW electric units have been used since the latter part of the 1940s. The 75 kW units were introduced in the early 1970s.

The zone of compaction around a single probe varies with the type of Vibroflot used. The cylindrical zone of compaction has a radius of about 2 m for a 23 kW unit. This radius can extend to about 3 m for a 75 kW unit.

Typical patterns of Vibroflot probe spacings are shown in Figure 6.33. Square and rectangular patterns generally are used to compact soil for isolated, shallow foundations. Equilateral triangular patterns generally are used to compact large areas. The capacity for successful densification of *in situ* soil depends on several factors, the most important of which is the grain-size distribution of the soil and the type of backfill used to fill the holes during the withdrawal period of the Vibroflot. The range of the grain-size distribution of *in situ* soil marked Zone 1 in Figure 6.34 is most suitable for compaction by vibroflotation. Soils that contain excessive amounts of fine sand and silt-size particles are difficult to compact, and considerable effort is needed to reach the

Table 6.4 Types of Vibroflot Units*

Motor type	75 kW electric and hydraulic	23 kW electric
a. Vibrating tip		
Length	2.1 m	1.86 m
Diameter	406 mm	381 mm
Weight	17.8 kN	17.8 kN
Maximum movement when full	12.5 mm	7.6 mm
Centrifugal force	160 kN	89 kN
b. Eccentric		
Weight	1.2 kN	0.76 kN
Offset	38 mm	32 mm
Length	610 mm	390 mm
Speed	1800 rpm	1800 rpm
c. Pump		
Operating flow rate	0–1.6 m^3/min	0–0.6 m^3/min
Pressure	700–1050 kN/m^2	700–1050 kN/m^2
d. Lower follow-up pipe and extensions		
Diameter	305 mm	305 mm
Weight	3.65 kN/m	3.65 kN/m

*After Brown (1977)

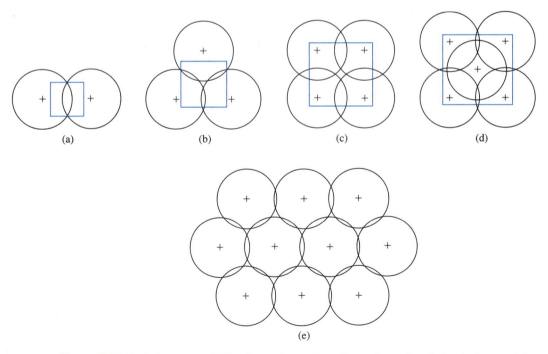

Figure 6.33 Typical patterns of Vibroflot probe spacings for a column foundation (a, b, c, and d) and for compaction over a large area (e)

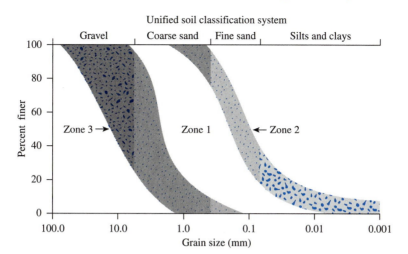

Figure 6.34 Effective range of grain-size distribution of soil for vibroflotation

proper relative density of compaction. Zone 2 in Figure 6.34 is the approximate lower limit of grain-size distribution for which compaction by vibroflotation is effective. Soil deposits whose grain-size distributions fall in Zone 3 contain appreciable amounts of gravel. For these soils, the rate of probe penetration may be slow and may prove uneconomical in the long run.

The grain-size distribution of the backfill material is an important factor that controls the rate of densification. Brown (1977) has defined a quantity called the *suitability number* for rating backfill as

$$S_N = 1.7\sqrt{\frac{3}{(D_{50})^2} + \frac{1}{(D_{20})^2} + \frac{1}{(D_{10})^2}}$$ (6.19)

where D_{50}, D_{20}, and D_{10} are the diameters (in mm) through which, respectively, 50, 20, and 10% of the material passes.

The smaller the value of S_N, the more desirable the backfill material. Following is a backfill rating system proposed by Brown:

Range of S_N	Rating as backfill
0–10	Excellent
10–20	Good
20–30	Fair
30–50	Poor
>50	Unsuitable

Dynamic Compaction

Dynamic compaction is a technique that has gained popularity in the United States for the densification of granular soil deposits. This process consists primarily of dropping a heavy weight repeatedly on the ground at regular intervals. The weight of the hammer used varies over a range of 80 to 360 kN and the height of the hammer drop varies between 7.5 and 30.5 m. The stress waves generated by the hammer drops aid in the densification. The degree of compaction achieved at a given site depends on the following three factors:

1. Weight of hammer
2. Height of hammer drop
3. Spacing of locations at which the hammer is dropped

Leonards, Cutter, and Holtz (1980) suggested that the significant depth of influence for compaction can be approximated by using the equation

$$D \simeq \left(\tfrac{1}{2}\right)\sqrt{W_H h} \tag{6.20}$$

where D = significant depth of densification (m)
$\quad W_H$ = dropping weight (metric ton)
$\quad h$ = height of drop (m)

In 1992, Poran and Rodriguez suggested a rational method for conducting dynamic compaction for granular soils in the field. According to their method, for a hammer of width D having a weight W_H and a drop h, the approximate shape of the densified area will be of the type shown in Figure 6.35 (i.e., a semiprolate spheroid). Note that in this figure $b = DI$ (where DI is the significant depth of densification). Figure 6.36 gives the design chart for a/D and b/D versus $NW_H h/Ab$ (D = width of the hammer if not circular in cross section; A = area of cross section of the hammer; and N = number of required hammer drops). This method uses the following steps.

Step 1: Determine the required significant depth of densification, $DI (= b)$.
Step 2: Determine the hammer weight (W_H), height of drop (h), dimensions of the cross section, and thus, the area A and the width D.
Step 3: Determine $DI/D = b/D$.
Step 4: Use Figure 6.36 and determine the magnitude of $NW_H h/Ab$ for the value of b/D obtained in Step 3.

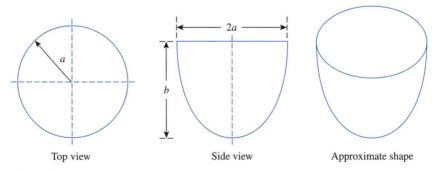

Figure 6.35 Approximate shape of the densified area due to dynamic compaction

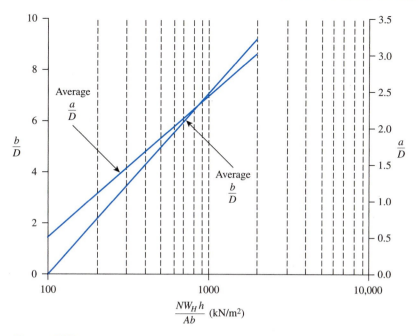

Figure 6.36 Poran and Rodriguez chart for a/D, b/D versus $NW_H h/Ab$

Step 5: Since the magnitudes of W_H, h, A, and b are known (or assumed) from Step 2, the number of hammer drops can be estimated from the value of $NW_H h/Ab$ obtained from Step 4.

Step 6: With known values of $NW_H h/Ab$, determine a/D and thus a from Figure 6.36.

Step 7: The grid spacing, S_g, for dynamic compaction may now be assumed to be equal to or somewhat less than a. (See Figure 6.37.)

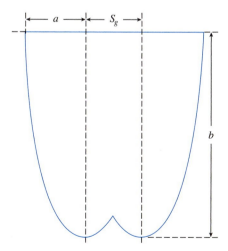

Figure 6.37 Approximate grid spacing for dynamic compaction

Blasting

Blasting is a technique that has been used successfully in many projects (Mitchell, 1970) for the densification of granular soils. The general soil grain sizes suitable for compaction by blasting are the same as those for compaction by vibroflotation. The process involves the detonation of explosive charges, such as 60% dynamite at a certain depth below the ground surface in saturated soil. The lateral spacing of the charges varies from about 3 to 9 m. Three to five successful detonations are usually necessary to achieve the desired compaction. Compaction (up to a relative density of about 80%) up to a depth of about 18 m over a large area can easily be achieved by using this process. Usually, the explosive charges are placed at a depth of about two-thirds of the thickness of the soil layer desired to be compacted. The sphere of influence of compaction by a 60% dynamite charge can be given as follows (Mitchell, 1970):

$$r = \sqrt{\frac{W_{\text{EX}}}{C}} \tag{6.21}$$

where r = sphere of influence
W_{EX} = weight of explosive—60% dynamite
C = 0.0122 when W_{EX} is in kg and r is in m

Figure 6.38 shows the test results of soil densification by blasting in an area measuring 15 m by 9 m (Mitchell, 1970). For these tests, twenty 2.09-kg charges of Gelamite No. 1 (Hercules Powder Company, Wilmington, Delaware) were used.

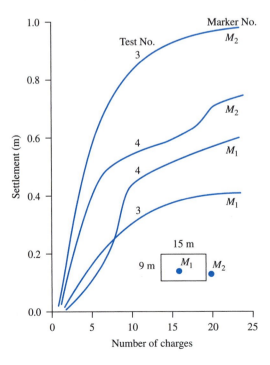

Figure 6.38 Ground settlement as a function of number of explosive charges

Example 6.4

Following are the details for the backfill material used in a vibroflotation project:

- $D_{10} = 0.36$ mm
- $D_{20} = 0.52$ mm
- $D_{50} = 1.42$ mm

Determine the suitability number S_N. What would be its rating as a backfill material?

Solution
From Eq. (6.19),

$$S_N = 1.7\sqrt{\frac{3}{(D_{50})^2} + \frac{1}{(D_{20})^2} + \frac{1}{(D_{10})^2}}$$

$$= 1.7\sqrt{\frac{3}{(1.42)^2} + \frac{1}{(0.52)^2} + \frac{1}{(0.36)^2}}$$

$$= 6.1$$

Rating: **Excellent**

6.12 *Summary and General Comments*

Laboratory standard and modified Proctor compaction tests described in this chapter are essentially for *impact* or *dynamic* compaction of soil; however, in the laboratory, *static compaction* and *kneading compaction* also can be used. It is important to realize that the compaction of clayey soils achieved by rollers in the field is essentially the kneading type. The relationships of dry unit weight (γ_d) and moisture content (w) obtained by dynamic and kneading compaction are not the same. Proctor compaction test results obtained in the laboratory are used primarily to determine whether the roller compaction in the field is sufficient. The structures of compacted cohesive soil at a similar dry unit weight obtained by dynamic and kneading compaction may be different. This difference, in turn, affects physical properties, such as hydraulic conductivity, compressibility, and strength.

For most fill operations, the final selection of the borrow site depends on such factors as the soil type and the cost of excavation and hauling.

Problems

6.1 Given $G_s = 2.75$, calculate the zero-air-void unit weight for a soil in kN/m^3 at $w = 5\%$, 8%, 10%, 12%, and 15%.

6.2 Repeat Problem 6.1 with $G_s = 2.65$.

6.3 Calculate the variation of dry density (kg/m^3) of a soil (G_s = 2.67) at w = 10% and 20% for degree of saturation (S) = 80%, 90%, and 100%.

6.4 The results of a standard Proctor test are given below. Determine the maximum dry unit weight of compaction and the optimum moisture content.

Volume of Proctor mold (cm³)	Weight of wet soil in the mold (N)	Moisture content (%)
943	14.5	8.4
943	18.46	10.2
943	20.77	12.3
943	17.88	14.6
943	16.15	16.8

6.5 For the soil described in Problem 6.4, if G_s = 2.72, determine the void ratio and the degree of saturation at optimum moisture content.

6.6 The results of a standard Proctor test are given in the following table. Determine the maximum dry density (kg/m^3) of compaction and the optimum moisture content.

Volume of Proctor mold (cm³)	Mass of wet soil in the mold (kg)	Moisture content (%)
943.3	1.68	9.9
943.3	1.71	10.6
943.3	1.77	12.1
943.3	1.83	13.8
943.3	1.86	15.1
943.3	1.88	17.4
943.3	1.87	19.4
943.3	1.85	21.2

6.7 A field unit weight determination test for the soil described in Problem 6.6 yielded the following data: moisture content = 10.5% and moist density = 1705 kg/m^3. Determine the relative compaction.

6.8 The *in situ* moisture content of a soil is 18% and the moist unit weight is 16.5 kN/m^3. The specific gravity of soil solids is 2.75. This soil is to be excavated and transported to a construction site for use in a compacted fill. If the specifications call for the soil to be compacted to a minimum dry unit weight of 16.27 kN/m^3 at the same moisture content of 18%, how many cubic meters of soil from the excavation site are needed to produce 7651 m^3 of compacted fill? How many 178 kN truckloads are needed to transport the excavated soil?

6.9 A proposed embankment fill requires 8000 m^3 of compacted soil. The void ratio of the compacted fill is specified as 0.7. Four borrow pits are available as described in the following table, which lists the respective void ratios of the soil and the cost per cubic meter for moving the soil to the proposed construction site. Make the necessary calculations to select the pit from which the soil should be bought to minimize the cost. Assume G_s to be the same at all pits.

Borrow pit	Void ratio	Cost ($/m³)
A	0.82	8
B	1.1	5
C	0.90	9
D	0.78	12

6.10 The maximum and minimum dry unit weights of a sand were determined in the laboratory to be 16.35 kN/m³ and 14.62 kN/m³, respectively. What would be the relative compaction in the field if the relative density is 78%?

6.11 The maximum and minimum dry densities of a sand were determined in the laboratory to be 1682 kg/m³ and 1510 kg/m³, respectively. In the field, if the relative density of compaction of the same sand is 70%, what are its relative compaction (%) and dry density (kg/m³)?

6.12 The relative compaction of a sand in the field is 90%. The maximum and minimum dry unit weights of the sand are 16.98 kN/m³ and 14.94 kN/m³, respectively. For the field condition, determine
  a. Dry unit weight
  b. Relative density of compaction
  c. Moist unit weight at a moisture content of 12%

6.13 Following are the results of a field unit weight determination test on a soil with the sand cone method:
  • Calibrated dry density of Ottawa sand = 1667 kg/m³
  • Calibrated mass of Ottawa sand to fill the cone = 0.117 kg
  • Mass of jar + cone + sand (before use) = 5.99 kg
  • Mass of jar + cone + sand (after use) = 2.81 kg
  • Mass of moist soil from hole = 3.331 kg
  • Moisture content of moist soil = 11.6%
  Determine the dry unit weight of compaction in the field.

6.14 The backfill material for a vibroflotation project has the following grain sizes:
  • D_{10} = 0.11 mm
  • D_{20} = 0.19 mm
  • D_{50} = 1.3 mm
  Determine the suitability number, S_N, for each.

6.15 Repeat Problem 6.14 using the following values:
  • D_{10} = 0.09 mm
  • D_{20} = 0.25 mm
  • D_{50} = 0.61 mm

References

AMERICAN ASSOCIATION OF STATE HIGHWAY AND TRANSPORTATION OFFICIALS (1982). *AASHTO Materials, Part II*, Washington, D.C.

AMERICAN SOCIETY FOR TESTING AND MATERIALS (2007). *Annual Book of ASTM Standards, Vol 04.08*, West Conshohocken, Pa.

D'APPOLONIA, E. (1953). "Loose Sands—Their Compaction by Vibroflotation," *Special Technical Publication No. 156,* ASTM, 138–154.

BROWN, E. (1977). "Vibroflotation Compaction of Cohesionless Soils," *Journal of the Geotechnical Engineering Division,* ASCE, Vol. 103, No. GT12, 1437–1451.

D'APPOLONIA, D. J., WHITMAN, R. V., and D'APPOLONIA, E. D. (1969). "Sand Compaction with Vibratory Rollers," *Journal of the Soil Mechanics and Foundations Division,* ASCE, Vol. 95, No. SM1, 263–284.

DAS, B. M., SELIM, A. A., and PFEIFLE, T. W. (1978). "Effective Use of Bottom Ash as a Geotechnical Material," *Proceedings,* 5th Annual UMR-DNR Conference and Exposition on Energy, University of Missouri, Rolla, 342–348.

DAS, B. M., TARQUIN, A. J., and JONES, A. D. (1983). "Geotechnical Properties of a Copper Slag," *Transportation Research Record No. 941,* National Research Council, Washington, D.C., 1–4.

FRANKLIN, A. F., OROZCO, L. F., and SEMRAU, R. (1973). "Compaction of Slightly Organic Soils," *Journal of the Soil Mechanics and Foundations Division,* ASCE, Vol. 99, No. SM7, 541–557.

GURTUG, Y., and SRIDHARAN, A. (2004). "Compaction Behaviour and Prediction of Its Characteristics of Fine Grained Soils with Particular Reference to Compaction Energy," *Soils and Foundations,* Vol. 44, No. 5, 27–36.

HOLTZ, R. D., and KOVACS, W. D. (1981). *An Introduction to Geotechnical Engineering,* Prentice-Hall, Englewood Cliffs, N.J.

JOHNSON, A. W., and SALLBERG, J. R. (1960). "Factors That Influence Field Compaction of Soil," Highway Research Board, *Bulletin No. 272.*

LAMBE, T. W. (1958a). "The Structure of Compacted Clay," *Journal of the Soil Mechanics and Foundations Division,* ASCE, Vol. 84, No. SM2, 1654–1 to 1654–35.

LAMBE, T. W. (1958b). "The Engineering Behavior of Compacted Clay," *Journal of the Soil Mechanics and Foundations Division,* ASCE, Vol. 84, No. SM2, 1655–1 to 1655–35.

LANCASTER, J., WACO, R., TOWLE, J., and CHANEY, R. (1996). "The Effect of Organic Content on Soil Compaction," *Proceedings,* 3rd International Symposium on Environmental Geotechnology, San Diego, 152–161.

LEE, K. W., and SINGH, A. (1971). "Relative Density and Relative Compaction," *Journal of the Soil Mechanics and Foundations Division,* ASCE, Vol. 97, No. SM7, 1049–1052.

LEE, P. Y., and SUEDKAMP, R. J. (1972). "Characteristics of Irregularly Shaped Compaction Curves of Soils," *Highway Research Record No. 381,* National Academy of Sciences, Washington, D.C., 1–9.

LEONARDS, G. A., CUTTER, W. A., and HOLTZ, R. D. (1980). "Dynamic Compaction of Granular Soils," *Journal of the Geotechnical Engineering Division,* ASCE, Vol. 106, No. GT1, 35–44.

MITCHELL, J. K. (1970). "In-Place Treatment of Foundation Soils," *Journal of the Soil Mechanics and Foundations Division,* ASCE, Vol. 96, No. SM1, 73–110.

OMAR, M., ABDALLAH, S., BASMA, A., and BARAKAT, S. (2003). "Compaction Characteristics of Granular Soils in United Arab Emirates," *Geotechnical and Geological Engineering,* Vol. 21, No. 3, 283–295.

PORAN, C. J., and RODRIGUEZ, J. A. (1992). "Design of Dynamic Compaction," *Canadian Geotechnical Journal,* Vol. 2, No. 5, 796–802.

PROCTOR, R. R. (1933). "Design and Construction of Rolled Earth Dams," *Engineering News Record,* Vol. 3, 245–248, 286–289, 348–351, 372–376.

SEALS, R. K. MOULTON, L. K., and RUTH, E. (1972). "Bottom Ash: An Engineering Material," *Journal of the Soil Mechanics and Foundations Division,* ASCE, Vol. 98, No. SM4, 311–325.

SEED, H. B. (1964). Lecture Notes, CE 271, Seepage and Earth Dam Design, University of California, Berkeley.

SEED, H. B., and CHAN, C. K. (1959). "Structure and Strength Characteristics of Compacted Clays," *Journal of the Soil Mechanics and Foundations Division,* ASCE, Vol. 85, No. SM5, 87–128.

U.S. DEPARTMENT OF NAVY (1971). "Design Manual—Soil Mechanics, Foundations, and Structures," *NAVFAC DM-7,* U.S. Government Printing Office, Washington, D.C.

7 Permeability

Soils are permeable due to the existence of interconnected voids through which water can flow from points of high energy to points of low energy. The study of the flow of water through permeable soil media is important in soil mechanics. It is necessary for estimating the quantity of underground seepage under various hydraulic conditions, for investigating problems involving the pumping of water for underground construction, and for making stability analyses of earth dams and earth-retaining structures that are subject to seepage forces.

7.1 Bernoulli's Equation

From fluid mechanics, we know that, according to Bernoulli's equation, the total head at a point in water under motion can be given by the sum of the pressure, velocity, and elevation heads, or

$$h = \underbrace{\frac{u}{\gamma_w}}_{\substack{\text{Pressure} \\ \text{head}}} + \underbrace{\frac{v^2}{2g}}_{\substack{\text{Velocity} \\ \text{head}}} + \underbrace{Z}_{\substack{\text{Elevation} \\ \text{head}}} \tag{7.1}$$

where h = total head
u = pressure
v = velocity
g = acceleration due to gravity
γ_w = unit weight of water

Note that the elevation head, Z, is the vertical distance of a given point above or below a datum plane. The pressure head is the water pressure, u, at that point divided by the unit weight of water, γ_w.

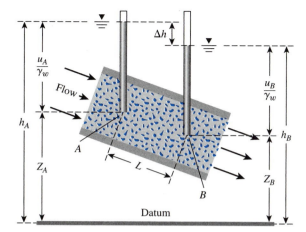

Figure 7.1 Pressure, elevation, and total heads for flow of water through soil

If Bernoulli's equation is applied to the flow of water through a porous soil medium, the term containing the velocity head can be neglected because the seepage velocity is small, and the total head at any point can be adequately represented by

$$h = \frac{u}{\gamma_w} + Z \tag{7.2}$$

Figure 7.1 shows the relationship among pressure, elevation, and total heads for the flow of water through soil. Open standpipes called *piezometers* are installed at points A and B. The levels to which water rises in the piezometer tubes situated at points A and B are known as the *piezometric levels* of points A and B, respectively. The pressure head at a point is the height of the vertical column of water in the piezometer installed at that point.

The loss of head between two points, A and B, can be given by

$$\Delta h = h_A - h_B = \left(\frac{u_A}{\gamma_w} + Z_A \right) - \left(\frac{u_B}{\gamma_w} + Z_B \right) \tag{7.3}$$

The head loss, Δh, can be expressed in a nondimensional form as

$$i = \frac{\Delta h}{L} \tag{7.4}$$

where i = hydraulic gradient

L = distance between points A and B—that is, the length of flow over
which the loss of head occurred

In general, the variation of the velocity v with the hydraulic gradient i is as shown in Figure 7.2. This figure is divided into three zones:

1. Laminar flow zone (Zone I)
2. Transition zone (Zone II)
3. Turbulent flow zone (Zone III)

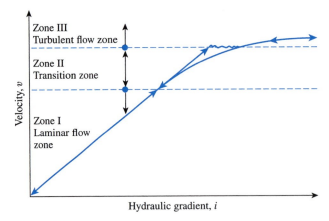

Figure 7.2 Nature of variation of v with hydraulic gradient, i

When the hydraulic gradient is increased gradually, the flow remains laminar in Zones I and II, and the velocity, v, bears a linear relationship to the hydraulic gradient. At a higher hydraulic gradient, the flow becomes turbulent (Zone III). When the hydraulic gradient is decreased, laminar flow conditions exist only in Zone I.

In most soils, the flow of water through the void spaces can be considered laminar; thus,

$$v \propto i \tag{7.5}$$

In fractured rock, stones, gravels, and very coarse sands, turbulent flow conditions may exist, and Eq. (7.5) may not be valid.

7.2   *Darcy's Law*

In 1856, Darcy published a simple equation for the discharge velocity of water through saturated soils, which may be expressed as

$$v = ki \tag{7.6}$$

where v = *discharge velocity,* which is the quantity of water flowing in unit time through a unit gross cross-sectional area of soil at right angles to the direction of flow

k = hydraulic conductivity (otherwise known as the coefficient of permeability)

This equation was based primarily on Darcy's observations about the flow of water through clean sands. Note that Eq. (7.6) is similar to Eq. (7.5); both are valid for laminar flow conditions and applicable for a wide range of soils.

In Eq. (7.6), v is the discharge velocity of water based on the gross cross-sectional area of the soil. However, the actual velocity of water (that is, the seepage velocity) through the

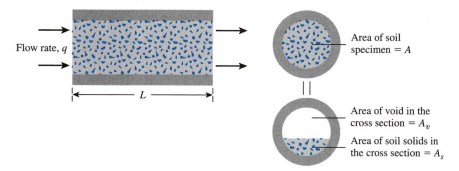

Figure 7.3 Derivation of Eq. (7.10)

void spaces is greater than v. A relationship between the discharge velocity and the seepage velocity can be derived by referring to Figure 7.3, which shows a soil of length L with a gross cross-sectional area A. If the quantity of water flowing through the soil in unit time is q, then

$$q = vA = A_v v_s \tag{7.7}$$

where v_s = seepage velocity
A_v = area of void in the cross section of the specimen

However,

$$A = A_v + A_s \tag{7.8}$$

where A_s = area of soil solids in the cross section of the specimen.
Combining Eqs. (7.7) and (7.8) gives

$$q = v(A_v + A_s) = A_v v_s$$

or

$$v_s = \frac{v(A_v + A_s)}{A_v} = \frac{v(A_v + A_s)L}{A_v L} = \frac{v(V_v + V_s)}{V_v} \tag{7.9}$$

where V_v = volume of voids in the specimen
V_s = volume of soil solids in the specimen

Equation (7.9) can be rewritten as

$$v_s = v \left[\frac{1 + \left(\dfrac{V_v}{V_s} \right)}{\dfrac{V_v}{V_s}} \right] = v \left(\frac{1 + e}{e} \right) = \frac{v}{n} \tag{7.10}$$

where e = void ratio
n = porosity

Darcy's law as defined by Eq. (7.6) implies that the discharge velocity v bears a linear relationship to the hydraulic gradient i and passes through the origin as shown in Figure 7.4. Hansbo (1960), however, reported the test results for four undisturbed

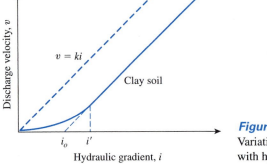

Figure 7.4
Variation of discharge velocity
with hydraulic gradient in clay

natural clays. On the basis of his results, a hydraulic gradient i' (see Figure 7.4) appears to exist, at which

$$v = k(i - i_0) \qquad (\text{for } i \geq i') \tag{7.11}$$

and

$$v = ki^m \qquad (\text{for } i < i') \tag{7.12}$$

The preceding equation implies that for very low hydraulic gradients, the relationship between v and i is nonlinear. The value of m in Eq. (7.12) for four Swedish clays was about 1.5. However, several other studies refute the preceding findings. Mitchell (1976) discussed these studies in detail. Taking all points into consideration, he concluded that Darcy's law is valid.

7.3 *Hydraulic Conductivity*

Hydraulic conductivity is generally expressed in cm/sec or m/sec in SI units.

The hydraulic conductivity of soils depends on several factors: fluid viscosity, pore-size distribution, grain-size distribution, void ratio, roughness of mineral particles, and degree of soil saturation. In clayey soils, structure plays an important role in hydraulic conductivity. Other major factors that affect the permeability of clays are the ionic concentration and the thickness of layers of water held to the clay particles.

The value of hydraulic conductivity (k) varies widely for different soils. Some typical values for saturated soils are given in Table 7.1. The hydraulic conductivity of unsaturated soils is lower and increases rapidly with the degree of saturation.

Table 7.1 Typical Values of Hydraulic Conductivity of Saturated Soils

Soil type	k (cm/sec)
Clean gravel	100−1.0
Coarse sand	1.0−0.01
Fine sand	0.01−0.001
Silty clay	0.001−0.00001
Clay	<0.000001

The hydraulic conductivity of a soil is also related to the properties of the fluid flowing through it by the equation

$$k = \frac{\gamma_w}{\eta} \overline{K}$$ (7.13)

where γ_w = unit weight of water
η = viscosity of water
$\overline{K}$ = absolute permeability

The *absolute permeability* $\overline{K}$ is expressed in units of L^2 (that is, cm², and so forth).

Equation (7.13) showed that hydraulic conductivity is a function of the unit weight and the viscosity of water, which is in turn a function of the temperature at which the test is conducted. So, from Eq. (7.13),

$$\frac{k_{T_1}}{k_{T_2}} = \left(\frac{\eta_{T_2}}{\eta_{T_1}}\right)\left(\frac{\gamma_{w(T_1)}}{\gamma_{w(T_2)}}\right)$$ (7.14)

where k_{T_1}, k_{T_2} = hydraulic conductivity at temperatures T_1 and T_2, respectively
η_{T_1}, η_{T_2} = viscosity of water at temperatures T_1 and T_2, respectively
$\gamma_{w(T_1)}, \gamma_{w(T_2)}$ = unit weight of water at temperatures T_1 and T_2, respectively

It is conventional to express the value of k at a temperature of 20°C. Within the range of test temperatures, we can assume that $\gamma_{w(T_1)} \simeq \gamma_{w(T_2)}$. So, from Eq. (7.14)

$$k_{20°C} = \left(\frac{\eta_{T°C}}{\eta_{20°C}}\right) k_{T°C}$$ (7.15)

The variation of $\eta_{T°C}/\eta_{20°C}$ with the test temperature T varying from 15 to 30°C is given in Table 7.2.

Table 7.2 Variation of $\eta_{T°C}/\eta_{20°C}$

Temperature, T (°C)	$\eta_{T°C}/\eta_{20°C}$	Temperature, T (°C)	$\eta_{T°C}/\eta_{20°C}$
15	1.135	23	0.931
16	1.106	24	0.910
17	1.077	25	0.889
18	1.051	26	0.869
19	1.025	27	0.850
20	1.000	28	0.832
21	0.976	29	0.814
22	0.953	30	0.797

Laboratory Determination of Hydraulic Conductivity

Two standard laboratory tests are used to determine the hydraulic conductivity of soil— the constant-head test and the falling-head test. A brief description of each follows.

Constant-Head Test

A typical arrangement of the constant-head permeability test is shown in Figure 7.5. In this type of laboratory setup, the water supply at the inlet is adjusted in such a way that the difference of head between the inlet and the outlet remains constant during the test period. After a constant flow rate is established, water is collected in a graduated flask for a known duration.

The total volume of water collected may be expressed as

$$Q = Avt = A(ki)t \tag{7.16}$$

where Q = volume of water collected
A = area of cross section of the soil specimen
t = duration of water collection

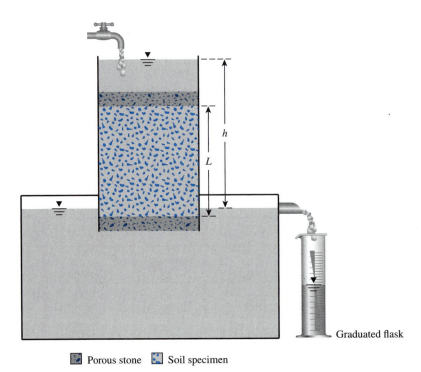

Graduated flask

Porous stone   Soil specimen

Figure 7.5 Constant-head permeability test

And because

$$i = \frac{h}{L} \tag{7.17}$$

where L = length of the specimen, Eq. (7.17) can be substituted into Eq. (7.16) to yield

$$Q = A\left(k\frac{h}{L}\right)t \tag{7.18}$$

or

$$k = \frac{QL}{Aht} \tag{7.19}$$

Falling-Head Test

A typical arrangement of the falling-head permeability test is shown in Figure 7.6. Water from a standpipe flows through the soil. The initial head difference h_1 at time $t = 0$ is recorded, and water is allowed to flow through the soil specimen such that the final head difference at time $t = t_2$ is h_2.

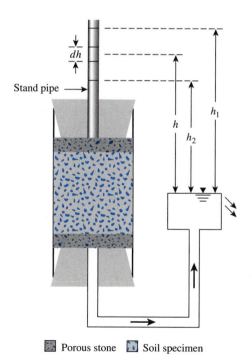

Porous stone   Soil specimen

Figure 7.6 Falling-head permeability test

The rate of flow of the water through the specimen at any time t can be given by

$$q = k\frac{h}{L}A = -a\frac{dh}{dt} \qquad (7.20)$$

where q = flow rate

a = cross-sectional area of the standpipe

A = cross-sectional area of the soil specimen

Rearrangement of Eq. (7.20) gives

$$dt = \frac{aL}{Ak}\left(-\frac{dh}{h}\right) \qquad (7.21)$$

Integration of the left side of Eq. (7.21) with limits of time from 0 to t and the right side with limits of head difference from h_1 to h_2 gives

$$t = \frac{aL}{Ak}\log_e \frac{h_1}{h_2}$$

or

$$k = 2.303\frac{aL}{At}\log_{10}\frac{h_1}{h_2} \qquad (7.22)$$

Example 7.1

Refer to the constant-head permeability test arrangement shown in Figure 7.5. A test gives these values:

- $L = 30$ cm
- A = area of the specimen = 177 cm^2
- Constant-head difference, $h = 50$ cm
- Water collected in a period of 5 min = 350 cm^3

Calculate the hydraulic conductivity in cm/sec.

Solution
From Eq. (7.19),

$$k = \frac{QL}{Aht}$$

Given $Q = 350$ cm^3, $L = 30$ cm, $A = 177$ cm^2, $h = 50$ cm, and $t = 5$ min, we have

$$k = \frac{(350)(30)}{(177)(50)(5)(60)} = \textbf{3.95} \times \textbf{10}^3 \textbf{ cm/sec} \qquad ∎$$

Example 7.2

For a falling-head permeability test, the following values are given:

- Length of specimen = 203 mm
- Area of soil specimen = 10.3 cm^2
- Area of standpipe = 0.39 cm^2
- Head difference at time $t = 0$ = 508 mm
- Head difference at time $t = 180$ sec = 305 mm

Determine the hydraulic conductivity of the soil in cm/sec.

Solution
From Eq. (7.22),

$$k = 2.303 \frac{aL}{At} \log_{10}\left(\frac{h_1}{h_2}\right)$$

We are given $a = 0.39$ cm^2, $L = 203$ mm, $A = 10.3$ cm^2, $t = 180$ sec, $h_1 = 508$ mm, and $h_2 = 305$ mm,

$$k = 2.303 \frac{(0.39)(20.3)}{(10.3)(180)} \log_{10}\left(\frac{508}{305}\right)$$

$$= \mathbf{2.18 \times 10^{-3}} \text{ cm/sec.} \qquad \blacksquare$$

Example 7.3

The hydraulic conductivity of a clayey soil is 3×10^{-7} cm/sec. The viscosity of water at 25°C is 0.0911×10^{-4} g · sec/cm^2. Calculate the absolute permeability $\overline{K}$ of the soil.

Solution
From Eq. (7.13),

$$k = \frac{\gamma_w}{\eta}\overline{K} = 3 \times 10^{-7} \text{ cm/sec}$$

so

$$3 \times 10^{-7} = \left(\frac{1 \text{ g/cm}^3}{0.0911 \times 10^{-4}}\right)\overline{K}$$

$$\overline{K} = \mathbf{0.2733 \times 10^{-11}} \text{ cm}^2 \qquad \blacksquare$$

Example 7.4

A permeable soil layer is underlain by an impervious layer, as shown in Figure 7.7a. With $k = 5.3 \times 10^{-5}$ m/sec for the permeable layer, calculate the rate of seepage through it in m^3/hr/m width if $H = 3$ m and $\alpha = 8°$.

Solution
From Figure 7.7b,

$$i = \frac{\text{head loss}}{\text{length}} = \frac{L \tan \alpha}{\left(\dfrac{L}{\cos \alpha}\right)} = \sin \alpha$$

$$q = kiA = (k)(\sin \alpha)(3 \cos \alpha)(1)$$

$$k = 5.3 \times 10^{-5}\,\text{m/sec}$$

$$q = (5.3 \times 10^{-5})(\sin 8°)(3 \cos 8°)(3600) = \mathbf{0.0789\ m^3/hr/m}$$

$$\uparrow$$

To change to
m/hr

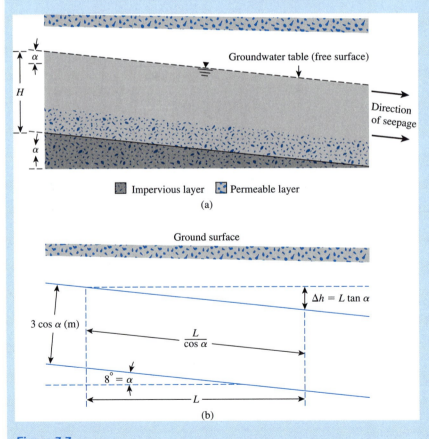

Impervious layer   Permeable layer
(a)

$\Delta h = L \tan \alpha$

(b)

Figure 7.7

Example 7.5

Find the flow rate in m^3/sec/m length (at right angles to the cross section shown) through the permeable soil layer shown in Figure 7.8 given $H = 8$ m, $H_1 = 3$ m, $h = 4$ m, $L = 50$ m, $\alpha = 8°$, and $k = 0.08$ cm/sec.

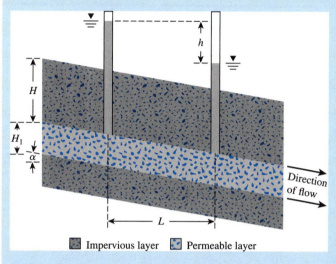

Impervious layer   Permeable layer

Figure 7.8 Flow through permeable layer

Solution

$$\text{Hydraulic gradient } (i) = \frac{h}{\dfrac{L}{\cos \alpha}}$$

From Eqs. (7.17) and (7.18),

$$q = kiA = k\left(\frac{h \cos \alpha}{L}\right)(H_1 \cos \alpha \times 1)$$

$$= (0.08 \times 10^{-2} \text{ m/sec})\left(\frac{4 \cos 8°}{50}\right)(3 \cos 8° \times 1)$$

$$= 0.19 \times 10^{-3} \text{ m}^3/\text{sec/m}$$

| 7.5 | *Relationships for Hydraulic Conductivity—Granular Soil* |

For fairly uniform sand (that is, sand with a small uniformity coefficient), Hazen (1930) proposed an empirical relationship for hydraulic conductivity in the form

$$k \text{ (cm/sec)} = c D_{10}^2 \qquad (7.23)$$

where c = a constant that varies from 1.0 to 1.5
D_{10} = the effective size, in mm

Equation (7.23) is based primarily on Hazen's observations of loose, clean, filter sands. A small quantity of silts and clays, when present in a sandy soil, may change the hydraulic conductivity substantially.

Over the last several years, experimental observations have shown that the magnitude of c for various types of granular soils may vary by three orders of magnitude (Carrier, 2003) and, hence, is not very reliable.

Another form of equation that gives fairly good results in estimating the hydraulic conductivity of sandy soils is based on the Kozeny-Carman equation (Kozeny, 1927; Carman, 1938, 1956). The derivation of this equation is not presented here. Interested readers are referred to any advanced soil mechanics book. According to the Kozeny-Carman equation

$$k = \frac{1}{C_s S_s^2 T^2} \frac{\gamma_w}{\eta} \frac{e^3}{1 + e} \qquad (7.24)$$

where C_s = shape factor, which is a function of the shape of flow channels
S_s = specific surface area per unit volume of particles
T = tortuosity of flow channels
γ_w = unit weight of water
η = viscosity of permeant
e = void ratio

For practical use, Carrier (2003) has modified Eq. (7.24) in the following manner. At 20°C, γ_w/η for water is about $9.93 \times 10^4 \left(\dfrac{1}{\text{cm} \cdot \text{s}} \right)$. Also, $(C_s T^2)$ is approximately equal to 5. Substituting these values in Eq. (7.24), we obtain

$$k = 1.99 \times 10^4 \left(\frac{1}{S_s} \right)^2 \frac{e^3}{1 + e} \qquad (7.25)$$

Again,

$$S_s = \frac{SF}{D_{\text{eff}}} \left(\frac{1}{\text{cm}} \right) \qquad (7.26)$$

with

$$D_{\text{eff}} = \frac{100\%}{\Sigma\left(\dfrac{f_i}{D_{(av)i}}\right)}$$ (7.27)

where f_i = fraction of particles between two sieve sizes, in percent
(*Note:* larger sieve, l; smaller sieve, s)
$$D_{(av)i}(\text{cm}) = [D_{li}(\text{cm})]^{0.5} \times [D_{si}(\text{cm})]^{0.5}$$ (7.28)
SF = shape factor

Combining Eqs. (7.25), (7.26), (7.27), and (7.28),

$$k = 1.99 \times 10^4 \left[\frac{100\%}{\Sigma \dfrac{f_i}{D_{li}^{0.5} \times D_{si}^{0.5}}}\right]^2 \left(\frac{1}{SF}\right)^2 \left(\frac{e^3}{1+e}\right)$$ (7.29)

The magnitude of SF may vary between 6 to 8, depending on the angularity of the soil particles.
Carrier (2003) further suggested a slight modification to Eq. (7.29), which can be written as

$$k = 1.99 \times 10^4 \left[\frac{100\%}{\Sigma \dfrac{f_i}{D_{li}^{0.404} \times D_{si}^{0.595}}}\right]^2 \left(\frac{1}{SF}\right)^2 \left(\frac{e^3}{1+e}\right)$$ (7.30)

Equation (7.30) suggests that

$$k \propto \frac{e^3}{1+e}$$ (7.31)

The author recommends the use of Eqs. (7.30) and (7.31). It is important to note that Eqs. (7.23) and (7.31) assume that laminar flow condition does exist.
More recently, Chapuis (2004) proposed an empirical relationship for k in conjunction with Eq. (7.31) as

$$k(\text{cm/s}) = 2.4622\left[D_{10}^2 \frac{e^3}{(1+e)}\right]^{0.7825}$$ (7.32)

where D_{10} = effective size (mm).
The preceding equation is valid for natural, uniform sand and gravel to predict k that is in the range of 10^{-1} to 10^{-3} cm/s. This can be extended to natural, silty sands without plasticity. It is not valid for crushed materials or silty soils with some plasticity.
Based on laboratory experimental results, Amer and Awad (1974) proposed the following relationship for k in granular soil:

$$k = 3.5 \times 10^{-4}\left(\frac{e^3}{1+e}\right)C_u^{0.6}D_{10}^{2.32}\left(\frac{\rho_w}{\eta}\right)$$ (7.33)

where k is in cm/sec

C_u = uniformity coefficient
D_{10} = effective size (mm)
ρ_W = density of water (g/cm³)
η = viscosity (g · s/cm²)

At 20 °C, $\rho_W = 1$ g/cm³ and $\eta \approx 0.1 \times 10^{-4}$ g · s/cm². So

$$k = 3.5 \times 10^{-4}\left(\frac{e^3}{1+e}\right)C_u^{0.6}D_{10}^{2.32}\left(\frac{1}{0.1 \times 10^{-4}}\right)$$

or

$$k(\text{cm/sec}) = 35\left(\frac{e^3}{1+e}\right)C_u^{0.6}(D_{10})^{2.32} \tag{7.34}$$

Mention was made at the end of Section 7.1 that turbulent flow conditions may exist in very coarse sands and gravels and that Darcy's law may not be valid for these materials. However, under a low hydraulic gradient, laminar flow conditions usually exist. Kenney, Lau, and Ofoegbu (1984) conducted laboratory tests on granular soils in which the particle sizes in various specimens ranged from 0.074 to 25.4 mm. The uniformity coefficients, C_u, of these specimens ranged from 1.04 to 12. All permeability tests were conducted at a relative density of 80% or more. These tests showed that for laminar flow conditions,

$$\overline{K}(\text{mm}^2) = (0.05 \text{ to } 1)D_5^2 \tag{7.35}$$

where D_5 = diameter (mm) through which 5% of soil passes.

Example 7.6

The hydraulic conductivity of a sand at a void ratio of 0.5 is 0.02 cm/sec. Estimate its hydraulic conductivity at a void ratio of 0.65.

Solution
From Eq. (7.31),

$$\frac{k_1}{k_2} = \frac{\dfrac{e_1^3}{1+e_1}}{\dfrac{e_2^3}{1+e_2}}$$

$$\frac{0.02}{k_2} = \frac{\dfrac{(0.5)^3}{1+0.5}}{\dfrac{(0.65)^3}{1+0.65}}$$

$$k_2 = \textbf{0.04 cm/sec} \qquad \blacksquare$$

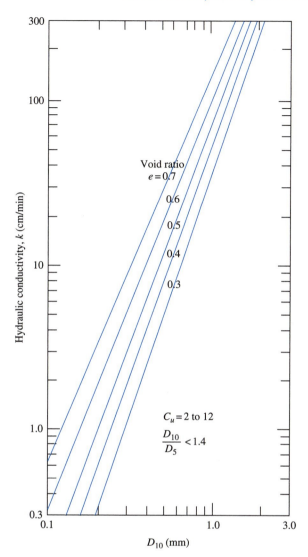

Figure 7.9 Hydraulic conductivity of granular soils (*After U.S. Department of Navy, 1971*)

On the basis of laboratory experiments, the U.S. Department of Navy (1971) provided an empirical correlation between k (cm/min) and D_{10} (mm) for granular soils with the
uniformity coefficient varying between 2 and 12 and $D_{10}/D_5 < 1.4$. This correlation is
shown in Figure 7.9.

Example 7.7

The grain-size distribution curve for a sand is shown in Figure 7.10. Estimate the
hydraulic conductivity using Eq. (7.30). Given: the void ratio of the sand is 0.6.
Use $SF = 7$.

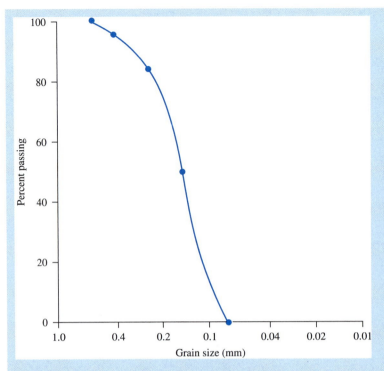

Figure 7.10

Solution

From Figure 7.10, the following table can be prepared.

Sieve no.	Sieve opening (cm)	Percent passing	Fraction of particles between two con-secutive sieves (%)
30	0.06	100	
40	0.0425	96	4
60	0.02	84	12
100	0.015	50	34
200	0.0075	0	50

For fraction between Nos. 30 and 40 sieves;

$$\frac{f_i}{D_{li}^{0.404} \times D_{si}^{0.595}} = \frac{4}{(0.06)^{0.404} \times (0.0425)^{0.595}} = 81.62$$

For fraction between Nos. 40 and 60 sieves;

$$\frac{f_i}{D_{li}^{0.404} \times D_{si}^{0.595}} = \frac{12}{(0.0425)^{0.404} \times (0.02)^{0.595}} = 440.76$$

Similarly, for fraction between Nos. 60 and 100 sieves;

$$\frac{f_i}{D_{li}^{0.404} \times D_{si}^{0.595}} = \frac{34}{(0.02)^{0.404} \times (0.015)^{0.595}} = 2009.5$$

And, for between Nos. 100 and 200 sieves;

$$\frac{f_i}{D_{li}^{0.404} \times D_{si}^{0.595}} = \frac{50}{(0.015)^{0.404} \times (0.0075)^{0.595}} = 5013.8$$

$$\frac{100\%}{\sum \dfrac{f_i}{D_{li}^{0.404} \times D_{si}^{0.595}}} = \frac{100}{81.62 + 440.76 + 2009.5 + 5013.8} \approx 0.0133$$

From Eq. (7.30),

$$k = (1.99 \times 10^4)(0.0133)^2 \left(\frac{1}{7}\right)^2 \left(\frac{0.6^3}{1 + 0.6}\right) = 0.0097 \text{ cm/s} \quad \blacksquare$$

Example 7.8

Solve Example 7.7 using Eq. (7.32).

Solution

From Fig. 7.10, $D_{10} = 0.09$ mm. From Eq. (7.32),

$$k = 2.4622 \left[D_{10}^2 \frac{e^3}{1 + e} \right]^{0.7825} = 2.4622 \left[(0.09)^2 \frac{0.6^3}{1 + 0.6} \right]^{0.7825} = \mathbf{0.0119\,cm/sec} \quad \blacksquare$$

Example 7.9

Solve Example 7.7 using Eq. (7.34).

Solution

From Figure 7.10, $D_{60} = 0.16$ mm and $D_{10} = 0.09$ mm. Thus,

$$C_u = \frac{D_{60}}{D_{10}} = \frac{0.16}{0.09} = 1.78$$

From Eq. (7.34),

$$k = 35 \left(\frac{e^3}{1 + e}\right) C_u^{0.6} (D_{10})^{2.32} = 35 \left(\frac{0.6^3}{1 + 0.6}\right) (1.78)^{0.6} (0.09)^{2.32} = \mathbf{0.025\,cm/sec} \quad \blacksquare$$

7.6   *Relationships for Hydraulic Conductivity—Cohesive Soils*

The Kozeny-Carman equation [Eq. (7.24)] has been used in the past to see if it will hold good for cohesive soil. Olsen (1961) conducted hydraulic conductivity tests on sodium illite and compared the results with Eq. (7.24). This comparison is shown in Figure 7.11. The marked degrees of variation between the theoretical and experimental values arise from several factors, including deviations from Darcy's law, high viscosity of the pore water, and unequal pore sizes.

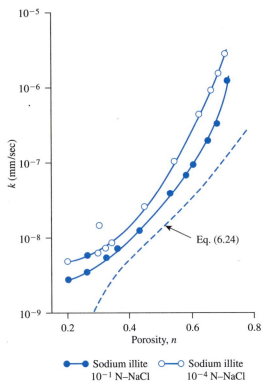

Figure 7.11 Coefficient of permeability for sodium illite (*Based on Olsen, 1961*)

Figure 7.12 Variation of hydraulic conductivity of sodium clay minerals (*Based on Mesri and Olson, 1971*)

Taylor (1948) proposed a linear relationship between the logarithm of k and the void ratio as

$$\log k = \log k_o - \frac{e_o - e}{C_k} \tag{7.36}$$

where k_o = *in situ* hydraulic conductivity at a void ratio e_o
$\quad\quad k$ = hydraulic conductivity at a void ratio e
$\quad\quad C_k$ = hydraulic conductivity change index

The preceding equation is a good correction for e_o, which is less than about 2.5. In this equation, the value of C_k may be taken to be about $0.5e_o$.

For a wide range of void ratio, Mesri and Olson (1971) suggested the use of a linear relationship between $\log k$ and $\log e$ in the form

$$\log k = A'\log e + B' \tag{7.37}$$

Figure 7.12 shows the plot of $\log k$ versus $\log e$ obtained in the laboratory based on which Eq. (7.37) was proposed.

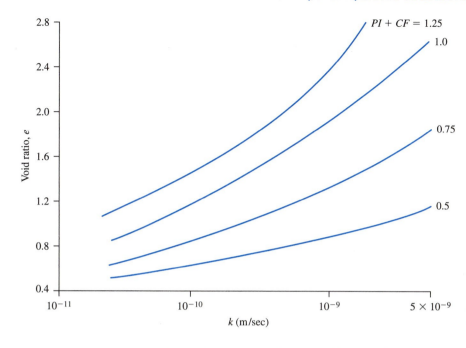

Figure 7.13 Variation of void ratio with hydraulic conductivity of clayey soils (*Based on Tavenas, et al., 1983*)

Samarasinghe, *et al.* (1982) conducted laboratory tests on New Liskeard clay and proposed that, for normally consolidated clays,

$$k = C\left(\frac{e^n}{1 + e}\right) \tag{7.38}$$

where C and n are constants to be determined experimentally.

Tavenas, *et al.* (1983) also gave a correlation between the void ratio and the hydraulic conductivity of clayey soil. This correlation is shown in Figure 7.13. An important point to note, however, is that in Figure 7.13, *PI*, the plasticity index, and *CF*, the clay-size fraction in the soil, are in *fraction* (decimal) form. One should keep in mind, however, that any empirical relationship of this type is for estimation only, because the magnitude of k is a highly variable parameter and depends on several factors.

Example 7.10

For a normally consolidated clay soil, the following values are given:

Void ratio	k (cm/sec)
1.1	0.302×10^{-7}
0.9	0.12×10^{-7}

Estimate the hydraulic conductivity of the clay at a void ratio of 0.75. Use Eq. (7.38).

Solution

From Eq. (7.38),

$$k = C\left(\frac{e^n}{1 + e}\right)$$

$$\frac{k_1}{k_2} = \frac{\left(\dfrac{e_1^n}{1 + e_1}\right)}{\left(\dfrac{e_2^n}{1 + e_2}\right)}$$

$$\frac{0.302 \times 10^{-7}}{0.12 \times 10^{-7}} = \frac{\dfrac{(1.1)^n}{1 + 1.1}}{\dfrac{(0.9)^n}{1 + 0.9}}$$

$$2.517 = \left(\frac{1.9}{2.1}\right)\left(\frac{1.1}{0.9}\right)^n$$

$$2.782 = (1.222)^n$$

$$n = \frac{\log(2.782)}{\log(1.222)} = \frac{0.444}{0.087} = 5.1$$

so

$$k = C\left(\frac{e^{5.1}}{1 + e}\right)$$

To find C,

$$0.302 \times 10^{-7} = C\left[\frac{(1.1)^{5.1}}{1 + 1.1}\right] = \left(\frac{1.626}{2.1}\right)C$$

$$C = \frac{(0.302 \times 10^{-7})(2.1)}{1.626} = 0.39 \times 10^{-7}$$

Hence,

$$k = (0.39 \times 10^{-7}\ \text{cm/sec})\left(\frac{e^n}{1 + e}\right)$$

At a void ratio of 0.75,

$$k = (0.39 \times 10^{-7})\left(\frac{0.75^{5.1}}{1 + 0.75}\right) = \textbf{0.514} \times \textbf{10}^{-8}\ \textbf{cm/sec} \qquad \blacksquare$$

7.7 *Directional Variation of Permeability*

Most soils are not isotropic with respect to permeability. In a given soil deposit, the magnitude of k changes with respect to the direction of flow. Figure 7.14 shows a soil layer through which water flows in a direction inclined at an angle α with the vertical. Let the hydraulic conductivity in the vertical ($\alpha = 0$) and horizontal ($\alpha = 90°$) directions be k_V and k_H, respectively. The magnitudes of k_V and k_H in a given soil depend on several factors, including the method of deposition in the field.

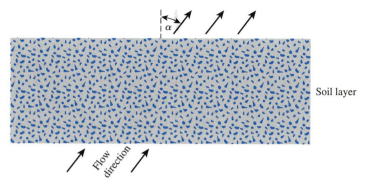

Figure 7.14 Directional variation of permeability

Figure 7.15 shows the laboratory test results obtained by Fukushima and Ishii (1986) related to k_V and k_H for compacted Masa-do soil (weathered granite). The soil specimens were initially compacted at a certain moisture content, and the hydraulic conductivity was determined at 100% saturation. Note that, for any given molding moisture content and confining pressure, k_H is larger than k_V.

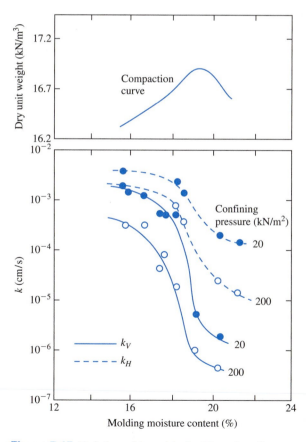

Figure 7.15 Variation of k_V and k_H for Masa-do soil compacted in the laboratory (*Based on the results of Fukushima and Ishii, 1986*)

There are several published results for fine-grained soils that show the ratio of k_H/k_V may vary over a wide range. Table 7.3 provided a summary of some of those studies.

Table 7.3 k_H/k_V for Fine-Grained Soils—Summary of Several Studies

Soil type	k_H/k_V	Reference
Organic silt with peat	1.2 to 1.7	Tsien (1955)
Plastic marine clay	1.2	Lumb and Holt (1968)
Soft clay	1.5	Basett and Brodie (1961)
Varved clay	1.5 to 1.7	Chan and Kenney (1973)
Varved clay	1.5	Kenney and Chan (1973)
Varved clay	3 to 15	Wu, *et al.* (1978)
Varved clay	4 to 40	Casagrande and Poulos (1969)

7.8 *Equivalent Hydraulic Conductivity in Stratified Soil*

In a stratified soil deposit where the hydraulic conductivity for flow in a given direction changes from layer to layer, an equivalent hydraulic conductivity can be computed to simplify calculations. The following derivations relate to the equivalent hydraulic conductivities for flow in vertical and horizontal directions through multilayered soils with horizontal stratification.

Figure 7.16 shows n layers of soil with flow in the *horizontal direction.* Let us consider a cross section of unit length passing through the n layer and perpendicular to the direction of flow. The total flow through the cross section in unit time can be written as

$$q = v \cdot 1 \cdot H$$
$$= v_1 \cdot 1 \cdot H_1 + v_2 \cdot 1 \cdot H_2 + v_3 \cdot 1 \cdot H_3 + \cdots + v_n \cdot 1 \cdot H_n \qquad (7.39)$$

where $\quad v$ = average discharge velocity

$v_1, v_2, v_3, \ldots, v_n$ = discharge velocities of flow in layers denoted by the subscripts

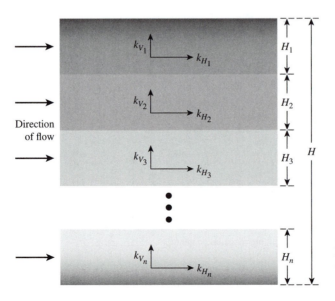

Figure 7.16 Equivalent hydraulic conductivity determination—horizontal flow in stratified soil

If $k_{H_1}, k_{H_2}, k_{H_3}, \ldots, k_{H_n}$ are the hydraulic conductivities of the individual layers in the horizontal direction and $k_{H(eq)}$ is the equivalent hydraulic conductivity in the horizontal direction, then, from Darcy's law,

$$v = k_{H(eq)}i_{eq}; \quad v_1 = k_{H_1}i_1; \quad v_2 = k_{H_2}i_2; \quad v_3 = k_{H_3}i_3; \quad \cdots \quad v_n = k_{H_n}i_n$$

Substituting the preceding relations for velocities into Eq. (7.39) and noting that $i_{eq} = i_1 = i_2 = i_3 = \cdots = i_n$ results in

$$k_{H(eq)} = \frac{1}{H}(k_{H_1}H_1 + k_{H_2}H_2 + k_{H_3}H_3 + \cdots + k_{H_n}H_n) \tag{7.40}$$

Figure 7.17 shows n layers of soil with flow in the vertical direction. In this case, the velocity of flow through all the layers is the same. However, the total head loss, h, is equal to the sum of the head losses in all layers. Thus,

$$v = v_1 = v_2 = v_3 = \cdots = v_n \tag{7.41}$$

and

$$h = h_1 + h_2 + h_3 + \cdots + h_n \tag{7.42}$$

Using Darcy's law, we can rewrite Eq. (7.41) as

$$k_{V(eq)}\left(\frac{h}{H}\right) = k_{V_1}i_1 = k_{V_2}i_2 = k_{V_3}i_3 = \cdots = k_{V_n}i_n \tag{7.43}$$

where $k_{V_1}, k_{V_2}, k_{V_3}, \ldots, k_{V_n}$ are the hydraulic conductivities of the individual layers in the vertical direction and $k_{V(eq)}$ is the equivalent hydraulic conductivity.

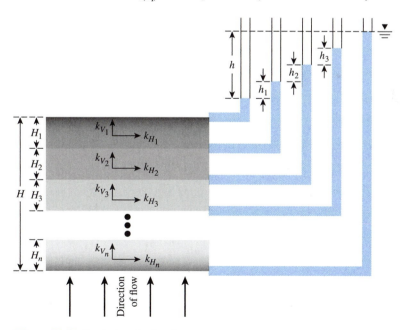

Figure 7.17 Equivalent hydraulic conductivity determination—vertical flow in stratified soil

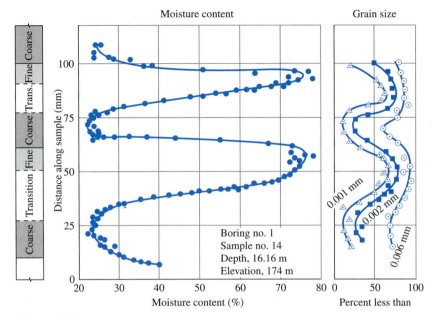

Figure 7.18 Variation of moisture content and grain-size distribution in New Liskeard varved soil. (*Source:* After "Laboratory Investigation of Permeability Ratio of New Liskeard Varved Clay," by H. T. Chan and T. C. Kenney, 1973, *Canadian Geotechnical Journal, 10*(3), p. 453–472. Copyright © 1973 National Research Council of Canada. Used by permission.)

Again, from Eq. (7.42),

$$h = H_1 i_1 + H_2 i_2 + H_3 i_3 + \cdots + H_n i_n \qquad (7.44)$$

Solving Eqs. (7.43) and (7.44) gives

$$k_{V(eq)} = \cfrac{H}{\left(\dfrac{H_1}{k_{V_1}}\right) + \left(\dfrac{H_2}{k_{V_2}}\right) + \left(\dfrac{H_3}{k_{V_3}}\right) + \cdots + \left(\dfrac{H_n}{k_{V_n}}\right)} \qquad (7.45)$$

An excellent example of naturally deposited layered soil is *varved soil,* which is a rhythmically layered sediment of coarse and fine minerals. Varved soils result from annual seasonal fluctuation of sediment conditions in glacial lakes. Figure 7.18 shows the variation of moisture content and grain-size distribution in New Liskeard, Canada, varved soil. Each varve is about 41 to 51 mm thick and consists of two homogeneous layers of soil—one coarse and one fine—with a transition layer between.

Example 7.11

A layered soil is shown in Figure 7.19. Given:

- $H_1 = 2$ m $k_1 = 10^{-4}$ cm/sec
- $H_2 = 3$ m $k_2 = 3.2 \times 10^{-2}$ cm/sec
- $H_3 = 4$ m $k_3 = 4.1 \times 10^{-5}$ cm/sec

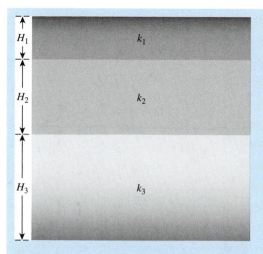

Figure 7.19 A layered soil profile

Estimate the ratio of equivalent hydraulic conductivity,

$$\frac{k_{H(\text{eq})}}{k_{V(\text{eq})}}$$

Solution

From Eq. (7.40),

$$k_{H(\text{eq})} = \frac{1}{H}\left(k_{H_1}H_1 + k_{H_2}H_2 + k_{H_3}H_3\right)$$

$$= \frac{1}{(2 + 3 + 4)}\left[(10^{-4})(2) + (3.2 \times 10^{-2})(3) + (4.1 \times 10^{-5})(4)\right]$$

$$= 107.07 \times 10^{-4}\ \text{cm/sec}$$

Again, from Eq. (7.45),

$$k_{V(\text{eq})} = \frac{H}{\left(\dfrac{H_1}{k_{V_1}}\right) + \left(\dfrac{H_2}{k_{V_2}}\right) + \left(\dfrac{H_3}{k_{V_3}}\right)}$$

$$= \frac{2 + 3 + 4}{\left(\dfrac{2}{10^{-4}}\right) + \left(\dfrac{3}{3.2 \times 10^{-2}}\right) + \left(\dfrac{4}{4.1 \times 10^{-5}}\right)}$$

$$= 0.765 \times 10^{-4}\ \text{cm/sec}$$

Hence,

$$\frac{k_{H(\text{eq})}}{k_{V(\text{eq})}} = \frac{107.07 \times 10^{-4}}{0.765 \times 10^{-4}} = \mathbf{139.96}$$

■

Example 7.12

Figure 7.20 shows the layers of soil in a tube that is 100 mm × 100 mm in cross section. Water is supplied to maintain a constant-head difference of 300 mm across the sample. The hydraulic conductivities of the soils in the direction of flow through them are as follows:

Soil	k (cm/sec)
A	10^{-2}
B	3×10^{-3}
C	4.9×10^{-4}

Find the rate of water supply in cm^3/hr.

Solution
From Eq. (7.45),

$$k_{V(eq)} = \frac{H}{\left(\dfrac{H_1}{k_1}\right) + \left(\dfrac{H_2}{k_2}\right) + \left(\dfrac{H_3}{k_3}\right)} = \frac{450}{\left(\dfrac{150}{10^{-2}}\right) + \left(\dfrac{150}{3 \times 10^{-3}}\right) + \left(\dfrac{150}{4.9 \times 10^{-4}}\right)}$$

$$= 0.001213 \text{ cm/sec}$$

$$q = k_{V(eq)}iA = (0.001213)\left(\frac{300}{450}\right)\left(\frac{100}{10} \times \frac{100}{10}\right)$$

$$= 0.0809 \text{ cm}^3/\text{sec} = \mathbf{291.24 \text{ cm}^3/\text{hr}}$$

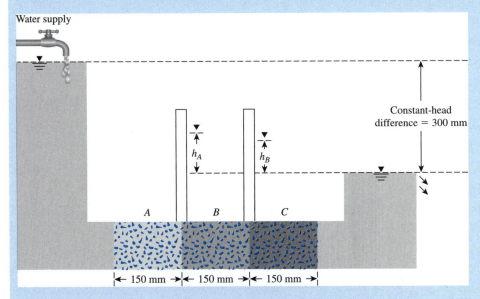

Figure 7.20 Layers of soil in a tube 100 mm × 100 mm in cross section

Permeability Test in the Field by Pumping from Wells

In the field, the average hydraulic conductivity of a soil deposit in the direction of flow can be determined by performing pumping tests from wells. Figure 7.21 shows a case where the top permeable layer, whose hydraulic conductivity has to be determined, is unconfined and underlain by an impermeable layer. During the test, water is pumped out at a constant rate from a test well that has a perforated casing. Several observation wells at various radial distances are made around the test well. Continuous observations of the water level in the test well and in the observation wells are made after the start of pumping, until a steady state is reached. The steady state is established when the water level in the test and observation wells becomes constant. The expression for the rate of flow of groundwater into the well, which is equal to the rate of discharge from pumping, can be written as

$$q = k\left(\frac{dh}{dr}\right)2\pi rh \qquad (7.46)$$

or

$$\int_{r_2}^{r_1} \frac{dr}{r} = \left(\frac{2\pi k}{q}\right) \int_{h_2}^{h_1} h \, dh$$

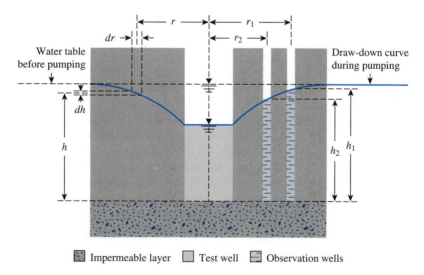

Figure 7.21 Pumping test from a well in an unconfined permeable layer underlain by an impermeable stratum.

Thus,

$$k = \frac{2.303q \log_{10}\left(\dfrac{r_1}{r_2}\right)}{\pi(h_1^2 - h_2^2)} \qquad (7.47)$$

From field measurements, if q, r_1, r_2, h_1, and h_2 are known, the hydraulic conductivity can be calculated from the simple relationship presented in Eq. (7.47).

The average hydraulic conductivity for a confined aquifer can also be determined by conducting a pumping test from a well with a perforated casing that penetrates the full depth of the aquifer and by observing the piezometric level in a number of observation wells at various radial distances (Figure 7.22). Pumping is continued at a uniform rate q until a steady state is reached.

Because water can enter the test well only from the aquifer of thickness H, the steady state of discharge is

$$q = k \left(\frac{dh}{dr}\right) 2\pi r H \qquad (7.48)$$

or

$$\int_{r_2}^{r_1} \frac{dr}{r} = \int_{h_2}^{h_1} \frac{2\pi k H}{q}\, dh$$

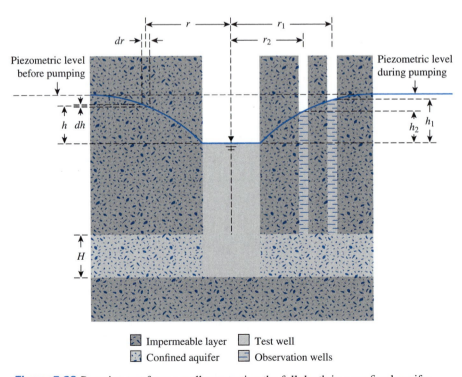

Figure 7.22 Pumping test from a well penetrating the full depth in a confined aquifer

This gives the hydraulic conductivity in the direction of flow as

$$k = \frac{q \log_{10}\left(\dfrac{r_1}{r_2}\right)}{2.727 H(h_1 - h_2)} \qquad (7.49)$$

7.10 In Situ *Hydraulic Conductivity of Compacted Clay Soils*

Daniel (1989) provided an excellent review of nine methods to estimate the *in situ* hydraulic conductivity of compacted clay layers. Three of these methods are described.

Boutwell Permeameter

A schematic diagram of the Boutwell permeameter is shown in Figure 7.23. A hole is first drilled and a casing is placed in it (Figure 7.23a). The casing is filled with water and a falling-head permeability test is conducted. Based on the test results, the hydraulic conductivity k_1 is calculated as

$$k_1 = \frac{\pi d^2}{\pi D(t_2 - t_1)} \ln\left(\frac{h_1}{h_2}\right) \qquad (7.50)$$

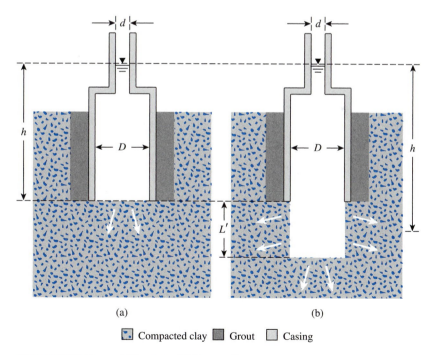

Compacted clay ▓ Grout ▢ Casing

Figure 7.23 Permeability test with Boutwell permeameter

where d = diameter of the standpipe
D = diameter of the casing
h_1 = head at time t_1
h_2 = head at time t_2

After the hydraulic conductivity is determined, the hole is deepened by augering, and the permeameter is reassembled as shown in Figure 7.23b. A falling-head hydraulic conductivity test is conducted again. The hydraulic conductivity is calculated as

$$k_2 = \frac{A'}{B'} \ln\left(\frac{h_1}{h_2}\right) \qquad (7.51)$$

where

$$A' = d^2\left\{\ln\left[\frac{L'}{D} + \sqrt{1 + \left(\frac{L'}{D}\right)^2}\right]\right\} \qquad (7.52)$$

$$B' = 8D\frac{L'}{D}(t_2 - t_1)\left\{1 - 0.562 \exp\left[-1.57\left(\frac{L'}{D}\right)\right]\right\} \qquad (7.53)$$

The anisotropy with respect to hydraulic conductivity is determined by referring to Figure 7.24, which is a plot of k_2/k_1 versus m ($m = \sqrt{k_H/k_V}$) for various values of L'/D. Figure 7.24 can be used to determine m using the experimental values of k_2/k_1 and L'/D. The plots in this figure are determined from

$$\frac{k_2}{k_1} = \frac{\ln[(L'/D) + \sqrt{1 + (L'/D)^2}]}{\ln[(mL'/D) + \sqrt{1 + (mL'/D)^2}]} m \qquad (7.54)$$

Once m is determined, we can calculate

$$k_H = mk_1 \qquad (7.55)$$

and

$$k_V = \frac{k_1}{m} \qquad (7.56)$$

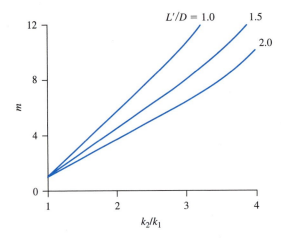

Figure 7.24
Variation of k_2/k_1 with m [Eq. (7.54)]

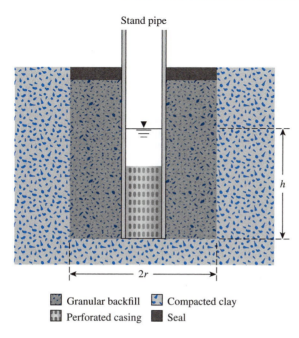

Stand pipe

2r

Granular backfill   Compacted clay
Perforated casing   Seal

h

Figure 7.25 Borehole test with constant water level

Constant-Head Borehole Permeameter

Figure 7.25 shows a constant-head borehole permeameter. In this arrangement a constant head h is maintained by supplying water, and the rate of flow q is measured. The hydraulic conductivity can be calculated as

$$k = \frac{q}{r^2 \sqrt{R^2 - 1}\,[F_1 + (F_2/A'')]} \tag{7.57}$$

where

$$R = \frac{h}{r} \tag{7.58}$$

$$F_1 = \frac{4.117(1 - R^2)}{\ln(R + \sqrt{R^2 - 1}) - [1 - (1/R^2)]^{0.5}} \tag{7.59}$$

$$F_2 = \frac{4.280}{\ln(R + \sqrt{R^2 - 1})} \tag{7.60}$$

$$A'' = \frac{1}{2}\,\alpha r \tag{7.61}$$

Typical values of α range from 0.002 to 0.01 cm^{-1} for fine-grained soil.

Porous Probes

Porous probes (Figure 7.26) are pushed or driven into the soil. Constant- or falling-head permeability tests are performed. The hydraulic conductivity is calculated as follows:

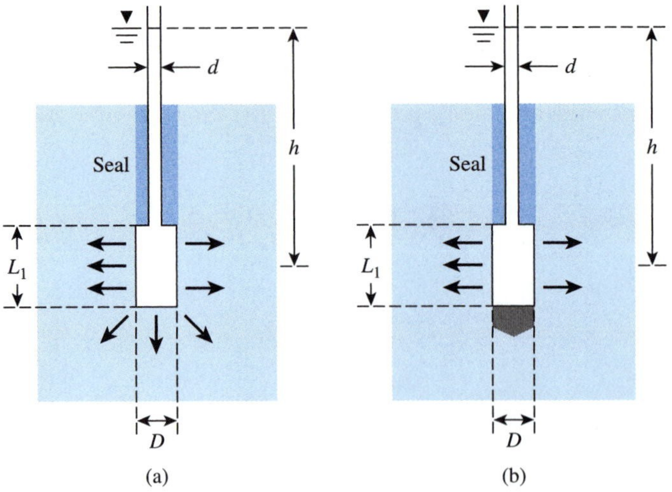

Figure 7.26 Porous probe: (a) test with permeable base; (b) test with impermeable base

The constant head is given by

$$k = \frac{q}{Fh} \tag{7.62}$$

The falling head is given by

$$k = \frac{\pi d^2/4}{F(t_2 - t_1)} \ln\left(\frac{h_1}{h_2}\right) \tag{7.63}$$

For probes with permeable bases (Figure 7.26a),

$$F = \frac{2\pi L_1}{\ln\left[(L_1/D) + \sqrt{1 + (L_1/D)^2}\right]} \tag{7.64}$$

For probes with impermeable bases (Figure 7.26b),

$$F = \frac{2\pi L_1}{\ln\left[(L_1/D) + \sqrt{1 + (L_1/D)^2}\right]} - 2.8D \tag{7.65}$$

7.11 Summary and General Comments

In this chapter, we discussed Darcy's law, the definition of hydraulic conductivity, laboratory determinations of hydraulic conductivity and the empirical relations for it, and field determinations of hydraulic conductivity of various types of soil. Hydraulic conductivity of various soil layers is highly variable. The empirical relations for hydraulic conductivity should be used as a general guide for all practical considerations. The accuracy of the values of k determined in the laboratory depends on several factors:

1. Temperature of the fluid
2. Viscosity of the fluid

3. Trapped air bubbles present in the soil specimen
4. Degree of saturation of the soil specimen
5. Migration of fines during testing
6. Duplication of field conditions in the laboratory

The hydraulic conductivity of saturated cohesive soils also can be determined by laboratory consolidation tests. The actual value of the hydraulic conductivity in the field also may be somewhat different than that obtained in the laboratory because of the non-homogeneity of the soil. Hence, proper care should be taken in assessing the order of the magnitude of k for all design considerations.

Problems

7.1 Refer to the constant-head arrangement shown in Figure 7.5. For a test, the following are given:
- $L = 457$ mm
- A = area of the specimen = 22.6 cm^2
- Constant-head difference = h = 711 mm
- Water collected in 3 min = 353.6 cm^3

Calculate the hydraulic conductivity (cm/sec).

7.2 Refer to Figure 7.5. For a constant-head permeability test in a sand, the following are given:
- $L = 300$ mm
- $A = 175$ cm^2
- $h = 500$ mm
- Water collected in 3 min = 620 cm^3
- Void ratio of sand = 0.58

Determine
a. Hydraulic conductivity, k (cm/sec)
b. Seepage velocity

7.3 In a constant-head permeability test in the laboratory, the following are given: $L = 305$ mm and $A = 96.8$ cm^2. If the value of $k = 0.015$ cm/sec and a flow rate of 7374 cm^3/hr must be maintained through the soil, what is the head difference, h, across the specimen? Also, determine the discharge velocity under the test conditions.

7.4 For a falling-head permeability test, the following are given:
- Length of the soil specimen = 508 mm
- Area of the soil specimen = 25.8 cm^2
- Area of the standpipe = 1.29 cm^2
- Head difference at time $t = 0$ is 762 mm
- Head difference at time $t = 10$ min is 305 mm

a. Determine the hydraulic conductivity of the soil (cm/min).
b. What was the head difference at time $t = 5$ min?

7.5 For a falling-head permeability test, the following are given: length of specimen = 380 mm; area of specimen = 6.5 cm^2; $k = 0.175$ cm/min. What should be the area of the standpipe for the head to drop from 650 cm to 300 cm in 8 min?

7.6 For a falling-head permeability test, the following are given:
- Length of soil specimen = 700 mm
- Area of the soil specimen = 20 cm²
- Area of the standpipe = 1.05 cm²
- Head difference at time $t = 0$ is 800 mm
- Head difference at time $t = 8$ min is 500 mm

a. Determine the absolute permeability of the soil.

b. What is the head difference at time $t = 6$ min?

Assume that the test was conducted at 20°C, and at 20°C, $\gamma_w = 9.789$ kN/m³ and $\eta = 1.005 \times 10^{-3}$ N·s/m².

7.7 A sand layer of the cross-sectional area shown in Fig. 7.27 has been determined to exist for a 800-m length of the levee. The hydraulic conductivity of the sand layer is 2.8 m/day. Determine the quantity of water which flows into the ditch in m³/min.

7.8 A permeable soil layer is underlain by an impervious layer, as shown in Figure 7.28. With $k = 5.2 \times 10^{-4}$ cm/sec for the permeable layer, calculate the rate of seepage through it in m³/hr/m length if $H = 3.8$ m and $\alpha = 8°$.

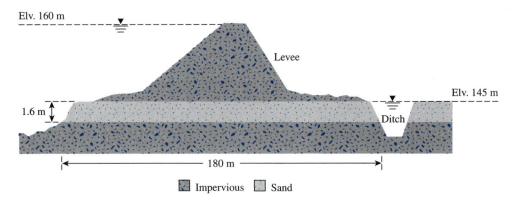

Elv. 160 m

Levee

Elv. 145 m

1.6 m

Ditch

180 m

Impervious Sand

Figure 7.27

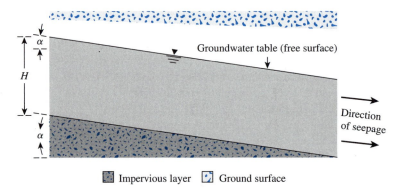

α

Groundwater table (free surface)

H

Direction of seepage

α

Impervious layer Ground surface

Figure 7.28

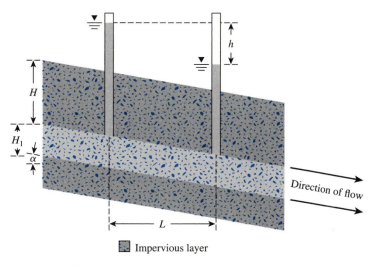

Figure 7.29

7.9 Refer to Figure 7.29. Find the flow rate in m^3/sec/m length (at right angles to the cross section shown) through the permeable soil layer. Given: $H = 5$ m, $H_1 = 2.8$ m, $h = 3.1$ m, $L = 60$ m, $\alpha = 5°$, $k = 0.05$ cm/sec.

7.10 The hydraulic conductivity of a sand at a void ratio of 0.5 is 0.022 cm/sec. Estimate its hydraulic conductivity at a void ratio of 0.7. Use Eq. (7.31).

7.11 For a sand, the following are given: porosity, $n = 0.31$ and $k = 6.1$ cm/min. Determine k when $n = 0.4$. Use Eq. (7.31).

7.12 The maximum dry density determined in the laboratory for a quartz sand is 1800 kg/m^3. In the field, if the relative compaction is 80%, determine the hydraulic conductivity of the sand in the field-compaction condition (given that $D_{10} = 0.15$ mm, $C_u = 2.2$ and $G_s = 2.66$). Use Eq. (7.34).

7.13 For a sandy soil, the following are given:
- Maximum void ratio = 0.7
- Minimum void ratio = 0.46
- $D_{10} = 0.2$ mm

Determine the hydraulic conductivity of the sand at a relative density of 60%. Use Eq. (7.32).

7.14 The sieve analysis for a sand is given in the following table. Estimate the hydraulic conductivity of the sand at a void ratio of 0.5. Use Eq. (7.30) and $SF = 6.5$.

U.S. Sieve No.	Percent passing
30	100
40	80
60	68
100	28
200	0

7.15 For a normally consolidated clay, the following are given:

Void ratio, e	k (cm/sec)
0.8	1.2×10^{-6}
1.4	3.6×10^{-6}

Estimate the hydraulic conductivity of the clay at a void ratio, $e = 0.9$. Use Eq. (7.38).

7.16 The *in situ* void ratio of a soft clay deposit is 2.1 and the hydraulic conductivity of the clay at this void ratio is 0.91×10^{-6} cm/sec. What is the hydraulic conductivity if the soil is compressed to have a void ratio of 1.1. Use Eq. (7.36).

7.17 A layered soil is shown in Figure 7.30. Given that
- $H_1 = 1.5$ m
- $k_1 = 10^{-5}$ cm/sec
- $H_2 = 2.5$ m
- $k_2 = 3.0 \times 10^{-3}$ cm/sec
- $H_3 = 3.0$ m
- $k_3 = 3.5 \times 10^{-5}$ cm/sec

Estimate the ratio of equivalent hydraulic conductivity, $k_{H(eq)}/k_{V(eq)}$.

7.18 A layered soil is shown in Figure 7.31. Estimate the ratio of equivalent hydraulic conductivity, $k_{H(eq)}/k_{V(eq)}$.

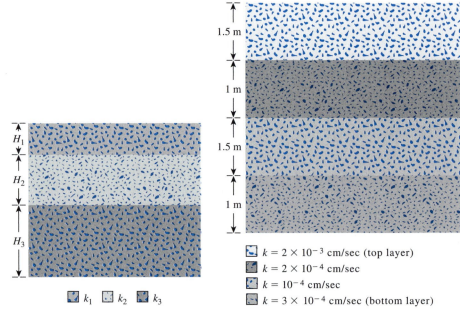

$k = 2 \times 10^{-3}$ cm/sec (top layer)
$k = 2 \times 10^{-4}$ cm/sec
$k = 10^{-4}$ cm/sec
$k = 3 \times 10^{-4}$ cm/sec (bottom layer)

k_1 k_2 k_3

Figure 7.30   **Figure 7.31**

References

AMER, A. M., and AWAD, A. A. (1974). "Permeability of Cohesionless Soils," *Journal of the Geotechnical Engineering Division*, ASCE, Vol. 100, No. GT12, 1309–1316.

BASETT, D. J., and BRODIE, A. F. (1961). "A Study of Matabitchual Varved Clay," *Ontario Hydroelectric Research News*, Vol. 13, No. 4, 1–6.

CARMAN, P. C. (1938). "The Determination of the Specific Surface of Powders." *J. Soc. Chem. Ind. Trans.*, Vol. 57, 225.

CARMAN, P. C. (1956). *Flow of Gases through Porous Media,* Butterworths Scientific Publications, London.

CARRIER III, W. D. (2003). "Goodbye, Hazen; Hello, Kozeny-Carman," *Journal of Geotechnical and Geoenvironmental Engineering,* ASCE, Vol. 129, No. 11, 1054–1056.

CASAGRANDE, L., and POULOS, S. J. (1969). "On the Effectiveness of Sand Drains," *Canadian Geotechnical Journal,* Vol. 6, No. 3, 287–326.

CHAN, H. T., and KENNEY, T. C. (1973). "Laboratory Investigation of Permeability Ratio of New Liskeard Varved Soil," *Canadian Geotechnical Journal,* Vol. 10, No. 3, 453–472.

CHAPUIS, R. P. (2004). "Predicting the Saturated Hydraulic Conductivity of Sand and Gravel Using Effective Diameter and Void Ratio," *Canadian Geotechnical Journal,* Vol. 41, No. 5, 787–795.

CHAPUIS, R. P., GILL, D. E., and BAASS, K. (1989). "Laboratory Permeability Tests on Sand: Influence of Compaction Method on Anisotropy," *Canadian Geotechnical Journal,* Vol. 26, 614–622.

DANIEL, D. E. (1989). "*In Situ* Hydraulic Conductivity Tests for Compacted Clay," *Journal of Geotechnical Engineering,* ASCE, Vol. 115, No. 9, 1205–1226.

DARCY, H. (1856). *Les Fontaines Publiques de la Ville de Dijon,* Dalmont, Paris.

FUKUSHIMA, S., and ISHII, T. (1986). "An Experimental Study of the Influence of Confining Pressure on Permeability Coefficients of Filldam Core Materials," *Soils and Foundations,* Vol. 26, No. 4, 32–46

HANSBO, S. (1960). "Consolidation of Clay with Special Reference to Influence of Vertical Sand Drains," Swedish Geotechnical Institute, *Proc. No. 18,* 41–61.

HAZEN, A. (1930). "Water Supply," in *American Civil Engineers Handbook,* Wiley, New York.

KENNEY, T. C., and CHAN, H. T. (1973). "Field Investigation of Permeability Ratio of New Liskeard Varved Soil," *Canadian Geotechnical Journal,* Vol. 10, No. 3, 473–488.

KENNEY, T. C., LAU, D., and OFOEGBU, G. I. (1984). "Permeability of Compacted Granular Materials," *Canadian Geotechnical Journal,* Vol. 21, No. 4, 726–729.

KOZENY, J. (1927). "Ueber kapillare Leitung des Wassers im Boden." *Wien, Akad. Wiss.,* Vol. 136, No. 2a, 271.

KRUMBEIN, W. C., and MONK, G. D. (1943). "Permeability as a Function of the Size Parameters of Unconsolidated Sand," *Transactions,* AIMME (Petroleum Division), Vol. 151, 153–163.

LUMB, P., and HOLT, J. K. (1968). "The Undrained Shear Strength of a Soft Marine Clay from Hong Kong," *Geotechnique,* Vol. 18, 25–36.

MESRI, G., and OLSON, R. E. (1971). "Mechanism Controlling the Permeability of Clays," *Clay and Clay Minerals,* Vol. 19, 151–158.

MITCHELL, J. K. (1976). *Fundamentals of Soil Behavior,* Wiley, New York.

OLSEN, H. W. (1961). "Hydraulic Flow Through Saturated Clay," Sc.D. Thesis, Massachusetts Institute of Technology.

SAMARASINGHE, A. M., HUANG, Y. H., and DRNEVICH, V. P. (1982). "Permeability and Consolidation of Normally Consolidated Soils," *Journal of the Geotechnical Engineering Division,* ASCE, Vol. 108, No. GT6, 835–850.

TAVENAS, F., JEAN, P., LEBLOND, F. T. P., and LEROUEIL, S. (1983). "The Permeability of Natural Soft Clays. Part II: Permeability Characteristics," *Canadian Geotechnical Journal,* Vol. 20, No. 4, 645–660.

TSIEN, S. I. (1955). "Stabilization of Marsh Deposit," *Highway Research Record,* Bulletin 115, 15–43.

U.S. DEPARTMENT OF NAVY (1971). "Design Manual—Soil Mechanics, Foundations, and Earth Structures," *NAVFAC DM-7,* U.S. Government Printing Office, Washington, D.C.

WU, T. H., CHANG, N. Y., and ALI, E. M. (1978). "Consolidation and Strength Properties of a Clay," *Journal of the Geotechnical Engineering Division,* ASCE, Vol. 104, No. GT7, 899–905.

8 Seepage

In the preceding chapter, we considered some simple cases for which direct application of Darcy's law was required to calculate the flow of water through soil. In many instances, the flow of water through soil is not in one direction only, nor is it uniform over the entire area perpendicular to the flow. In such cases, the groundwater flow is generally calculated by the use of graphs referred to as *flow nets*. The concept of the flow net is based on *Laplace's equation of continuity,* which governs the steady flow condition for a given point in the soil mass. In the following sections of this chapter, the derivation of Laplace's equation of continuity will be presented along with its application to seepage problems.

8.1 Laplace's Equation of Continuity

To derive the Laplace differential equation of continuity, let us consider a single row of sheet piles that have been driven into a permeable soil layer, as shown in Figure 8.1a. The row of sheet piles is assumed to be impervious. The steady state flow of water from the upstream to the downstream side through the permeable layer is a two-dimensional flow. For flow at a point A, we consider an elemental soil block. The block has dimensions dx, dy, and dz (length dy is perpendicular to the plane of the paper); it is shown in an enlarged scale in Figure 8.1b. Let v_x and v_z be the components of the discharge velocity in the horizontal and vertical directions, respectively. The rate of flow of water into the elemental block in the horizontal direction is equal to $v_x\, dz\, dy$, and in the vertical direction it is $v_z\, dx\, dy$. The rates of outflow from the block in the horizontal and vertical directions are, respectively,

$$\left(v_x + \frac{\partial v_x}{\partial x}\, dx \right) dz\, dy$$

and

$$\left(v_z + \frac{\partial v_z}{\partial z}\, dz \right) dx\, dy$$

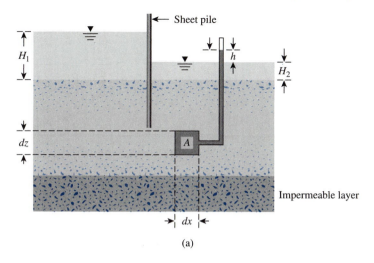

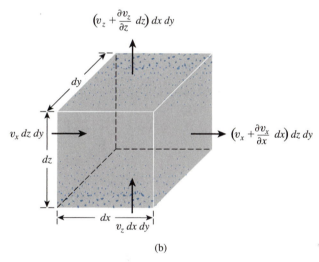

Figure 8.1 (a) Single-row sheet piles driven into permeable layer; (b) flow at A

Assuming that water is incompressible and that no volume change in the soil mass occurs, we know that the total rate of inflow should equal the total rate of outflow. Thus,

$$\left[\left(v_x + \frac{\partial v_x}{\partial x}\,dx\right)dz\,dy + \left(v_z + \frac{\partial v_z}{\partial z}\,dz\right)dx\,dy\right] - [v_x\,dz\,dy + v_z\,dx\,dy] = 0$$

or

$$\frac{\partial v_x}{\partial x} + \frac{\partial v_z}{\partial z} = 0 \qquad (8.1)$$

With Darcy's law, the discharge velocities can be expressed as

$$v_x = k_x i_x = k_x \frac{\partial h}{\partial x} \tag{8.2}$$

and

$$v_z = k_z i_z = k_z \frac{\partial h}{\partial z} \tag{8.3}$$

where k_x and k_z are the hydraulic conductivities in the vertical and horizontal directions, respectively.

From Eqs. (8.1), (8.2), and (8.3), we can write

$$k_x \frac{\partial^2 h}{\partial x^2} + k_z \frac{\partial^2 h}{\partial z^2} = 0 \tag{8.4}$$

If the soil is isotropic with respect to the hydraulic conductivity—that is, $k_x = k_z$—the preceding continuity equation for two-dimensional flow simplifies to

$$\frac{\partial^2 h}{\partial x^2} + \frac{\partial^2 h}{\partial z^2} = 0 \tag{8.5}$$

8.2 Continuity Equation for Solution of Simple Flow Problems

The continuity equation given in Eq. (8.5) can be used in solving some simple flow problems. To illustrate this, let us consider a one-dimensional flow problem, as shown in Figure 8.2, in which a constant head is maintained across a two-layered soil for the flow of water. The head difference between the top of soil layer no. 1 and the bottom of soil layer no. 2 is h_1. Because the flow is in only the z direction, the continuity equation [Eq. (8.5)] is simplified to the form

$$\frac{\partial^2 h}{\partial z^2} = 0 \tag{8.6}$$

or

$$h = A_1 z + A_2 \tag{8.7}$$

where A_1 and A_2 are constants.

To obtain A_1 and A_2 for flow through soil layer no. 1, we must know the boundary conditions, which are as follows:

Condition 1: At $z = 0$, $h = h_1$.
Condition 2: At $z = H_1$, $h = h_2$.

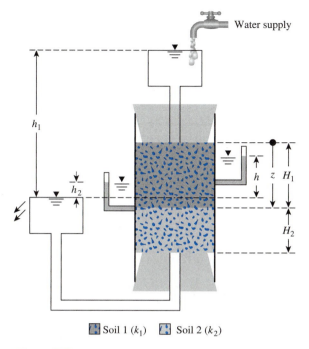

Figure 8.2 Flow through a two-layered soil

Combining Eq. (8.7) and Condition 1 gives

$$A_2 = h_1 \qquad (8.8)$$

Similarly, combining Eq. (8.7) and Condition 2 with Eq. (8.8) gives

$$h_2 = A_1 H_1 + h_1$$

or

$$A_1 = -\left(\frac{h_1 - h_2}{H_1}\right) \qquad (8.9)$$

Combining Eqs. (8.7), (8.8), and (8.9), we obtain

$$h = -\left(\frac{h_1 - h_2}{H_1}\right)z + h_1 \qquad \text{(for } 0 \le z \le H_1) \qquad (8.10)$$

For flow through soil layer no. 2, the boundary conditions are

Condition 1: At $z = H_1$, $h = h_2$.
Condition 2: At $z = H_1 + H_2$, $h = 0$.

From Condition 1 and Eq. (8.7),

$$A_2 = h_2 - A_1 H_1 \qquad (8.11)$$

Also, from Condition 2 and Eqs. (8.7) and (8.11),

$$0 = A_1(H_1 + H_2) + (h_2 - A_1 H_1)$$
$$A_1 H_1 + A_1 H_2 + h_2 - A_1 H_1 = 0$$

or

$$A_1 = -\frac{h_2}{H_2} \tag{8.12}$$

So, from Eqs. (8.7), (8.11), and (8.12),

$$h = -\left(\frac{h_2}{H_2}\right) z + h_2 \left(1 + \frac{H_1}{H_2}\right) \qquad \text{(for } H_1 \leq z \leq H_1 + H_2) \tag{8.13}$$

At any given time, flow through soil layer no. 1 equals flow through soil layer no. 2, so

$$q = k_1 \left(\frac{h_1 - h_2}{H_1}\right) A = k_2 \left(\frac{h_2 - 0}{H_2}\right) A$$

where A = area of cross section of the soil
  k_1 = hydraulic conductivity of soil layer no. 1
  k_2 = hydraulic conductivity of soil layer no. 2

or

$$h_2 = \frac{h_1 k_1}{H_1 \left(\dfrac{k_1}{H_1} + \dfrac{k_2}{H_2}\right)} \tag{8.14}$$

Substituting Eq. (8.14) into Eq. (8.10), we obtain

$$h = h_1 \left(1 - \frac{k_2 z}{k_1 H_2 + k_2 H_1}\right) \qquad \text{(for } 0 \leq z \leq H_1) \tag{8.15}$$

Similarly, combining Eqs. (8.13) and (8.14) gives

$$h = h_1 \left[\left(\frac{k_1}{k_1 H_2 + k_2 H_1}\right)(H_1 + H_2 - z)\right] \qquad \text{(for } H_1 \leq z \leq H_1 + H_2) \tag{8.16}$$

Example 8.1

Refer to Fig. 8.2. Given: $H_1 = 305$ mm; $H_2 = 508$ mm; $h_1 = 610$ mm; $h = 508$ mm; $z = 203$ mm; $k_1 = 0.066$ cm/sec, and diameter of the soil specimen, $D = 76$ mm. Determine the rate of flow of water through the two-layered soil (cm^3/hr).

Solution
Since $z = 203$ mm is located in soil layer 1, Eq. (8.15) is valid. Thus,

$$h = h_1 \left(1 - \frac{k_2 z}{k_1 H_2 + k_2 H_1} \right) = h_1 \left[1 - \frac{z}{\left(\frac{k_1}{k_2}\right) H_2 + H_1} \right]$$

$$508 = 610 \left[1 - \frac{203}{\left(\frac{k_1}{k_2}\right) 508 + 305} \right]$$

$$\frac{k_1}{k_2} = 1.795 \approx 1.8$$

Given $k_1 = 0.066$ cm/sec. So

$$k_2 = \frac{k_1}{1.8} = \frac{0.066}{1.8} = 0.037 \, \text{cm/sec}$$

The rate of flow is

$$q = k_{eq} i A$$

$$i = \frac{h_1}{H_1 + H_2} = \frac{610}{305 + 508} = 0.75$$

$$A = \frac{\pi}{4} D^2 = \frac{\pi}{4} (7.6)^2 = 45.36 \, \text{cm}^2$$

$$k_{eq} = \frac{H_1 + H_2}{\frac{H_1}{k_1} + \frac{H_2}{k_2}} = \frac{30.5 \, \text{cm} + 50.8 \, \text{cm}}{\frac{30.5}{0.066} + \frac{50.8}{0.037}} = 0.0443 \, \text{cm/sec} = 159.48 \, \text{cm/hr}$$

Thus,

$$q = k_{eq} \, i \, A = (159.48)(0.75)(45.36) \approx \textbf{5426 cm}^3\textbf{/hr}$$

8.3   Flow Nets

The continuity equation [Eq. (8.5)] in an isotropic medium represents two orthogonal families of curves—that is, the flow lines and the equipotential lines. A *flow line* is a line along which a water particle will travel from upstream to the downstream side in the permeable soil medium. An *equipotential line* is a line along which the potential head at all points is equal. Thus, if piezometers are placed at different points along an equipotential line, the water level will rise to the same elevation in all of them. Figure 8.3a demonstrates the

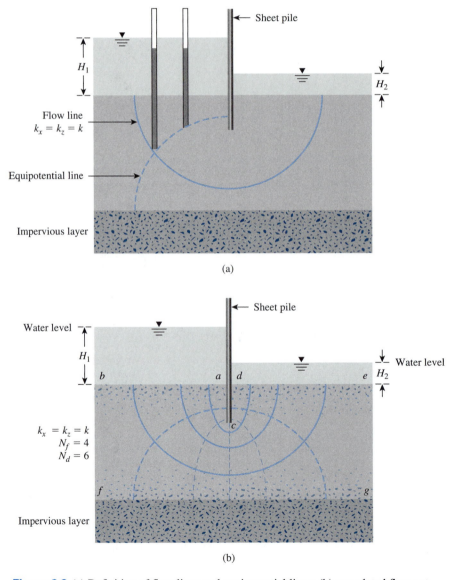

(a)

(b)

Figure 8.3 (a) Definition of flow lines and equipotential lines; (b) completed flow net

definition of flow and equipotential lines for flow in the permeable soil layer around the row of sheet piles shown in Figure 8.1 (for $k_x = k_z = k$).

A combination of a number of flow lines and equipotential lines is called a *flow net.* As mentioned in the introduction, flow nets are constructed for the calculation of ground-water flow and the evaluation of heads in the media. To complete the graphic construction of a flow net, one must draw the flow and equipotential lines in such a way that

1. The equipotential lines intersect the flow lines at right angles.
2. The flow elements formed are approximate squares.

Figure 8.3b shows an example of a completed flow net. One more example of flow net in isotropic permeable layer are given in Figure 8.4. In these figures, N_f is the number of flow channels in the flow net, and N_d is the number of potential drops (defined later in this chapter).

Drawing a flow net takes several trials. While constructing the flow net, keep the boundary conditions in mind. For the flow net shown in Figure 8.3b, the following four boundary conditions apply:

Condition 1: The upstream and downstream surfaces of the permeable layer (lines *ab* and *de*) are equipotential lines.

Condition 2: Because *ab* and *de* are equipotential lines, all the flow lines intersect them at right angles.

Condition 3: The boundary of the impervious layer—that is, line *fg*—is a flow line, and so is the surface of the impervious sheet pile, line *acd*.

Condition 4: The equipotential lines intersect *acd* and *fg* at right angles.

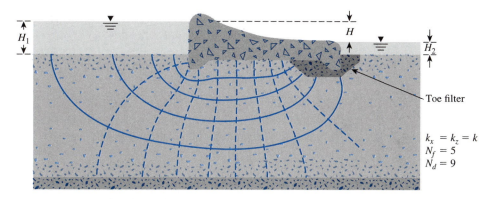

Figure 8.4 Flow net under a dam with toe filter

8.4 *Seepage Calculation from a Flow Net*

In any flow net, the strip between any two adjacent flow lines is called a *flow channel.* Figure 8.5 shows a flow channel with the equipotential lines forming square elements. Let $h_1, h_2, h_3, h_4, \ldots, h_n$ be the piezometric levels corresponding to the equipotential lines. The rate of seepage through the flow channel per unit length (perpendicular to the vertical section through the permeable layer) can be calculated as follows. Because there is no flow across the flow lines,

$$\Delta q_1 = \Delta q_2 = \Delta q_3 = \cdots = \Delta q \qquad (8.17)$$

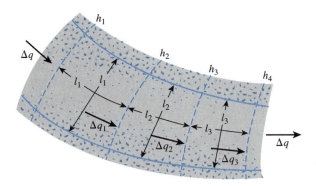

Figure 8.5 Seepage through a flow channel with square elements

From Darcy's law, the flow rate is equal to kiA. Thus, Eq. (8.17) can be written as

$$\Delta q = k\left(\frac{h_1 - h_2}{l_1}\right)l_1 = k\left(\frac{h_2 - h_3}{l_2}\right)l_2 = k\left(\frac{h_3 - h_4}{l_3}\right)l_3 = \cdots \qquad (8.18)$$

Eq. (8.18) shows that if the flow elements are drawn as approximate squares, the drop in the piezometric level between any two adjacent equipotential lines is the same. This is called the *potential drop*. Thus,

$$h_1 - h_2 = h_2 - h_3 = h_3 - h_4 = \cdots = \frac{H}{N_d} \qquad (8.19)$$

and

$$\Delta q = k\frac{H}{N_d} \qquad (8.20)$$

where H = head difference between the upstream and downstream sides
  N_d = number of potential drops

In Figure 8.3b, for any flow channel, $H = H_1 - H_2$ and $N_d = 6$.

If the number of flow channels in a flow net is equal to N_f, the total rate of flow through all the channels per unit length can be given by

$$q = k\frac{HN_f}{N_d} \qquad (8.21)$$

Although drawing square elements for a flow net is convenient, it is not always necessary. Alternatively, one can draw a rectangular mesh for a flow channel, as shown in Figure 8.6, provided that the width-to-length ratios for all the rectangular elements in the flow net are the same. In this case, Eq. (8.18) for rate of flow through the channel can be modified to

$$\Delta q = k\left(\frac{h_1 - h_2}{l_1}\right)b_1 = k\left(\frac{h_2 - h_3}{l_2}\right)b_2 = k\left(\frac{h_3 - h_4}{l_3}\right)b_3 = \cdots \qquad (8.22)$$

If $b_1/l_1 = b_2/l_2 = b_3/l_3 = \cdots = n$ (i.e., the elements are not square), Eqs. (8.20) and (8.21) can be modified to

$$\Delta q = kH\left(\frac{n}{N_d}\right) \qquad (8.23)$$

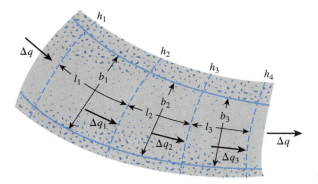

Figure 8.6 Seepage through a flow channel with rectangular elements

and

$$q = kH\left(\frac{N_f}{N_d}\right)n \qquad (8.24)$$

Figure 8.7 shows a flow net for seepage around a single row of sheet piles. Note that flow channels 1 and 2 have square elements. Hence, the rate of flow through these two channels can be obtained from Eq. (8.20):

$$\Delta q_1 + \Delta q_2 = \frac{k}{N_d}H + \frac{k}{N_d}H = \frac{2kH}{N_d}$$

However, flow channel 3 has rectangular elements. These elements have a width-to-length ratio of about 0.38; hence, from Eq. (8.23)

$$\Delta q_3 = \frac{k}{N_d}H(0.38)$$

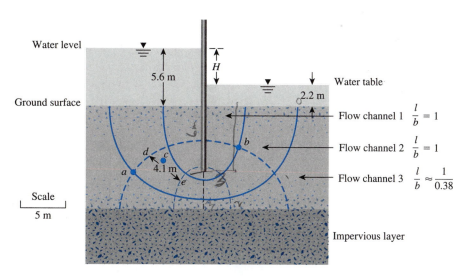

Figure 8.7 Flow net for seepage around a single row of sheet piles

So, the total rate of seepage can be given as

$$q = \Delta q_1 + \Delta q_2 + \Delta q_3 = 2.38\frac{kH}{N_d} \qquad (8.25)$$

Example 8.2

A flow net for flow around a single row of sheet piles in a permeable soil layer is shown in Figure 8.7. Given that $k_x = k_z = k = 5 \times 10^{-3}$ cm/sec, determine

    a. How high (above the ground surface) the water will rise if piezometers are placed at points a and b.
    b. The total rate of seepage through the permeable layer per unit length
    c. The approximate average hydraulic gradient at c.

Solution
Part a
From Figure 8.7, we have $N_d = 6$, $H_1 = 5.6$ m, and $H_2 = 2.2$ m. So the head loss of each potential drop is

$$\Delta H = \frac{H_1 - H_2}{N_d} = \frac{5.6 - 2.2}{6} = 0.567\,\text{m}$$

At point a, we have gone through one potential drop. So the water in the piezometer will rise to an elevation of

$$(5.6 - 0.567) = \textbf{5.033 m above the ground surface}$$

At point b, we have five potential drops. So the water in the piezometer will rise to an elevation of

$$[5.6 - (5)(0.567)] = \textbf{2.765 m above the ground surface}$$

Part b
From Eq. (8.25),

$$q = 2.38\,\frac{k(H_1 - H_2)}{N_d} = \frac{(2.38)(5 \times 10^{-5}\text{m/sec})(5.6 - 2.2)}{6}$$

$$= \textbf{6.74} \times \textbf{10}^{-5}\textbf{m}^3\textbf{/sec/m}$$

Part c
The average hydraulic gradient at c can be given as

$$i = \frac{\text{head loss}}{\text{average length of flow between } d \text{ and } e} = \frac{\Delta H}{\Delta L} = \frac{0.567\,\text{m}}{4.1\,\text{m}} = \textbf{0.138}$$

(*Note*: The average length of flow has been scaled.)   ∎

Flow Nets in Anisotropic Soil

The flow-net construction described thus far and the derived Eqs. (8.21) and (8.24) for seepage calculation have been based on the assumption that the soil is isotropic. However, in nature, most soils exhibit some degree of anisotropy. To account for soil anisotropy with respect to hydraulic conductivity, we must modify the flow net construction.

The differential equation of continuity for a two-dimensional flow [Eq. (8.4)] is

$$k_x \frac{\partial^2 h}{\partial x^2} + k_z \frac{\partial^2 h}{\partial z^2} = 0$$

For anisotropic soils, $k_x \neq k_z$. In this case, the equation represents two families of curves that do not meet at 90°. However, we can rewrite the preceding equation as

$$\frac{\partial^2 h}{(k_z/k_x)\,\partial x^2} + \frac{\partial^2 h}{\partial z^2} = 0 \tag{8.26}$$

Substituting $x' = \sqrt{k_z/k_x}\,x$, we can express Eq. (8.26) as

$$\frac{\partial^2 h}{\partial x'^2} + \frac{\partial^2 h}{\partial z^2} = 0 \tag{8.27}$$

Now Eq. (8.27) is in a form similar to that of Eq. (8.5), with x replaced by x', which is the new transformed coordinate. To construct the flow net, use the following procedure:

Step 1: Adopt a vertical scale (that is, z axis) for drawing the cross section.
Step 2: Adopt a horizontal scale (that is, x axis) such that horizontal scale = $\sqrt{k_z/k_x} \times$ vertical scale.
Step 3: With scales adopted as in Steps 1 and 2, plot the vertical section through the permeable layer parallel to the direction of flow.
Step 4: Draw the flow net for the permeable layer on the section obtained from Step 3, with flow lines intersecting equipotential lines at right angles and the elements as approximate squares.

The rate of seepage per unit length can be calculated by modifying Eq. (8.21) to

$$q = \sqrt{k_x k_z}\,\frac{H N_f}{N_d} \tag{8.28}$$

where H = total head loss
N_f and N_d = number of flow channels and potential drops, respectively (from flow net drawn in Step 4)

Note that when flow nets are drawn in transformed sections (in anisotropic soils), the flow lines and the equipotential lines are orthogonal. However, when they are redrawn in a true section, these lines are not at right angles to each other. This fact is shown in Figure 8.8. In this figure, it is assumed that $k_x = 6k_z$. Figure 8.8a shows a flow element in a transformed section. The flow element has been redrawn in a true section in Figure 8.8b.

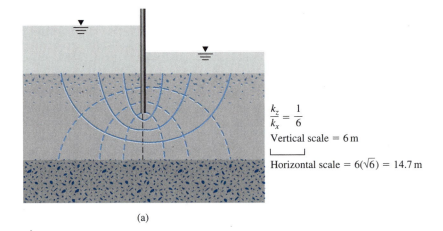

$$\frac{k_z}{k_x} = \frac{1}{6}$$

Vertical scale = 6 m

Horizontal scale = $6(\sqrt{6})$ = 14.7 m

(a)

Scale 6 m

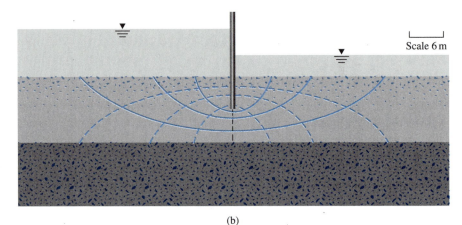

Figure 8.8
A flow element
in anisotropic soil:
(a) in transformed
section; (b) in true
section

(b)

Example 8.3

A dam section is shown in Figure 8.9a. The hydraulic conductivity of the permeable layer in the vertical and horizontal directions are 2×10^{-2} mm/s and 4×10^{-2} mm/s, respectively. Draw a flow net and calculate the seepage loss of the dam in m³/day/m.

Solution
From the given data,

$$k_z = 2 \times 10^{-2} \text{ mm/s} = 1.728 \text{ m/day}$$

$$k_x = 4 \times 10^{-2} \text{ mm/s} = 3.456 \text{ m/day}$$

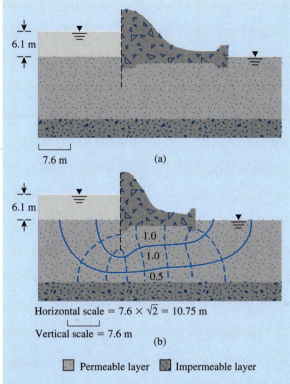

6.1 m

7.6 m   (a)

6.1 m

1.0

1.0

0.5

Horizontal scale = 7.6 × √2 = 10.75 m

Vertical scale = 7.6 m

(b)

☐ Permeable layer ☐ Impermeable layer

Figure 8.9

and $h = 6.1$ m. For drawing the flow net,

$$\text{Horizontal scale} = \sqrt{\frac{2 \times 10^{-2}}{4 \times 10^{-2}}}(\text{vertical scale})$$

$$= \frac{1}{\sqrt{2}}(\text{vertical scale})$$

On the basis of this, the dam section is replotted, and the flow net drawn as in Figure 8.9b. The rate of seepage is given by $q = \sqrt{k_x k_z}\, H(N_f/N_d)$. From Figure 8.9b, $N_d = 8$ and $N_f = 2.5$ (the lowermost flow channel has a width-to-length ratio of 0.5). So,

$$q = \sqrt{(1.728)(3.456)}\,(6.1)(2.5/8) = \textbf{4.66 m}^3\textbf{/day/m}$$ ∎

8.6   *Mathematical Solution for Seepage*

The seepage under several simple hydraulic structures can be solved mathematically. Harr (1962) has analyzed many such conditions. Figure 8.10 shows a nondimensional plot for the rate of seepage around a single row of sheet piles. In a similar manner, Figure 8.11 is a nondimensional plot for the rate of seepage under a dam. In Figure 8.10, the depth of penetration of the sheet pile is S, and the thickness of the permeable soil layer is T'.

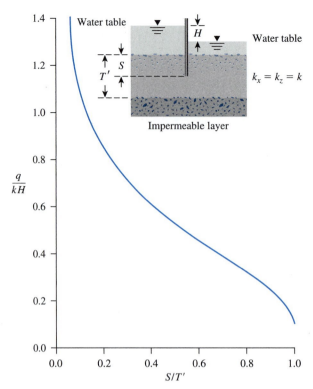

Figure 8.10 Plot of q/kH against S/T' for flow around a single row of sheet piles (*After Harr, 1962*)

Example 8.4

Refer to Figure 8.11. Given; the width of the dam, $B = 6$ m; length of the dam, $L = 120$ m; $S = 3$ m; $T' = 6$ m; $x = 2.4$ m; and $H_1 - H_2 = 5$ m. If the hydraulic conductivity of the permeable layer is 0.008 cm/sec, estimate the seepage under the dam (Q) in m³/day/m.

Solution
Given that $B = 6$ m, $T' = 6$ m, and $S = 3$ m, so $b = B/2 = 3$ m.

$$\frac{b}{T'} = \frac{3}{6} = 0.5$$

$$\frac{S}{T'} = \frac{3}{6} = 0.5$$

$$\frac{x}{b} = \frac{2.4}{3} = 0.8$$

From Figure 8.11, for $b/T' = 0.5$, $S/T' = 0.5$, and $x/b = 0.8$, the value of $q/kH \approx 0.378$. Thus,

$$Q = q L = k HL = (0.008 \times 10^{-2} \times 60 \times 60 \times 24 \text{ m/day})(5)(120) = \textbf{4147.2 m}^3\textbf{/day} \blacksquare$$

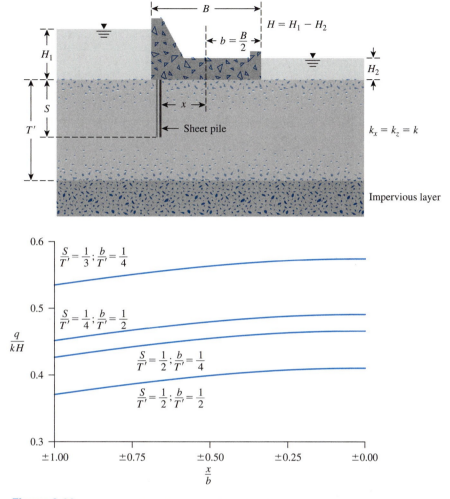

Figure 8.11 Seepage under a dam (*After Harr, 1962*)

8.7 *Uplift Pressure Under Hydraulic Structures*

Flow nets can be used to determine the uplift pressure at the base of a hydraulic structure. This general concept can be demonstrated by a simple example. Figure 8.12a shows a weir, the base of which is 2 m below the ground surface. The necessary flow net also has been drawn (assuming that $k_x = k_z = k$). The pressure distribution diagram at the base of the weir can be obtained from the equipotential lines as follows.

There are seven equipotential drops (N_d) in the flow net, and the difference in the water levels between the upstream and downstream sides is $H = 7$ m. The head loss for each potential drop is $H/7 = 7/7 = 1$ m. The uplift pressure at

$$a \text{ (left corner of the base)} = (\text{Pressure head at } a) \times (\gamma_w)$$
$$= [(7 + 2) - 1]\gamma_w = 8\gamma_w$$

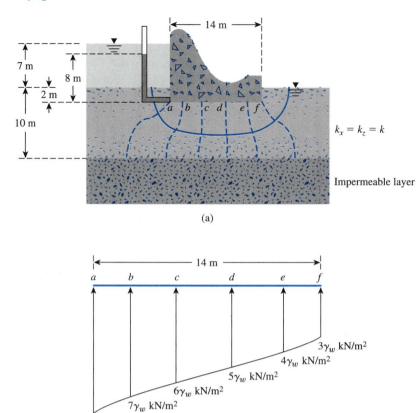

Figure 8.12 (a) A weir; (b) uplift force under a hydraulic structure

Similarly, the uplift pressure at

$$b = [9 - (2)(1)]\gamma_w = 7\gamma_w$$

and at

$$f = [9 - (6)(1)]\gamma_w = 3\gamma_w$$

The uplift pressures have been plotted in Figure 8.12b. The uplift force per unit length measured along the axis of the weir can be calculated by finding the area of the pressure diagram.

8.8   ## *Seepage Through an Earth Dam on an Impervious Base*

Figure 8.13 shows a homogeneous earth dam resting on an impervious base. Let the hydraulic conductivity of the compacted material of which the earth dam is made be equal to k. The free surface of the water passing through the dam is given by *abcd*. It is assumed

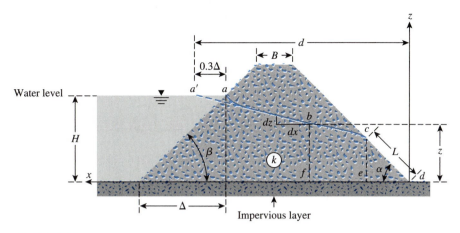

Figure 8.13 Flow through an earth dam constructed over an impervious base

that $a'bc$ is parabolic. The slope of the free surface can be assumed to be equal to the hydraulic gradient. It also is assumed that, because this hydraulic gradient is constant with depth (Dupuit, 1863),

$$i \simeq \frac{dz}{dx} \tag{8.29}$$

Considering the triangle cde, we can give the rate of seepage per unit length of the dam (at right angles to the cross section shown in Figure 8.13) as

$$q = kiA$$

$$i = \frac{dz}{dx} = \tan \alpha$$

$$A = (\overline{ce})(1) = L \sin \alpha$$

So

$$q = k(\tan \alpha)(L \sin \alpha) = kL \tan \alpha \sin \alpha \tag{8.30}$$

Again, the rate of seepage (per unit length of the dam) through the section bf is

$$q = kiA = k\left(\frac{dz}{dx}\right)(z \times 1) = kz \frac{dz}{dx} \tag{8.31}$$

For continuous flow,

$$q_{Eq.\ (8.30)} = q_{Eq.\ (8.31)}$$

or

$$kz\frac{dz}{dx} = kL\tan\alpha\sin\alpha$$

or

$$\int_{z=L\sin\alpha}^{z=H} kz\ dz = \int_{x=L\cos\alpha}^{x=d} (kL\tan\alpha\sin\alpha)\ dx$$

$$\tfrac{1}{2}(H^2 - L^2\sin^2\alpha) = L\tan\alpha\sin\alpha(d - L\cos\alpha)$$

$$\frac{H^2}{2} - \frac{L^2\sin^2\alpha}{2} = Ld\left(\frac{\sin^2\alpha}{\cos\alpha}\right) - L^2\sin^2\alpha$$

$$\frac{H^2\cos\alpha}{2\sin^2\alpha} - \frac{L^2\cos\alpha}{2} = Ld - L^2\cos\alpha$$

or

$$L^2\cos\alpha - 2Ld + \frac{H^2\cos\alpha}{\sin^2\alpha} = 0$$

So,

$$L = \frac{d}{\cos\alpha} - \sqrt{\frac{d^2}{\cos^2\alpha} - \frac{H^2}{\sin^2\alpha}} \qquad (8.32)$$

Following is a step-by-step procedure to obtain the seepage rate q (per unit length of the dam):

Step 1: Obtain α.
Step 2: Calculate Δ (see Figure 8.13) and then 0.3Δ.
Step 3: Calculate d.
Step 4: With known values of α and d, calculate L from Eq. (8.32).
Step 5: With known value of L, calculate q from Eq. (8.30).

The preceding solution generally is referred to as Schaffernak's solution (1917) with Casagrande's correction, since Casagrande experimentally showed that the parabolic free surface starts from a', not a (Figure 8.13).

Example 8.5

Refer to the earth dam shown in Figure 8.13. Given that $\beta = 45°$, $\alpha = 30°$, $B = 3$ m, $H = 6$ m, height of dam $= 7.6$ m, and $k = 61 \times 10^{-6}$ m/min, calculate the seepage rate, q, in m³/day/m length.

Solution
We know that $\beta = 45°$ and $\alpha = 30°$. Thus,

$$\Delta = \frac{H}{\tan \beta} = \frac{6}{\tan 45°} = 6 \text{ m} \quad 0.3\Delta = (0.3)(6) = 1.8 \text{ m}$$

$$d = 0.3\Delta + \frac{(7.6 - 6)}{\tan \beta} + B + \frac{7.6}{\tan \alpha}$$

$$= 1.8 + \frac{(7.6 - 6)}{\tan 45°} + 3 + \frac{7.6}{\tan 30} = 23.36 \text{ m}$$

From Eq. (8.32),

$$L = \frac{d}{\cos \alpha} - \sqrt{\frac{d^2}{\cos^2 \alpha} - \frac{H^2}{\sin^2 \alpha}}$$

$$= \frac{23.36}{\cos 30} - \sqrt{\left(\frac{23.36}{\cos 30}\right)^2 - \left(\frac{6}{\sin 30}\right)^2} = 2.81 \text{ m}$$

From Eq. (8.30)

$$q = kL \tan \alpha \sin \alpha = (61 \times 10^{-6})(\tan 30)(\sin 30)$$

$$= 49.48 \times 10^{-6} \text{ m}^3/\text{min/m} = \mathbf{7.13 \times 10^{-2} \text{ m}^3/\text{day/m}} \quad \blacksquare$$

8.9 ## L. Casagrande's Solution for Seepage Through an Earth Dam

Equation (8.32) is derived on the basis of Dupuit's assumption (i.e., $i \approx dz/dx$). It was shown by Casagrande (1932) that, when the downstream slope angle α in Figure 8.13 becomes greater than 30°, deviations from Dupuit's assumption become more noticeable. Thus (see Figure 8.13), L. Casagrande (1932) suggested that

$$i = \frac{dz}{ds} = \sin \alpha \tag{8.33}$$

where $ds = \sqrt{dx^2 + dz^2}$

So Eq. (8.30) can now be modified as

$$q = kiA = k \sin \alpha (L \sin \alpha) = kL \sin^2 \alpha \tag{8.34}$$

Again,

$$q = kiA = k\left(\frac{dz}{ds}\right)(1 \times z) \tag{8.35}$$

Combining Eqs. (8.34) and (8.35) yields

$$\int_{L \sin \alpha}^{H} z \, dz = \int_{L}^{s} L \sin^2 \alpha \, ds \tag{8.36}$$

where s = length of curve $a'bc$

$$\frac{1}{2}(H^2 - L^2 \sin^2 \alpha) = L \sin^2 \alpha(s - L)$$

or

$$L = s - \sqrt{s^2 - \frac{H^2}{\sin^2 \alpha}} \tag{8.37}$$

With about 4 to 5% error, we can write

$$s = \sqrt{d^2 + H^2} \tag{8.38}$$

Combining Eqs. (8.37) and (8.38) yields

$$L = \sqrt{d^2 + H^2} - \sqrt{d^2 - H^2 \cot^2 \alpha} \tag{8.39}$$

Once the magnitude of L is known, the rate of seepage can be calculated from Eq. (8.34) as

$$q = kL \sin^2 \alpha$$

In order to avoid the approximation introduced in Eqs. (8.38) and (8.39), a solution was provided by Gilboy (1934). This is shown in a graphical form in Figure 8.14. Note, in this graph,

$$m = \frac{L \sin \alpha}{H} \tag{8.40}$$

In order to use the graph,

Step 1: Determine d/H.
Step 2: For a given d/H and α, determine m.
Step 3: Calculate $L = \dfrac{mH}{\sin \alpha}$.
Step 4: Calculate $kL \sin^2 \alpha$.

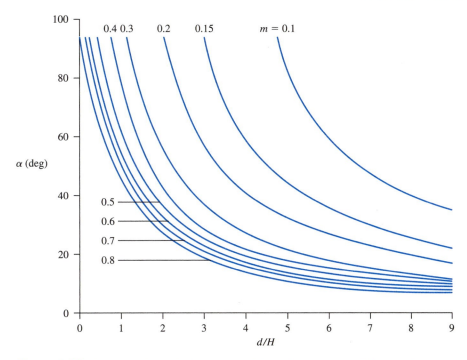

Figure 8.14 Chart for solution by L. Casagrande's method based on Gilboy's solution

8.10 Filter Design

When seepage water flows from a soil with relatively fine grains into a coarser material, there is danger that the fine soil particles may wash away into the coarse material. Over a period of time, this process may clog the void spaces in the coarser material. Hence, the grain-size distribution of the coarse material should be properly manipulated to avoid this situation. A properly designed coarser material is called a *filter*. Figure 8.15 shows the steady-state seepage condition in an earth dam which has a toe filter. For proper selection of the filter material, two conditions should be kept in mind:

Condition 1: The size of the voids in the filter material should be small enough to hold the larger particles of the protected material in place.

Condition 2: The filter material should have a high hydraulic conductivity to prevent buildup of large seepage forces and hydrostatic pressures in the filters.

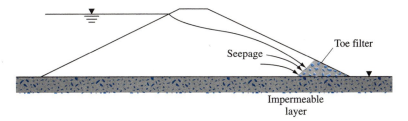

Figure 8.15 Steady-state seepage in an earth dam with a toe filter

It can be shows that, if three perfect spheres have diameters greater than 6.5 times the diameter of a smaller sphere, the small sphere can move through the void spaces of the larger ones (Figure 8.16a). Generally speaking, in a given soil, the sizes of the grains vary over a wide range. If the pore spaces in a filter are small enough to hold D_{85} of the soil to be protected, then the finer soil particles also will be protected (Figure 8.16b). This means that the effective diameter of the pore spaces in the filter should be less than D_{85} of the soil to be protected. The effective pore diameter is about $\frac{1}{5} D_{15}$ of the filter. With this in mind and based on the experimental investigation of filters, Terzaghi and Peck (1948) provided the following criteria to satisfy Condition 1:

$$\frac{D_{15(F)}}{D_{85(S)}} \leq 4 \text{ to } 5 \qquad \text{(to satisfy Condition 1)} \qquad (8.41)$$

In order to satisfy Condition 2, they suggested that

$$\frac{D_{15(F)}}{D_{15(S)}} \geq 4 \text{ to } 5 \qquad \text{(to satisfy Condition 2)} \qquad (8.42)$$

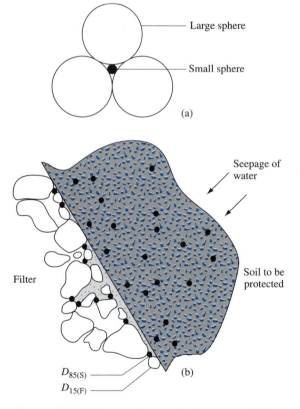

Figure 8.16 (a) Large spheres with diameters of 6.5 times the diameter of the small sphere; (b) boundary between a filter and the soil to be protected

where $D_{15(F)}$ = diameter through which 15% of filter material will pass
$\quad\quad D_{15(S)}$ = diameter through which 15% of soil to be protected will pass
$\quad\quad D_{85(S)}$ = diameter through which 85% of soil to be protected will pass

The proper use of Eqs. (8.41) and (8.42) to determine the grain-size distribution of soils used as filters is shown in Figure 8.17. Consider the soil used for the construction of the earth dam shown in Figure 8.15. Let the grain-size distribution of this soil be given by curve *a* in Figure 8.17. We can now determine $5D_{85(S)}$ and $5D_{15(S)}$ and plot them as shown in Figure 8.17. The acceptable grain-size distribution of the filter material will have to lie in the shaded zone. (*Note:* The shape of curves *b* and *c* are approximately the same as curve *a*.)

The U.S. Navy (1971) requires the following conditions for the design of filters.

Condition 1: For avoiding the movement of the particles of the protected soil:

$$\frac{D_{15(F)}}{D_{85(S)}} < 5$$

$$\frac{D_{50(F)}}{D_{50(S)}} < 25$$

$$\frac{D_{15(F)}}{D_{15(S)}} < 20$$

If the uniformity coefficient C_u of the protected soil is less than 1.5, $D_{15(F)}/D_{85(S)}$ may be increased to 6. Also, if C_u of the protected soil is greater than 4, $D_{15(F)}/D_{15(S)}$ may be increased to 40.

Condition 2: For avoiding buildup of large seepage force in the filter:

$$\frac{D_{15(F)}}{D_{15(S)}} > 4$$

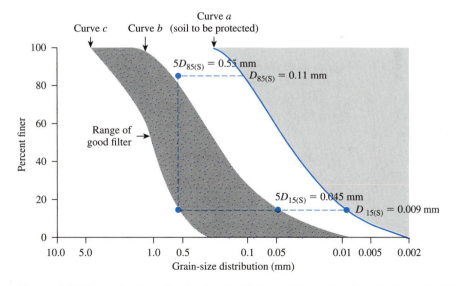

Figure 8.17 Determination of grain-size distribution of filter using Eqs. (8.41) and (8.42)

Condition 3:   The filter material should not have grain sizes greater than 76.2 mm. (This is to avoid segregation of particles in the filter.)

Condition 4:   To avoid internal movement of fines in the filter, it should have no more than 5% passing a No. 200 sieve.

Condition 5:   When perforated pipes are used for collecting seepage water, filters also are used around the pipes to protect the fine-grained soil from being washed into the pipes. To avoid the movement of the filter material into the drain-pipe perforations, the following additional conditions should be met:

$$\frac{D_{85(F)}}{\text{slot width}} > 1.2 \text{ to } 1.4$$

$$\frac{D_{85(F)}}{\text{hole diameter}} > 1.0 \text{ to } 1.2$$

8.11   *Summary*

In this chapter, we studied Laplace's equation of continuity and its application in solving problems related to seepage calculation. The continuity equation is the fundamental basis on which the concept of drawing flow nets is derived. Flow nets are very powerful tools for calculation of seepage as well as uplift pressure under various hydraulic structures.

Also discussed in this chapter (Sections 8.8 and 8.9) is the procedure to calculate seepage through an earth dam constructed over an impervious base. Section 8.8 derives the relationship for seepage based on Dupuit's assumption that the hydraulic gradient is constant with depth. An improved procedure (L. Casagrande's solution) for seepage calculation is provided in Section 8.9. The principles of designing a protective filter are given in Section 8.10.

Problems

8.1   Refer to the constant-head permeability test arrangement in a two-layered soil as shown in Figure 8.2. During the test, it was seen that when a constant head of $h_1 = 200$ mm was maintained, the magnitude of h_2 was 80 mm. If k_1 is 0.004 cm/sec, determine the value of k_2 given $H_1 = 100$ mm and $H_2 = 150$ mm.

8.2   Refer to Figure 8.18. Given:
   • $H_1 = 6$ m   • $D = 3$ m
   • $H_2 = 1.5$ m   • $D_1 = 6$ m

   draw a flow net. Calculate the seepage loss per meter length of the sheet pile (at a right angle to the cross section shown).

8.3   Draw a flow net for the single row of sheet piles driven into a permeable layer as shown in Figure 8.18. Given:
   • $H_1 = 3$ m   • $D = 1.5$ m
   • $H_2 = 0.5$ m   • $D_1 = 3.75$ m

   calculate the seepage loss per meter length of the sheet pile (at right angles to the cross section shown).

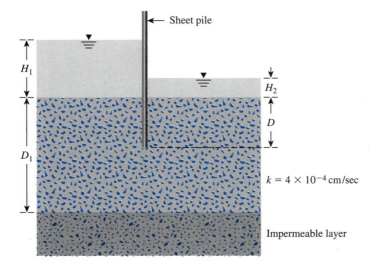

Figure 8.18

8.4 Refer to Figure 8.18. Given:
- $H_1 = 4$ m
- $H_2 = 1.5$ m
- $D_1 = 6$ m
- $D = 3.6$ m

calculate the seepage loss in m³/day per meter length of the sheet pile (at right angles to the cross section shown). Use Figure 8.10.

8.5 For the hydraulic structure shown in Figure 8.19, draw a flow net for flow through the permeable layer and calculate the seepage loss in m³/day/m.

8.6 Refer to Problem 8.5. Using the flow net drawn, calculate the hydraulic uplift force at the base of the hydraulic structure per meter length (measured along the axis of the structure).

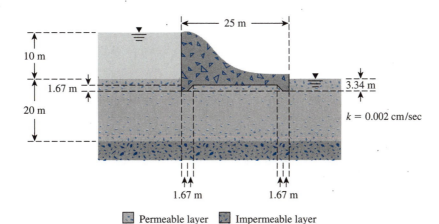

Figure 8.19

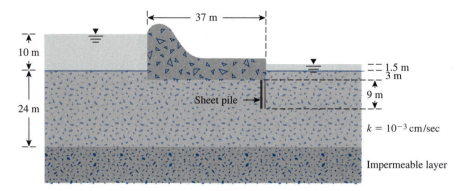

Figure 8.20

8.7 Draw a flow net for the weir shown in Figure 8.20. Calculate the rate of seepage under the weir.

8.8 For the weir shown in Figure 8.21, calculate the seepage in the permeable layer in m^3/day/m for (a) $x' = 1$ m and (b) $x' = 2$ m. Use Figure 8.11.

8.9 An earth dam is shown in Figure 8.22. Determine the seepage rate, q, in m^3/day/m length. Given: $\alpha_1 = 35°$, $\alpha_2 = 40°$, $L_1 = 5$ m, $H = 7$ m, $H_1 = 10$ m, and $k = 3×10^{-4}$ cm/sec. Use Schaffernak's solution.

8.10 Repeat Problem 8.9 using L. Casagrande's method.

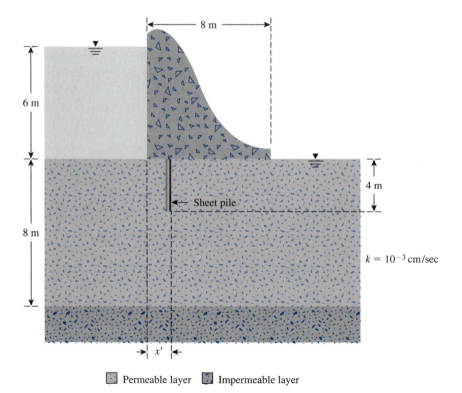

Figure 8.21

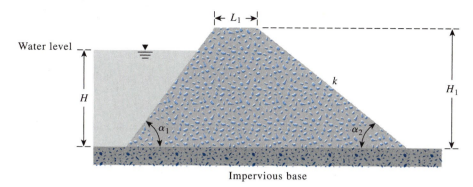

Figure 8.22

References

CASAGRANDE, L. (1932). "Naeherungsmethoden zur Bestimmurg von Art und Menge der Sickerung durch geschuettete Daemme," Thesis, Technische Hochschule, Vienna.

DUPUIT, J. (1863). *Etudes Theoriques et Practiques sur le Mouvement des Eaux dans les Canaux Decouverts et a Travers les Terrains Permeables,* Dunod, Paris.

GILBOY, G. (1934). "Mechanics of Hydraulic Fill Dams," in *Contributions to Soil Mechanics 1925–1940,* Boston Society of Civil Engineers, Boston.

HARR, M. E. (1962). *Ground Water and Seepage,* McGraw-Hill, New York.

SCHAFFERNAK, F. (1917). "Über die Standicherheit durchlaessiger geschuetteter Dämme," *Allgem. Bauzeitung.*

TERZAGHI, K., and PECK, R. B. (1948). *Soil Mechanics in Engineering Practice,* Wiley, New York.

U.S. DEPARTMENT OF THE NAVY, Naval Facilities Engineering Command (1971). "Design Manual—Soil Mechanics, Foundations, and Earth Structures," *NAVFAC DM-7,* Washington, D.C.

9 *In Situ* Stresses

As described in Chapter 3, soils are multiphase systems. In a given volume of soil, the solid particles are distributed randomly with void spaces between. The void spaces are continuous and are occupied by water and/or air. To analyze problems (such as compressibility of soils, bearing capacity of foundations, stability of embankments, and lateral pressure on earth-retaining structures), we need to know the nature of the distribution of stress along a given cross section of the soil profile. We can begin the analysis by considering a saturated soil with no seepage.

9.1 Stresses in Saturated Soil without Seepage

Figure 9.1a shows a column of saturated soil mass with no seepage of water in any direction. The total stress at the elevation of point A can be obtained from the saturated unit weight of the soil and the unit weight of water above it. Thus,

$$\sigma = H\gamma_w + (H_A - H)\gamma_{sat} \tag{9.1}$$

where σ = total stress at the elevation of point A
γ_w = unit weight of water
γ_{sat} = saturated unit weight of the soil
H = height of water table from the top of the soil column
H_A = distance between point A and the water table

The total stress, σ, given by Eq. (9.1) can be divided into two parts:

1. A portion is carried by water in the continuous void spaces. This portion acts with equal intensity in all directions.
2. The rest of the total stress is carried by the soil solids at their points of contact. The sum of the vertical components of the forces developed at the points of contact of the solid particles per unit cross-sectional area of the soil mass is called the *effective stress*.

This can be seen by drawing a wavy line, *a–a,* through point A that passes only through the points of contact of the solid particles. Let $P_1, P_2, P_3, \ldots, P_n$ be the forces that act at the points of contact of the soil particles (Figure 9.1b). The sum of the vertical

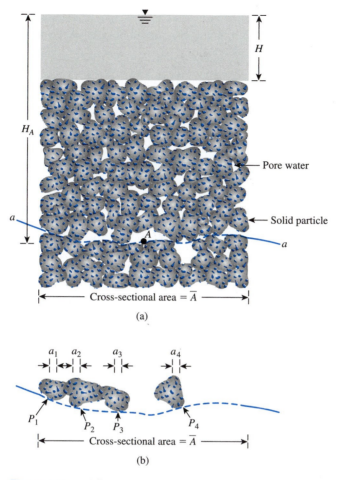

Figure 9.1 (a) Effective stress consideration for a saturated soil column without seepage; (b) forces acting at the points of contact of soil particles at the level of point A

components of all such forces over the unit cross-sectional area is equal to the effective stress σ', or

$$\sigma' = \frac{P_{1(v)} + P_{2(v)} + P_{3(v)} + \cdots + P_{n(v)}}{\overline{A}} \tag{9.2}$$

where $P_{1(v)}$, $P_{2(v)}$, $P_{3(v)}$, . . ., $P_{n(v)}$ are the vertical components of P_1, P_2, P_3, . . ., P_n, respectively, and $\overline{A}$ is the cross-sectional area of the soil mass under consideration.

Again, if a_s is the cross-sectional area occupied by solid-to-solid contacts (that is, $a_s = a_1 + a_2 + a_3 + \cdots + a_n$), then the space occupied by water equals $(\overline{A} - a_s)$. So we can write

$$\sigma = \sigma' + \frac{u(\overline{A} - a_s)}{\overline{A}} = \sigma' + u(1 - a_s') \tag{9.3}$$

where $u = H_A\gamma_w$ = pore water pressure (that is, the hydrostatic pressure at A)

$a'_s = a_s/\overline{A}$ = fraction of unit cross-sectional area of the soil mass occupied by solid-to-solid contacts

The value of a'_s is extremely small and can be neglected for pressure ranges generally encountered in practical problems. Thus, Eq. (9.3) can be approximated by

$$\sigma = \sigma' + u \qquad (9.4)$$

where u is also referred to as *neutral stress*. Substitution of Eq. (9.1) for σ in Eq. (9.4) gives

$$\sigma' = [H\gamma_w + (H_A - H)\gamma_{sat}] - H_A\gamma_w$$
$$= (H_A - H)(\gamma_{sat} - \gamma_w)$$
$$= (\text{Height of the soil column}) \times \gamma' \qquad (9.5)$$

where $\gamma' = \gamma_{sat} - \gamma_w$ equals the submerged unit weight of soil. Thus, we can see that the effective stress at any point A is independent of the depth of water, H, above the submerged soil.

Figure 9.2a shows a layer of submerged soil in a tank where there is no seepage. Figures 9.2b through 9.2d show plots of the variations of the total stress, pore water pressure, and effective stress, respectively, with depth for a submerged layer of soil placed in a tank with no seepage.

The principle of effective stress [Eq. (9.4)] was first developed by Terzaghi (1925, 1936). Skempton (1960) extended the work of Terzaghi and proposed the relationship between total and effective stress in the form of Eq. (9.3).

In summary, effective stress is approximately the force per unit area carried by the soil skeleton. The effective stress in a soil mass controls its volume change and strength. Increasing the effective stress induces soil to move into a denser state of packing.

The effective stress principle is probably the most important concept in geotechnical engineering. The compressibility and shearing resistance of a soil depend to a great extent on the effective stress. Thus, the concept of effective stress is significant in solving geotechnical engineering problems, such as the lateral earth pressure on retaining structures, the load-bearing capacity and settlement of foundations, and the stability of earth slopes.

In Eq. (9.2), the effective stress, σ', is defined as the sum of the vertical components of all intergranular *contact* forces over a unit gross cross-sectional area. This definition is mostly true for granular soils; however, for fine-grained soils, intergranular contact may not physically be there, because the clay particles are surrounded by tightly held water film. In a more general sense, Eq. (9.3) can be rewritten as

$$\sigma = \sigma_{ig} + u(1 - a'_s) - A' + R' \qquad (9.6)$$

where σ_{ig} = intergranular stress

A' = electrical attractive force per unit cross-sectional area of soil

R' = electrical repulsive force per unit cross-sectional area of soil

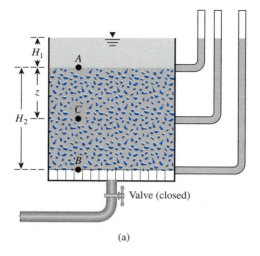

(a)

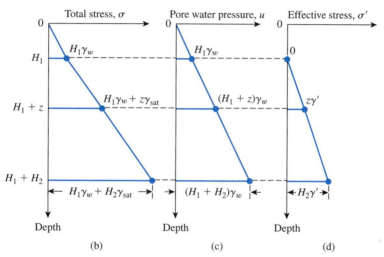

(b)   (c)   (d)

Figure 9.2 (a) Layer of soil in a tank where there is no seepage; Variation of (b) total stress, (c) pore water pressure, and (d) effective stress with depth for a submerged soil layer without seepage

For granular soils, silts, and clays of low plasticity, the magnitudes of A' and R' are small. Hence, for all practical purposes,

$$\sigma_{ig} = \sigma' \approx \sigma - u$$

However, if $A' - R'$ is large, then $\sigma_{ig} \neq \sigma'$. Such situations can be encountered in highly plastic, dispersed clay. Many interpretations have been made in the past to distinguish between the intergranular stress and effective stress. In any case, the effective stress principle is an excellent approximation used in solving engineering problems.

Example 9.1

A soil profile is shown in Figure 9.3. Calculate the total stress, pore water pressure, and effective stress at points *A, B,* and *C.*

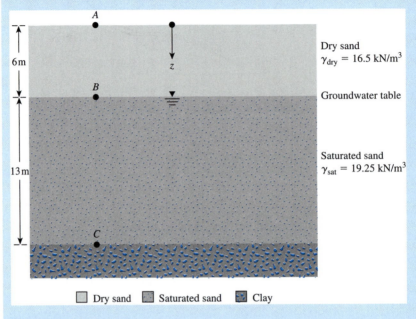

Figure 9.3 Soil profile

Solution
At Point *A,*

$$\text{Total stress: } \sigma_A = 0$$
$$\text{Pore water pressure: } u_A = 0$$
$$\text{Effective stress: } \sigma'_A = 0$$

At Point *B,*

$$\sigma_B = 6\gamma_{\text{dry(sand)}} = 6 \times 16.5 = 99 \text{ kN/m}^2$$
$$u_B = 0 \text{ kN/m}^2$$
$$\sigma'_B = 99 - 0 = 99 \text{ kN/m}^2$$

At Point *C,*

$$\sigma_C = 6\gamma_{\text{dry(sand)}} + 13\gamma_{\text{sat(clay)}}$$
$$= 6 \times 16.5 + 13 \times 19.25$$
$$= 99 + 250.25 = 349.25 \text{ kN/m}^2$$
$$u_C = 13\gamma_w = 13 \times 9.81 = 127.53 \text{ kN/m}^2$$
$$\sigma'_C = 349.25 - 127.53 = 221.72 \text{ kN/m}^2$$

Stresses in Saturated Soil with Upward Seepage

If water is seeping, the effective stress at any point in a soil mass will differ from that in the static case. It will increase or decrease, depending on the direction of seepage.

Figure 9.4a shows a layer of granular soil in a tank where upward seepage is caused by adding water through the valve at the bottom of the tank. The rate of water supply is kept constant. The loss of head caused by upward seepage between the levels of A and B is h. Keeping in mind that the total stress at any point in the soil mass is due solely to the

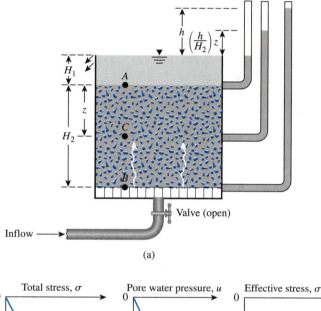

(a)

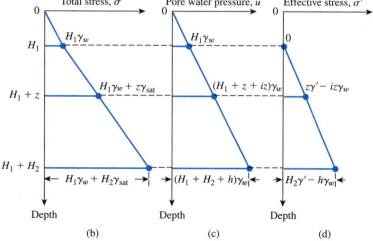

(b) (c) (d)

Figure 9.4 (a) Layer of soil in a tank with upward seepage. Variation of (b) total stress; (c) pore water pressure; and (d) effective stress with depth for a soil layer with upward seepage

weight of soil and water above it, we find that the effective stress calculations at points A and B are as follows:

At A,

- Total stress: $\sigma_A = H_1 \gamma_w$
- Pore water pressure: $u_A = H_1 \gamma_w$
- Effective stress: $\sigma'_A = \sigma_A - u_A = 0$

At B,

- Total stress: $\sigma_B = H_1 \gamma_w + H_2 \gamma_{\text{sat}}$
- Pore water pressure: $u_B = (H_1 + H_2 + h) \gamma_w$
- Effective stress: $\sigma'_B = \sigma_B - u_B$
$$= H_2(\gamma_{\text{sat}} - \gamma_w) - h\gamma_w$$
$$= H_2 \gamma' - h\gamma_w$$

Similarly, the effective stress at a point C located at a depth z below the top of the soil surface can be calculated as follows:

At C,

- Total stress: $\sigma_C = H_1 \gamma_w + z\gamma_{\text{sat}}$
- Pore water pressure: $u_C = \left(H_1 + z + \dfrac{h}{H_2} z \right) \gamma_w$
- Effective stress: $\sigma'_C = \sigma_C - u_C$
$$= z(\gamma_{\text{sat}} - \gamma_w) - \frac{h}{H_2} z\gamma_w$$
$$= z\gamma' - \frac{h}{H_2} z\gamma_w$$

Note that h/H_2 is the hydraulic gradient i caused by the flow, and therefore,

$$\sigma'_C = z\gamma' - iz\gamma_w \tag{9.7}$$

The variations of total stress, pore water pressure, and effective stress with depth are plotted in Figures 9.4b through 9.4d, respectively. A comparison of Figures 9.2d and 9.3d shows that the effective stress at a point located at a depth z measured from the surface of a soil layer is reduced by an amount $iz\gamma_w$ because of upward seepage of water. If the rate of seepage and thereby the hydraulic gradient gradually are increased, a limiting condition will be reached, at which point

$$\sigma'_C = z\gamma' - i_{\text{cr}} z\gamma_w = 0 \tag{9.8}$$

where i_{cr} = critical hydraulic gradient (for zero effective stress).

Under such a situation, soil stability is lost. This situation generally is referred to as *boiling,* or a *quick condition.*

From Eq. (9.8),

$$i_{\text{cr}} = \frac{\gamma'}{\gamma_w} \tag{9.9}$$

For most soils, the value of i_{cr} varies from 0.9 to 1.1, with an average of 1.

Example 9.2

A 6 m thick layer of stiff saturated clay is underlain by a layer of sand (Figure 9.5). The sand is under artesian pressure. Calculate the maximum depth of cut H that can be made in the clay.

Solution

Due to excavation, there will be unloading of the overburden pressure. Let the depth of the cut be H, at which point the bottom will heave. Let us consider the stability of point A at that time:

$$\sigma_A = (6 - H)\gamma_{sat(clay)}$$
$$u_A = 3.66$$

For heave to occur, σ'_A should be 0. So

$$\sigma_A - u_A = (6 - H)\gamma_{sat(clay)} - 3.66\gamma_w$$

or

$$(6 - H)18.87 - (3.66)9.81 = 0$$

$$H = \frac{(6)18.87 - (3.66)9.81}{18.87} = \textbf{4.62 m}$$

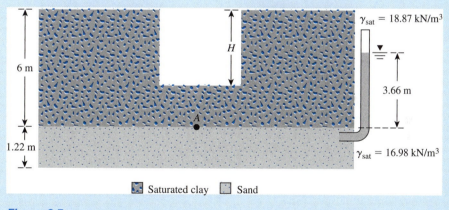

Saturated clay   Sand

Figure 9.5

Stresses in Saturated Soil with Downward Seepage

The condition of downward seepage is shown in Figure 9.6a. The water level in the soil tank is held constant by adjusting the supply from the top and the outflow at the bottom.

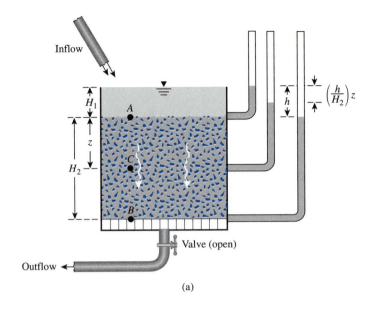

(a)

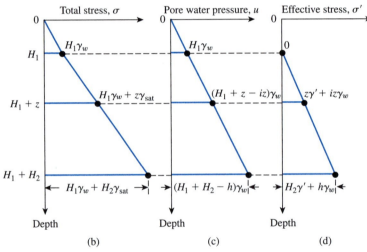

(b)   (c)   (d)

Figure 9.6
(a) Layer of soil in
a tank with down-
ward seepage;
variation of
(b) total stress;
(c) pore water
pressure; (d) effec-
tive stress with
depth for a soil
layer with down-
ward seepage

The hydraulic gradient caused by the downward seepage equals $i = h/H_2$. The total stress, pore water pressure, and effective stress at any point C are, respectively,

$$\sigma_C = H_1\gamma_w + z\gamma_{sat}$$

$$u_C = (H_1 + z - iz)\gamma_w$$

$$\sigma'_C = (H_1\gamma_w + z\gamma_{sat}) - (H_1 + z - iz)\gamma_w$$

$$= z\gamma' + iz\gamma_w$$

The variations of total stress, pore water pressure, and effective stress with depth also are shown graphically in Figures 9.6b through 9.6d.

9.4 Seepage Force

The preceding section showed that the effect of seepage is to increase or decrease the effective stress at a point in a layer of soil. Often, expressing the seepage force per unit volume of soil is convenient.

In Figure 9.2, it was shown that, with no seepage, the effective stress at a depth z measured from the surface of the soil layer in the tank is equal to $z\gamma'$. Thus, the effective force on an area A is

$$P_1' = z\gamma' A \tag{9.10}$$

(The direction of the force P_1' is shown in Figure 9.7a.)

Again, if there is an upward seepage of water in the vertical direction through the same soil layer (Figure 9.4), the effective force on an area A at a depth z can be given by

$$P_2' = (z\gamma' - iz\gamma_w)A \tag{9.11}$$

Hence, the decrease in the total force because of seepage is

$$P_1' - P_2' = iz\gamma_w A \tag{9.12}$$

The volume of the soil contributing to the effective force equals zA, so the seepage force per unit volume of soil is

$$\frac{P_1' - P_2'}{(\text{Volume of soil})} = \frac{iz\gamma_w A}{zA} = i\gamma_w \tag{9.13}$$

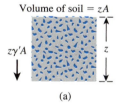

Volume of soil = zA

$z\gamma'A$

z

(a)

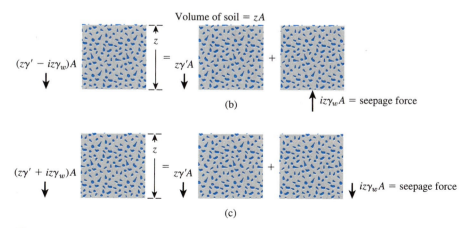

Volume of soil = zA

$(z\gamma' - iz\gamma_w)A$ = $z\gamma'A$ + $i z\gamma_w A$ = seepage force

z

(b)

$(z\gamma' + iz\gamma_w)A$ = $z\gamma'A$ + $iz\gamma_w A$ = seepage force

z

(c)

Figure 9.7 Force due to (a) no seepage; (b) upward seepage; (c) downward seepage on a volume of soil

The force per unit volume, $i\gamma_w$, for this case acts in the upward direction—that is, in the direction of flow. This upward force is demonstrated in Figure 9.7b. Similarly, for downward seepage, it can be shown that the seepage force in the downward direction per unit volume of soil is $i\gamma_w$ (Figure 9.7c).

From the preceding discussions, we can conclude that the seepage force per unit volume of soil is equal to $i\gamma_w$, and in isotropic soils the force acts in the same direction as the direction of flow. This statement is true for flow in any direction. Flow nets can be used to find the hydraulic gradient at any point and, thus, the seepage force per unit volume of soil.

Example 9.3

Consider the upward flow of water through a layer of sand in a tank as shown in Figure 9.8. For the sand, the following are given: void ratio $(e) = 0.52$ and specific gravity of solids $= 2.67$.

   a. Calculate the total stress, pore water pressure, and effective stress at points A and B.
   b. What is the upward seepage force per unit volume of soil?

Solution
Part a
The saturated unit weight of sand is calculated as follows:

$$\gamma_{sat} = \frac{(G_s + e)\gamma_w}{1 + e} = \frac{(2.67 + 0.52)9.81}{1 + 0.52} = 20.59 \text{ kN/m}^3$$

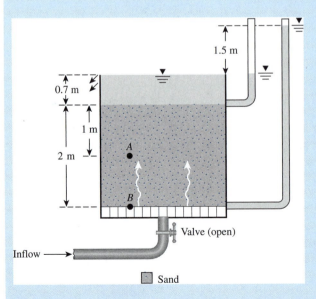

Figure 9.8 Upward flow of water through a layer of sand in a tank

Now, the following table can be prepared:

Point	Total stress, σ (kN/m²)	Pore water pressure, u (kN/m²)	Effective stress, $\sigma' = \sigma - u$ (kN/m²)
A	$0.7\gamma_w + 1\gamma_{sat} = (0.7)(9.81)$ $+ (1)(20.59) = \textbf{27.46}$	$\left[(1 + 0.7) + \left(\dfrac{1.5}{2}\right)(1)\right]\gamma_w$ $= (2.45)(9.81) = \textbf{24.03}$	**3.43**
B	$0.7\gamma_w + 2\gamma_{sat} = (0.7)(9.81)$ $+ (2)(20.59) = \textbf{48.05}$	$(2 + 0.7 + 1.5)\gamma_w$ $= (4.2)(9.81) = \textbf{41.2}$	**6.85**

Part b

Hydraulic gradient (i) = 1.5/2 = 0.75. Thus, the seepage force per unit volume can be calculated as

$$i\gamma_w = (0.75)(9.81) = \textbf{7.36} \textbf{ kN/m}^3$$ ∎

9.5 Heaving in Soil Due to Flow Around Sheet Piles

Seepage force per unit volume of soil can be used for checking possible failure of sheet-pile structures where underground seepage may cause heaving of soil on the downstream side (Figure 9.9a). After conducting several model tests, Terzaghi (1922) concluded that heaving generally occurs within a distance of $D/2$ from the sheet piles (when D equals depth of embedment of sheet piles into the permeable layer). Therefore, we need to

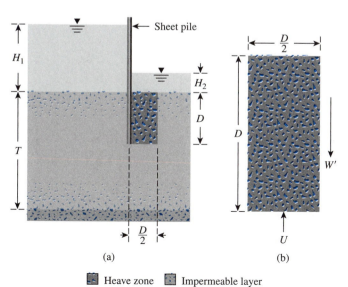

(a)

(b)

■ Heave zone   ▨ Impermeable layer

Figure 9.9
(a) Check for heaving on the downstream side for a row of sheet piles driven into a permeable layer; (b) enlargement of heave zone

investigate the stability of soil in a zone measuring D by $D/2$ in cross-section as shown in Figure 9.9b.

The factor of safety against heaving can be given by

$$FS = \frac{W'}{U} \tag{9.14}$$

where FS = factor of safety

W' = submerged weight of soil in the heave zone per unit length of sheet pile = $D(D/2)(\gamma_{sat} - \gamma_w) = (\frac{1}{2})D^2\gamma'$

U = uplifting force caused by seepage on the same volume of soil

From Eq. (9.13),

$$U = (\text{Soil volume}) \times (i_{av}\gamma_w) = \frac{1}{2}D^2 i_{av}\gamma_w$$

where i_{av} = average hydraulic gradient at the bottom of the block of soil (see Example 9.4). Substituting the values of W' and U in Eq. (9.14), we can write

$$FS = \frac{\gamma'}{i_{av}\gamma_w} \tag{9.15}$$

For the case of *flow around a sheet pile* in a *homogeneous soil,* as shown in Figure 9.9, it can be demonstrated that

$$\frac{U}{0.5\gamma_w D(H_1 - H_2)} = C_o \tag{9.16}$$

where C_o is a function of D/T (see Table 9.1). Hence, from Eq. (9.14),

$$FS = \frac{W'}{U} = \frac{0.5D^2\gamma'}{0.5C_o\gamma_w D(H_1 - H_2)} = \frac{D\gamma'}{C_o\gamma_w(H_1 - H_2)} \tag{9.17}$$

Table 9.1 Variation of C_o with D/T

D/T	C_o
0.1	0.385
0.2	0.365
0.3	0.359
0.4	0.353
0.5	0.347
0.6	0.339
0.7	0.327
0.8	0.309
0.9	0.274

Example 9.4

Figure 9.10 shows the flow net for seepage of water around a single row of sheet piles driven into a permeable layer. Calculate the factor of safety against downstream heave, given that γ_{sat} for the permeable layer $= 17.6$ kN/m^3. (*Note:* thickness of permeable layer $T = 18.3$ m)

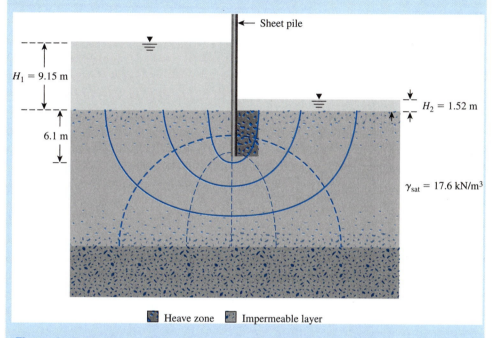

Sheet pile

$H_1 = 9.15$ m

6.1 m

$H_2 = 1.52$ m

$\gamma_{sat} = 17.6$ kN/m^3

Heave zone   Impermeable layer

Figure 9.10 Flow net for seepage of water around sheet piles driven into permeable layer

Solution

From the dimensions given in Figure 9.10, the soil prism to be considered is 6.1 m $\times$ 3.05 m in cross section.

The soil prism is drawn to an enlarged scale in Figure 9.11. By use of the flow net, we can calculate the head loss through the prism as follows:

- At b, the driving head $= \frac{3}{6}(H_1 - H_2)$

- At c, the driving head $\approx \frac{1.6}{6}(H_1 - H_2)$

Similarly, for other intermediate points along bc, the approximate driving heads have been calculated and are shown in Figure 9.11.

The average value of the head loss in the prism is $0.36(H_1 - H_2)$, and the average hydraulic gradient is

$$i_{av} = \frac{0.36(H_1 - H_2)}{D}$$

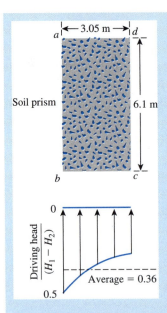

Figure 9.11 Soil prism—enlarged scale

Thus, the factor of safety [Eq. (9.15)] is

$$FS = \frac{\gamma'}{i_{av}\gamma_w} = \frac{\gamma'D}{0.36(H_1 - H_2)\gamma_w} = \frac{(17.6 - 9.81)\,6.1}{0.36(9.15 - 1.52) \times 9.81} = \mathbf{1.76}$$

Alternate Solution
For this case, $D/T = 1/3$. From Table 9.1, for $D/T = 1/3$, the value of $C_o = 0.357$. Thus, from Eq. (9.17),

$$FS = \frac{D\gamma'}{C_o\gamma_w(H_1 - H_2)} = \frac{(6.1)(17.6 - 9.81)}{(0.357)(9.81)(9.15 - 1.52)} = \mathbf{1.78} \quad \blacksquare$$

9.6   *Use of Filters to Increase the Factor of Safety Against Heave*

The factor of safety against heave as calculated in Example 9.4 is low. In practice, a minimum factor of safety of about 4 to 5 is required for the safety of the structure. Such a high factor of safety is recommended primarily because of the inaccuracies inherent in the analysis. One way to increase the factor of safety against heave is to use a *filter* in the downstream side of the sheet-pile structure (Figure 9.12a). A filter is a granular material with openings small enough to prevent the movement of the soil particles upon which it is placed and, at the same time, is pervious enough to offer little resistance to seepage through it (see Section 8.10). In Figure 9.12a, the thickness of the filter material is D_1. In this case, the factor of safety against heave can be calculated as follows (Figure 9.12b).

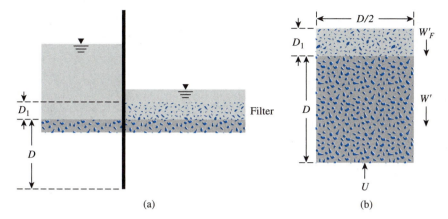

Figure 9.12 Factor of safety against heave, with a filter

The submerged weight of the soil and the filter in the heave zone per unit length of sheet pile $= W' + W'_F$, where

$$W' = (D)\left(\frac{D}{2}\right)(\gamma_{sat} - \gamma_w) = \frac{1}{2}D^2\gamma'$$

$$W'_F = (D_1)\left(\frac{D}{2}\right)(\gamma'_F) = \frac{1}{2}D_1 D\gamma'_F$$

in which γ'_F = effective unit weight of the filter.

The uplifting force caused by seepage on the same volume of soil is given by

$$U = \tfrac{1}{2}D^2 i_{av}\gamma_w$$

The preceding relationship was derived in Section 9.5.

The factor of safety against heave is thus

$$FS = \frac{W' + W'_F}{U} = \frac{\dfrac{1}{2}D^2\gamma' + \dfrac{1}{2}D_1 D\gamma'_F}{\dfrac{1}{2}D^2 i_{av}\gamma_w} = \frac{\gamma' + \left(\dfrac{D_1}{D}\right)\gamma'_F}{i_{av}\gamma_w} \qquad (9.18)$$

The principles for selection of filter materials were given in Section 8.10.

If Eq. (9.16) is used,

$$FS = \frac{\dfrac{1}{2}D^2\gamma' + \dfrac{1}{2}D_1 D\gamma'_F}{0.5C_o\gamma_w D(H_1 - H_2)} = \frac{D\gamma' + D_1\gamma'_F}{C_o\gamma_w(H_1 - H_2)} \qquad (9.19)$$

The values of C_o are given in Table 9.1.

9.7   *Effective Stress in Partially Saturated Soil*

In partially saturated soil, water in the void spaces is not continuous, and it is a three-phase system—that is, solid, pore water, and pore air (Figure 9.13). Hence, the total stress at any point in a soil profile consists of intergranular, pore air, and pore water pressures. From laboratory test results, Bishop, *et al.* (1960) gave the following equation for effective stress in partially saturated soils

$$\sigma' = \sigma - u_a + \chi(u_a - u_w)$$ (9.20)

where σ' = effective stress
σ = total stress
u_a = pore air pressure
u_w = pore water pressure

In Eq. (9.20), χ represents the fraction of a unit cross-sectional area of the soil occupied by water. For dry soil $\chi = 0$, and for saturated soil $\chi = 1$.

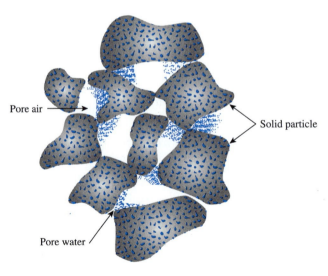

Pore air

Solid particle

Pore water

Figure 9.13 Partially saturated soil

Bishop, *et al.* have pointed out that the intermediate values of χ will depend primarily on the degree of saturation S. However, these values also will be influenced by factors such as soil structure. The nature of variation of χ with the degree of saturation for a silt is shown in Figure 9.14.

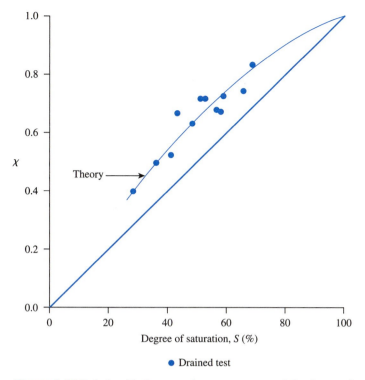

• Drained test

Figure 9.14 Relationship between the parameter χ and the degree of saturation for Bearhead silt *(After Bishop et al, 1960. With permission from ASCE.)*

9.8 Capillary Rise in Soils

The continuous void spaces in soil can behave as bundles of capillary tubes of variable cross section. Because of surface tension force, water may rise above the phreatic surface.

Figure 9.15 shows the fundamental concept of the height of rise in a capillary tube. The height of rise of water in the capillary tube can be given by summing the forces in the vertical direction, or

$$\left(\frac{\pi}{4}d^2\right)h_c\gamma_w = \pi dT \cos \alpha$$

$$h_c = \frac{4T \cos \alpha}{d\gamma_w} \tag{9.21}$$

where T = surface tension (force/length)
α = angle of contact
d = diameter of capillary tube
γ_w = unit weight of water

For pure water and clean glass, $\alpha = 0$. Thus, Eq. (9.21) becomes

$$h_c = \frac{4T}{d\gamma_w} \tag{9.22}$$

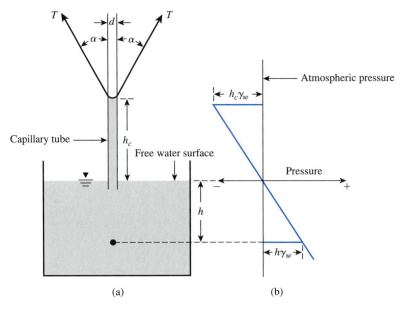

Figure 9.15 (a) Rise of water in the capillary tube; (b) pressure within the height of rise in the capillary tube (atmospheric pressure taken as datum)

For water, $T = 72$ mN/m. From Eq. (9.22), we see that the height of capillary rise

$$h_c \propto \frac{1}{d} \qquad (9.23)$$

Thus, the smaller the capillary tube diameter, the larger the capillary rise.

Although the concept of capillary rise as demonstrated for an ideal capillary tube can be applied to soils, one must realize that the capillary tubes formed in soils because of the continuity of voids have variable cross sections. The results of the nonuniformity on capillary rise can be seen when a dry column of sandy soil is placed in contact with water (Figure 9.16). After the lapse of a given amount of time, the variation of the degree of saturation with the height of the soil column caused by capillary rise is approximately as shown in Figure 9.16b. The degree of saturation is about 100% up to a height of h_2, and this corresponds to the largest voids. Beyond the height h_2, water can occupy only the smaller voids; hence, the degree of saturation is less than 100%. The maximum height of capillary rise corresponds to the smallest voids. Hazen (1930) gave a formula for the approximation of the height of capillary rise in the form.

$$h_1 \,(\text{mm}) = \frac{C}{e D_{10}} \qquad (9.24)$$

where D_{10} = effective size (mm)
      e = void ratio
      C = a constant that varies from 10 to 50 mm^2

Equation (9.24) has an approach similar to that of Eq. (9.23). With the decrease of D_{10}, the pore size in soil decreases, which causes higher capillary rise. Table 9.2 shows the approximate range of capillary rise that is encountered in various types of soils.

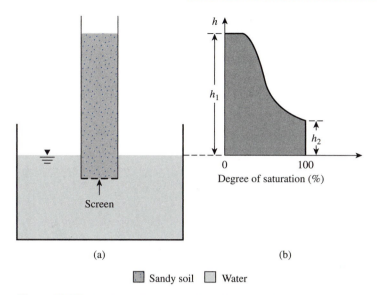

Sandy soil ▢ Water

Figure 9.16 Capillary effect in sandy soil: (a) a soil column in contact with water; (b) variation of degree of saturation in the soil column

Table 9.2 Approximate Range of Capillary Rise in Soils

Soil type	Range of capillary rise
	m
Coarse sand	0.1–0.2
Fine sand	0.3–1.2
Silt	0.75–7.5
Clay	7.5–23

Capillary rise is important in the formation of some types of soils such as *caliche,* which can be found in the desert Southwest of the United States. Caliche is a mixture of sand, silt, and gravel bonded by calcareous deposits. These deposits are brought to the surface by a net upward migration of water by capillary action. The water evaporates in the high local temperature. Because of sparse rainfall, the carbonates are not washed out of the top soil layer.

9.9   *Effective Stress in the Zone of Capillary Rise*

The general relationship among total stress, effective stress, and pore water pressure was given in Eq. (9.4) as

$$\sigma = \sigma' + u$$

The pore water pressure u at a point in a layer of soil fully saturated by capillary rise is equal to $-\gamma_w h$ (h = height of the point under consideration measured from the groundwater table) with the atmospheric pressure taken as datum. If partial saturation is caused by capillary action, it can be approximated as

$$u = -\left(\frac{S}{100}\right)\gamma_w h \qquad (9.25)$$

where S = degree of saturation, in percent.

Example 9.5

A soil profile is shown in Figure 9.17. Given: $H_1 = 1.83$ m, $H_2 = 0.91$ m, $H_3 = 1.83$ m. Plot the variation of σ, u, and σ' with depth.

Solution
Determination of Unit Weight

Dry sand:

$$\gamma_{d(sand)} = \frac{G_s \gamma_w}{1 + e} = \frac{(2.65)(9.81)}{1 + 0.5} = 17.33 \text{ kN/m}^3$$

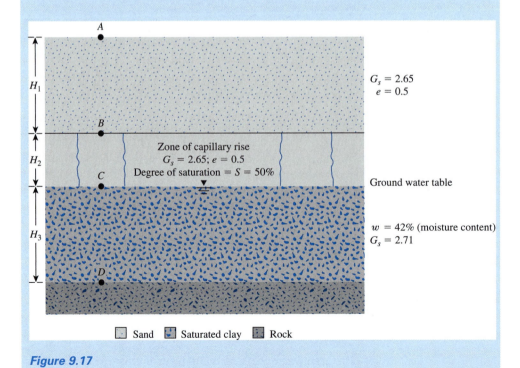

Sand   Saturated clay   Rock

Figure 9.17

Moist sand:

$$\gamma_{\text{sand}} = \frac{(G_s + Se)\gamma_w}{1 + e} = \frac{[2.65 + (0.5)(0.5)]9.81}{1 + 0.5} = 18.97 \text{ kN/m}^3$$

Saturated clay:

$$e = \frac{G_s w}{S} = \frac{(2.71)(0.42)}{1.0} = 1.1382$$

$$\gamma_{\text{sat(clay)}} = \frac{(G_s + e)\gamma_w}{1 + e} = \frac{(2.71 + 1.1382)9.81}{1 + 1.1382} = 17.66 \text{ kN/m}^3$$

Calculation of Stress

At the ground surface (i.e., point A):

$$\sigma = 0$$
$$u = 0$$
$$\sigma' = \sigma - u = 0$$

At depth H_1 (i.e., point B):

$$\sigma = \gamma_{d(\text{sand})}(1.83) = (17.33)(1.83) = \textbf{31.71 kN/m}^2$$

$u = \textbf{0}$ (immediately above)

$u = -(S\gamma_w H_2) = -(0.5)(9.81)(0.91) = \textbf{−4.46 kN/m}^2$ (immediately below)

$\sigma' = 31.71 - 0 = \textbf{31.71 kN/m}^2$ (immediately above)

$\sigma' = 31.71 - (-4.46) = \textbf{36.17 kN/m}^2$ (immediately below)

At depth $H_1 + H_2$ (i.e., at point C):

$$\sigma = (17.33)(1.83) + (18.97)(0.91) = \textbf{48.97 kN/m}^2$$

$$u = \textbf{0}$$

$$\sigma' = 48.97 - 0 = \textbf{48.97 kN/m}^2$$

At depth $H_1 + H_2 + H_3$ (i.e., at point D):

$$\sigma = 48.97 + (17.66)(1.83) = \textbf{81.29 kN/m}^2$$

$$u = 1.83\gamma_w = (1.83)(9.81) = \textbf{17.95 kN/m}^2$$

$$\sigma' = 81.29 - 17.95 = \textbf{63.34 kN/m}^2$$

The plot of the stress variation with depth is shown in Figure 9.18.

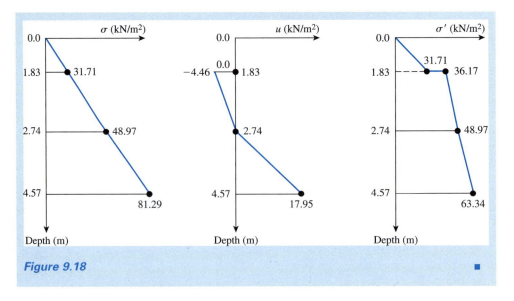

Figure 9.18

9.10   *Summary and General Comments*

The effective stress principle is probably the most important concept in geotechnical engineering. The compressibility and shearing resistance of a soil depend to a great extent on the effective stress. Thus, the concept of effective stress is significant in solving geotechnical engineering problems, such as the lateral earth pressure on retaining structures, the load-bearing capacity and settlement of foundations, and the stability of earth slopes.

In Eq. (9.2), the effective stress σ' is defined as the sum of the vertical components of all intergranular *contact* forces over a unit gross cross-sectional area. This definition is mostly true for granular soils; however, for fine-grained soils, intergranular contact may not be there physically, because the clay particles are surrounded by tightly held water film. In a more general sense, Eq. (9.3) can be rewritten as

$$\sigma = \sigma_{ig} + u(1 - a_s') - A' + R' \tag{9.26}$$

where σ_{ig} = intergranular stress
$\quad\quad A'$ = electrical attractive force per unit cross-sectional area of soil
$\quad\quad R'$ = electrical repulsive force per unit cross-sectional area of soil

For granular soils, silts, and clays of low plasticity, the magnitudes of A' and R' are small. Hence, for all practical purposes,

$$\sigma_{ig} = \sigma' \approx \sigma - u$$

However, if $A' - R'$ is large, then $\sigma_{ig} \neq \sigma'$. Such situations can be encountered in highly plastic, dispersed clay. Many interpretations have been made in the past to distinguish between the intergranular stress and effective stress. In any case, the effective stress principle is an excellent approximation used in solving engineering problems.

Problems

9.1 through 9.5   Refer to Figure 9.19. Calculate σ, u, and σ' at A, B, C, and D for the following cases and plot the variations with depth. (*Note: e* = void ratio, w = moisture content, G_s = specific gravity of soil solids, γ_d = dry unit weight, and γ_{sat} = saturated unit weight.)

Problem	I	II	III
	Details of soil layer		
9.1	$H_1 = 1.5$ m $\gamma_d = 17.6$ kN/m³	$H_2 = 1.83$ m $\gamma_{sat} = 18.87$ kN/m³	$H_3 = 2.44$ m $\gamma_{sat} = 19.65$ kN/m³
9.2	$H_1 = 1.5$ m $\gamma_d = 15.72$ kN/m³	$H_2 = 3.05$ m $\gamma_{sat} = 18.24$ kN/m³	$H_3 = 2.74$ m $\gamma_{sat} = 19.18$ kN/m³
9.3	$H_1 = 3$ m $\gamma_d = 15$ kN/m³	$H_2 = 4$ m $\gamma_{sat} = 16$ kN/m³	$H_3 = 5$ m $\gamma_{sat} = 18$ kN/m³
9.4	$H_1 = 4$ m $e = 0.4$ $G_s = 2.62$	$H_2 = 5$ m $e = 0.6$ $G_s = 2.68$	$H_3 = 3$ m $e = 0.81$ $G_s = 2.73$
9.5	$H_1 = 4$ m $e = 0.6$ $G_s = 2.65$	$H_2 = 3$ m $e = 0.52$ $G_s = 2.68$	$H_3 = 1.5$ m $w = 40\%$ $e = 1.1$

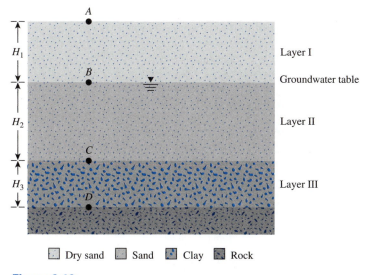

Figure 9.19

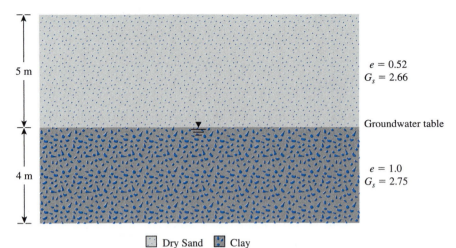

Dry Sand   Clay

Figure 9.20

9.6 Refer to the soil profile shown in Figure 9.20.
  a. Calculate the variation of σ, u, and σ' with depth.
  b. If the water table rises to the top of the ground surface, what is the change in the effective stress at the bottom of the clay layer?
  c. How many meters must the groundwater table rise to decrease the effective stress by 15 kN/m^2 at the bottom of the clay layer?

9.7 An exploratory drill hole was made in a stiff saturated clay (see Figure 9.21). The sand layer underlying the clay was observed to be under artesian pressure. Water in the drill hole rose to a height of 3.66 m above the top of the sand layer. If an open excavation is to be made in the clay, how deep can the excavation proceed before the bottom heaves?

9.8 A cut is made in a stiff saturated clay that is underlain by a layer of sand. (See Figure 9.22.) What should be the height of the water, h, in the cut so that the stability of the saturated clay is not lost?

9.9 Refer to Figure 9.4a. If $H_1 = 1.5$ m, $H_2 = 2.5$ m, $h = 1.5$ m, $\gamma_{\text{sat}} = 18.6$ kN/m^3, hydraulic conductivity of sand (k) = 0.12 cm/sec, and area of tank = 0.45 m^2, what is the rate of upward seepage of water (m^3/min)?

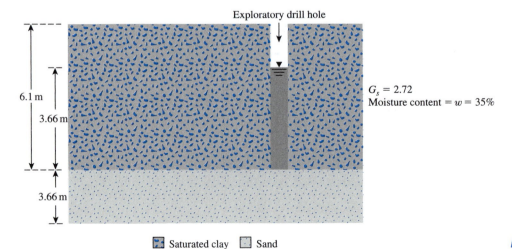

Saturated clay   Sand

Figure 9.21

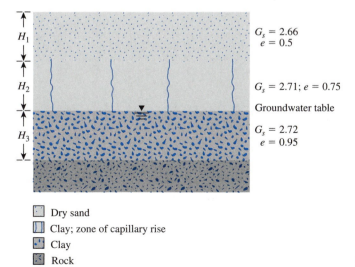

Figure 9.22

9.10 Refer to Figure 9.4a. Given: $H_1 = 1$ m, $H_2 = 2$ m, $h = 1.2$ m, void ratio of sand $(e) = 0.55$, specific gravity of soil solids $(G_s) = 2.68$, area of the tank $= 0.5$ m^2, and hydraulic conductivity of sand $= 0.1$ cm/sec.
   a. What is the rate of upward seepage?
   b. If $h = 1.2$ m, will boiling occur? Why?
   c. What should be the value of h to cause boiling?

9.11 Find the factor of safety against heave on the downstream side of the single-row sheet pile structure shown in Figure 8.7. (*Note:* Thickness of permeable layer $= 10$ m and depth of penetration of sheet pile $= 6$ m.) Assume $\gamma_{sat} = 19$ kN/m^3.

9.12 through 9.13 Refer to Figure 9.23. For the following variables, calculate and plot σ, u, and σ' with depth.

Problem	H_1	H_2	H_3	Degree of saturation in capillary rise zone, S (%)
9.12	2.44 m	1.22 m	2.13 m	
9.13	5 m	3 m	3.5 m	65

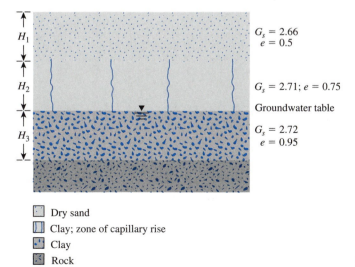

☐ Dry sand
▥ Clay; zone of capillary rise
▨ Clay
▧ Rock

Figure 9.23

References

BISHOP, A. W., ALPAN, I., BLIGHT, G. E., and DONALD, I. B. (1960). "Factors Controlling the Strength of Partially Saturated Cohesive Soils," *Proceedings,* Research Conference on Shear Strength of Cohesive Soils, ASCE, 502–532.

HAZEN, A. (1930). "Water Supply," in *American Civil Engineering Handbook,* Wiley, New York.

SKEMPTON, A. W. (1960). "Correspondence," *Geotechnique,* Vol. 10, No. 4, 186.

TERZAGHI, K. (1922). "Der Grundbruch an Stauwerken und seine Verhütung," *Die Wasserkraft,* Vol. 17, 445–449.

TERZAGHI, K. (1925). *Erdbaumechanik auf Bodenphysikalischer Grundlage,* Dueticke, Vienna.

TERZAGHI, K. (1936). "Relation Between Soil Mechanics and Foundation Engineering: Presidential Address," *Proceedings,* First International Conference on Soil Mechanics and Foundation Engineering, Boston, Vol. 3, 13–18.

TERZAGHI, K., and PECK, R. B. (1948). *Soil Mechanics in Engineering Practice,* Wiley, New York.

10 | Stresses in a Soil Mass

Construction of a foundation causes changes in the stress, usually a net increase. The net stress increase in the soil depends on the load per unit area to which the foundation is subjected, the depth below the foundation at which the stress estimation is desired, and other factors. It is necessary to estimate the net increase of vertical stress in soil that occurs as a result of the construction of a foundation so that settlement can be calculated. The settlement calculation procedure is discussed in more detail in Chapter 11. This chapter discusses the principles of estimation of vertical stress increase in soil caused by various types of loading, based on the theory of elasticity. Although natural soil deposits, in most cases, are not fully elastic, isotropic, or homogeneous materials, calculations for estimating increases in vertical stress yield fairly good results for practical work.

10.1 Normal and Shear Stresses on a Plane

Students in a soil mechanics course are familiar with the fundamental principles of the mechanics of deformable solids. This section is a brief review of the basic concepts of normal and shear stresses on a plane that can be found in any course on the mechanics of materials.

Figure 10.1a shows a two-dimensional soil element that is being subjected to normal and shear stresses ($\sigma_y > \sigma_x$). To determine the normal stress and the shear stress on a plane EF that makes an angle θ with the plane AB, we need to consider the free body diagram of EFB shown in Figure 10.1b. Let σ_n and τ_n be the normal stress and the shear stress, respectively, on the plane EF. From geometry, we know that

$$\overline{EB} = \overline{EF} \cos \theta \tag{10.1}$$

and

$$\overline{FB} = \overline{EF} \sin \theta \tag{10.2}$$

Summing the components of forces that act on the element in the direction of N and T, we have

$$\sigma_n(\overline{EF}) = \sigma_x(\overline{EF}) \sin^2 \theta + \sigma_y(\overline{EF}) \cos^2 \theta + 2\tau_{xy}(\overline{EF}) \sin \theta \cos \theta$$

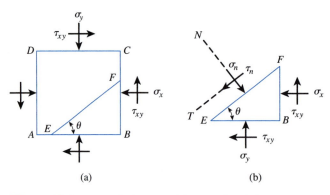

Figure 10.1 (a) A soil element with normal and shear stresses acting on it; (b) free body diagram of *EFB* as shown in (a)

or

$$\sigma_n = \sigma_x \sin^2 \theta + \sigma_y \cos^2 \theta + 2\tau_{xy} \sin \theta \cos \theta$$

or

$$\sigma_n = \frac{\sigma_y + \sigma_x}{2} + \frac{\sigma_y - \sigma_x}{2} \cos 2\theta + \tau_{xy} \sin 2\theta \qquad (10.3)$$

Again,

$$\tau_n(\overline{EF}) = -\sigma_x(\overline{EF}) \sin \theta \cos \theta + \sigma_y(\overline{EF}) \sin \theta \cos \theta$$
$$- \tau_{xy}(\overline{EF}) \cos^2 \theta + \tau_{xy}(\overline{EF}) \sin^2 \theta$$

or

$$\tau_n = \sigma_y \sin \theta \cos \theta - \sigma_x \sin \theta \cos \theta - \tau_{xy}(\cos^2 \theta - \sin^2 \theta)$$

or

$$\tau_n = \frac{\sigma_y - \sigma_x}{2} \sin 2\theta - \tau_{xy} \cos 2\theta \qquad (10.4)$$

From Eq. (10.4), we can see that we can choose the value of θ in such a way that τ_n will be equal to zero. Substituting $\tau_n = 0$, we get

$$\tan 2\theta = \frac{2\tau_{xy}}{\sigma_y - \sigma_x} \qquad (10.5)$$

For given values of τ_{xy}, σ_x, and σ_y, Eq. (10.5) will give two values of θ that are 90° apart. This means that there are two planes that are at right angles to each other on which the shear stress is zero. Such planes are called *principal planes*. The normal stresses that act on the principal planes are referred to as *principal stresses*. The values of principal stresses can be found by substituting Eq. (10.5) into Eq. (10.3), which yields

Major principal stress:

$$\sigma_n = \sigma_1 = \frac{\sigma_y + \sigma_x}{2} + \sqrt{\left[\frac{(\sigma_y - \sigma_x)}{2}\right]^2 + \tau_{xy}^2} \qquad (10.6)$$

Minor principal stress:

$$\sigma_n = \sigma_3 = \frac{\sigma_y + \sigma_x}{2} - \sqrt{\left[\frac{(\sigma_y - \sigma_x)}{2}\right]^2 + \tau_{xy}^2} \qquad (10.7)$$

The normal stress and shear stress that act on any plane can also be determined by plotting a Mohr's circle, as shown in Figure 10.2. The following sign conventions are used in Mohr's circles: compressive normal stresses are taken as positive, and shear stresses are considered positive if they act on opposite faces of the element in such a way that they tend to produce a counterclockwise rotation.

For plane *AD* of the soil element shown in Figure 10.1a, normal stress equals $+\sigma_x$ and shear stress equals $+\tau_{xy}$. For plane *AB*, normal stress equals $+\sigma_y$ and shear stress equals $-\tau_{xy}$.

The points *R* and *M* in Figure 10.2 represent the stress conditions on planes *AD* and *AB*, respectively. *O* is the point of intersection of the normal stress axis with the line *RM*. The circle *MNQRS* drawn with *O* as the center and *OR* as the radius is the

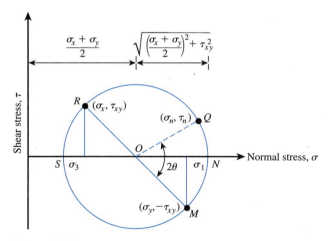

Figure 10.2 Principles of the Mohr's circle

Mohr's circle for the stress conditions considered. The radius of the Mohr's circle is equal to

$$\sqrt{\left[\frac{(\sigma_y - \sigma_x)}{2}\right]^2 + \tau_{xy}^2}$$

The stress on plane EF can be determined by moving an angle 2θ (which is twice the angle that the plane EF makes in a counterclockwise direction with plane AB in Figure 10.1a) in a counterclockwise direction from point M along the circumference of the Mohr's circle to reach point Q. The abscissa and ordinate of point Q, respectively, give the normal stress σ_n and the shear stress τ_n on plane EF.

Because the ordinates (that is, the shear stresses) of points N and S are zero, they represent the stresses on the principal planes. The abscissa of point N is equal to σ_1 [Eq. (10.6)], and the abscissa for point S is σ_3 [Eq. (10.7)].

As a special case, if the planes AB and AD were major and minor principal planes, the normal stress and the shear stress on plane EF could be found by substituting $\tau_{xy} = 0$. Equations (10.3) and (10.4) show that $\sigma_y = \sigma_1$ and $\sigma_x = \sigma_3$ (Figure 10.3a). Thus,

$$\sigma_n = \frac{\sigma_1 + \sigma_3}{2} + \frac{\sigma_1 - \sigma_3}{2} \cos 2\theta \tag{10.8}$$

$$\tau_n = \frac{\sigma_1 - \sigma_3}{2} \sin 2\theta \tag{10.9}$$

The Mohr's circle for such stress conditions is shown in Figure 10.3b. The abscissa and the ordinate of point Q give the normal stress and the shear stress, respectively, on the plane EF.

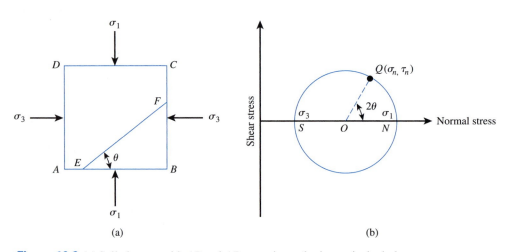

(a) (b)

Figure 10.3 (a) Soil element with AB and AD as major and minor principal planes; (b) Mohr's circle for soil element shown in (a)

Example 10.1

A soil element is shown in Figure 10.4. The magnitudes of stresses are $\sigma_x = 96$ kN/m^2, $\tau = 38$ kN/m^2, $\sigma_y = 120$ kN/m^2, and $\theta = 20°$. Determine

    a. Magnitudes of the principal stresses
    b. Normal and shear stresses on plane *AB*. Use Eqs. (10.3), (10.4), (10.6), and (10.7).

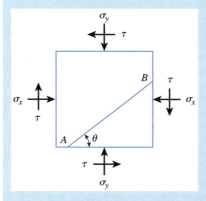

Figure 10.4 Soil element with stresses acting on it

Solution
Part a
From Eqs. (10.6) and (10.7),

$$\left.\begin{array}{r}\sigma_3\\\sigma_1\end{array}\right\} = \frac{\sigma_y + \sigma_x}{2} \pm \sqrt{\left[\frac{\sigma_y - \sigma_x}{2}\right]^2 + \tau_{xy}^2}$$

$$= \frac{120 + 96}{2} \pm \sqrt{\left[\frac{120 - 96}{2}\right]^2 + (-38)^2}$$

$$\sigma_1 = \textbf{147.85 kN/m}^2$$

$$\sigma_3 = \textbf{68.15 kN/m}^2$$

Part b
From Eq. (10.3),

$$\sigma_n = \frac{\sigma_y + \sigma_x}{2} + \frac{\sigma_y - \sigma_x}{2}\cos 2\theta + \tau \sin 2\theta$$

$$= \frac{120 + 96}{2} + \frac{120 - 96}{2}\cos(2 \times 20) + (-38)\sin(2 \times 20)$$

$$= \textbf{92.76 kN/m}^2$$

From Eq. (10.4),

$$\tau_n = \frac{\sigma_y - \sigma_x}{2} \sin 2\theta - \tau \cos 2\theta$$

$$= \frac{120 - 96}{2} \sin (2 \times 20) - (-38) \cos (2 \times 20)$$

$$= \textbf{36.82 kN/m}^2 \qquad \blacksquare$$

10.2 The Pole Method of Finding Stresses Along a Plane

Another important technique of finding stresses along a plane from a Mohr's circle is the *pole method*, or the method of *origin of planes*. This is demonstrated in Figure 10.5. Figure 10.5a is the same stress element that is shown in Figure 10.1a; Figure 10.5b is the Mohr's circle for the stress conditions indicated. According to the pole method, we draw a line from a known point on the Mohr's circle parallel to the plane on which the state of stress acts. The point of intersection of this line with the Mohr's circle is called the *pole*. This is a unique point for the state of stress under consideration. For example, the point M on the Mohr's circle in Figure 10.5b represents the stresses on the plane AB. The line MP is drawn parallel to AB. So point P is the pole (origin of planes) in this case. If we need to find the stresses on a plane EF, we draw a line from the pole parallel to EF. The point of intersection of this line with the Mohr's circle is Q. The coordinates of Q give the stresses on the plane EF. (*Note*: From geometry, angle QOM is twice the angle QPM.)

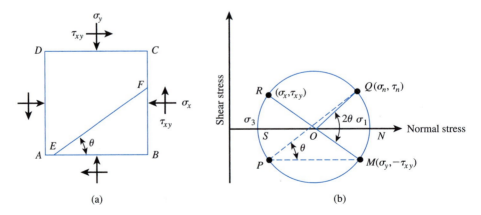

Figure 10.5 (a) Soil element with normal and shear stresses acting on it; (b) use of pole method to find the stresses along a plane

Example 10.2

For the stressed soil element shown in Figure 10.6a, determine

    a. Major principal stress
    b. Minor principal stress
    c. Normal and shear stresses on the plane AE

Use the pole method.

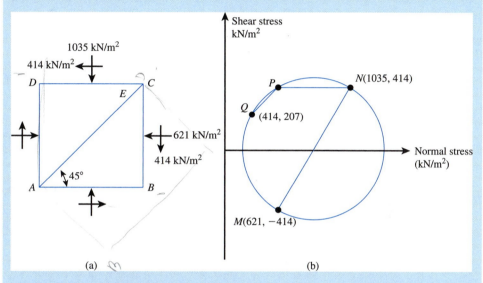

Figure 10.6 (a) Soil element with stresses acting on it; (b) Mohr's circle

Solution
On plane AD:

$$\text{Normal stress} = 621 \text{ kN/m}^2$$

$$\text{Shear stress} = -414 \text{ kN/m}^2$$

On plane AB:

$$\text{Normal stress} = 1035 \text{ kN/m}^2$$

$$\text{Shear stress} = 414 \text{ kN/m}^2$$

The Mohr's circle is plotted in Figure 10.6b. From the plot,

Part a

$$\text{Major principal stress} = \mathbf{1291 \text{ kN/m}^2}$$

Part b

$$\text{Minor principal stress} = \mathbf{365 \text{ kN/m}^2}$$

Part c

NP is the line drawn parallel to the plane *CD*. *P* is the pole. *PQ* is drawn parallel to *AE* (see Figure 10.6a). The coordinates of point *Q* give the stresses on the plane *AE*. Thus,

$$\text{Normal stress} = \textbf{414 kN/m}^2$$

$$\text{Shear stress} = \textbf{207 kN/m}^2$$

10.3 *Stresses Caused by a Point Load*

Boussinesq (1883) solved the problem of stresses produced at any point in a homogeneous, elastic, and isotropic medium as the result of a point load applied on the surface of an infinitely large half-space. According to Figure 10.7, Boussinesq's solution for normal stresses at a point caused by the point load *P* is

$$\Delta\sigma_x = \frac{P}{2\pi}\left\{\frac{3x^2z}{L^5} - (1 - 2\mu)\left[\frac{x^2 - y^2}{Lr^2(L + z)} + \frac{y^2z}{L^3r^2}\right]\right\} \tag{10.10}$$

$$\Delta\sigma_y = \frac{P}{2\pi}\left\{\frac{2y^2z}{L^5} - (1 - 2\mu)\left[\frac{y^2 - x^2}{Lr^2(L + z)} + \frac{x^2z}{L^3r^2}\right]\right\} \tag{10.11}$$

and

$$\Delta\sigma_z = \frac{3P}{2\pi}\frac{z^3}{L^5} = \frac{3P}{2\pi}\frac{z^3}{(r^2 + z^2)^{5/2}} \tag{10.12}$$

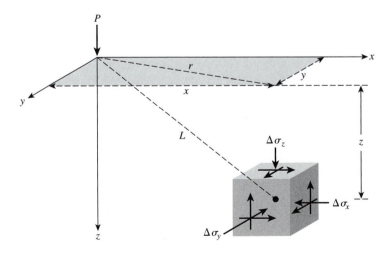

Figure 10.7
Stresses in an elastic medium caused by a point load

where $r = \sqrt{x^2 + y^2}$

$$L = \sqrt{x^2 + y^2 + z^2} = \sqrt{r^2 + z^2}$$

μ = Poisson's ratio

Note that Eqs. (10.10) and (10.11), which are the expressions for horizontal normal stresses, depend on the Poisson's ratio of the medium. However, the relationship for the vertical normal stress, $\Delta\sigma_z$, as given by Eq. (10.12), is independent of Poisson's ratio. The relationship for $\Delta\sigma_z$ can be rewritten as

$$\Delta\sigma_z = \frac{P}{z^2}\left\{\frac{3}{2\pi}\frac{1}{[(r/z)^2 + 1]^{5/2}}\right\} = \frac{P}{z^2}I_1 \tag{10.13}$$

where

$$I_1 = \frac{3}{2\pi}\frac{1}{[(r/z)^2 + 1]^{5/2}} \tag{10.14}$$

The variation of I_1 for various values of r/z is given in Table 10.1.

Table 10.1 Variation of I_1 for Various Values of r/z [Eq. (10.14)]

r/z	I_1	r/z	I_1	r/z	I_1
0	0.4775	0.36	0.3521	1.80	0.0129
0.02	0.4770	0.38	0.3408	2.00	0.0085
0.04	0.4765	0.40	0.3294	2.20	0.0058
0.06	0.4723	0.45	0.3011	2.40	0.0040
0.08	0.4699	0.50	0.2733	2.60	0.0029
0.10	0.4657	0.55	0.2466	2.80	0.0021
0.12	0.4607	0.60	0.2214	3.00	0.0015
0.14	0.4548	0.65	0.1978	3.20	0.0011
0.16	0.4482	0.70	0.1762	3.40	0.00085
0.18	0.4409	0.75	0.1565	3.60	0.00066
0.20	0.4329	0.80	0.1386	3.80	0.00051
0.22	0.4242	0.85	0.1226	4.00	0.00040
0.24	0.4151	0.90	0.1083	4.20	0.00032
0.26	0.4050	0.95	0.0956	4.40	0.00026
0.28	0.3954	1.00	0.0844	4.60	0.00021
0.30	0.3849	1.20	0.0513	4.80	0.00017
0.32	0.3742	1.40	0.0317	5.00	0.00014
0.34	0.3632	1.60	0.0200		

Example 10.3

Consider a point load $P = 5$ kN (Figure 10.7). Calculate the vertical stress increase ($\Delta\sigma_z$) at $z = 0$, 2 m, 4 m, 6 m, 10 m, and 20 m. Given $x = 3$ m and $y = 4$ m.

Solution

$$r = \sqrt{x^2 + y^2} = \sqrt{3^2 + 4^2} = 5\,\text{m}$$

The following table can now be prepared.

r (m)	z (m)	$\dfrac{r}{z}$	I_1	$\Delta\sigma_z = \left(\dfrac{P}{z^2}\right)I_1$ (kN/m²)
5	0	∞	0	0
	2	2.5	0.0034	0.0043
	4	1.25	0.0424	0.0133
	6	0.83	0.1295	0.0180
	10	0.5	0.2733	0.0137
	20	0.25	0.4103	0.0051

■

10.4 *Vertical Stress Caused by a Vertical Line Load*

Figure 10.8 shows a vertical flexible line load of infinite length that has an intensity q/unit length on the surface of a semi-infinite soil mass. The vertical stress increase, $\Delta\sigma_z$, inside the soil mass can be determined by using the principles of the theory of elasticity, or

$$\Delta\sigma_z = \frac{2qz^3}{\pi(x^2 + z^2)^2} \tag{10.15}$$

This equation can be rewritten as

$$\Delta\sigma_z = \frac{2q}{\pi z[(x/z)^2 + 1]^2}$$

or

$$\frac{\Delta\sigma_z}{(q/z)} = \frac{2}{\pi[(x/z)^2 + 1]^2} \tag{10.16}$$

Note that Eq. (10.16) is in a nondimensional form. Using this equation, we can calculate the variation of $\Delta\sigma_z/(q/z)$ with x/z. This is given in Table 10.2. The value of $\Delta\sigma_z$ calculated by using Eq. (10.16) is the additional stress on soil caused by the line load. The value of $\Delta\sigma_z$ does not include the overburden pressure of the soil above point A.

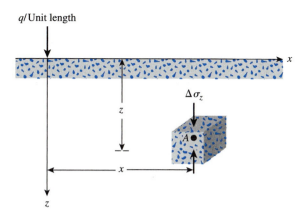

q/Unit length

Figure 10.8 Line load over the surface of a semi-infinite soil mass

Table 10.2 Variation of $\Delta\sigma_z/(q/z)$ with x/z [Eq. (10.16)]

x/z	$\Delta\sigma_z/(q/z)$	x/z	$\Delta\sigma_z/(q/z)$
0	0.637	1.3	0.088
0.1	0.624	1.4	0.073
0.2	0.589	1.5	0.060
0.3	0.536	1.6	0.050
0.4	0.473	1.7	0.042
0.5	0.407	1.8	0.035
0.6	0.344	1.9	0.030
0.7	0.287	2.0	0.025
0.8	0.237	2.2	0.019
0.9	0.194	2.4	0.014
1.0	0.159	2.6	0.011
1.1	0.130	2.8	0.008
1.2	0.107	3.0	0.006

Example 10.4

Figure 10.9a shows two line loads on the ground surface. Determine the increase of stress at point A.

Solution
Refer to Figure 10.9b. The total stress at A is

$$\Delta\sigma_z = \Delta\sigma_{z(1)} + \Delta\sigma_{z(2)}$$

$$\Delta\sigma_{z(1)} = \frac{2q_1 z^3}{\pi(x_1^2 + z^2)^2} = \frac{(2)(7.3)(1.2)^3}{\pi(1.5^2 + 1.2^2)^2} = 0.59 \ \text{kN/m}^2$$

$$\Delta\sigma_{z(2)} = \frac{2q_2 z^3}{\pi(x_2^2 + z^2)^2} = \frac{(2)(14.6)(1.2)^3}{\pi(3^2 + 1.2^2)^2} = 0.147 \text{ kN/m}^2$$

$$\Delta\sigma_z = 0.59 + 0.147 = 0.737 \text{ kN/m}^2$$

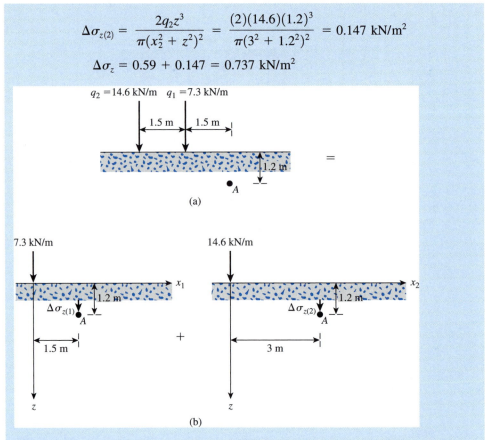

Figure 10.9 (a) Two line loads on the ground surface; (b) use of superposition principle to obtain stress at point A

10.5 Vertical Stress Caused by a Horizontal Line Load

Figure 10.10 shows a horizontal flexible line load on the surface on a semi-infinite soil mass. The vertical stress increase at point A in the soil mass can be given as

$$\Delta\sigma_z = \frac{2q}{\pi} \frac{xz^2}{(x^2 + z^2)^2} \tag{10.17}$$

Table 10.3 gives the variation of $\Delta\sigma_z/(q/z)$ with x/z.

q/Unit length

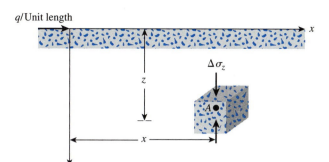

Figure 10.10 Horizontal line load over the surface of a semi-infinite soil mass

Table 10.3 Variation of $\Delta\sigma_z/(q/z)$ with x/z

x/z	$\Delta\sigma_z/(q/z)$	x/z	$\Delta\sigma_z/(q/z)$
0	0	0.7	0.201
0.1	0.062	0.8	0.189
0.2	0.118	0.9	0.175
0.3	0.161	1.0	0.159
0.4	0.189	1.5	0.090
0.5	0.204	2.0	0.051
0.6	0.207	3.0	0.019

Example 10.5

An inclined line load with a magnitude of 14.6 kN/m is shown in Figure 10.11. Determine the increase of vertical stress $\Delta\sigma_z$ at point A due to the line load.

Solution

The vertical component of the inclined load $q_V = 14.6 \cos 20 = 13.72$ kN/m, and the horizontal component $q_H = 14.6 \sin 20 = 5$ kN/m. For point A, $x/z = 1.5/1.2 = 1.25$. Using Table 10.2, the vertical stress increase at point A due to q_V is

$$\frac{\Delta\sigma_{z(V)}}{\left(\frac{q_v}{z}\right)} = 0.098$$

$$\Delta\sigma_{z(V)} = (0.098)\left(\frac{q_v}{z}\right) = (0.098)\left(\frac{13.72}{1.2}\right) = 1.12 \text{ kN/m}^2$$

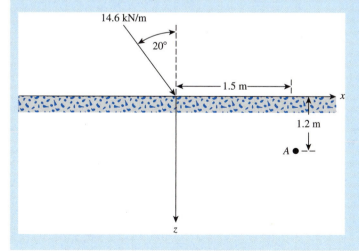

Figure 10.11

Similarly, using Table 10.3, the vertical stress increase at point A due to q_H is

$$\frac{\Delta\sigma_{z(H)}}{\left(\dfrac{q_H}{z}\right)} = 0.125$$

$$\Delta\sigma_{z(V)} = (0.125)\left(\frac{5}{1.2}\right) = 0.52 \text{ kN/m}^2$$

Thus, the total is

$$\Delta\sigma_z = \Delta\sigma_{z(V)} + \Delta\sigma_{z(H)} = 1.12 + 0.52 = \textbf{1.64 kN/m}^2$$

■

10.6 *Vertical Stress Caused by a Vertical Strip Load (Finite Width and Infinite Length)*

The fundamental equation for the vertical stress increase at a point in a soil mass as the result of a line load (Section 10.4) can be used to determine the vertical stress at a point caused by a flexible strip load of width B. (See Figure 10.12.) Let the load per unit area of the strip shown in Figure 10.12 be equal to q. If we consider an elemental strip of width dr,

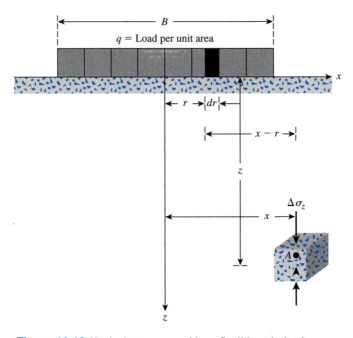

Figure 10.12 Vertical stress caused by a flexible strip load

the load per unit length of this strip is equal to $q\ dr$. This elemental strip can be treated as a line load. Equation (10.15) gives the vertical stress increase $d\sigma_z$ at point A inside the soil mass caused by this elemental strip load. To calculate the vertical stress increase, we need to substitute $q\ dr$ for q and $(x - r)$ for x. So,

$$d\sigma_z = \frac{2(q\ dr)z^3}{\pi[(x - r)^2 + z^2]^2} \tag{10.18}$$

The total increase in the vertical stress $(\Delta\sigma_z)$ at point A caused by the entire strip load of width B can be determined by integration of Eq. (10.18) with limits of r from $-B/2$ to $+B/2$, or

$$\Delta\sigma_z = \int d\sigma_z = \int_{-B/2}^{+B/2} \left(\frac{2q}{\pi}\right)\left\{\frac{z^3}{[(x - r)^2 + z^2]^2}\right\} dr$$

$$= \frac{q}{\pi}\left\{\tan^{-1}\left[\frac{z}{x - (B/2)}\right] - \tan^{-1}\left[\frac{z}{x + (B/2)}\right]\right. \tag{10.19}$$

$$\left. - \frac{Bz[x^2 - z^2 - (B^2/4)]}{[x^2 + z^2 - (B^2/4)]^2 + B^2z^2}\right\}$$

Table 10.4 shows the variation of $\Delta\sigma_z/q$ with $2z/B$ for $2x/B$. This table can be used conveniently for the calculation of vertical stress at a point caused by a flexible strip load.

10.7   *Vertical Stress Due to Embankment Loading*

Figure 10.14 shows the cross section of an embankment of height H. For this two-dimensional loading condition the vertical stress increase may be expressed as

$$\Delta\sigma_z = \frac{q_o}{\pi}\left[\left(\frac{B_1 + B_2}{B_2}\right)(\alpha_1 + \alpha_2) - \frac{B_1}{B_2}(\alpha_2)\right] \tag{10.20}$$

where $q_o = \gamma H$

γ = unit weight of the embankment soil
H = height of the embankment

$$\alpha_1\ (\text{radians}) = \tan^{-1}\left(\frac{B_1 + B_2}{z}\right) - \tan^{-1}\left(\frac{B_1}{z}\right) \tag{10.21}$$

$$\alpha_2 = \tan^{-1}\left(\frac{B_1}{z}\right) \tag{10.22}$$

Table 10.4 Variation of $\Delta\sigma_z/q$ with $2z/B$ and $2x/B$ [Eq. (10.19)]

2z/B	2x/B										
	0.0	0.1	0.2	0.3	0.4	0.5	0.6	0.7	0.8	0.9	1.0
0.00	1.000	1.000	1.000	1.000	1.000	1.000	1.000	1.000	1.000	1.000	0.000
0.10	1.000	1.000	0.999	0.999	0.999	0.998	0.997	0.993	0.980	0.909	0.500
0.20	0.997	0.997	0.996	0.995	0.992	0.988	0.979	0.959	0.909	0.775	0.500
0.30	0.990	0.989	0.987	0.984	0.978	0.967	0.947	0.908	0.833	0.697	0.499
0.40	0.977	0.976	0.973	0.966	0.955	0.937	0.906	0.855	0.773	0.651	0.498
0.50	0.959	0.958	0.953	0.943	0.927	0.902	0.864	0.808	0.727	0.620	0.497
0.60	0.937	0.935	0.928	0.915	0.896	0.866	0.825	0.767	0.691	0.598	0.495
0.70	0.910	0.908	0.899	0.885	0.863	0.831	0.788	0.732	0.662	0.581	0.492
0.80	0.881	0.878	0.869	0.853	0.829	0.797	0.755	0.701	0.638	0.566	0.489
0.90	0.850	0.847	0.837	0.821	0.797	0.765	0.724	0.675	0.617	0.552	0.485
1.00	0.818	0.815	0.805	0.789	0.766	0.735	0.696	0.650	0.598	0.540	0.480
1.10	0.787	0.783	0.774	0.758	0.735	0.706	0.670	0.628	0.580	0.529	0.474
1.20	0.755	0.752	0.743	0.728	0.707	0.679	0.646	0.607	0.564	0.517	0.468
1.30	0.725	0.722	0.714	0.699	0.679	0.654	0.623	0.588	0.548	0.506	0.462
1.40	0.696	0.693	0.685	0.672	0.653	0.630	0.602	0.569	0.534	0.495	0.455
1.50	0.668	0.666	0.658	0.646	0.629	0.607	0.581	0.552	0.519	0.484	0.448
1.60	0.642	0.639	0.633	0.621	0.605	0.586	0.562	0.535	0.506	0.474	0.440
1.70	0.617	0.615	0.608	0.598	0.583	0.565	0.544	0.519	0.492	0.463	0.433
1.80	0.593	0.591	0.585	0.576	0.563	0.546	0.526	0.504	0.479	0.453	0.425
1.90	0.571	0.569	0.564	0.555	0.543	0.528	0.510	0.489	0.467	0.443	0.417
2.00	0.550	0.548	0.543	0.535	0.524	0.510	0.494	0.475	0.455	0.433	0.409
2.10	0.530	0.529	0.524	0.517	0.507	0.494	0.479	0.462	0.443	0.423	0.401
2.20	0.511	0.510	0.506	0.499	0.490	0.479	0.465	0.449	0.432	0.413	0.393
2.30	0.494	0.493	0.489	0.483	0.474	0.464	0.451	0.437	0.421	0.404	0.385
2.40	0.477	0.476	0.473	0.467	0.460	0.450	0.438	0.425	0.410	0.395	0.378
2.50	0.462	0.461	0.458	0.452	0.445	0.436	0.426	0.414	0.400	0.386	0.370
2.60	0.447	0.446	0.443	0.439	0.432	0.424	0.414	0.403	0.390	0.377	0.363
2.70	0.433	0.432	0.430	0.425	0.419	0.412	0.403	0.393	0.381	0.369	0.355
2.80	0.420	0.419	0.417	0.413	0.407	0.400	0.392	0.383	0.372	0.360	0.348
2.90	0.408	0.407	0.405	0.401	0.396	0.389	0.382	0.373	0.363	0.352	0.341
3.00	0.396	0.395	0.393	0.390	0.385	0.379	0.372	0.364	0.355	0.345	0.334
3.10	0.385	0.384	0.382	0.379	0.375	0.369	0.363	0.355	0.347	0.337	0.327
3.20	0.374	0.373	0.372	0.369	0.365	0.360	0.354	0.347	0.339	0.330	0.321
3.30	0.364	0.363	0.362	0.359	0.355	0.351	0.345	0.339	0.331	0.323	0.315
3.40	0.354	0.354	0.352	0.350	0.346	0.342	0.337	0.331	0.324	0.316	0.308
3.50	0.345	0.345	0.343	0.341	0.338	0.334	0.329	0.323	0.317	0.310	0.302
3.60	0.337	0.336	0.335	0.333	0.330	0.326	0.321	0.316	0.310	0.304	0.297
3.70	0.328	0.328	0.327	0.325	0.322	0.318	0.314	0.309	0.304	0.298	0.291
3.80	0.320	0.320	0.319	0.317	0.315	0.311	0.307	0.303	0.297	0.292	0.285
3.90	0.313	0.313	0.312	0.310	0.307	0.304	0.301	0.296	0.291	0.286	0.280
4.00	0.306	0.305	0.304	0.303	0.301	0.298	0.294	0.290	0.285	0.280	0.275
4.10	0.299	0.299	0.298	0.296	0.294	0.291	0.288	0.284	0.280	0.275	0.270
4.20	0.292	0.292	0.291	0.290	0.288	0.285	0.282	0.278	0.274	0.270	0.265
4.30	0.286	0.286	0.285	0.283	0.282	0.279	0.276	0.273	0.269	0.265	0.260
4.40	0.280	0.280	0.279	0.278	0.276	0.274	0.271	0.268	0.264	0.260	0.256
4.50	0.274	0.274	0.273	0.272	0.270	0.268	0.266	0.263	0.259	0.255	0.251
4.60	0.268	0.268	0.268	0.266	0.265	0.263	0.260	0.258	0.254	0.251	0.247
4.70	0.263	0.263	0.262	0.261	0.260	0.258	0.255	0.253	0.250	0.246	0.243
4.80	0.258	0.258	0.257	0.256	0.255	0.253	0.251	0.248	0.245	0.242	0.239
4.90	0.253	0.253	0.252	0.251	0.250	0.248	0.246	0.244	0.241	0.238	0.235
5.00	0.248	0.248	0.247	0.246	0.245	0.244	0.242	0.239	0.237	0.234	0.231

Table 10.4 (*continued*)

2z/B	2x/B									
	1.1	1.2	1.3	1.4	1.5	1.6	1.7	1.8	1.9	2.0
0.00	0.000	0.000	0.000	0.000	0.000	0.000	0.000	0.000	0.000	0.000
0.10	0.091	0.020	0.007	0.003	0.002	0.001	0.001	0.000	0.000	0.000
0.20	0.225	0.091	0.040	0.020	0.011	0.007	0.004	0.003	0.002	0.002
0.30	0.301	0.165	0.090	0.052	0.031	0.020	0.013	0.009	0.007	0.005
0.40	0.346	0.224	0.141	0.090	0.059	0.040	0.027	0.020	0.014	0.011
0.50	0.373	0.267	0.185	0.128	0.089	0.063	0.046	0.034	0.025	0.019
0.60	0.391	0.298	0.222	0.163	0.120	0.088	0.066	0.050	0.038	0.030
0.70	0.403	0.321	0.250	0.193	0.148	0.113	0.087	0.068	0.053	0.042
0.80	0.411	0.338	0.273	0.218	0.173	0.137	0.108	0.086	0.069	0.056
0.90	0.416	0.351	0.291	0.239	0.195	0.158	0.128	0.104	0.085	0.070
1.00	0.419	0.360	0.305	0.256	0.214	0.177	0.147	0.122	0.101	0.084
1.10	0.420	0.366	0.316	0.271	0.230	0.194	0.164	0.138	0.116	0.098
1.20	0.419	0.371	0.325	0.282	0.243	0.209	0.178	0.152	0.130	0.111
1.30	0.417	0.373	0.331	0.291	0.254	0.221	0.191	0.166	0.143	0.123
1.40	0.414	0.374	0.335	0.298	0.263	0.232	0.203	0.177	0.155	0.135
1.50	0.411	0.374	0.338	0.303	0.271	0.240	0.213	0.188	0.165	0.146
1.60	0.407	0.373	0.339	0.307	0.276	0.248	0.221	0.197	0.175	0.155
1.70	0.402	0.370	0.339	0.309	0.281	0.254	0.228	0.205	0.183	0.164
1.80	0.396	0.368	0.339	0.311	0.284	0.258	0.234	0.212	0.191	0.172
1.90	0.391	0.364	0.338	0.312	0.286	0.262	0.239	0.217	0.197	0.179
2.00	0.385	0.360	0.336	0.311	0.288	0.265	0.243	0.222	0.203	0.185
2.10	0.379	0.356	0.333	0.311	0.288	0.267	0.246	0.226	0.208	0.190
2.20	0.373	0.352	0.330	0.309	0.288	0.268	0.248	0.229	0.212	0.195
2.30	0.366	0.347	0.327	0.307	0.288	0.268	0.250	0.232	0.215	0.199
2.40	0.360	0.342	0.323	0.305	0.287	0.268	0.251	0.234	0.217	0.202
2.50	0.354	0.337	0.320	0.302	0.285	0.268	0.251	0.235	0.220	0.205
2.60	0.347	0.332	0.316	0.299	0.283	0.267	0.251	0.236	0.221	0.207
2.70	0.341	0.327	0.312	0.296	0.281	0.266	0.251	0.236	0.222	0.208
2.80	0.335	0.321	0.307	0.293	0.279	0.265	0.250	0.236	0.223	0.210
2.90	0.329	0.316	0.303	0.290	0.276	0.263	0.249	0.236	0.223	0.211
3.00	0.323	0.311	0.299	0.286	0.274	0.261	0.248	0.236	0.223	0.211
3.10	0.317	0.306	0.294	0.283	0.271	0.259	0.247	0.235	0.223	0.212
3.20	0.311	0.301	0.290	0.279	0.268	0.256	0.245	0.234	0.223	0.212
3.30	0.305	0.296	0.286	0.275	0.265	0.254	0.243	0.232	0.222	0.211
3.40	0.300	0.291	0.281	0.271	0.261	0.251	0.241	0.231	0.221	0.211
3.50	0.294	0.286	0.277	0.268	0.258	0.249	0.239	0.229	0.220	0.210
3.60	0.289	0.281	0.273	0.264	0.255	0.246	0.237	0.228	0.218	0.209
3.70	0.284	0.276	0.268	0.260	0.252	0.243	0.235	0.226	0.217	0.208
3.80	0.279	0.272	0.264	0.256	0.249	0.240	0.232	0.224	0.216	0.207
3.90	0.274	0.267	0.260	0.253	0.245	0.238	0.230	0.222	0.214	0.206
4.00	0.269	0.263	0.256	0.249	0.242	0.235	0.227	0.220	0.212	0.205
4.10	0.264	0.258	0.252	0.246	0.239	0.232	0.225	0.218	0.211	0.203
4.20	0.260	0.254	0.248	0.242	0.236	0.229	0.222	0.216	0.209	0.202
4.30	0.255	0.250	0.244	0.239	0.233	0.226	0.220	0.213	0.207	0.200
4.40	0.251	0.246	0.241	0.235	0.229	0.224	0.217	0.211	0.205	0.199
4.50	0.247	0.242	0.237	0.232	0.226	0.221	0.215	0.209	0.203	0.197
4.60	0.243	0.238	0.234	0.229	0.223	0.218	0.212	0.207	0.201	0.195
4.70	0.239	0.235	0.230	0.225	0.220	0.215	0.210	0.205	0.199	0.194
4.80	0.235	0.231	0.227	0.222	0.217	0.213	0.208	0.202	0.197	0.192
4.90	0.231	0.227	0.223	0.219	0.215	0.210	0.205	0.200	0.195	0.190
5.00	0.227	0.224	0.220	0.216	0.212	0.207	0.203	0.198	0.193	0.188

Example 10.6

With reference to Figure 10.12, we are given $q = 200$ kN/m², $B = 6$ m, and $z = 3$ m. Determine the vertical stress increase at $x = \pm 9, \pm 6, \pm 3$, and 0 m. Plot a graph of $\Delta \sigma_z$ against x.

Solution

The following table can be made:

x(m)	$2x/B$	$2z/B$	$\Delta \sigma_z / q^a$	$\Delta \sigma_z{}^b$ (kN/m²)
±9	±3	1	0.017	3.4
±6	±2	1	0.084	16.8
±3	±1	1	0.480	96.0
0	0	1	0.818	163.6

^{a}From Table 10.4
$^b q = 200$ kN/m²

The plot of $\Delta \sigma_z$ against x is given in Figure 10.13.

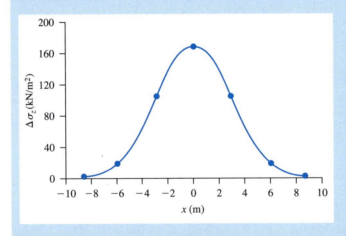

Figure 10.13 Plot of $\Delta \sigma_z$ against distance x

For a detailed derivation of the equation, see Das (2008). A simplified form of Eq. (10.20) is

$$\Delta \sigma_z = q_o I_2 \tag{10.23}$$

where $I_2 =$ a function of B_1/z and B_2/z.

The variation of I_2 with B_1/z and B_2/z is shown in Figure 10.15 (Osterberg, 1957).

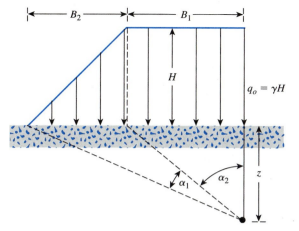

Figure 10.14
Embankment loading

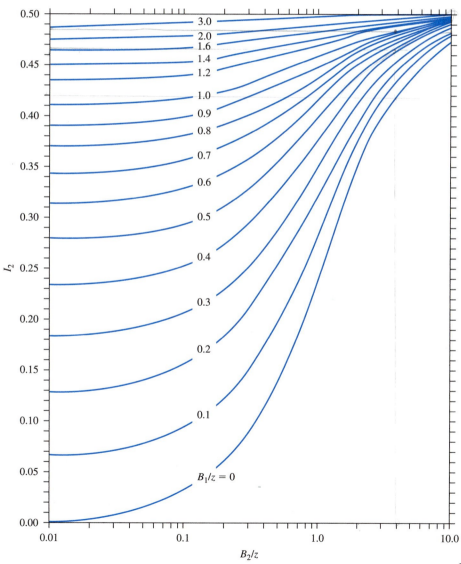

Figure 10.15
Osterberg's chart for determination of vertical stress due to embankment loading

Example 10.7

An embankment is shown in Figure 10.16a. Determine the stress increase under the embankment at point A.

Solution

Refer to Figures 10.16b and c.

At point A:

$$\sigma_z = \sigma_{z(1)} - \sigma_{z(2)}$$

$$q_{o(1)} = q_{o(2)} = (10)(18) = 180 \text{ kN/m}^2$$

From Fig. 10.16b, $B_1 = 21$ m and $B_2 = 20$ m. So,

$$\frac{B_1}{z} = \frac{21}{6} = 3.5$$

$$\frac{B_2}{z} = \frac{20}{6} = 3.33$$

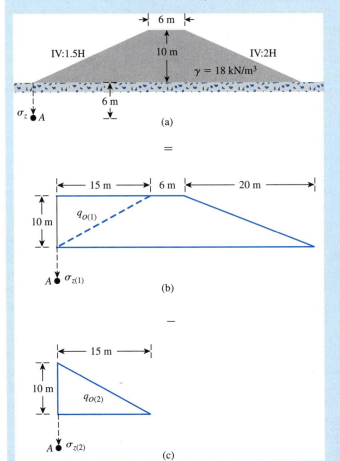

Figure 10.16

From Fig. 10.15, $I_{2(1)} \approx 0.495$. Again, from Fig. 10.16c, $B_1 = 0$ and $B_2 = 15$ m. So,

$$\frac{B_1}{z} = \frac{0}{6} = 0$$

$$\frac{B_2}{z} = \frac{15}{6} = 2.5$$

From Fig. 10.15, $I_{2(2)} \approx 0.39$. Therefore,

$$\sigma_z = \sigma_{z(1)} - \sigma_{z(2)} = q_{o(1)}I_{2(1)} - q_{o(2)}I_{2(2)} = 180[0.490 - 0.39] = \textbf{18.9 kN/m}^2 \quad \blacksquare$$

10.8 *Vertical Stress Below the Center of a Uniformly Loaded Circular Area*

Using Boussinesq's solution for vertical stress $\Delta\sigma_z$ caused by a point load [Eq. (10.12)], one also can develop an expression for the vertical stress below the center of a uniformly loaded flexible circular area.

From Figure 10.17, let the intensity of pressure on the circular area of radius R be equal to q. The total load on the elemental area (shaded in the figure) is equal to $qr\, dr\, d\alpha$. The vertical stress, $d\sigma_z$, at point A caused by the load on the elemental area (which may be assumed to be a concentrated load) can be obtained from Eq. (10.12):

$$d\sigma_z = \frac{3(qr\, dr\, d\alpha)}{2\pi} \frac{z^3}{(r^2 + z^2)^{5/2}} \tag{10.24}$$

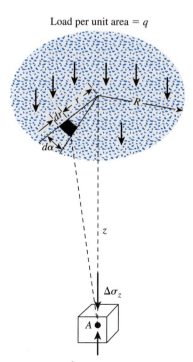

Load per unit area = q

Figure 10.17
Vertical stress below the center of a uniformly loaded flexible circular area

The increase in the stress at point A caused by the entire loaded area can be found by integrating Eq. (10.24):

$$\Delta\sigma_z = \int d\sigma_z = \int_{\alpha=0}^{\alpha=2\pi} \int_{r=0}^{r=R} \frac{3q}{2\pi} \frac{z^3 r}{(r^2 + z^2)^{5/2}} \, dr \, d\alpha$$

So,

$$\Delta\sigma_z = q\left\{1 - \frac{1}{[(R/z)^2 + 1]^{3/2}}\right\} \tag{10.25}$$

The variation of $\Delta\sigma_z/q$ with z/R as obtained from Eq. (10.25) is given in Table 10.5. A plot of this also is shown in Figure 10.18. The value of $\Delta\sigma_z$ decreases rapidly with depth, and at $z = 5R$, it is about 6% of q, which is the intensity of pressure at the ground surface.

Table 10.5 Variation of $\Delta\sigma_z/q$ with z/R [Eq. (10.25)]

z/R	$\Delta\sigma_z/q$	z/R	$\Delta\sigma_z/q$
0	1	1.0	0.6465
0.02	0.9999	1.5	0.4240
0.05	0.9998	2.0	0.2845
0.10	0.9990	2.5	0.1996
0.2	0.9925	3.0	0.1436
0.4	0.9488	4.0	0.0869
0.5	0.9106	5.0	0.0571
0.8	0.7562		

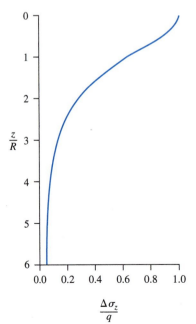

Figure 10.18 Stress under the center of a uniformly loaded flexible circular area

Vertical Stress at Any Point Below a Uniformly Loaded Circular Area

A detailed tabulation for calculation of vertical stress below a uniformly loaded flexible circular area was given by Ahlvin and Ulery (1962). Referring to Figure 10.19, we find that $\Delta \sigma_z$ at any point A located at a depth z at any distance r from the center of the loaded area can be given as

$$\Delta \sigma_z = q(A' + B') \qquad (10.26)$$

where A' and B' are functions of z/R and r/R. (See Tables 10.6 and 10.7.)

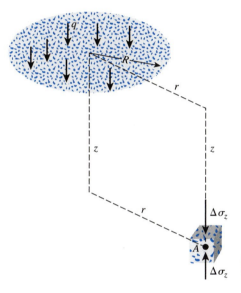

Figure 10.19 Vertical stress at any point below a uniformly loaded circular area

---

Example 10.8

Consider a uniformly loaded flexible circular area on the ground surface, as shown in Fig. 10.19. Given: $R = 3$ m and uniform load $q = 100$ kN/m².

Calculate the increase in vertical stress at depths of 1.5 m, 3 m, 4.5 m, 6 m, and 12 m below the ground surface for points at (a) $r = 0$ and (b) $r = 4.5$ m.

Solution
From Eq. (10.26),

$$\Delta \sigma_z = q\,(A' + B')$$

Given $R = 3$ m and $q = 100$ kN/m².

(continued on page 278)

Table 10.6 Variation of A' with z/R and r/R^*

z/R									r/R
	0	0.2	0.4	0.6	0.8	1	1.2	1.5	2
0	1.0	1.0	1.0	1.0	1.0	0.5	0	0	0
0.1	0.90050	0.89748	0.88679	0.86126	0.78797	0.43015	0.09645	0.02787	0.00856
0.2	0.80388	0.79824	0.77884	0.73483	0.63014	0.38269	0.15433	0.05251	0.01680
0.3	0.71265	0.70518	0.68316	0.62690	0.52081	0.34375	0.17964	0.07199	0.02440
0.4	0.62861	0.62015	0.59241	0.53767	0.44329	0.31048	0.18709	0.08593	0.03118
0.5	0.55279	0.54403	0.51622	0.46448	0.38390	0.28156	0.18556	0.09499	0.03701
0.6	0.48550	0.47691	0.45078	0.40427	0.33676	0.25588	0.17952	0.10010	
0.7	0.42654	0.41874	0.39491	0.35428	0.29833	0.21727	0.17124	0.10228	0.04558
0.8	0.37531	0.36832	0.34729	0.31243	0.26581	0.21297	0.16206	0.10236	
0.9	0.33104	0.32492	0.30669	0.27707	0.23832	0.19488	0.15253	0.10094	
1	0.29289	0.28763	0.27005	0.24697	0.21468	0.17868	0.14329	0.09849	0.05185
1.2	0.23178	0.22795	0.21662	0.19890	0.17626	0.15101	0.12570	0.09192	0.05260
1.5	0.16795	0.16552	0.15877	0.14804	0.13436	0.11892	0.10296	0.08048	0.05116
2	0.10557	0.10453	0.10140	0.09647	0.09011	0.08269	0.07471	0.06275	0.04496
2.5	0.07152	0.07098	0.06947	0.06698	0.06373	0.05974	0.05555	0.04880	0.03787
3	0.05132	0.05101	0.05022	0.04886	0.04707	0.04487	0.04241	0.03839	0.03150
4	0.02986	0.02976	0.02907	0.02802	0.02832	0.02749	0.02651	0.02490	0.02193
5	0.01942	0.01938				0.01835			0.01573
6	0.01361					0.01307			0.01168
7	0.01005					0.00976			0.00894
8	0.00772					0.00755			0.00703
9	0.00612					0.00600			0.00566
10								0.00477	0.00465

*After Ahlvin and Ulery (1962)

Table 10.7 Variation of B' with z/R and r/R^*

z/R									r/R
	0	0.2	0.4	0.6	0.8	1	1.2	1.5	2
0	0	0	0	0	0	0	0	0	0
0.1	0.09852	0.10140	0.11138	0.13424	0.18796	0.05388	−0.07899	−0.02672	−0.00845
0.2	0.18857	0.19306	0.20772	0.23524	0.25983	0.08513	−0.07759	−0.04448	−0.01593
0.3	0.26362	0.26787	0.28018	0.29483	0.27257	0.10757	−0.04316	−0.04999	−0.02166
0.4	0.32016	0.32259	0.32748	0.32273	0.26925	0.12404	−0.00766	−0.04535	−0.02522
0.5	0.35777	0.35752	0.35323	0.33106	0.26236	0.13591	0.02165	−0.03455	−0.02651
0.6	0.37831	0.37531	0.36308	0.32822	0.25411	0.14440	0.04457	−0.02101	
0.7	0.38487	0.37962	0.36072	0.31929	0.24638	0.14986	0.06209	−0.00702	−0.02329
0.8	0.38091	0.37408	0.35133	0.30699	0.23779	0.15292	0.07530	0.00614	
0.9	0.36962	0.36275	0.33734	0.29299	0.22891	0.15404	0.08507	0.01795	
1	0.35355	0.34553	0.32075	0.27819	0.21978	0.15355	0.09210	0.02814	−0.01005
1.2	0.31485	0.30730	0.28481	0.24836	0.20113	0.14915	0.10002	0.04378	0.00023
1.5	0.25602	0.25025	0.23338	0.20694	0.17368	0.13732	0.10193	0.05745	0.01385
2	0.17889	0.18144	0.16644	0.15198	0.13375	0.11331	0.09254	0.06371	0.02836
2.5	0.12807	0.12633	0.12126	0.11327	0.10298	0.09130	0.07869	0.06022	0.03429
3	0.09487	0.09394	0.09099	0.08635	0.08033	0.07325	0.06551	0.05354	0.03511
4	0.05707	0.05666	0.05562	0.05383	0.05145	0.04773	0.04532	0.03995	0.03066
5	0.03772	0.03760				0.03384			0.02474
6	0.02666					0.02468			0.01968
7	0.01980					0.01868			0.01577
8	0.01526					0.01459			0.01279
9	0.01212					0.01170			0.01054
10								0.00924	0.00879

Source: From "Tabulated Values for Determining the Complete Pattern of Stresses, Strains, and Deflections Beneath a Uniform Circular Load on a Homogeneous Half Space," by R. G Ahlvin and H. H. Ulery. In *Highway Research Bulletin 342,* Transportation Research Board, National Research Council, Washington, D.C., 1962.

Table 10.6 (*continued*)

3	4	5	6	7	8	10	12	14
0	0	0	0	0	0	0	0	0
0.00211	0.00084	0.00042						
0.00419	0.00167	0.00083	0.00048	0.00030	0.00020			
0.00622	0.00250							
0.01013	0.00407	0.00209	0.00118	0.00071	0.00053	0.00025	0.00014	0.00009
0.01742	0.00761	0.00393	0.00226	0.00143	0.00097	0.00050	0.00029	0.00018
0.01935	0.00871	0.00459	0.00269	0.00171	0.00115			
0.02142	0.01013	0.00548	0.00325	0.00210	0.00141	0.00073	0.00043	0.00027
0.02221	0.01160	0.00659	0.00399	0.00264	0.00180	0.00094	0.00056	0.00036
0.02143	0.01221	0.00732	0.00463	0.00308	0.00214	0.00115	0.00068	0.00043
0.01980	0.01220	0.00770	0.00505	0.00346	0.00242	0.00132	0.00079	0.00051
0.01592	0.01109	0.00768	0.00536	0.00384	0.00282	0.00160	0.00099	0.00065
0.01249	0.00949	0.00708	0.00527	0.00394	0.00298	0.00179	0.00113	0.00075
0.00983	0.00795	0.00628	0.00492	0.00384	0.00299	0.00188	0.00124	0.00084
0.00784	0.00661	0.00548	0.00445	0.00360	0.00291	0.00193	0.00130	0.00091
0.00635	0.00554	0.00472	0.00398	0.00332	0.00276	0.00189	0.00134	0.00094
0.00520	0.00466	0.00409	0.00353	0.00301	0.00256	0.00184	0.00133	0.00096
0.00438	0.00397	0.00352	0.00326	0.00273	0.00241			

Table 10.7 (*continued*)

3	4	5	6	7	8	10	12	14
0	0	0	0	0	0	0	0	0
−0.00210	−0.00084	−0.00042						
−0.00412	−0.00166	−0.00083	−0.00024	−0.00015	−0.00010			
−0.00599	−0.00245							
−0.00991	−0.00388	−0.00199	−0.00116	−0.00073	−0.00049	−0.00025	−0.00014	−0.00009
−0.01115	−0.00608	−0.00344	−0.00210	−0.00135	−0.00092	−0.00048	−0.00028	−0.00018
−0.00995	−0.00632	−0.00378	−0.00236	−0.00156	−0.00107			
−0.00669	−0.00600	−0.00401	−0.00265	−0.00181	−0.00126	−0.00068	−0.00040	−0.00026
0.00028	−0.00410	−0.00371	−0.00278	−0.00202	−0.00148	−0.00084	−0.00050	−0.00033
0.00661	−0.00130	−0.00271	−0.00250	−0.00201	−0.00156	−0.00094	−0.00059	−0.00039
0.01112	0.00157	−0.00134	−0.00192	−0.00179	−0.00151	−0.00099	−0.00065	−0.00046
0.01515	0.00595	0.00155	−0.00029	−0.00094	−0.00109	−0.00094	−0.00068	−0.00050
0.01522	0.00810	0.00371	0.00132	0.00013	−0.00043	−0.00070	−0.00061	−0.00049
0.01380	0.00867	0.00496	0.00254	0.00110	0.00028	−0.00037	−0.00047	−0.00045
0.01204	0.00842	0.00547	0.00332	0.00185	0.00093	−0.00002	−0.00029	−0.00037
0.01034	0.00779	0.00554	0.00372	0.00236	0.00141	0.00035	−0.00008	−0.00025
0.00888	0.00705	0.00533	0.00386	0.00265	0.00178	0.00066	0.00012	−0.00012
0.00764	0.00631	0.00501	0.00382	0.00281	0.00199			

Part (a)

We can prepare the following table: (*Note: r/R = 0.A'* and *B'* values are from Tables 10.6 and 10.7.)

Depth, z (m)	z/R	A'	B'	$\Delta\sigma_z$ (kN/m²)
1.5	0.5	0.553	0.358	91.1
3	1.0	0.293	0.354	64.7
4.5	1.5	0.168	0.256	42.4
6	2.0	0.106	0.179	28.5
12	4.0	0.03	0.057	8.7

Part (b)

$$r/R = 4.5/3 = 1.5$$

Depth, z (m)	z/R	A'	B'	$\Delta\sigma_z$ (kN/m²)
1.5	0.5	0.095	−0.035	6.0
3	1.0	0.098	0.028	12.6
4.5	1.5	0.08	0.057	13.7
6	2.0	0.063	0.064	12.7
12	4.0	0.025	0.04	6.5

■

10.10 *Vertical Stress Caused by a Rectangularly Loaded Area*

Boussinesq's solution also can be used to calculate the vertical stress increase below a flexible rectangular loaded area, as shown in Figure 10.20. The loaded area is located at the ground surface and has length L and width B. The uniformly distributed load per unit area is equal to q. To determine the increase in the vertical stress ($\Delta\sigma_z$) at point A, which is located at depth z below the corner of the rectangular area, we need to consider a small elemental area $dx\,dy$ of the rectangle. (This is shown in Figure 10.20.) The load on this elemental area can be given by

$$dq = q\,dx\,dy \tag{10.27}$$

The increase in the stress ($d\sigma_z$) at point A caused by the load dq can be determined by using Eq. (10.12). However, we need to replace P with $dq = q\,dx\,dy$ and r^2 with $x^2 + y^2$. Thus,

$$d\sigma_z = \frac{3q\,dx\,dy\,z^3}{2\pi(x^2 + y^2 + z^2)^{5/2}} \tag{10.28}$$

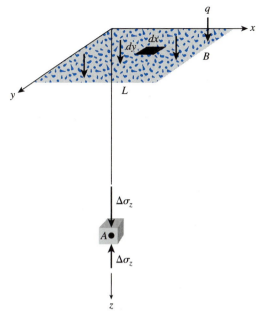

Figure 10.20
Vertical stress below the corner
of a uniformly loaded flexible
rectangular area

The increase in the stress, at point A caused by the entire loaded area can now be determined by integrating the preceding equation. We obtain

$$\Delta\sigma_z = \int d\sigma_z = \int_{y=0}^{B} \int_{x=0}^{L} \frac{3qz^3(dx\,dy)}{2\pi(x^2 + y^2 + z^2)^{5/2}} = qI_3 \qquad (10.29)$$

where

$$I_3 = \frac{1}{4\pi}\left[\frac{2mn\sqrt{m^2 + n^2 + 1}}{m^2 + n^2 + m^2n^2 + 1}\left(\frac{m^2 + n^2 + 2}{m^2 + n^2 + 1}\right) + \tan^{-1}\left(\frac{2mn\sqrt{m^2 + n^2 + 1}}{m^2 + n^2 - m^2n^2 + 1}\right)\right] \qquad (10.30)$$

$$m = \frac{B}{z} \qquad (10.31)$$

$$n = \frac{L}{z} \qquad (10.32)$$

The variation of I_3 with m and n is shown in Table 10.8 and Figure 10.21.

Table 10.8 Variation of I_3 with m and n [Eq. (10.30)]

										m
n	0.1	0.2	0.3	0.4	0.5	0.6	0.7	0.8	0.9	1.0
0.1	0.0047	0.0092	0.0132	0.0168	0.0198	0.0222	0.0242	0.0258	0.0270	0.0279
0.2	0.0092	0.0179	0.0259	0.0328	0.0387	0.0435	0.0474	0.0504	0.0528	0.0547
0.3	0.0132	0.0259	0.0374	0.0474	0.0559	0.0629	0.0686	0.0731	0.0766	0.0794
0.4	0.0168	0.0328	0.0474	0.0602	0.0711	0.0801	0.0873	0.0931	0.0977	0.1013
0.5	0.0198	0.0387	0.0559	0.0711	0.0840	0.0947	0.1034	0.1104	0.1158	0.1202
0.6	0.0222	0.0435	0.0629	0.0801	0.0947	0.1069	0.1168	0.1247	0.1311	0.1361
0.7	0.0242	0.0474	0.0686	0.0873	0.1034	0.1169	0.1277	0.1365	0.1436	0.1491
0.8	0.0258	0.0504	0.0731	0.0931	0.1104	0.1247	0.1365	0.1461	0.1537	0.1598
0.9	0.0270	0.0528	0.0766	0.0977	0.1158	0.1311	0.1436	0.1537	0.1619	0.1684
1.0	0.0279	0.0547	0.0794	0.1013	0.1202	0.1361	0.1491	0.1598	0.1684	0.1752
1.2	0.0293	0.0573	0.0832	0.1063	0.1263	0.1431	0.1570	0.1684	0.1777	0.1851
1.4	0.0301	0.0589	0.0856	0.1094	0.1300	0.1475	0.1620	0.1739	0.1836	0.1914
1.6	0.0306	0.0599	0.0871	0.1114	0.1324	0.1503	0.1652	0.1774	0.1874	0.1955
1.8	0.0309	0.0606	0.0880	0.1126	0.1340	0.1521	0.1672	0.1797	0.1899	0.1981
2.0	0.0311	0.0610	0.0887	0.1134	0.1350	0.1533	0.1686	0.1812	0.1915	0.1999
2.5	0.0314	0.0616	0.0895	0.1145	0.1363	0.1548	0.1704	0.1832	0.1938	0.2024
3.0	0.0315	0.0618	0.0898	0.1150	0.1368	0.1555	0.1711	0.1841	0.1947	0.2034
4.0	0.0316	0.0619	0.0901	0.1153	0.1372	0.1560	0.1717	0.1847	0.1954	0.2042
5.0	0.0316	0.0620	0.0901	0.1154	0.1374	0.1561	0.1719	0.1849	0.1956	0.2044
6.0	0.0316	0.0620	0.0902	0.1154	0.1374	0.1562	0.1719	0.1850	0.1957	0.2045

The increase in the stress at any point below a rectangularly loaded area can be found by using Eq. (10.29). This can be explained by reference to Figure 10.22. Let us determine the stress at a point below point A' at depth z. The loaded area can be divided into four rectangles as shown. The point A' is the corner common to all four rectangles. The increase in the stress at depth z below point A' due to each rectangular area can now be calculated by using Eq. (10.29). The total stress increase caused by the entire loaded area can be given by

$$\Delta\sigma_z = q[I_{3(1)} + I_{3(2)} + I_{3(3)} + I_{3(4)}] \tag{10.33}$$

where $I_{3(1)}$, $I_{3(2)}$, $I_{3(3)}$, and $I_{3(4)}$ = values of I_3 for rectangles 1, 2, 3, and 4, respectively.

In most cases the vertical stress increase below the center of a rectangular area (Figure 10.23) is important. This stress increase can be given by the relationship

$$\Delta\sigma_z = qI_4 \tag{10.34}$$

Table 10.8 (*continued*)

1.2	1.4	1.6	1.8	2.0	2.5	3.0	4.0	5.0	6.0
0.0293	0.0301	0.0306	0.0309	0.0311	0.0314	0.0315	0.0316	0.0316	0.0316
0.0573	0.0589	0.0599	0.0606	0.0610	0.0616	0.0618	0.0619	0.0620	0.0620
0.0832	0.0856	0.0871	0.0880	0.0887	0.0895	0.0898	0.0901	0.0901	0.0902
0.1063	0.1094	0.1114	0.1126	0.1134	0.1145	0.1150	0.1153	0.1154	0.1154
0.1263	0.1300	0.1324	0.1340	0.1350	0.1363	0.1368	0.1372	0.1374	0.1374
0.1431	0.1475	0.1503	0.1521	0.1533	0.1548	0.1555	0.1560	0.1561	0.1562
0.1570	0.1620	0.1652	0.1672	0.1686	0.1704	0.1711	0.1717	0.1719	0.1719
0.1684	0.1739	0.1774	0.1797	0.1812	0.1832	0.1841	0.1847	0.1849	0.1850
0.1777	0.1836	0.1874	0.1899	0.1915	0.1938	0.1947	0.1954	0.1956	0.1957
0.1851	0.1914	0.1955	0.1981	0.1999	0.2024	0.2034	0.2042	0.2044	0.2045
0.1958	0.2028	0.2073	0.2103	0.2124	0.2151	0.2163	0.2172	0.2175	0.2176
0.2028	0.2102	0.2151	0.2184	0.2206	0.2236	0.2250	0.2260	0.2263	0.2264
0.2073	0.2151	0.2203	0.2237	0.2261	0.2294	0.2309	0.2320	0.2323	0.2325
0.2103	0.2183	0.2237	0.2274	0.2299	0.2333	0.2350	0.2362	0.2366	0.2367
0.2124	0.2206	0.2261	0.2299	0.2325	0.2361	0.2378	0.2391	0.2395	0.2397
0.2151	0.2236	0.2294	0.2333	0.2361	0.2401	0.2420	0.2434	0.2439	0.2441
0.2163	0.2250	0.2309	0.2350	0.2378	0.2420	0.2439	0.2455	0.2461	0.2463
0.2172	0.2260	0.2320	0.2362	0.2391	0.2434	0.2455	0.2472	0.2479	0.2481
0.2175	0.2263	0.2324	0.2366	0.2395	0.2439	0.2460	0.2479	0.2486	0.2489
0.2176	0.2264	0.2325	0.2367	0.2397	0.2441	0.2463	0.2482	0.2489	0.2492

where

$$I_4 = \frac{2}{\pi}\left[\frac{m_1 n_1}{\sqrt{1 + m_1^2 + n_1^2}}\frac{1 + m_1^2 + 2n_1^2}{(1 + n_1^2)(m_1^2 + n_1^2)} + \sin^{-1}\frac{m_1}{\sqrt{m_1^2 + n_1^2}\sqrt{1 + n_1^2}}\right] \quad (10.35)$$

$$m_1 = \frac{L}{B} \quad (10.36)$$

$$n_1 = \frac{z}{b} \quad (10.37)$$

$$b = \frac{B}{2} \quad (10.38)$$

The variation of I_4 with m_1 and n_1 is given in Table 10.9.

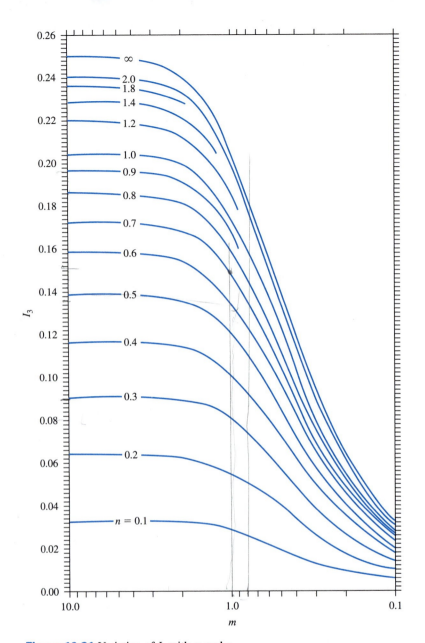

Figure 10.21 Variation of I_3 with m and n

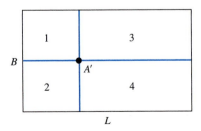

Figure 10.22 Increase of stress at any point below a rectangularly loaded flexible area

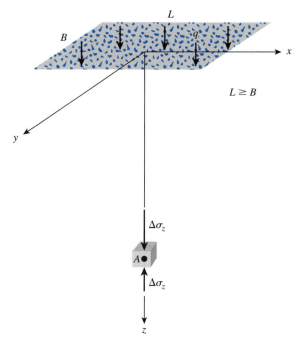

Figure 10.23 Vertical stress below the center of a uniformly loaded flexible rectangular area

Table 10.9 Variation of I_4 with m_1 and n_1 [Eq. (10.35)]

					m_1					
n_1	1	2	3	4	5	6	7	8	9	10
0.20	0.994	0.997	0.997	0.997	0.997	0.997	0.997	0.997	0.997	0.997
0.40	0.960	0.976	0.977	0.977	0.977	0.977	0.977	0.977	0.977	0.977
0.60	0.892	0.932	0.936	0.936	0.937	0.937	0.937	0.937	0.937	0.937
0.80	0.800	0.870	0.878	0.880	0.881	0.881	0.881	0.881	0.881	0.881
1.00	0.701	0.800	0.814	0.817	0.818	0.818	0.818	0.818	0.818	0.818
1.20	0.606	0.727	0.748	0.753	0.754	0.755	0.755	0.755	0.755	0.755
1.40	0.522	0.658	0.685	0.692	0.694	0.695	0.695	0.696	0.696	0.696
1.60	0.449	0.593	0.627	0.636	0.639	0.640	0.641	0.641	0.641	0.642
1.80	0.388	0.534	0.573	0.585	0.590	0.591	0.592	0.592	0.593	0.593
2.00	0.336	0.481	0.525	0.540	0.545	0.547	0.548	0.549	0.549	0.549
3.00	0.179	0.293	0.348	0.373	0.384	0.389	0.392	0.393	0.394	0.395
4.00	0.108	0.190	0.241	0.269	0.285	0.293	0.298	0.301	0.302	0.303
5.00	0.072	0.131	0.174	0.202	0.219	0.229	0.236	0.240	0.242	0.244
6.00	0.051	0.095	0.130	0.155	0.172	0.184	0.192	0.197	0.200	0.202
7.00	0.038	0.072	0.100	0.122	0.139	0.150	0.158	0.164	0.168	0.171
8.00	0.029	0.056	0.079	0.098	0.113	0.125	0.133	0.139	0.144	0.147
9.00	0.023	0.045	0.064	0.081	0.094	0.105	0.113	0.119	0.124	0.128
10.00	0.019	0.037	0.053	0.067	0.079	0.089	0.097	0.103	0.108	0.112

Example 10.9

The plan of a uniformly loaded rectangular area is shown in Figure 10.24a. Determine the vertical stress increase $\Delta\sigma_z$ below point A' at a depth of $z = 4$ m.

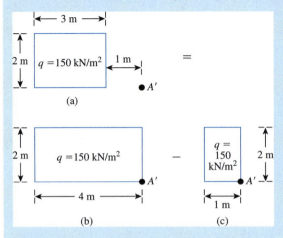

(a)

(b)   (c)   *Figure 10.24*

Solution

The stress increase $\Delta\sigma_z$ can be written as

$$\Delta\sigma_z = \Delta\sigma_{z(1)} - \Delta\sigma_{z(2)}$$

where

$\Delta\sigma_{z(1)}$ = stress increase due to the loaded area shown in Figure 10.24b
$\Delta\sigma_{z(2)}$ = stress increase due to the loaded area shown in Figure 10.24c

For the loaded area shown in Figure 10.24b:

$$m = \frac{B}{z} = \frac{2}{4} = 0.5$$

$$n = \frac{L}{z} = \frac{4}{4} = 1$$

From Figure 10.21 for $m = 0.5$ and $n = 1$, the value of $I_3 = 0.1225$. So

$$\Delta\sigma_{z(1)} = qI_3 = (150)(0.1202) = 18.38 \text{ kN/m}^2$$

Similarly, for the loaded area shown in Figure 10.24c:

$$m = \frac{B}{z} = \frac{1}{4} = 0.25$$

$$n = \frac{L}{z} = \frac{2}{4} = 0.5$$

Thus, $I_3 = 0.0473$. Hence

$$\Delta\sigma_{z(2)} = (150)(0.0473) = 7.1 \text{ kN/m}^2$$

So

$$\Delta\sigma_z = \Delta\sigma_{z(1)} - \Delta\sigma_{z(2)} = 18.38 - 7.1 = \mathbf{11.28 \text{ kN/m}^2}$$

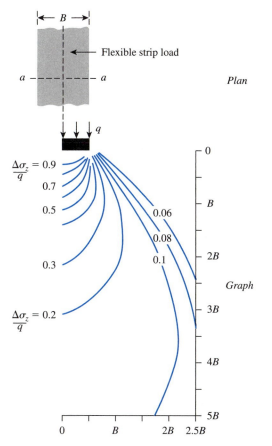

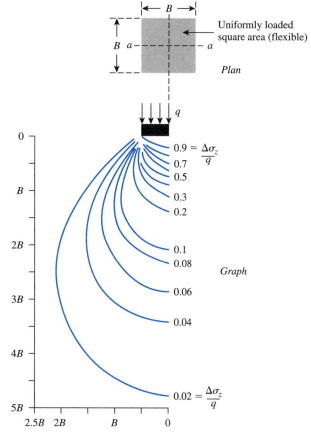

Figure 10.25 Vertical pressure isobars under a flexible strip load (*Note:* Isobars are for line *a–a* as shown on the plan)

Figure 10.26 Vertical pressure isobars under a uniformly loaded square area (*Note:* Isobars are for line *a–a* as shown on the plan)

10.11 Stress Isobars

In Section 10.6, we developed the relationship to estimate $\Delta\sigma_z$ at any point due to a vertical strip loading. Also, Section 10.10 provides the relationships to calculate $\Delta\sigma_z$ at any point due to a vertically and uniformly loaded rectangular area. These relationships for $\Delta\sigma_z$ can be used to calculate the stress increase at various grid points below the loaded area. Based on those calculated stress increases, stress isobars can be plotted. Figures 10.25 and 10.26 show such stress isobars under uniformly loaded (vertically) strip and square areas.

10.12 Influence Chart for Vertical Pressure

Equation (10.25) can be rearranged and written in the form

$$\frac{R}{z} = \sqrt{\left(1 - \frac{\Delta\sigma_z}{q}\right)^{-2/3} - 1} \qquad (10.39)$$

Table 10.10 Values of R/z for Various Pressure Ratios [Eq. (10.39)]

$\Delta\sigma_z/q$	R/z	$\Delta\sigma_z/q$	R/z
0	0	0.55	0.8384
0.05	0.1865	0.60	0.9176
0.10	0.2698	0.65	1.0067
0.15	0.3383	0.70	1.1097
0.20	0.4005	0.75	1.2328
0.25	0.4598	0.80	1.3871
0.30	0.5181	0.85	1.5943
0.35	0.5768	0.90	1.9084
0.40	0.6370	0.95	2.5232
0.45	0.6997	1.00	∞
0.50	0.7664		

Note that R/z and $\Delta\sigma_z/q$ in this equation are nondimensional quantities. The values of R/z that correspond to various pressure ratios are given in Table 10.10.

Using the values of R/z obtained from Eq. (10.39) for various pressure ratios, Newmark (1942) presented an influence chart that can be used to determine the vertical pressure at any point below a uniformly loaded flexible area of any shape.

Figure 10.27 shows an influence chart that has been constructed by drawing concentric circles. The radii of the circles are equal to the R/z values corresponding to $\Delta\sigma_z/q =$ 0, 0.1, 0.2, . . . , 1. (*Note:* For $\Delta\sigma_z/q = 0$, $R/z = 0$, and for $\Delta\sigma_z/q = 1$, $R/z = \infty$, so nine

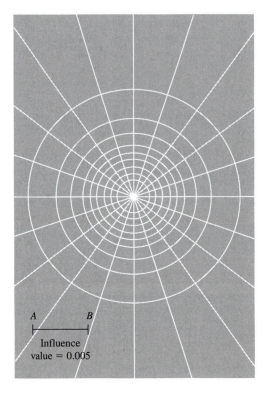

A   *B*

Influence value = 0.005

Figure 10.27
Influence chart for vertical pressure based on Boussinesq's theory (*Bulletin No. 338. Influence Charts for Computation of Stresses in Elastic Foundations, by Nathan M. Newmark University of Illinois, 1942.*)

circles are shown.) The unit length for plotting the circles is $\overline{AB}$. The circles are divided by several equally spaced radial lines. The influence value of the chart is given by $1/N$, where N is equal to the number of elements in the chart. In Figure 10.27, there are 200 elements; hence, the influence value is 0.005.

The procedure for obtaining vertical pressure at any point below a loaded area is as follows:

1. Determine the depth z below the uniformly loaded area at which the stress increase is required.
2. Plot the plan of the loaded area with a scale of z equal to the unit length of the chart $(\overline{AB})$.
3. Place the plan (plotted in step 2) on the influence chart in such a way that the point below which the stress is to be determined is located at the center of the chart.
4. Count the number of elements (M) of the chart enclosed by the plan of the loaded area.

The increase in the pressure at the point under consideration is given by

$$\Delta\sigma_z = (IV)qM \tag{10.40}$$

where IV = influence value
$\quad\;\; q$ = pressure on the loaded area

Example 10.10

The cross section and plan of a column footing are shown in Figure 10.28. Find the increase in vertical stress produced by the column footing at point A.

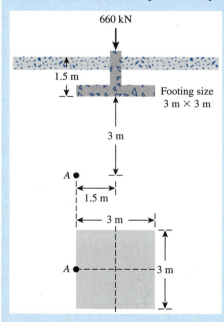

Figure 10.28
Cross section and plan
of a column footing

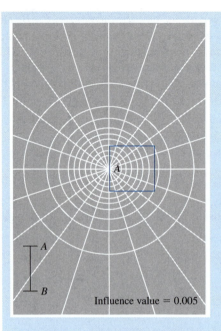

Figure 10.29 Determination of stress at a point by use of Newmark's influence chart

Solution
Point A is located at à depth 3 m below the bottom of the footing. The plan of the square footing has been replotted to a scale of $\overline{AB} = 3$ m and placed on the influence chart (Figure 10.29) in such a way that point A on the plan falls directly over the center of the chart. The number of elements inside the outline of the plan is about 48.5. Hence,

$$\Delta\sigma_z = (IV)qM = 0.005\left(\frac{660}{3 \times 3}\right)48.5 = \textbf{17.78 kN/m}^2 \qquad \blacksquare$$

10.13   *Summary and General Comments*

This chapter presents the relationships for determining vertical stress at a point due to the application of various types of loading on the surface of a soil mass. The types of loading considered here are point, line, strip, embankment, circular, and rectangular. These relationships are derived by integration of Boussinesq's equation for a point load.

The equations and graphs presented in this chapter are based entirely on the principles of the theory of elasticity; however, one must realize the limitations of these theories when they are applied to a soil medium. This is because soil deposits, in general, are not homogeneous, perfectly elastic, and isotropic. Hence, some deviations from the theoretical stress calculations can be expected in the field. Only a limited number of field observations are available in the literature for comparison purposes. On the basis of these results, it appears that one could expect a difference of ±25 to 30% between theoretical estimates and actual field values.

Problems

10.1 A soil element is shown in Figure 10.30. Determine the following:
  a. Maximum and minimum principal stresses
  b. Normal and shear stresses on the plane *AB*
  Use Eqs. (10.3), (10.4), (10.6), and (10.7).

10.2 Repeat Problem 10.1 for the element shown in Figure 10.31.

10.3 Using the principles of Mohr's circles for the soil element shown in Figure 10.32, determine the following:
  a. Maximum and minimum principal stresses
  b. Normal and shear stresses on the plane *AB*

10.4 Repeat Problem 10.3 for the soil element shown in Figure 10.33.

10.5 A soil element is shown in Figure 10.34. Determine the following:
  a. Maximum and minimum principal stresses
  b. Normal and shear stresses on the plane *AB*
  Use the pole method.

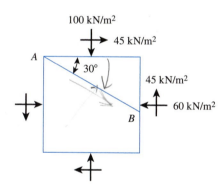

Figure 10.30

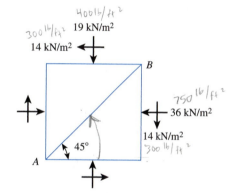

Figure 10.31

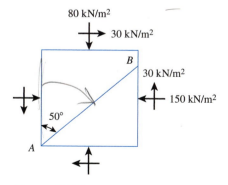

Figure 10.32

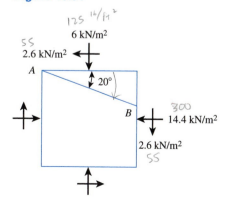

Figure 10.33

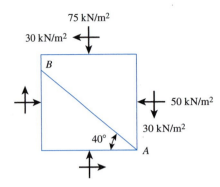

Figure 10.34

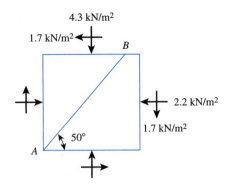

Figure 10.35

10.6 Repeat Problem 10.5 for the soil element shown in Figure 10.35.

10.7 Point loads of magnitude 8.9, 17.8, and 26.7 kN act at *A*, *B*, and *C*, respectively (Figure 10.36). Determine the increase in vertical stress at a depth of 3 m below point *D*. Use Boussinesq's equation.

10.8 Refer to Figure 10.37. Determine the vertical stress increase, $\Delta \sigma_z$, at point *A* with the following values:

- $q_1 = 75$ kN/m   • $x_1 = 2$ m
- $q_2 = 300$ kN/m   • $x_2 = 3$ m
- $z = 2$ m

10.9 Repeat Problem 10.8 with the following data:

- $q_1 = 300$ kN/m   • $x_1 = 4$ m
- $q_2 = 260$ kN/m   • $x_2 = 3$ m
- $z = 3$ m

10.10 Refer to Figure 10.37. Given: $q_1 = 10.9$ kN/m, $x_1 = 2.45$ m, $x_2 = 1.22$ m, and $z = 0.9$ m. If the vertical stress increase at point *A* due to the loading is 1.7 kN/m², determine the magnitude of q_2.

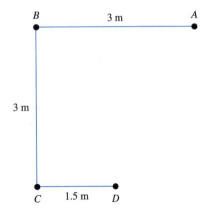

Figure 10.36

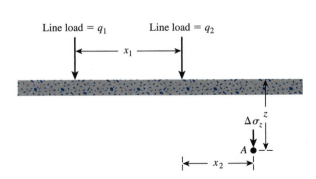

Figure 10.37

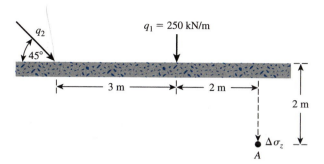

Figure 10.38

10.11 Refer to Figure 10.38. Due to the application of line loads q_1 and q_2, the vertical stress increase, $\Delta \sigma_z$, at A is 30 kN/m². Determine the magnitude of q_2.

10.12 Refer to Figure 10.12. Given: $B = 3.7$ m, $q = 16.8$ kN/m², $x = 2.7$ m, and $z = 1.5$ m. Determine the vertical stress increase, $\Delta \sigma_z$, at point A.

10.13 Repeat Problem 10.12 for $B = 3$ m, $q = 60$ kN/m², $x = 1.5$ m, and $z = 3$ m.

10.14 An earth embankment diagram is shown in Figure 10.39. Determine the stress increase at point A due to the embankment load.

10.15 Figure 10.40 shows an embankment load for a silty clay soil layer. Determine the vertical stress increase at points A, B, and C.

10.16 Consider a circularly loaded flexible area on the ground surface. Given that the radius of the circular area $R = 4$ m and that the uniformly distributed load $q = 200$ kN/m², calculate the vertical stress increase, $\Delta \sigma_z$, at points 1.5, 3, 6, 9, and 12 m below the ground surface (immediately below the center of the circular area).

10.17 Figure 10.19 shows a flexible circular area of radius $R = 3$ m. The uniformly distributed load on the circular area is 96 kN/m². Calculate the vertical stress increase at $r = 0, 0.6, 1.2, 2.4$, and 3.6 and $z = 1.5$ m.

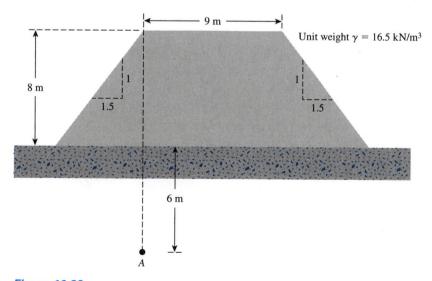

Figure 10.39

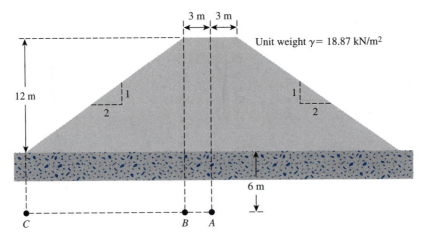

Figure 10.40

10.18 Refer to Figure 10.41. The circular flexible area is uniformly loaded. Given $q = 300$ kN/m^2 and using Newmark's chart, determine the vertical stress increase $\Delta \sigma_z$ at point A.

10.19 The plan of a flexible rectangular loaded area is shown in Figure 10.42. The uniformly distributed load on the flexible area, q, is 100 kN/m^2. Determine the increase in the vertical stress, $\Delta \sigma_z$, at a depth of $z = 2$ m below

   a. Point A
   b. Point B
   c. Point C

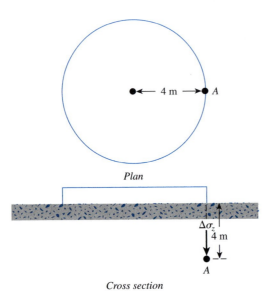

Plan

Cross section

Figure 10.41

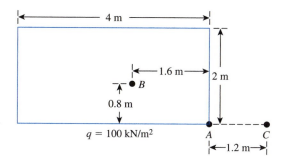

Figure 10.42

10.20 Refer to the flexible loaded rectangular area shown in Figure 10.42. Using Eq. (10.34), determine the vertical stress increase below the center of the area at a depth of 3.5 m.

References

AHLVIN, R. G., and ULERY, H. H. (1962). "Tabulated Values for Determining the Complete Pattern of Stresses, Strains, and Deflections Beneath a Uniform Circular Load on a Homogeneous Half Space," in *Highway Research Bulletin 342,* Transportation Research Board, National Research Council, Washington, D.C., 1–13.

BOUSSINESQ, J. (1883). *Application des Potentials à L'Etude de L'Equilibre et du Mouvement des Solides Elastiques,* Gauthier-Villars, Paris.

DAS, B. (2008). *Advanced Soil Mechanics,* 3rd ed., Taylor and Francis, London.

NEWMARK, N. M. (1942). "Influence Charts for Computation of Stresses in Elastic Soil," University of Illinois Engineering Experiment Station, *Bulletin No. 338.*

OSTERBERG, J. O. (1957). "Influence Values for Vertical Stresses in Semi-Infinite Mass Due to Embankment Loading," *Proceedings,* Fourth International Conference on Soil Mechanics and Foundation Engineering, London, Vol. 1, 393–396.

11 Compressibility of Soil

A stress increase caused by the construction of foundations or other loads compresses soil layers. The compression is caused by (a) deformation of soil particles, (b) relocations of soil particles, and (c) expulsion of water or air from the void spaces. In general, the soil settlement caused by loads may be divided into three broad categories:

1. *Elastic settlement* (or *immediate settlement*), which is caused by the elastic deformation of dry soil and of moist and saturated soils without any change in the moisture content. Elastic settlement calculations generally are based on equations derived from the theory of elasticity.
2. *Primary consolidation settlement,* which is the result of a volume change in saturated cohesive soils because of expulsion of the water that occupies the void spaces.
3. *Secondary consolidation settlement,* which is observed in saturated cohesive soils and is the result of the plastic adjustment of soil fabrics. It is an additional form of compression that occurs at constant effective stress.

This chapter presents the fundamental principles for estimating the elastic and consolidation settlements of soil layers under superimposed loadings.

The total settlement of a foundation can then be given as

$$S_T = S_c + S_s + S_e$$

where S_T = total settlement
S_c = primary consolidation settlement
S_s = secondary consolidation settlement
S_e = elastic settlement

When foundations are constructed on very compressible clays, the consolidation settlement can be several times greater than the elastic settlement.

ELASTIC SETTLEMENT

11.1 Contact Pressure and Settlement Profile

Elastic, or immediate, settlement of foundations (S_e) occurs directly after the application of a load without a change in the moisture content of the soil. The magnitude of the contact

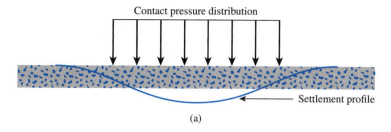

(a)

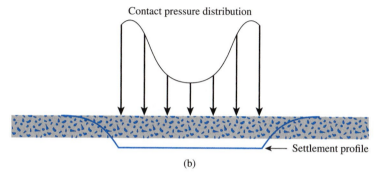

(b)

Figure 11.1
Elastic settlement
profile and contact
pressure in clay:
(a) flexible foun-
dation; (b) rigid foun-
dation

settlement will depend on the flexibility of the foundation and the type of material on which it is resting.

In Chapter 10, the relationships for determining the increase in stress (which causes elastic settlement) due to the application of line load, strip load, embankment load, circular load, and rectangular load were based on the following assumptions:

* The load is applied at the ground surface.
* The loaded area is *flexible.*
* The soil medium is homogeneous, elastic, isotropic, and extends to a great depth.

In general, foundations are not perfectly flexible and are embedded at a certain depth below the ground surface. It is instructive, however, to evaluate the distribution of the contact pressure under a foundation along with the settlement profile under idealized conditions. Figure 11.1a shows a *perfectly flexible* foundation resting on an elastic material such as saturated clay. If the foundation is subjected to a uniformly distributed load, the contact pressure will be uniform and the foundation will experience a sagging profile. On the other hand, if we consider a *perfectly rigid* foundation resting on the ground surface subjected to a uniformly distributed load, the contact pressure and foundation settlement profile will be as shown in Figure 11.1b: the foundation will undergo a uniform settlement and the contact pressure will be redistributed.

The settlement profile and contact pressure distribution described are true for soils in which the modulus of elasticity is fairly constant with depth. In the case of cohesionless sand, the modulus of elasticity increases with depth. Additionally, there is a lack of lateral confinement on the edge of the foundation at the ground surface. The sand at the edge of a flexible foundation is pushed outward, and the deflection curve of the foundation takes a concave downward shape. The distributions of contact pressure and the settlement profiles of a flexible and a rigid foundation resting on sand and subjected to uniform loading are shown in Figures 11.2a and 11.2b, respectively.

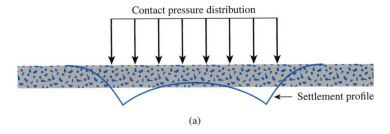

Contact pressure distribution

Settlement profile

(a)

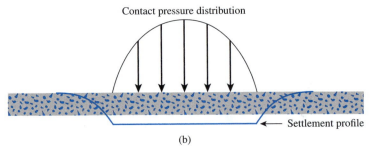

Contact pressure distribution

Settlement profile

(b)

Figure 11.2
Elastic settlement profile and contact pressure in sand: (a) flexible foundation; (b) rigid foundation

11.2 Relations for Elastic Settlement Calculation

Figure 11.3 shows a shallow foundation subjected to a net force per unit area equal to $\Delta\sigma$. Let the Poisson's ratio and the modulus of elasticity of the soil supporting it be μ_s and E_s, respectively. Theoretically, if the foundation is perfectly flexible, the settlement may be expressed as

$$S_e = \Delta\sigma(\alpha B')\frac{1 - \mu_s^2}{E_s}I_sI_f \qquad (11.1)$$

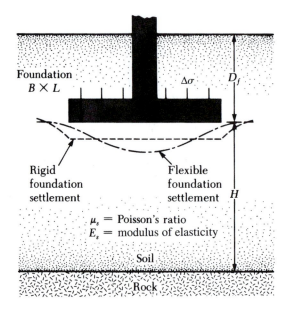

Foundation
$B \times L$

$\Delta\sigma$

D_f

Rigid foundation settlement

Flexible foundation settlement

H

μ_s = Poisson's ratio
E_s = modulus of elasticity

Soil

Rock

Figure 11.3 Elastic settlement of flexible and rigid foundations

where $\Delta\sigma$ = net applied pressure on the foundation

μ_s = Poisson's ratio of soil

E_s = average modulus of elasticity of the soil under the foundation measured from $z = 0$ to about $z = 4B$

$B' = B/2$ for center of foundation

$= B$ for corner of foundation

I_s = shape factor (Steinbrenner, 1934)

$$= F_1 + \frac{1 - 2\mu_s}{1 - \mu_s} F_2 \tag{11.2}$$

$$F_1 = \frac{1}{\pi}(A_0 + A_1) \tag{11.3}$$

$$F_2 = \frac{n'}{2\pi} \tan^{-1} A_2 \tag{11.4}$$

$$A_0 = m' \ln \frac{(1 + \sqrt{m'^2 + 1})\sqrt{m'^2 + n'^2}}{m'(1 + \sqrt{m'^2 + n'^2 + 1})} \tag{11.5}$$

$$A_1 = \ln \frac{(m' + \sqrt{m'^2 + 1})\sqrt{1 + n'^2}}{m' + \sqrt{m'^2 + n'^2 + 1}} \tag{11.6}$$

$$A_2 = \frac{m'}{n'\sqrt{m'^2 + n'^2 + 1}} \tag{11.7}$$

$$I_f = \text{depth factor (Fox, 1948)} = f\left(\frac{D_f}{B}, \mu_s, \text{ and } \frac{L}{B}\right) \tag{11.8}$$

α = factor that depends on the location on the foundation where settlement is being calculated

• For calculation of settlement at the *center* of the foundation:

$$\alpha = 4$$
$$m' = \frac{L}{B}$$
$$n' = \frac{H}{\left(\frac{B}{2}\right)}$$

• For calculation of settlement at a *corner* of the foundation:

$$\alpha = 1$$
$$m' = \frac{L}{B}$$
$$n' = \frac{H}{B}$$

The variations of F_1 and F_2 [Eqs. (11.3) and (11.4)] with m' and n' given in Tables 11.1 and 11.2. Also the variation of I_f with D_f/B and μ_s is given in Table 11.3. Note that *when $D_f = 0$, the value of $I_f = 1$ in all cases.*

Table 11.1 Variation of F_1 with m' and n'

n'	m'									
	1.0	1.2	1.4	1.6	1.8	2.0	2.5	3.0	3.5	4.0
0.25	0.014	0.013	0.012	0.011	0.011	0.011	0.010	0.010	0.010	0.010
0.50	0.049	0.046	0.044	0.042	0.041	0.040	0.038	0.038	0.037	0.037
0.75	0.095	0.090	0.087	0.084	0.082	0.080	0.077	0.076	0.074	0.074
1.00	0.142	0.138	0.134	0.130	0.127	0.125	0.121	0.118	0.116	0.115
1.25	0.186	0.183	0.179	0.176	0.173	0.170	0.165	0.161	0.158	0.157
1.50	0.224	0.224	0.222	0.219	0.216	0.213	0.207	0.203	0.199	0.197
1.75	0.257	0.259	0.259	0.258	0.255	0.253	0.247	0.242	0.238	0.235
2.00	0.285	0.290	0.292	0.292	0.291	0.289	0.284	0.279	0.275	0.271
2.25	0.309	0.317	0.321	0.323	0.323	0.322	0.317	0.313	0.308	0.305
2.50	0.330	0.341	0.347	0.350	0.351	0.351	0.348	0.344	0.340	0.336
2.75	0.348	0.361	0.369	0.374	0.377	0.378	0.377	0.373	0.369	0.365
3.00	0.363	0.379	0.389	0.396	0.400	0.402	0.402	0.400	0.396	0.392
3.25	0.376	0.394	0.406	0.415	0.420	0.423	0.426	0.424	0.421	0.418
3.50	0.388	0.408	0.422	0.431	0.438	0.442	0.447	0.447	0.444	0.441
3.75	0.399	0.420	0.436	0.447	0.454	0.460	0.467	0.458	0.466	0.464
4.00	0.408	0.431	0.448	0.460	0.469	0.476	0.484	0.487	0.486	0.484
4.25	0.417	0.440	0.458	0.472	0.481	0.484	0.495	0.514	0.515	0.515
4.50	0.424	0.450	0.469	0.484	0.495	0.503	0.516	0.521	0.522	0.522
4.75	0.431	0.458	0.478	0.494	0.506	0.515	0.530	0.536	0.539	0.539
5.00	0.437	0.465	0.487	0.503	0.516	0.526	0.543	0.551	0.554	0.554
5.25	0.443	0.472	0.494	0.512	0.526	0.537	0.555	0.564	0.568	0.569
5.50	0.448	0.478	0.501	0.520	0.534	0.546	0.566	0.576	0.581	0.584
5.75	0.453	0.483	0.508	0.527	0.542	0.555	0.576	0.588	0.594	0.597
6.00	0.457	0.489	0.514	0.534	0.550	0.563	0.585	0.598	0.606	0.609
6.25	0.461	0.493	0.519	0.540	0.557	0.570	0.594	0.609	0.617	0.621
6.50	0.465	0.498	0.524	0.546	0.563	0.577	0.603	0.618	0.627	0.632
6.75	0.468	0.502	0.529	0.551	0.569	0.584	0.610	0.627	0.637	0.643
7.00	0.471	0.506	0.533	0.556	0.575	0.590	0.618	0.635	0.646	0.653
7.25	0.474	0.509	0.538	0.561	0.580	0.596	0.625	0.643	0.655	0.662
7.50	0.477	0.513	0.541	0.565	0.585	0.601	0.631	0.650	0.663	0.671
7.75	0.480	0.516	0.545	0.569	0.589	0.606	0.637	0.658	0.671	0.680
8.00	0.482	0.519	0.549	0.573	0.594	0.611	0.643	0.664	0.678	0.688
8.25	0.485	0.522	0.552	0.577	0.598	0.615	0.648	0.670	0.685	0.695
8.50	0.487	0.524	0.555	0.580	0.601	0.619	0.653	0.676	0.692	0.703
8.75	0.489	0.527	0.558	0.583	0.605	0.623	0.658	0.682	0.698	0.710
9.00	0.491	0.529	0.560	0.587	0.609	0.627	0.663	0.687	0.705	0.716
9.25	0.493	0.531	0.563	0.589	0.612	0.631	0.667	0.693	0.710	0.723
9.50	0.495	0.533	0.565	0.592	0.615	0.634	0.671	0.697	0.716	0.719
9.75	0.496	0.536	0.568	0.595	0.618	0.638	0.675	0.702	0.721	0.735
10.00	0.498	0.537	0.570	0.597	0.621	0.641	0.679	0.707	0.726	0.740
20.00	0.529	0.575	0.614	0.647	0.677	0.702	0.756	0.797	0.830	0.858
50.00	0.548	0.598	0.640	0.678	0.711	0.740	0.803	0.853	0.895	0.931
100.00	0.555	0.605	0.649	0.688	0.722	0.753	0.819	0.872	0.918	0.956

Table 11.1 (*continued*)

					m′					
n′	4.5	5.0	6.0	7.0	8.0	9.0	10.0	25.0	50.0	100.0
0.25	0.010	0.010	0.010	0.010	0.010	0.010	0.010	0.010	0.010	0.010
0.50	0.036	0.036	0.036	0.036	0.036	0.036	0.036	0.036	0.036	0.036
0.75	0.073	0.073	0.072	0.072	0.072	0.072	0.071	0.071	0.071	0.071
1.00	0.114	0.113	0.112	0.112	0.112	0.111	0.111	0.110	0.110	0.110
1.25	0.155	0.154	0.153	0.152	0.152	0.151	0.151	0.150	0.150	0.150
1.50	0.195	0.194	0.192	0.191	0.190	0.190	0.189	0.188	0.188	0.188
1.75	0.233	0.232	0.229	0.228	0.227	0.226	0.225	0.223	0.223	0.223
2.00	0.269	0.267	0.264	0.262	0.261	0.260	0.259	0.257	0.256	0.256
2.25	0.302	0.300	0.296	0.294	0.293	0.291	0.291	0.287	0.287	0.287
2.50	0.333	0.331	0.327	0.324	0.322	0.321	0.320	0.316	0.315	0.315
2.75	0.362	0.359	0.355	0.352	0.350	0.348	0.347	0.343	0.342	0.342
3.00	0.389	0.386	0.382	0.378	0.376	0.374	0.373	0.368	0.367	0.367
3.25	0.415	0.412	0.407	0.403	0.401	0.399	0.397	0.391	0.390	0.390
3.50	0.438	0.435	0.430	0.427	0.424	0.421	0.420	0.413	0.412	0.411
3.75	0.461	0.458	0.453	0.449	0.446	0.443	0.441	0.433	0.432	0.432
4.00	0.482	0.479	0.474	0.470	0.466	0.464	0.462	0.453	0.451	0.451
4.25	0.516	0.496	0.484	0.473	0.471	0.471	0.470	0.468	0.462	0.460
4.50	0.520	0.517	0.513	0.508	0.505	0.502	0.499	0.489	0.487	0.487
4.75	0.537	0.535	0.530	0.526	0.523	0.519	0.517	0.506	0.504	0.503
5.00	0.554	0.552	0.548	0.543	0.540	0.536	0.534	0.522	0.519	0.519
5.25	0.569	0.568	0.564	0.560	0.556	0.553	0.550	0.537	0.534	0.534
5.50	0.584	0.583	0.579	0.575	0.571	0.568	0.585	0.551	0.549	0.548
5.75	0.597	0.597	0.594	0.590	0.586	0.583	0.580	0.565	0.583	0.562
6.00	0.611	0.610	0.608	0.604	0.601	0.598	0.595	0.579	0.576	0.575
6.25	0.623	0.623	0.621	0.618	0.615	0.611	0.608	0.592	0.589	0.588
6.50	0.635	0.635	0.634	0.631	0.628	0.625	0.622	0.605	0.601	0.600
6.75	0.646	0.647	0.646	0.644	0.641	0.637	0.634	0.617	0.613	0.612
7.00	0.656	0.658	0.658	0.656	0.653	0.650	0.647	0.628	0.624	0.623
7.25	0.666	0.669	0.669	0.668	0.665	0.662	0.659	0.640	0.635	0.634
7.50	0.676	0.679	0.680	0.679	0.676	0.673	0.670	0.651	0.646	0.645
7.75	0.685	0.688	0.690	0.689	0.687	0.684	0.681	0.661	0.656	0.655
8.00	0.694	0.697	0.700	0.700	0.698	0.695	0.692	0.672	0.666	0.665
8.25	0.702	0.706	0.710	0.710	0.708	0.705	0.703	0.682	0.676	0.675
8.50	0.710	0.714	0.719	0.719	0.718	0.715	0.713	0.692	0.686	0.684
8.75	0.717	0.722	0.727	0.728	0.727	0.725	0.723	0.701	0.695	0.693
9.00	0.725	0.730	0.736	0.737	0.736	0.735	0.732	0.710	0.704	0.702
9.25	0.731	0.737	0.744	0.746	0.745	0.744	0.742	0.719	0.713	0.711
9.50	0.738	0.744	0.752	0.754	0.754	0.753	0.751	0.728	0.721	0.719
9.75	0.744	0.751	0.759	0.762	0.762	0.761	0.759	0.737	0.729	0.727
10.00	0.750	0.758	0.766	0.770	0.770	0.770	0.768	0.745	0.738	0.735
20.00	0.878	0.896	0.925	0.945	0.959	0.969	0.977	0.982	0.965	0.957
50.00	0.962	0.989	1.034	1.070	1.100	1.125	1.146	1.265	1.279	1.261
100.00	0.990	1.020	1.072	1.114	1.150	1.182	1.209	1.408	1.489	1.499

Table 11.2 Variation of F_2 with m' and n'

n'	1.0	1.2	1.4	1.6	1.8	2.0	2.5	3.0	3.5	4.0
0.25	0.049	0.050	0.051	0.051	0.051	0.052	0.052	0.052	0.052	0.052
0.50	0.074	0.077	0.080	0.081	0.083	0.084	0.086	0.086	0.0878	0.087
0.75	0.083	0.089	0.093	0.097	0.099	0.101	0.104	0.106	0.107	0.108
1.00	0.083	0.091	0.098	0.102	0.106	0.109	0.114	0.117	0.119	0.120
1.25	0.080	0.089	0.096	0.102	0.107	0.111	0.118	0.122	0.125	0.127
1.50	0.075	0.084	0.093	0.099	0.105	0.110	0.118	0.124	0.128	0.130
1.75	0.069	0.079	0.088	0.095	0.101	0.107	0.117	0.123	0.128	0.131
2.00	0.064	0.074	0.083	0.090	0.097	0.102	0.114	0.121	0.127	0.131
2.25	0.059	0.069	0.077	0.085	0.092	0.098	0.110	0.119	0.125	0.130
2.50	0.055	0.064	0.073	0.080	0.087	0.093	0.106	0.115	0.122	0.127
2.75	0.051	0.060	0.068	0.076	0.082	0.089	0.102	0.111	0.119	0.125
3.00	0.048	0.056	0.064	0.071	0.078	0.084	0.097	0.108	0.116	0.122
3.25	0.045	0.053	0.060	0.067	0.074	0.080	0.093	0.104	0.112	0.119
3.50	0.042	0.050	0.057	0.064	0.070	0.076	0.089	0.100	0.109	0.116
3.75	0.040	0.047	0.054	0.060	0.067	0.073	0.086	0.096	0.105	0.113
4.00	0.037	0.044	0.051	0.057	0.063	0.069	0.082	0.093	0.102	0.110
4.25	0.036	0.042	0.049	0.055	0.061	0.066	0.079	0.090	0.099	0.107
4.50	0.034	0.040	0.046	0.052	0.058	0.063	0.076	0.086	0.096	0.104
4.75	0.032	0.038	0.044	0.050	0.055	0.061	0.073	0.083	0.093	0.101
5.00	0.031	0.036	0.042	0.048	0.053	0.058	0.070	0.080	0.090	0.098
5.25	0.029	0.035	0.040	0.046	0.051	0.056	0.067	0.078	0.087	0.095
5.50	0.028	0.033	0.039	0.044	0.049	0.054	0.065	0.075	0.084	0.092
5.75	0.027	0.032	0.037	0.042	0.047	0.052	0.063	0.073	0.082	0.090
6.00	0.026	0.031	0.036	0.040	0.045	0.050	0.060	0.070	0.079	0.087
6.25	0.025	0.030	0.034	0.039	0.044	0.048	0.058	0.068	0.077	0.085
6.50	0.024	0.029	0.033	0.038	0.042	0.046	0.056	0.066	0.075	0.083
6.75	0.023	0.028	0.032	0.036	0.041	0.045	0.055	0.064	0.073	0.080
7.00	0.022	0.027	0.031	0.035	0.039	0.043	0.053	0.062	0.071	0.078
7.25	0.022	0.026	0.030	0.034	0.038	0.042	0.051	0.060	0.069	0.076
7.50	0.021	0.025	0.029	0.033	0.037	0.041	0.050	0.059	0.067	0.074
7.75	0.020	0.024	0.028	0.032	0.036	0.039	0.048	0.057	0.065	0.072
8.00	0.020	0.023	0.027	0.031	0.035	0.038	0.047	0.055	0.063	0.071
8.25	0.019	0.023	0.026	0.030	0.034	0.037	0.046	0.054	0.062	0.069
8.50	0.018	0.022	0.026	0.029	0.033	0.036	0.045	0.053	0.060	0.067
8.75	0.018	0.021	0.025	0.028	0.032	0.035	0.043	0.051	0.059	0.066
9.00	0.017	0.021	0.024	0.028	0.031	0.034	0.042	0.050	0.057	0.064
9.25	0.017	0.020	0.024	0.027	0.030	0.033	0.041	0.049	0.056	0.063
9.50	0.017	0.020	0.023	0.026	0.029	0.033	0.040	0.048	0.055	0.061
9.75	0.016	0.019	0.023	0.026	0.029	0.032	0.039	0.047	0.054	0.060
10.00	0.016	0.019	0.022	0.025	0.028	0.031	0.038	0.046	0.052	0.059
20.00	0.008	0.010	0.011	0.013	0.014	0.016	0.020	0.024	0.027	0.031
50.00	0.003	0.004	0.004	0.005	0.006	0.006	0.008	0.010	0.011	0.013
100.00	0.002	0.002	0.002	0.003	0.003	0.003	0.004	0.005	0.006	0.006

Table 11.2 (*continued*)

n'	4.5	5.0	6.0	7.0	8.0	9.0	10.0	25.0	50.0	100.0
0.25	0.053	0.053	0.053	0.053	0.053	0.053	0.053	0.053	0.053	0.053
0.50	0.087	0.087	0.088	0.088	0.088	0.088	0.088	0.088	0.088	0.088
0.75	0.109	0.109	0.109	0.110	0.110	0.110	0.110	0.111	0.111	0.111
1.00	0.121	0.122	0.123	0.123	0.124	0.124	0.124	0.125	0.125	0.125
1.25	0.128	0.130	0.131	0.132	0.132	0.133	0.133	0.134	0.134	0.134
1.50	0.132	0.134	0.136	0.137	0.138	0.138	0.139	0.140	0.140	0.140
1.75	0.134	0.136	0.138	0.140	0.141	0.142	0.142	0.144	0.144	0.145
2.00	0.134	0.136	0.139	0.141	0.143	0.144	0.145	0.147	0.147	0.148
2.25	0.133	0.136	0.140	0.142	0.144	0.145	0.146	0.149	0.150	0.150
2.50	0.132	0.135	0.139	0.142	0.144	0.146	0.147	0.151	0.151	0.151
2.75	0.130	0.133	0.138	0.142	0.144	0.146	0.147	0.152	0.152	0.153
3.00	0.127	0.131	0.137	0.141	0.144	0.145	0.147	0.152	0.153	0.154
3.25	0.125	0.129	0.135	0.140	0.143	0.145	0.147	0.153	0.154	0.154
3.50	0.122	0.126	0.133	0.138	0.142	0.144	0.146	0.153	0.155	0.155
3.75	0.119	0.124	0.131	0.137	0.141	0.143	0.145	0.154	0.155	0.155
4.00	0.116	0.121	0.129	0.135	0.139	0.142	0.145	0.154	0.155	0.156
4.25	0.113	0.119	0.127	0.133	0.138	0.141	0.144	0.154	0.156	0.156
4.50	0.110	0.116	0.125	0.131	0.136	0.140	0.143	0.154	0.156	0.156
4.75	0.107	0.113	0.123	0.130	0.135	0.139	0.142	0.154	0.156	0.157
5.00	0.105	0.111	0.120	0.128	0.133	0.137	0.140	0.154	0.156	0.157
5.25	0.102	0.108	0.118	0.126	0.131	0.136	0.139	0.154	0.156	0.157
5.50	0.099	0.106	0.116	0.124	0.130	0.134	0.138	0.154	0.156	0.157
5.75	0.097	0.103	0.113	0.122	0.128	0.133	0.136	0.154	0.157	0.157
6.00	0.094	0.101	0.111	0.120	0.126	0.131	0.135	0.153	0.157	0.157
6.25	0.092	0.098	0.109	0.118	0.124	0.129	0.134	0.153	0.157	0.158
6.50	0.090	0.096	0.107	0.116	0.122	0.128	0.132	0.153	0.157	0.158
6.75	0.087	0.094	0.105	0.114	0.121	0.126	0.131	0.153	0.157	0.158
7.00	0.085	0.092	0.103	0.112	0.119	0.125	0.129	0.152	0.157	0.158
7.25	0.083	0.090	0.101	0.110	0.117	0.123	0.128	0.152	0.157	0.158
7.50	0.081	0.088	0.099	0.108	0.115	0.121	0.126	0.152	0.156	0.158
7.75	0.079	0.086	0.097	0.106	0.114	0.120	0.125	0.151	0.156	0.158
8.00	0.077	0.084	0.095	0.104	0.112	0.118	0.124	0.151	0.156	0.158
8.25	0.076	0.082	0.093	0.102	0.110	0.117	0.122	0.150	0.156	0.158
8.50	0.074	0.080	0.091	0.101	0.108	0.115	0.121	0.150	0.156	0.158
8.75	0.072	0.078	0.089	0.099	0.107	0.114	0.119	0.150	0.156	0.158
9.00	0.071	0.077	0.088	0.097	0.105	0.112	0.118	0.149	0.156	0.158
9.25	0.069	0.075	0.086	0.096	0.104	0.110	0.116	0.149	0.156	0.158
9.50	0.068	0.074	0.085	0.094	0.102	0.109	0.115	0.148	0.156	0.158
9.75	0.066	0.072	0.083	0.092	0.100	0.107	0.113	0.148	0.156	0.158
10.00	0.065	0.071	0.082	0.091	0.099	0.106	0.112	0.147	0.156	0.158
20.00	0.035	0.039	0.046	0.053	0.059	0.065	0.071	0.124	0.148	0.156
50.00	0.014	0.016	0.019	0.022	0.025	0.028	0.031	0.071	0.113	0.142
100.00	0.007	0.008	0.010	0.011	0.013	0.014	0.016	0.039	0.071	0.113

Header spans *m'* over columns 4.5–100.0.

Table 11.3 Variation of I_f with L/B and D_f/B

		I_f		
L/B	D_f/B	$\mu_s = 0.3$	$\mu_s = 0.4$	$\mu_s = 0.5$
1	0.5	0.77	0.82	0.85
	0.75	0.69	0.74	0.77
	1	0.65	0.69	0.72
2	0.5	0.82	0.86	0.89
	0.75	0.75	0.79	0.83
	1	0.71	0.75	0.79
5	0.5	0.87	0.91	0.93
	0.75	0.81	0.86	0.89
	1	0.78	0.82	0.85

The elastic settlement of a *rigid foundation* can be estimated as

$$S_{e(\text{rigid})} \approx 0.93 S_{e(\text{flexible, center})} \tag{11.9}$$

Due to the nonhomogeneous nature of soil deposits, the magnitude of E_s may vary with depth. For that reason, Bowles (1987) recommended using a weighted average value of E_s in Eq. (11.1) or

$$E_s = \frac{\Sigma E_{s(i)} \Delta z}{\bar{z}} \tag{11.10}$$

Where $E_{s(i)}$ = soil modulus of elasticity within a depth Δz
$\bar{z} = H$ or $5B$, whichever is smaller

Representative values of the modulus of elasticity and Poisson's ratio for different types of soils are given in Tables 11.4 and 11.5, respectively.

Table 11.4 Representative Values of the Modulus of Elasticity of Soil

Soil type	E_s (kN/m^2)
Soft clay	1,800–3,500
Hard clay	6,000–14,000
Loose sand	10,000–28,000
Dense sand	35,000–70,000

Table 11.5 Representative Values of Poisson's Ratio

Type of soil	Poisson's ratio, μ_s
Loose sand	0.2–0.4
Medium sand	0.25–0.4
Dense sand	0.3–0.45
Silty sand	0.2–0.4
Soft clay	0.15–0.25
Medium clay	0.2–0.5

Example 11.1

A rigid shallow foundation 1 m × 2 m is shown in Figure 11.4. Calculate the elastic settlement at the center of the foundation.

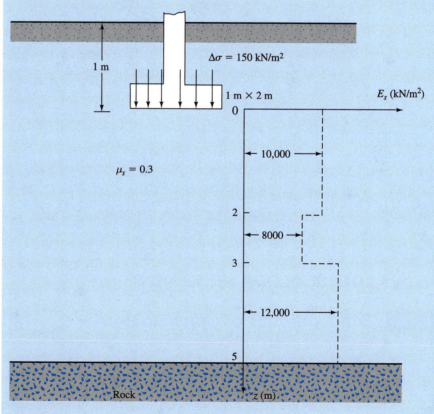

Figure 11.4

Solution

Given: $B = 1$ m and $L = 2$ m. Note that $\bar{z} = 5$ m $= 5B$. From Eq. (11.10),

$$E_s = \frac{\Sigma E_{s(i)}\Delta z}{\bar{z}}$$

$$= \frac{(10,000)(2) + (8,000)(1) + (12,000)(2)}{5} = 10,400\,\text{kN/m}^2$$

For the *center of the foundation*,

$$\alpha = 4$$

$$m' = \frac{L}{B} = \frac{2}{1} = 2$$

$$n' = \frac{H}{\left(\dfrac{B}{2}\right)} = \frac{5}{\left(\dfrac{1}{2}\right)} = 10$$

From Tables 11.1 and 11.2, $F_1 = 0.641$ and $F_2 = 0.031$. From Eq. (11.2).

$$I_s = F_1 + \frac{2 - \mu_s}{1 - \mu_s}F_2$$

$$= 0.641 + \frac{2 - 0.3}{1 - 0.3}(0.031) = 0.716$$

Again, $\dfrac{D_f}{B} = \dfrac{1}{1} = 1, \dfrac{L}{B} = 2, \mu_s = 0.3$. From Table 11.3, $I_f = 0.71$. Hence,

$$S_{e(\text{flexible})} = \Delta\sigma(\alpha B')\frac{1 - \mu_s^2}{E_s}I_s I_f$$

$$= (150)\left(4 \times \frac{1}{2}\right)\left(\frac{1 - 0.3^2}{10,400}\right)(0.716)(0.71) = 0.0133\,\text{m} = 13.3\,\text{mm}$$

Since the foundation is rigid, from Eq. (11.9),

$$S_e(\text{rigid}) = (0.93)(13.3) = \mathbf{12.4\ mm}$$

■

CONSOLIDATION SETTLEMENT

Fundamentals of Consolidation

When a saturated soil layer is subjected to a stress increase, the pore water pressure is increased suddenly. In sandy soils that are highly permeable, the drainage caused by the increase in the pore water pressure is completed immediately. Pore water drainage is accompanied by a reduction in the volume of the soil mass, which results in settlement. Because of rapid drainage of the pore water in sandy soils, elastic settlement and consolidation occur simultaneously.

When a saturated compressible clay layer is subjected to a stress increase, elastic settlement occurs immediately. Because the hydraulic conductivity of clay is significantly smaller than that of sand, the excess pore water pressure generated by loading gradually dissipates over a long period. Thus, the associated volume change (that is, the consolidation) in the clay may continue long after the elastic settlement. The settlement caused by consolidation in clay may be several times greater than the elastic settlement.

The time-dependent deformation of saturated clayey soil best can be understood by considering a simple model that consists of a cylinder with a spring at its center. Let the inside area of the cross section of the cylinder be equal to A. The cylinder is filled with water and has a frictionless watertight piston and valve as shown in Figure 11.5a. At this time, if we place a load P on the piston (Figure 11.5b) and keep the valve closed, the entire load will be taken by the water in the cylinder because water is *incompressible*. The spring will not go through any deformation. The excess hydrostatic pressure at this time can be given as

$$\Delta u = \frac{P}{A} \qquad (11.11)$$

This value can be observed in the pressure gauge attached to the cylinder.

In general, we can write

$$P = P_s + P_w \qquad (11.12)$$

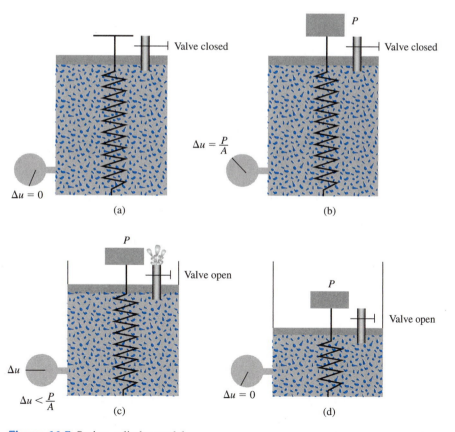

Figure 11.5 Spring-cylinder model

where P_s = load carried by the spring and P_w = load carried by the water.

From the preceding discussion, we can see that when the valve is closed after the placement of the load P,

$$P_s = 0 \quad \text{and} \quad P_w = P$$

Now, if the valve is opened, the water will flow outward (Figure 11.5c). This flow will be accompanied by a reduction of the excess hydrostatic pressure and an increase in the compression of the spring. So, at this time, Eq. (11.12) will hold. However,

$$P_s > 0 \quad \text{and} \quad P_w < P \quad \text{(that is, } \Delta u < P/A)$$

After some time, the excess hydrostatic pressure will become zero and the system will reach a state of equilibrium, as shown in Figure 11.5d. Now we can write

$$P_s = P \quad \text{and} \quad P_w = 0$$

and

$$P = P_s + P_w$$

With this in mind, we can analyze the strain of a saturated clay layer subjected to a stress increase (Figure 11.6a). Consider the case where a layer of saturated clay of thickness H that is confined between two layers of sand is being subjected to an instantaneous increase of *total stress* of $\Delta\sigma$. This incremental total stress will be transmitted to the pore water and the soil solids. This means that the total stress, $\Delta\sigma$, will be divided in some proportion between effective stress and pore water pressure. The behavior of the effective stress change will be similar to that of the spring in Figure 11.5, and the behavior of the pore water pressure change will be similar to that of the excess hydrostatic pressure in Figure 11.5. From the principle of effective stress (Chapter 9), it follows that

$$\Delta\sigma = \Delta\sigma' + \Delta u \tag{11.13}$$

where $\Delta\sigma'$ = increase in the effective stress
Δu = increase in the pore water pressure

Because clay has a very low hydraulic conductivity and water is incompressible as compared with the soil skeleton, at time $t = 0$, the entire incremental stress, $\Delta\sigma$, will be carried by water ($\Delta\sigma = \Delta u$) at all depths (Figure 11.6b). None will be carried by the soil skeleton—that is, incremental effective stress ($\Delta\sigma'$) = 0.

After the application of incremental stress, $\Delta\sigma$, to the clay layer, the water in the void spaces will start to be squeezed out and will drain in both directions into the sand layers. By this process, the excess pore water pressure at any depth in the clay layer

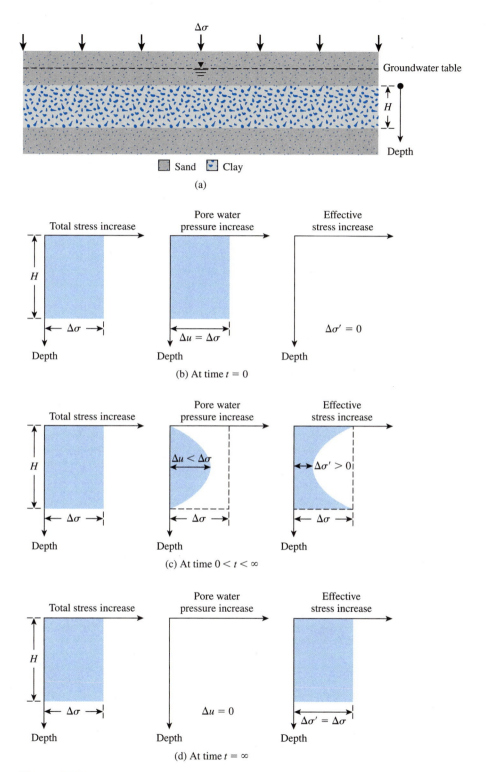

Figure 11.6 Variation of total stress, pore water pressure, and effective stress in a clay layer drained at top and bottom as the result of an added stress, $\Delta\sigma$

307

gradually will decrease, and the stress carried by the soil solids (effective stress) will increase. Thus, at time $0 < t < \infty$,

$$\Delta\sigma = \Delta\sigma' + \Delta u \qquad (\Delta\sigma' > 0 \quad \text{and} \quad \Delta u < \Delta\sigma)$$

However, the magnitudes of $\Delta\sigma'$ and Δu at various depths will change (Figure 11.6c), depending on the minimum distance of the drainage path to either the top or bottom sand layer.

Theoretically, at time $t = \infty$, the entire excess pore water pressure would be dissipated by drainage from all points of the clay layer; thus, $\Delta u = 0$. Now the total stress increase, $\Delta\sigma$, will be carried by the soil structure (Figure 11.6d). Hence,

$$\Delta\sigma = \Delta\sigma'$$

This gradual process of drainage under an additional load application and the associated transfer of excess pore water pressure to effective stress cause the time-dependent settlement in the clay soil layer.

11.4   *One-Dimensional Laboratory Consolidation Test*

The one-dimensional consolidation testing procedure was first suggested by Terzaghi. This test is performed in a consolidometer (sometimes referred to as an *oedometer*). The schematic diagram of a consolidometer is shown in Figure 11.7a. Figure 11.7b. shows a photograph of a consolidometer. The soil specimen is placed inside a metal ring with two porous stones, one at the top of the specimen and another at the bottom. The specimens are usually 64 mm in diameter and 25 mm thick. The load on the specimen is applied through a lever arm, and compression is measured by a micrometer dial gauge. The specimen is kept under water during the test. Each load usually is kept for 24 hours. After that, the load usually is doubled, which doubles the pressure on the specimen, and the compression measurement is continued. At the end of the test, the dry weight of the test specimen is determined. Figure 11.7c shows a consolidation test in progress (right-hand side).

The general shape of the plot of deformation of the specimen against time for a given load increment is shown in Figure 11.8. From the plot, we can observe three distinct stages, which may be described as follows:

Stage I: Initial compression, which is caused mostly by preloading.
Stage II: Primary consolidation, during which excess pore water pressure gradually is transferred into effective stress because of the expulsion of pore water.
Stage III: Secondary consolidation, which occurs after complete dissipation of the excess pore water pressure, when some deformation of the specimen takes place because of the plastic readjustment of soil fabric.

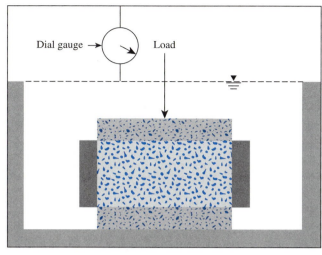

Dial gauge → Load

Porous stone   Soil specimen   Specimen ring

(a)

(b)

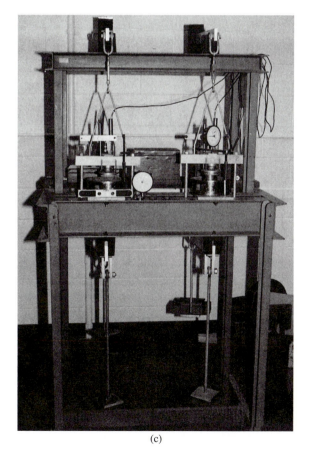

(c)

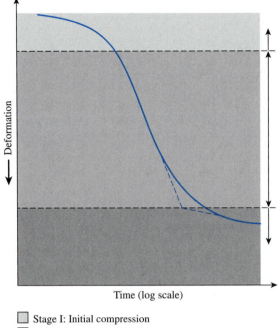

Stage I: Initial compression
Stage II: Primary consolidation
Stage III: Secondary consolidation

Figure 11.8
Time–deformation plot during consolidation for a given load increment

11.5   *Void Ratio–Pressure Plots*

After the time–deformation plots for various loadings are obtained in the laboratory, it is necessary to study the change in the void ratio of the specimen with pressure. Following is a step-by-step procedure for doing so:

Step 1:   Calculate the height of solids, H_s, in the soil specimen (Figure 11.9) using the equation

$$H_s = \frac{W_s}{AG_s\gamma_w} = \frac{M_s}{AG_s\rho_w}$$ (11.14)

where W_s = dry weight of the specimen
M_s = dry mass of the specimen
A = area of the specimen
G_s = specific gravity of soil solids
γ_w = unit weight of water
ρ_w = density of water

Step 2:   Calculate the initial height of voids as

$$H_v = H - H_s$$ (11.15)

where H = initial height of the specimen.

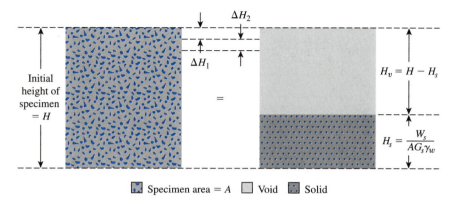

Figure 11.9 Change of height of specimen in one-dimensional consolidation test

Step 3: Calculate the initial void ratio, e_O, of the specimen, using the equation

$$e_O = \frac{V_v}{V_s} = \frac{H_v A}{H_s A} = \frac{H_v}{H_s} \qquad (11.16)$$

Step 4: For the first incremental loading, σ_1 (total load/unit area of specimen), which causes a deformation ΔH_1, calculate the change in the void ratio as

$$\Delta e_1 = \frac{\Delta H_1}{H_s} \qquad (11.17)$$

(ΔH_1 is obtained from the initial and the final dial readings for the loading).

It is important to note that, at the end of consolidation, total stress σ_1 is equal to effective stress σ_1'.

Step 5: Calculate the new void ratio after consolidation caused by the pressure increment as

$$e_1 = e_O - \Delta e_1 \qquad (11.18)$$

For the next loading, σ_2 (*note:* σ_2 equals the cumulative load per unit area of specimen), which causes additional deformation ΔH_2, the void ratio at the end of consolidation can be calculated as

$$e_2 = e_1 - \frac{\Delta H_2}{H_s} \qquad (11.19)$$

At this time, $\sigma_2 =$ effective stress, σ_2'. Proceeding in a similar manner, one can obtain the void ratios at the end of the consolidation for all load increments.

The effective stress σ' and the corresponding void ratios (e) at the end of consolidation are plotted on semilogarithmic graph paper. The typical shape of such a plot is shown in Figure 11.10.

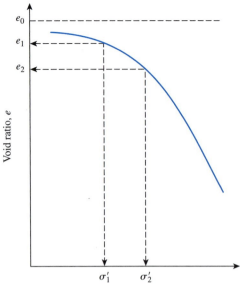

Figure 11.10 Typical plot of *e* against log σ'

Example 11.2

Following are the results of a laboratory consolidation test on a soil specimen obtained from the field: Dry mass of specimen = 128 g, height of specimen at the beginning of the test = 2.54 cm, G_s = 2.75, and area of the specimen = 30.68 cm².

Effective pressure, σ' (kN/m²)	Final height of specimen at the end of consolidation (cm)
0	2.540
50	2.488
100	2.465
200	2.431
400	2.389
800	2.324
1600	2.225
3200	2.115

Make necessary calculations and draw an *e* versus log σ' curve.

Solution

From Eq. (11.14),

$$H_s = \frac{W_s}{AG_s\gamma_w} = \frac{M_s}{AG_s\rho_w} = \frac{128\,\text{g}}{(30.68\,\text{cm}^2)(2.75)(1\,\text{g/cm}^3)} = 1.52\,\text{cm}$$

Now the following table can be prepared.

Effective pressure, σ' (kN/m²)	Height at the end of consolidation, H (cm)	$H_v = H - H_s$ (cm)	$e = H_v/H_s$
0	2.540	1.02	0.671
50	2.488	0.968	0.637
100	2.465	0.945	0.622
200	2.431	0.911	0.599
400	2.389	0.869	0.572
800	2.324	0.804	0.529
1600	2.225	0.705	0.464
3200	2.115	0.595	0.390

The e versus log σ' plot is shown in Figure 11.11

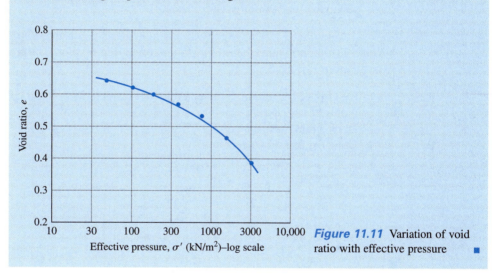

Figure 11.11 Variation of void ratio with effective pressure ∎

11.6   *Normally Consolidated and Overconsolidated Clays*

Figure 11.10 shows that the upper part of the e–log σ' plot is somewhat curved with a flat slope, followed by a linear relationship for the void ratio with log σ' having a steeper slope. This phenomenon can be explained in the following manner:

A soil in the field at some depth has been subjected to a certain maximum effective past pressure in its geologic history. This maximum effective past pressure may be equal to or less than the existing effective overburden pressure at the time of sampling. The reduction of effective pressure in the field may be caused by natural geologic processes or human processes. During the soil sampling, the existing effective overburden pressure is also released, which results in some expansion. When this specimen is subjected to a consolidation test, a small amount of compression (that is, a small change in void ratio) will occur when the effective pressure applied is less than the maximum effective overburden pressure in the field to which the soil has been subjected in the past. When the effective pressure on the specimen becomes greater than the maximum effective past pressure, the change in the void ratio is much larger, and the e–log σ' relationship is practically linear with a steeper slope.

This relationship can be verified in the laboratory by loading the specimen to exceed the maximum effective overburden pressure, and then unloading and reloading again. The e–log σ' plot for such cases is shown in Figure 11.12, in which cd represents unloading and dfg represents the reloading process.

This leads us to the two basic definitions of clay based on stress history:

1. *Normally consolidated,* whose present effective overburden pressure is the maximum pressure that the soil was subjected to in the past.
2. *Overconsolidated,* whose present effective overburden pressure is less than that which the soil experienced in the past. The maximum effective past pressure is called the *preconsolidation pressure.*

Casagrande (1936) suggested a simple graphic construction to determine the preconsolidation pressure σ'_c from the laboratory e–log σ' plot. The procedure is as follows (see Figure 11.13):

1. By visual observation, establish point a, at which the e–log σ' plot has a minimum radius of curvature.
2. Draw a horizontal line ab.
3. Draw the line ac tangent at a.

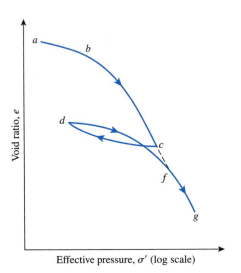

Figure 11.12 Plot of e against log σ' showing loading, unloading, and reloading branches

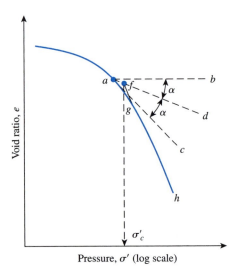

Figure 11.13 Graphic procedure for determining preconsolidation pressure

4. Draw the line *ad,* which is the bisector of the angle *bac.*
5. Project the straight-line portion *gh* of the *e*–log σ' plot back to intersect line *ad* at *f.* The abscissa of point *f* is the preconsolidation pressure, σ'_c.

The overconsolidation ratio (*OCR*) for a soil can now be defined as

$$OCR = \frac{\sigma'_c}{\sigma'} \tag{11.20}$$

where σ'_c = preconsolidation pressure of a specimen
σ' = present effective vertical pressure

In the literature, some empirical relationships are available to predict the preconsolidation pressure. Some examples are given below.

• Nagaraj and Murty (1985):

$$\log \sigma'_c = \frac{1.112 - \left(\dfrac{e_o}{e_L}\right)0.0463\sigma'_o}{0.188} \tag{11.21}$$

where e_o = *in situ* void ratio
$$e_L = \text{void ratio at liquid limit} = \left[\frac{LL(\%)}{100}\right]G_s \tag{11.22}$$

G_s = specific gravity of soil solids
σ'_o = *in situ* effective overburden pressure

(*Note:* σ'_c and σ'_o are in kN/m²)

• Stas and Kulhawy (1984):

$$\frac{\sigma'_c}{p_a} = 10^{[1.11-1.62(LI)]} \tag{11.23}$$

where p_a = atmospheric pressure ($\approx$100 kN/m²)
LI = liquidity index

• Hansbo (1957):

$$\sigma'_c = \alpha_{(VST)}\, c_{u(VST)} \tag{11.24}$$

where $\alpha_{(VST)}$ = an empirical coefficient = $\dfrac{222}{LL(\%)}$

$c_{u(VST)}$ = undrained shear strength obtained from vane shear test (Chapter 12)

In any case, these above relationships may change from soil to soil. They may be taken as an initial approximation.

11.7   *Effect of Disturbance on Void Ratio–Pressure Relationship*

A soil specimen will be remolded when it is subjected to some degree of disturbance. This remolding will result in some deviation of the e–log σ' plot as observed in the laboratory from the actual behavior in the field. The field e–log σ' plot can be reconstructed from the laboratory test results in the manner described in this section (Terzaghi and Peck, 1967).

Normally Consolidated Clay of Low to Medium Plasticity (Figure 11.14)

1. In Figure 11.14, curve 2 is the laboratory e–log σ' plot. From this plot, determine the preconsolidation pressure $(\sigma_c') = \sigma_0'$ (that is, the present effective overburden pressure). Knowing where $\sigma_c' = \sigma_0'$, draw vertical line *ab*.
2. Calculate the void ratio in the field, e_0 [Section 11.5, Eq. (11.16)]. Draw horizontal line *cd*.
3. Calculate $0.4e_0$ and draw line *ef*. (*Note: f* is the point of intersection of the line with curve 2.)
4. Join points *f* and *g*. Note that *g* is the point of intersection of lines *ab* and *cd*. This is the *virgin compression curve*.

It is important to point out that if a soil is remolded completely, the general position of the e–log σ' plot will be as represented by curve 3.

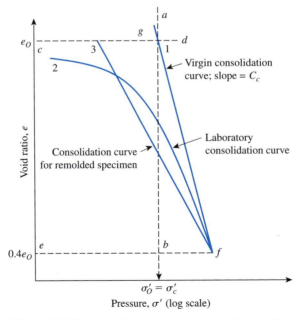

Figure 11.14 Consolidation characteristics of normally consolidated clay of low to medium sensitivity

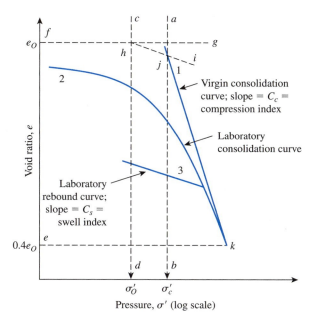

Figure 11.15 Consolidation characteristics of overconsolidated clay of low to medium sensitivity

Overconsolidated Clay of Low to Medium Plasticity (Figure 11.15)

1. In Figure 11.15, curve 2 is the laboratory e–log σ' plot (loading), and curve 3 is the laboratory unloading, or rebound, curve. From curve 2, determine the preconsolidation pressure σ_c'. Draw the vertical line ab.
2. Determine the field effective overburden pressure σ_0'. Draw vertical line cd.
3. Determine the void ratio in the field, e_0. Draw the horizontal line fg. The point of intersection of lines fg and cd is h.
4. Draw a line hi, which is parallel to curve 3 (which is practically a straight line). The point of intersection of lines hi and ab is j.
5. Join points j and k. Point k is on curve 2, and its ordinate is $0.4e_0$.

The field consolidation plot will take a path hjk. The recompression path in the field is hj and is parallel to the laboratory rebound curve (Schmertmann, 1953).

11.8   Calculation of Settlement from One-Dimensional Primary Consolidation

With the knowledge gained from the analysis of consolidation test results, we can now proceed to calculate the probable settlement caused by primary consolidation in the field, assuming one-dimensional consolidation.

Let us consider a saturated clay layer of thickness H and cross-sectional area A under an existing average effective overburden pressure, σ_0'. Because of an increase of effective pressure, $\Delta\sigma'$, let the primary settlement be S_c. Thus, the change in volume (Figure 11.16) can be given by

$$\Delta V = V_0 - V_1 = HA - (H - S_c)A = S_c A \qquad (11.25)$$

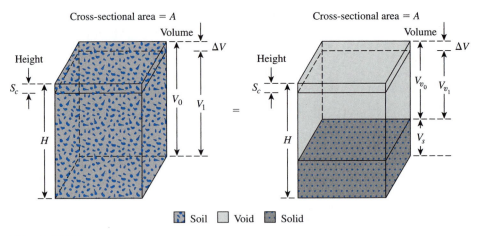

Figure 11.16 Settlement caused by one-dimensional consolidation

where V_0 and V_1 are the initial and final volumes, respectively. However, the change in the total volume is equal to the change in the volume of voids, ΔV_v. Hence,

$$\Delta V = S_c A = V_{v_0} - V_{v_1} = \Delta V_v \tag{11.26}$$

where V_{v_0} and V_{v_1} are the initial and final void volumes, respectively. From the definition of void ratio, it follows that

$$\Delta V_v = \Delta e V_s \tag{11.27}$$

where Δe = change of void ratio. But

$$V_s = \frac{V_0}{1 + e_o} = \frac{AH}{1 + e_o} \tag{11.28}$$

where e_O = initial void ratio at volume V_0. Thus, from Eqs. (11.25) through (11.28),

$$\Delta V = S_c A = \Delta e V_s = \frac{AH}{1 + e_o}\Delta e$$

or

$$S_c = H\frac{\Delta e}{1 + e_o} \tag{11.29}$$

For normally consolidated clays that exhibit a linear e–log σ' relationship (see Figure 11.14),

$$\Delta e = C_c[\log(\sigma_o' + \Delta\sigma') - \log \sigma_o'] \tag{11.30}$$

where C_c = slope of the e–log σ' plot and is defined as the compression index. Substitution of Eq. (11.30) into Eq. (11.29) gives

$$S_c = \frac{C_c H}{1 + e_o} \log\left(\frac{\sigma_o' + \Delta\sigma'}{\sigma_o'}\right) \tag{11.31}$$

In overconsolidated clays (see Figure 11.15), for $\sigma'_o + \Delta\sigma' \leq \sigma'_c$, field e–log σ' variation will be along the line *hj*, the slope of which will be approximately equal to that for the laboratory rebound curve. The slope of the rebound curve C_s is referred to as the *swell index;* so

$$\Delta e = C_s[\log(\sigma'_o + \Delta\sigma') - \log\sigma'_o] \tag{11.32}$$

From Eqs. (11.29) and (11.32), we obtain

$$S_c = \frac{C_s H}{1 + e_o}\log\left(\frac{\sigma'_o + \Delta\sigma'}{\sigma'_o}\right) \tag{11.33}$$

If $\sigma'_o + \Delta\sigma' > \sigma'_c$, then

$$S_c = \frac{C_s H}{1 + e_o}\log\frac{\sigma'_c}{\sigma'_o} + \frac{C_c H}{1 + e_o}\log\left(\frac{\sigma'_o + \Delta\sigma'}{\sigma'_c}\right) \tag{11.34}$$

However, if the e–log σ' curve is given, one can simply pick Δe off the plot for the appropriate range of pressures. This number may be substituted into Eq. (11.29) for the calculation of settlement, S_c.

11.9 *Compression Index (C$_c$)*

The compression index for the calculation of field settlement caused by consolidation can be determined by graphic construction (as shown in Figure 11.14) after one obtains the laboratory test results for void ratio and pressure.

Skempton (1944) suggested the following empirical expression for the compression index for undisturbed clays:

$$C_c = 0.009(LL - 10) \tag{11.35}$$

where LL = liquid limit.

Several other correlations for the compression index are also available. They have been developed by tests on various clays. Some of these correlations are given in Table 11.6.

On the basis of observations on several natural clays, Rendon-Herrero (1983) gave the relationship for the compression index in the form

$$C_c = 0.141 G_s^{1.2}\left(\frac{1 + e_o}{G_s}\right)^{2.38} \tag{11.36}$$

Table 11.6 Correlations for Compression Index, C_c*

Equation	Reference	Region of applicability
$C_c = 0.007(LL - 7)$	Skempton (1944)	Remolded clays
$C_c = 0.01w_N$		Chicago clays
$C_c = 1.15(e_O - 0.27)$	Nishida (1956)	All clays
$C_c = 0.30(e_O - 0.27)$	Hough (1957)	Inorganic cohesive soil: silt, silty clay, clay
$C_c = 0.0115w_N$		Organic soils, peats, organic silt, and clay
$C_c = 0.0046(LL - 9)$		Brazilian clays
$C_c = 0.75(e_O - 0.5)$		Soils with low plasticity
$C_c = 0.208e_O + 0.0083$		Chicago clays
$C_c = 0.156e_O + 0.0107$		All clays

* After Rendon-Herrero (1980)

Note: e_O = *in situ* void ratio; w_N = *in situ* water content.

Nagaraj and Murty (1985) expressed the compression index as

$$C_c = 0.2343 \left[\frac{LL\,(\%)}{100} \right] G_s \tag{11.37}$$

Based on the modified Cam clay model, Wroth and Wood (1978) have shown that

$$C_c \approx 0.5 G_s \frac{[PI(\%)]}{100} \tag{11.38}$$

where PI = plasticity index.

If an average value of G_s is taken to be about 2.7 (Kulhawy and Mayne, 1990)

$$C_c \approx \frac{PI}{74} \tag{11.39}$$

More recently, Park and Koumoto (2004) expressed the compression index by the following relationship.

$$C_c = \frac{n_o}{371.747 - 4.275n_o} \tag{11.40}$$

where n_o = *in situ* porosity of the soil

11.10 *Swell Index (C_s)*

The swell index is appreciably smaller in magnitude than the compression index and generally can be determined from laboratory tests. In most cases,

$$C_s \approx \tfrac{1}{5} \text{ to } \tfrac{1}{10} C_c$$

The swell index was expressed by Nagaraj and Murty (1985) as

$$C_s = 0.0463 \left[\frac{LL(\%)}{100} \right] G_s \qquad (11.41)$$

Based on the modified Cam clay model, Kulhawy and Mayne (1990) have shown that

$$C_c \approx \frac{PI}{370} \qquad (11.42)$$

Typical values of the liquid limit, plastic limit, virgin compression index, and swell index for some natural soils are given in Table 11.7.

Table 11.7 Compression and Swell of Natural Soils

Soil	Liquid limit	Plastic limit	Compression index, C_c	Swell index, C_s
Boston blue clay	41	20	0.35	0.07
Chicago clay	60	20	0.4	0.07
Ft. Gordon clay, Georgia	51	26	0.12	—
New Orleans clay	80	25	0.3	0.05
Montana clay	60	28	0.21	0.05

Example 11.3

The following are the results of a laboratory consolidation test:

Pressure, σ' (kN/m²)	Void ratio, e	Remarks	Pressure, σ' (kN/m²)	Void ratio, e	Remarks
25	1.03	Loading	800	0.71	Loading
50	1.02		1600	0.62	
100	0.98		800	0.635	Unloading
200	0.91		400	0.655	
400	0.79		200	0.67	

a. Draw an e-log σ'_O graph and determine the preconsolidation pressure, σ'_c
b. Calculate the compression index and the ratio of C_s/C_c
c. On the basis of the average e-log σ' plot, calculate the void ratio at $\sigma'_O = 1200$ kN/m²

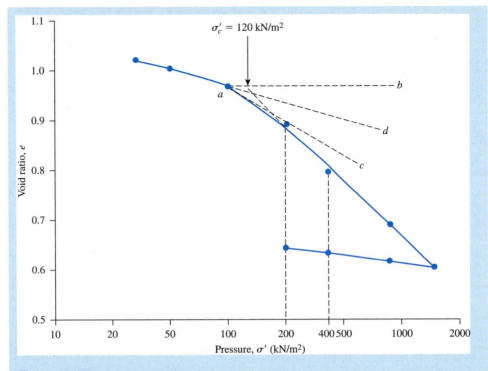

Figure 11.17 Plot of *e* versus log σ'

Solution
Part a
The *e* versus log σ' plot is shown in Figure 11.17. Casagrande's graphic procedure is used to determine the preconsolidation pressure:

$$\sigma'_O = 120 \text{ kN/m}^2$$

Part b
From the average *e*-log σ' plot, for the loading and unloading branches, the following values can be determined:

Branch	e	σ'_0(kN/m²)
Loading	0.9	200
	0.8	400
Unloading	0.67	200
	0.655	400

From the loading branch,

$$C_c = \frac{e_1 - e_2}{\log \dfrac{\sigma'_2}{\sigma'_1}} = \frac{0.9 - 0.8}{\log\left(\dfrac{400}{200}\right)} = \mathbf{0.33}$$

From the unloading branch,

$$C_s = \frac{e_1 - e_2}{\log \dfrac{\sigma'_2}{\sigma'_1}} = \frac{0.67 - 0.655}{\log\left(\dfrac{400}{200}\right)} = \mathbf{0.05}$$

$$\frac{C_s}{C_c} = \frac{0.05}{0.33} = \mathbf{0.15}$$

Part c

$$C_C = \frac{e_1 - e_3}{\log \dfrac{\sigma'_3}{\sigma'_1}}$$

We know that $e_1 = 0.9$ at $\sigma'_1 = 200$ kN/m² and that $C_c = 0.33$ [part (b)]. Let $\sigma'_3 = 1200$ kN/m². So,

$$0.33 = \frac{0.9 - e_3}{\log\left(\dfrac{1200}{200}\right)}$$

$$e_3 = 0.9 - 0.33 \log\left(\frac{1200}{200}\right) = \mathbf{0.64}$$ ∎

Example 11.4

A soil profile is shown in Figure 11.18. If a uniformly distributed load, $\Delta\sigma$, is applied at the ground surface, what is the settlement of the clay layer caused by primary consolidation if

    a. The clay is normally consolidated
    b. The preconsolidation pressure (σ'_c) = 190 kN/m²
    c. $\sigma'_c = 170$ kN/m²

Use $C_s \approx \frac{1}{6}C_c$.

Solution
Part a
The average effective stress at the middle of the clay layer is

$$\sigma'_o = 2\gamma_{dry} + 4[\gamma_{sat(sand)} - \gamma_w] + \tfrac{4}{2}[\gamma_{sat(clay)} - \gamma_w]$$

$$\sigma'_o = (2)(14) + 4(18 - 9.81) + 2(19 - 9.81) = 79.14 \text{ kN/m}^2$$

From Eq. (11.31),

$$S_c = \frac{C_c H}{1 + e_o} \log\left(\frac{\sigma'_o + \Delta\sigma'}{\sigma'_o}\right)$$

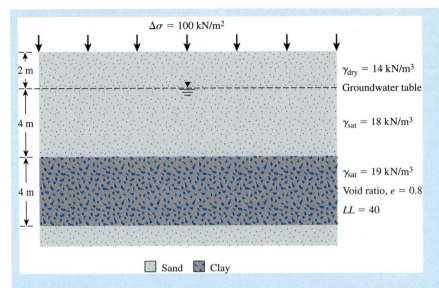

$\Delta\sigma = 100 \text{ kN/m}^2$

2 m — $\gamma_{dry} = 14 \text{ kN/m}^3$
Groundwater table

4 m — $\gamma_{sat} = 18 \text{ kN/m}^3$

4 m — $\gamma_{sat} = 19 \text{ kN/m}^3$
Void ratio, $e = 0.8$
$LL = 40$

Sand   Clay

Figure 11.18

From Eq. (11.35),

$$C_c = 0.009(LL - 10) = 0.009(40 - 10) = 0.27$$

So,

$$S_c = \frac{(0.27)(4)}{1 + 0.8} \log\left(\frac{79.14 + 100}{79.14}\right) = 0.213 \text{ m} = \mathbf{213 \text{ mm}}$$

Part b

$$\sigma_O' + \Delta\sigma' = 79.14 + 100 = 179.14 \text{ kN/m}^2$$

$$\sigma_c' = 190 \text{ kN/m}^2$$

Because $\sigma_O' + \Delta\sigma' > \sigma_c'$, use Eq. (11.33):

$$S_c = \frac{C_s H}{1 + e_O} \log\left(\frac{\sigma_O' + \Delta\sigma'}{\sigma_O'}\right)$$

$$C_s = \frac{C_c}{6} = \frac{0.27}{6} = 0.045$$

$$S_c = \frac{(0.045)(4)}{1 + 0.8} \log\left(\frac{79.14 + 100}{79.14}\right) = 0.036 \text{ m} = \mathbf{36 \text{ mm}}$$

Part c

$$\sigma_o' = 79.14 \text{ kN/m}^2$$

$$\sigma_o' + \Delta\sigma' = 179.14 \text{ kN/m}^2$$

$$\sigma_c' = 170 \text{ kN/m}^2$$

Because $\sigma'_o < \sigma'_c < \sigma'_o + \Delta\sigma'$, use Eq. 11.34:

$$S_c = \frac{C_s H}{1 + e_o} \log \frac{\sigma'_c}{\sigma'_o} + \frac{C_c H}{1 + e_o} \log\left(\frac{\sigma'_o + \Delta\sigma'}{\sigma'_c}\right)$$

$$= \frac{(0.045)(4)}{1.8} \log\left(\frac{170}{79.14}\right) + \frac{(0.27)(4)}{1.8} \log\left(\frac{179.14}{170}\right)$$

$$\approx 0.0468 \text{ m} = \textbf{46.8 mm}$$

Example 11.5

A soil profile is shown in Figure 11.19a. Laboratory consolidation tests were conducted on a specimen collected from the middle of the clay layer. The field consolidation curve interpolated from the laboratory test results is shown in Figure 11.19b. Calculate the settlement in the field caused by primary consolidation for a surcharge of 48 kN/m² applied at the ground surface.

Solution

$$\sigma'_O = (5)(\gamma_{sat} - \gamma_w) = 5(18.0 - 9.81)$$

$$= 40.95 \text{ kN/m}^2$$

$$e_O = 1.1$$

$$\Delta\sigma' = 48 \text{ kN/m}^2$$

$$\sigma'_O + \Delta\sigma' = 40.95 + 48 = 88.95 \text{ kN/m}^2$$

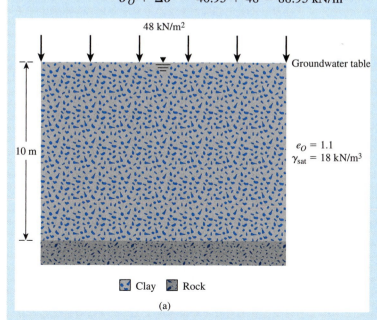

Clay Rock

(a)

Figure 11.19
(a) Soil profile

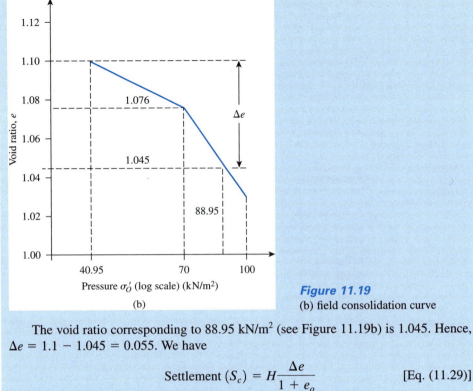

Figure 11.19
(b) field consolidation curve

(b)

The void ratio corresponding to 88.95 kN/m² (see Figure 11.19b) is 1.045. Hence, $\Delta e = 1.1 - 1.045 = 0.055$. We have

$$\text{Settlement } (S_c) = H\frac{\Delta e}{1 + e_o} \qquad\qquad \text{[Eq. (11.29)]}$$

so,

$$S_c = 10\frac{(0.055)}{1 + 1.1} = 0.262 \text{ m} = \textbf{262 mm} \qquad\qquad \blacksquare$$

11.11 *Secondary Consolidation Settlement*

Section 11.4 showed that at the end of primary consolidation (that is, after complete dissipation of excess pore water pressure) some settlement is observed because of the plastic adjustment of soil fabrics. This stage of consolidation is called *secondary consolidation*. During secondary consolidation the plot of deformation against the log of time is practically linear (see Figure 11.8). The variation of the void ratio, e, with time t for a given load increment will be similar to that shown in Figure 11.8. This variation is shown in Figure 11.20. From Figure 11.20, the secondary compression index can be defined as

$$C_\alpha = \frac{\Delta e}{\log t_2 - \log t_1} = \frac{\Delta e}{\log(t_2/t_1)} \qquad\qquad (11.43)$$

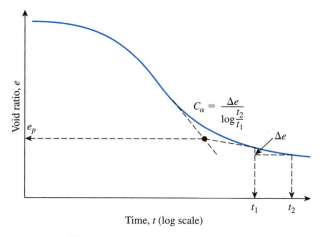

Figure 11.20 Variation of e with log t under a given load increment and definition of secondary consolidation index

where C_α = secondary compression index
Δe = change of void ratio
t_1, t_2 = time

The magnitude of the secondary consolidation can be calculated as

$$S_s = C'_\alpha H \log\left(\frac{t_2}{t_1}\right)$$ (11.44)

and

$$C'_\alpha = \frac{C_\alpha}{1 + e_p}$$ (11.45)

where e_p = void ratio at the end of primary consolidation (see Figure 11.20)
H = thickness of clay layer

The general magnitudes of C'_α as observed in various natural deposits (also see Figure 11.21) are as follows:

- Overconsolidated clays = 0.001 or less
- Normally consolidated clays = 0.005 to 0.03
- Organic soil = 0.04 or more

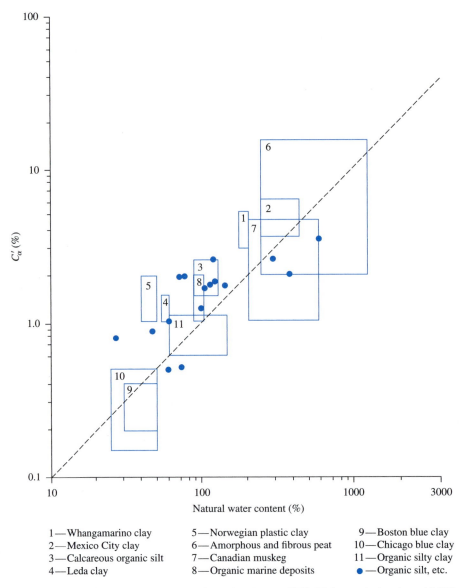

1—Whangamarino clay   5—Norwegian plastic clay   9—Boston blue clay
2—Mexico City clay   6—Amorphous and fibrous peat   10—Chicago blue clay
3—Calcareous organic silt 7—Canadian muskeg   11—Organic silty clay
4—Leda clay   8—Organic marine deposits   ● —Organic silt, etc.

Figure 11.21 C_α' for natural soil deposits *(After Mesri, 1973. With permission from ASCE.)*

Mersri and Godlewski (1977) compiled the ratio of C_α'/C_c for a number of natural clays. From this study, it appears that C_α'/C_c for

- Inorganic clays and silts $\approx 0.04 \pm 0.01$
- Organic clays and silts $\approx 0.05 \pm 0.01$
- Peats $\approx 0.075 \pm 0.01$

Secondary consolidation settlement is more important than primary consolidation in organic and highly compressible inorganic soils. In overconsolidated inorganic clays, the secondary compression index is very small and of less practical significance.

Example 11.6

For a normally consolidated clay layer in the field, the following values are given:

- Thickness of clay layer = 2.6 m
- Void ratio (e_O) = 0.8
- Compression index (C_c) = 0.28
- Average effective pressure on the clay layer (σ'_o) = 127 kN/m²
- $\Delta\sigma'$ = 46.5 kN/m²
- Secondary compression index (C_α) = 0.02

What is the total consolidation settlement of the clay layer five years after the completion of primary consolidation settlement? (*Note:* Time for completion of primary settlement = 1.5 years.)

Solution
From Eq. (11.45),

$$C'_\alpha = \frac{C_\alpha}{1 + e_p}$$

The value of e_p can be calculated as

$$e_p = e_o - \Delta e_{\text{primary}}$$

Combining Eqs. (11.29) and (11.30), we find that

$$\Delta e = C_c \log\left(\frac{\sigma'_o + \Delta\sigma'}{\sigma'_o}\right) = 0.28 \log\left(\frac{127 + 46.5}{127}\right)$$

$$= 0.038$$

$$\text{Primary consolidation, } S_c = \frac{\Delta e H}{1 + e_o} = \frac{(0.038)(2.6 \times 1000)}{1 + 0.8} = 54.9 \text{ mm}$$

It is given that e_O = 0.8, and thus,

$$e_p = 0.8 - 0.038 = 0.762$$

Hence,

$$C'_\alpha = \frac{0.02}{1 + 0.762} = 0.011$$

From Eq. (11.44),

$$S_s = C'_\alpha H \log\left(\frac{t_2}{t_1}\right) = (0.011)(2.6 \times 1000) \log\left(\frac{5}{1.5}\right) \approx 14.95 \text{ mm}$$

Total consolidation settlement = primary consolidation (S_c) + secondary settlement (S_s). So

$$\text{Total consolidation settlement} = 54.9 + 14.95 = \textbf{69.85 mm}$$

■

11.12 *Time Rate of Consolidation*

The total settlement caused by primary consolidation resulting from an increase in the stress on a soil layer can be calculated by the use of one of the three equations—(11.31), (11.33), or (11.34)—given in Section 11.8. However, they do not provide any information regarding the rate of primary consolidation. Terzaghi (1925) proposed the first theory to consider the rate of one-dimensional consolidation for saturated clay soils. The mathematical derivations are based on the following six assumptions (also see Taylor, 1948):

1. The clay–water system is homogeneous.
2. Saturation is complete.
3. Compressibility of water is negligible.
4. Compressibility of soil grains is negligible (but soil grains rearrange).
5. The flow of water is in one direction only (that is, in the direction of compression).
6. Darcy's law is valid.

Figure 11.22a shows a layer of clay of thickness $2H_{dr}$ (*Note: H_{dr}* = length of maximum drainage path) that is located between two highly permeable sand layers. If the clay layer is subjected to an increased pressure of $\Delta\sigma$, the pore water pressure at any point A in the clay layer will increase. For one-dimensional consolidation, water will be squeezed out in the vertical direction toward the sand layer.

Figure 11.22b shows the flow of water through a prismatic element at A. For the soil element shown,

$$
\text{Rate of outflow} \atop \text{of water} \quad - \quad {\text{Rate of inflow} \atop \text{of water}} \quad = \quad {\text{Rate of} \atop \text{volume change}}
$$

Thus,

$$
\left(v_z + \frac{\partial v_z}{\partial z}\, dz \right) dx\, dy - v_z\, dx\, dy = \frac{\partial V}{\partial t}
$$

where V = volume of the soil element
$\quad\ v_z$ = velocity of flow in z direction

or

$$
\frac{\partial v_z}{\partial z}\, dx\, dy\, dz = \frac{\partial V}{\partial t} \tag{11.46}
$$

Using Darcy's law, we have

$$
v_z = ki = -k\frac{\partial h}{\partial z} = -\frac{k}{\gamma_w}\frac{\partial u}{\partial z} \tag{11.47}
$$

where u = excess pore water pressure caused by the increase of stress.

From Eqs. (11.46) and (11.47),

$$
-\frac{k}{\gamma_w}\frac{\partial^2 u}{\partial z^2} = \frac{1}{dx\, dy\, dz}\frac{\partial V}{\partial t} \tag{11.48}
$$

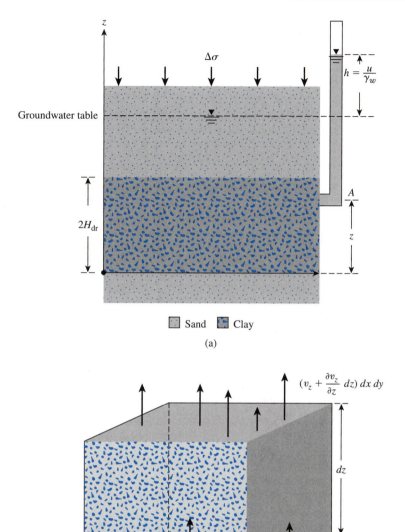

Figure 11.22
(a) Clay layer undergoing consolidation; (b) flow of water at A during consolidation

During consolidation, the rate of change in the volume of the soil element is equal to the rate of change in the volume of voids. Thus,

$$\frac{\partial V}{\partial t} = \frac{\partial V_v}{\partial t} = \frac{\partial (V_s + eV_s)}{\partial t} = \frac{\partial V_s}{\partial t} + V_s \frac{\partial e}{\partial t} + e \frac{\partial V_s}{\partial t} \qquad (11.49)$$

where V_s = volume of soil solids
$\quad\quad V_v$ = volume of voids

But (assuming that soil solids are incompressible)

$$\frac{\partial V_s}{\partial t} = 0$$

and

$$V_s = \frac{V}{1 + e_o} = \frac{dx \, dy \, dz}{1 + e_o}$$

Substitution for $\partial V_s / \partial t$ and V_s in Eq. (11.49) yields

$$\frac{\partial V}{\partial t} = \frac{dx \, dy \, dz}{1 + e_o} \frac{\partial e}{\partial t} \tag{11.50}$$

where e_O = initial void ratio.

Combining Eqs. (11.48) and (11.50) gives

$$-\frac{k}{\gamma_w} \frac{\partial^2 u}{\partial z^2} = \frac{1}{1 + e_o} \frac{\partial e}{\partial t} \tag{11.51}$$

The change in the void ratio is caused by the increase of effective stress (i.e., a decrease of excess pore water pressure). Assuming that they are related linearly, we have

$$\partial e = a_v \partial (\Delta \sigma') = -a_v \partial u \tag{11.52}$$

where $\partial(\Delta \sigma')$ = change in effective pressure
$\quad a_v$ = coefficient of compressibility (a_v can be considered constant for a narrow range of pressure increase)

Combining Eqs. (11.51) and (11.52) gives

$$-\frac{k}{\gamma_w} \frac{\partial^2 u}{\partial z^2} = -\frac{a_v}{1 + e_o} \frac{\partial u}{\partial t} = -m_v \frac{\partial u}{\partial t}$$

where

$$m_v = \text{coefficient of volume compressibility} = a_v/(1 + e_o) \tag{11.53}$$

or,

$$\frac{\partial u}{\partial t} = c_v \frac{\partial^2 u}{\partial z^2} \tag{11.54}$$

where

$$c_v = \text{coefficient of consolidation} = k/(\gamma_w m_v) \tag{11.55}$$

Thus,

$$c_v = \frac{k}{\gamma_w m_v} = \frac{k}{\gamma_w \left(\dfrac{a_v}{1 + e_o} \right)} \tag{11.56}$$

Eq. (11.54) is the basic differential equation of Terzaghi's consolidation theory and can be solved with the following boundary conditions:

$$z = 0, \quad u = 0$$

$$z = 2H_{dr}, \quad u = 0$$

$$t = 0, \quad u = u_O$$

The solution yields

$$u = \sum_{m=0}^{m=\infty} \left[\frac{2u_o}{M} \sin\left(\frac{Mz}{H_{dr}}\right) \right] e^{-M^2 T_v} \tag{11.57}$$

where m = an integer
$M = (\pi/2)(2m + 1)$
u_O = initial excess pore water pressure

$$T_v = \frac{c_v t}{H_{dr}^2} = \text{time factor} \tag{11.58}$$

The time factor is a nondimensional number.

Because consolidation progresses by the dissipation of excess pore water pressure, the degree of consolidation at a distance z at any time t is

$$U_z = \frac{u_o - u_z}{u_o} = 1 - \frac{u_z}{u_o} \tag{11.59}$$

where u_z = excess pore water pressure at time t.

Equations (11.57) and (11.59) can be combined to obtain the degree of consolidation at any depth z. This is shown in Figure 11.23.

The average degree of consolidation for the entire depth of the clay layer at any time t can be written from Eq. (11.59) as

$$U = \frac{S_{c(t)}}{S_c} = 1 - \frac{\left(\dfrac{1}{2H_{dr}}\right) \displaystyle\int_0^{2H_{dr}} u_z \, dz}{u_o} \tag{11.60}$$

where U = average degree of consolidation
$S_{c(t)}$ = settlement of the layer at time t
S_c = ultimate settlement of the layer from primary consolidation

Substitution of the expression for excess pore water pressure u_z given in Eq. (11.57) into Eq. (11.60) gives

$$U = 1 - \sum_{m=0}^{m=\infty} \frac{2}{M^2} e^{-M^2 T_v} \tag{11.61}$$

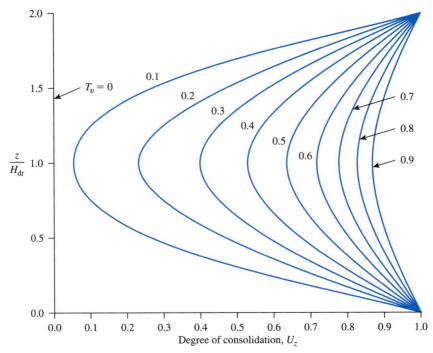

Figure 11.23
Variation of U_z with T_v and z/H_{dr}

The variation in the average degree of consolidation with the nondimensional time factor, T_v, is given in Figure 11.24, which represents the case where u_O is the same for the entire depth of the consolidating layer.

The values of the time factor and their corresponding average degrees of consolidation for the case presented in Figure 11.24 may also be approximated by the following simple relationship:

$$\text{For } U = 0 \text{ to } 60\%, \quad T_v = \frac{\pi}{4}\left(\frac{U\%}{100}\right)^2 \tag{11.62}$$

$$\text{For } U > 60\%, \quad T_v = 1.781 - 0.933 \log(100 - U\%) \tag{11.63}$$

Table 11.8 gives the variation of T_v with U on the basis of Eqs. (11.62) and (11.63).

Sivaram and Swamee (1977) gave the following equation for U varying from 0 to 100%:

$$\frac{U\%}{100} = \frac{(4T_v/\pi)^{0.5}}{[1 + (4T_v/\pi)^{2.8}]^{0.179}} \tag{11.64}$$

or

$$T_v = \frac{(\pi/4)(U\%/100)^2}{[1 - (U\%/100)^{5.6}]^{0.357}} \tag{11.65}$$

Equations (11.64) and (11.65) give an error in T_v of less than 1% for 0% < U < 90% and less than 3% for 90% < U < 100%.

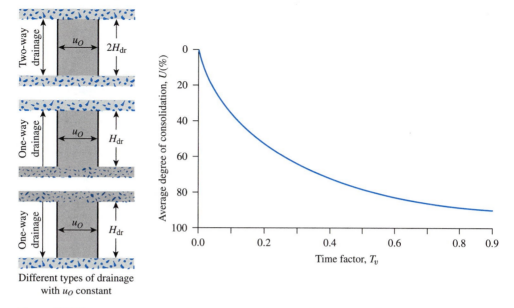

Different types of drainage
with u_O constant

Figure 11.24 Variation of average degree of consolidation with time factor, T_v (u_O constant with depth)

Table 11.8 Variation of T_v with U

U (%)	T_v	U (%)	T_v	U (%)	T_v	U (%)	T_v
0	0	26	0.0531	52	0.212	78	0.529
1	0.00008	27	0.0572	53	0.221	79	0.547
2	0.0003	28	0.0615	54	0.230	80	0.567
3	0.00071	29	0.0660	55	0.239	81	0.588
4	0.00126	30	0.0707	56	0.248	82	0.610
5	0.00196	31	0.0754	57	0.257	83	0.633
6	0.00283	32	0.0803	58	0.267	84	0.658
7	0.00385	33	0.0855	59	0.276	85	0.684
8	0.00502	34	0.0907	60	0.286	86	0.712
9	0.00636	35	0.0962	61	0.297	87	0.742
10	0.00785	36	0.102	62	0.307	88	0.774
11	0.0095	37	0.107	63	0.318	89	0.809
12	0.0113	38	0.113	64	0.329	90	0.848
13	0.0133	39	0.119	65	0.304	91	0.891
14	0.0154	40	0.126	66	0.352	92	0.938
15	0.0177	41	0.132	67	0.364	93	0.993
16	0.0201	42	0.138	68	0.377	94	1.055
17	0.0227	43	0.145	69	0.390	95	1.129
18	0.0254	44	0.152	70	0.403	96	1.219
19	0.0283	45	0.159	71	0.417	97	1.336
20	0.0314	46	0.166	72	0.431	98	1.500
21	0.0346	47	0.173	73	0.446	99	1.781
22	0.0380	48	0.181	74	0.461	100	∞
23	0.0415	49	0.188	75	0.477		
24	0.0452	50	0.197	76	0.493		
25	0.0491	51	0.204	77	0.511		

Example 11.7

The time required for 50% consolidation of a 25-mm-thick clay layer (drained at both top and bottom) in the laboratory is 2 min. 20 sec. How long (in days) will it take for a 3-m-thick clay layer of the same clay in the field under the same pressure increment to reach 50% consolidation? In the field, there is a rock layer at the bottom of the clay.

Solution

$$T_{50} = \frac{c_v t_{lab}}{H^2_{dr(lab)}} = \frac{c_v t_{field}}{H^2_{dr(field)}}$$

or

$$\frac{t_{lab}}{H^2_{dr(lab)}} = \frac{t_{field}}{H^2_{dr(field)}}$$

$$\frac{140\,\text{sec}}{\left(\dfrac{0.025\,\text{m}}{2}\right)^2} = \frac{t_{field}}{(3\,\text{m})^2}$$

$$t_{field} = 8{,}064{,}000\,\text{sec} = \textbf{93.33 days} \qquad \blacksquare$$

Example 11.8

Refer to Example 11.7. How long (in days) will it take in the field for 30% primary consolidation to occur? Use Eq. (11.62).

Solution
From Eq. (11.62)

$$\frac{c_v t_{field}}{H^2_{dr(lab)}} = T_v \propto U^2$$

So

$$t \propto U^2$$

$$\frac{t_1}{t_2} = \frac{U_1^2}{U_2^2}$$

or

$$\frac{93.33\,\text{days}}{t_2} = \frac{50^2}{30^2}$$

$$t_2 = \textbf{33.6 days} \qquad \blacksquare$$

Example 11.9

A 3-m-thick layer (double drainage) of saturated clay under a surcharge loading underwent 90% primary consolidation in 75 days. Find the coefficient of consolidation of clay for the pressure range.

Solution

$$T_{90} = \frac{c_v t_{90}}{H_{dr}^2}$$

Because the clay layer has two-way drainage, $H_{dr} = 3 \text{ m}/2 = 1.5 \text{ m}$. Also, $T_{90} = 0.848$ (see Table 11.8). So,

$$0.848 = \frac{c_v(75 \times 24 \times 60 \times 60)}{(1.5 \times 100)^2}$$

$$c_v = \frac{0.848 \times 2.25 \times 10^4}{75 \times 24 \times 60 \times 60} = \mathbf{0.00294 \text{ cm}^2/\text{sec}} \qquad \blacksquare$$

Example 11.10

For a normally consolidated laboratory clay specimen drained on both sides, the following are given:

- $\sigma'_o = 150 \text{ kN/m}^2$, $e = e_o = 1.1$
- $\sigma'_o + \Delta\sigma' = 300 \text{ kN/m}^2$, $e = 0.9$
- Thickness of clay specimen = 25 mm.
- Time for 50% consolidation = 2 min

a. Determine the hydraulic conductivity (min) of the clay for the loading range.
b. How long (in days) will it take for a 1.8 m clay layer in the field (drained on one side) to reach 60% consolidation?

Solution
Part a

The coefficient of compressibility is

$$m_v = \frac{a_v}{1 + e_{av}} = \frac{\left(\dfrac{\Delta e}{\Delta\sigma'}\right)}{1 + e_{av}}$$

$$\Delta e = 1.1 - 0.9 = 0.2$$

$$\Delta\sigma' = 300 - 150 = 150 \text{ kN/m}^2$$

$$e_{av} = \frac{1.1 + 0.9}{2} = 1.0$$

So

$$m_v = \frac{\dfrac{0.2}{150}}{1 + 1.0} = 6.35 \times 10^{-4} \text{ m}^2/\text{kN}$$

From Table 11.8, for $U = 50\%$, $T_v = 0.197$; thus,

$$c_v = \frac{(0.197)\left(\dfrac{25}{2 \times 1000}\right)^2}{2} = 1.53 \times 10^{-5} \text{ m}^2/\text{min}$$

$$k = c_v m_v \gamma_w = (1.53 \times 10^{-5})(6.35 \times 10^{-4})(9.81)$$

$$= 95.3 \times 10^{-9} \text{ m/min} = 95.3 \times 10^{-7} \text{ cm/min}$$

Part b

$$T_{60} = \frac{c_v t_{60}}{H_{dr}^2}$$

$$t_{60} = \frac{T_{60} H_{dr}^2}{c_v}$$

From Table 11.8, for $U = 60\%$, $T_v = 0.286$,

$$t_{60} = \frac{(0.286)(1.8)^2}{1.53 \times 10^{-5}} = 60{,}565 \text{ min} = \textbf{42.06 days} \qquad \blacksquare$$

11.13 *Coefficient of Consolidation*

The coefficient of consolidation c_v generally decreases as the liquid limit of soil increases. The range of variation of c_v for a given liquid limit of soil is wide.

For a given load increment on a specimen, two graphical methods commonly are used for determining c_v from laboratory one-dimensional consolidation tests. The first is the *logarithm-of-time method* proposed by Casagrande and Fadum (1940), and the other is the *square-root-of-time method* given by Taylor (1942). More recently, at least two other methods were proposed. They are the *hyperbola method* (Sridharan and Prakash, 1985) and the *early stage log-t method* (Robinson and Allam, 1996). The general procedures for obtaining c_v by these methods are described in this section.

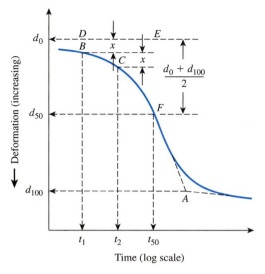

Figure 11.25
Logarithm-of-time method for determining coefficient of consolidation

Logarithm-of-Time Method

For a given incremental loading of the laboratory test, the specimen deformation against log-of-time plot is shown in Figure 11.25. The following constructions are needed to determine c_v.

Step 1: Extend the straight-line portions of primary and secondary consolidations to intersect at A. The ordinate of A is represented by d_{100}—that is, the deformation at the end of 100% primary consolidation.

Step 2: The initial curved portion of the plot of deformation versus log t is approximated to be a parabola on the natural scale. Select times t_1 and t_2 on the curved portion such that $t_2 = 4t_1$. Let the difference of specimen deformation during time ($t_2 - t_1$) be equal to x.

Step 3: Draw a horizontal line DE such that the vertical distance BD is equal to x. The deformation corresponding to the line DE is d_0 (that is, deformation at 0% consolidation).

Step 4: The ordinate of point F on the consolidation curve represents the deformation at 50% primary consolidation, and its abscissa represents the corresponding time (t_{50}).

Step 5: For 50% average degree of consolidation, $T_v = 0.197$ (see Table 11.8), so,

$$T_{50} = \frac{c_v t_{50}}{H_{dr}^2}$$

or

$$c_v = \frac{0.197 H_{dr}^2}{t_{50}} \tag{11.66}$$

where H_{dr} = average longest drainage path during consolidation.

For specimens drained at both top and bottom, H_{dr} equals one-half the average height of the specimen during consolidation. For specimens drained on only one side, H_{dr} equals the average height of the specimen during consolidation.

Square-Root-of-Time Method

In the square-root-of-time method, a plot of deformation against the square root of time is made for the incremental loading (Figure 11.26). Other graphic constructions required are as follows:

Step 1: Draw a line AB through the early portion of the curve.

Step 2: Draw a line AC such that $\overline{OC} = 1.15\overline{OB}$. The abscissa of point D, which is the intersection of AC and the consolidation curve, gives the square root of time for 90% consolidation ($\sqrt{t_{90}}$).

Step 3: For 90% consolidation, $T_{90} = 0.848$ (see Table 11.8), so

$$T_{90} = 0.848 = \frac{c_v t_{90}}{H_{dr}^2}$$

or

$$c_v = \frac{0.848 H_{dr}^2}{t_{90}} \tag{11.67}$$

H_{dr} in Eq. (11.67) is determined in a manner similar to that in the logarithm-of-time method.

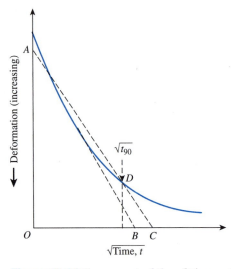

Figure 11.26 Square-root-of-time fitting method

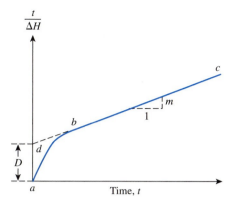

Figure 11.27
Hyperbola method for determination of c_v

Hyperbola Method

In the hyperbola method, the following procedure is recommended for the determination of c_v.

Step 1: Obtain the time t and the specimen deformation (ΔH) from the laboratory consolidation test.
Step 2: Plot the graph of $t/\Delta H$ against t as shown in Figure 11.27.
Step 3: Identify the straight-line portion bc and project it back to point d. Determine the intercept D.
Step 4: Determine the slope m of the line bc.
Step 5: Calculate c_v as

$$c_v = 0.3\left(\frac{mH_{dr}^2}{D}\right) \tag{11.68}$$

Note that because the unit of D is time/length and the unit of m is (time/length)/time = 1/length, the unit of c_v is

$$\frac{\left(\dfrac{1}{\text{length}}\right)(\text{length})^2}{\left(\dfrac{\text{time}}{\text{length}}\right)} = \frac{(\text{length})^2}{\text{time}}$$

The hyperbola method is fairly simple to use, and it gives good results for $U = 60$ to 90%.

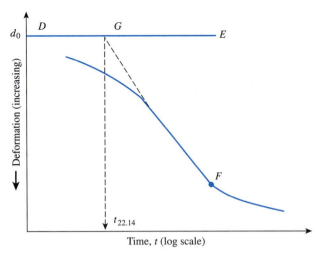

Figure 11.28 Early stage log-*t* method

Early Stage log-t Method

The early stage log-*t* method, an extension of the logarithm-of-time method, is based on specimen deformation against log-of-time plot as shown in Figure 11.28. According to this method, follow Steps 2 and 3 described for the logarithm-of-time method to determine d_0. Draw a horizontal line *DE* through d_0. Then draw a tangent through the point of inflection, *F*. The tangent intersects line *DE* at point *G*. Determine the time *t* corresponding to *G*, which is the time at $U = 22.14\%$. So

$$c_v = \frac{0.0385 H_{dr}^2}{t_{22.14}} \qquad (11.69)$$

In most cases, for a given soil and pressure range, the magnitude of c_v determined by using the *logarithm-of-time method* provides *lowest value*. The *highest value* is obtained from the *early stage log-t method*. The primary reason is because the early stage log-*t* method uses the earlier part of the consolidation curve, whereas the logarithm-of-time method uses the lower portion of the consolidation curve. When the lower portion of the consolidation curve is taken into account, the effect of secondary consolidation plays a role in the magnitude of c_v. This fact is demonstrated for several soils in Table 11.9.

Several investigators also have reported that the c_v value obtained from the field is substantially higher than that obtained from laboratory tests conducted by using conventional testing methods (that is, logarithm-of-time and square-root-of-time methods). This finding is shown in Table 11.10 (Leroueil, 1988). Hence, the early stage log-*t* method may provide a more realistic value of fieldwork.

Table 11.9 Comparison of c_v Obtained from Various Methods*

Soil	Range of pressure σ' (kN/m^2)	$c_v \times 10^4$ cm^2/sec		
		Logarithm-of-time method	Square-root-of-time method	Early stage log-t method
Red earth	25–50	4.63	5.45	6.12
	50–100	6.43	7.98	9.00
	100–200	7.32	9.99	11.43
	200–400	8.14	10.90	12.56
	400–800	8.10	11.99	12.80
Brown soil	25–50	3.81	4.45	5.42
	50–100	3.02	3.77	3.80
	100–200	2.86	3.40	3.52
	200–400	2.09	2.21	2.74
	400–800	1.30	1.45	1.36
Black cotton soil	25–50	5.07	6.55	9.73
	50–100	3.06	3.69	4.78
	100–200	2.00	2.50	3.45
	200–400	1.15	1.57	2.03
	400–800	0.56	0.64	0.79
Illite	25–50	1.66	2.25	2.50
	50–100	1.34	3.13	3.32
	100–200	2.20	3.18	3.65
	200–400	3.15	4.59	5.14
	400–800	4.15	5.82	6.45
Bentonite	25–50	0.063	0.130	0.162
	50–100	0.046	0.100	0.130
	100–200	0.044	0.052	0.081
	200–400	0.021	0.022	0.040
	400–800	0.015	0.017	0.022
Chicago clay (Taylor, 1948)	12.5–25	25.10	45.50	46.00
	25–50	20.10	23.90	31.50
	50–100	13.70	17.40	20.20
	100–200	3.18	4.71	4.97
	200–400	4.56	4.40	4.91
	400–800	6.05	6.44	7.41
	800–1600	7.09	8.62	9.09

*After a table from "Determination of Coefficient of Consolidation from Early Stage of Log t Plot," by R. G. Robinson and M. M. Allam, 1996, *Geotechnical Testing Journal, 19*(3) pp. 316–320. Copyright © 1996 American Society for Testing and Materials. Reprinted with permission.

Table 11.10 Comparison Between the Coefficients of Consolidation Determined in the Laboratory and Those Deduced from Embankment Settlement Analysis, as Observed by Leroueil (1988)

Site	$c_{v(lab)}$ (m²/sec)	$c_{v(in\ situ)}$ (m²/sec)	$c_{v(in\ situ)}/c_{v(lab)}$	Reference
Ska-Edeby IV	5.0×10^{-9}	1.0×10^{-7}	20	Holtz and Broms (1972)
Oxford (1)			4–57	Lewis, Murray, and Symons (1975)
Donnington			4–7	Lewis, Murray, and Symons (1975)
Oxford (2)			3–36	Lewis, Murray, and Symons (1975)
Avonmouth			6–47	Lewis, Murray, and Symons (1975)
Tickton			7–47	Lewis, Murray, and Symons (1975)
Over causeway			3–12	Lewis, Murray, and Symons (1975)
Melbourne			200	Walker and Morgan (1977)
Penang	1.6×10^{-8}	1.1×10^{-6}	70	Adachi and Todo (1979)
Cubzac B	2.0×10^{-8}	2.0×10^{-7}	10	Magnan, *et al.* (1983)
Cubzac C	1.4×10^{-8}	4.3×10^{-7}	31	Leroueil, Magnan, and Tavenas (1985)
A-64	7.5×10^{-8}	2.0×10^{-6}	27	Leroueil, Magnan, and Tavenas (1985)
Saint-Alban	1.0×10^{-8}	8.0×10^{-8}	8	Leroueil, Magnan, and Tavenas (1985)
R-7	6.0×10^{-9}	2.8×10^{-7}	47	Leroueil, Magnan, and Tavenas (1985)
Matagami	8.0×10^{-9}	8.5×10^{-8}	10	Leroueil, Magnan, and Tavenas (1985)
Berthierville		4.0×10^{-8}	3–10	Kabbaj (1985)

Example 11.11

During a laboratory consolidation test, the time and dial gauge readings obtained from an increase of pressure on the specimen from 50 kN/m² to 100 kN/m² are given here.

Time (min)	Dial gauge reading (cm × 10⁴)	Time (min)	Dial gauge reading (cm × 10⁴)
0	3975	16.0	4572
0.1	4082	30.0	4737
0.25	4102	60.0	4923
0.5	4128	120.0	5080
1.0	4166	240.0	5207
2.0	4224	480.0	5283
4.0	4298	960.0	5334
8.0	4420	1440.0	5364

Using the logarithm-of-time method, determine c_v. The average height of the specimen during consolidation was 2.24 cm, and it was drained at the top and bottom.

Solution

The semi-logarithmic plot of dial reading versus time is shown in Figure 11.29. For this, $t_1 = 0.1$ min, $t_2 = 0.4$ min to determine d_0. Following the procedure outlined in Figure 11.25, $t_{50} \approx 19$ min. From Eq. (11.66)

$$C_v = \frac{0.197 H_{dr}^2}{t_{50}} = \frac{0.197 \left(\frac{2.24}{2}\right)^2}{19} = 0.013\,\text{cm}^2/\text{min} = \mathbf{2.17 \times 10^{-4}\,\text{cm}^2/\text{sec}}$$

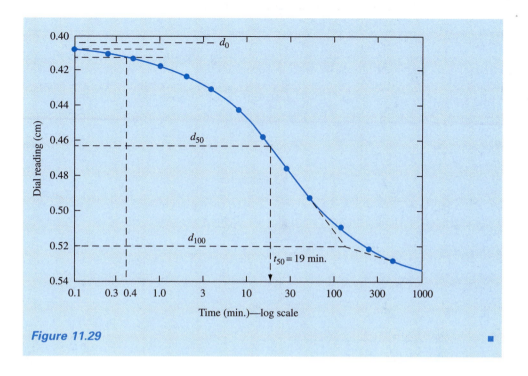

Figure 11.29

11.14 Calculation of Consolidation Settlement Under a Foundation

Chapter 10 showed that the increase in the vertical stress in soil caused by a load applied over a limited area decreases with depth z measured from the ground surface downward. Hence to estimate the one-dimensional settlement of a foundation, we can use Eq. (11.31), (11.33), or (11.34). However, the increase of effective stress, $\Delta\sigma'$, in these equations should be the average increase in the pressure below the center of the foundation. The values can be determined by using the procedure described in Chapter 10.

Assuming that the pressure increase varies parabolically, using Simpson's rule, we can estimate the value of $\Delta\sigma'_{av}$ as

$$\Delta\sigma'_{av} = \frac{\Delta\sigma'_t + 4\Delta\sigma'_m + \Delta\sigma'_b}{6} \qquad (11.70)$$

where $\Delta\sigma'_t$, $\Delta\sigma'_m$, and $\Delta\sigma'_b$ represent the increase in the effective pressure at the top, middle, and bottom of the layer, respectively.

Example 11.12

Calculate the settlement of the 3 m-thick clay layer (Figure 11.30) that will result from the load carried by a 1.5-m-square footing. The clay is normally consolidated. Use the weighted average method [Eq. (11.70)] to calculate the average increase of effective pressure in the clay layer.

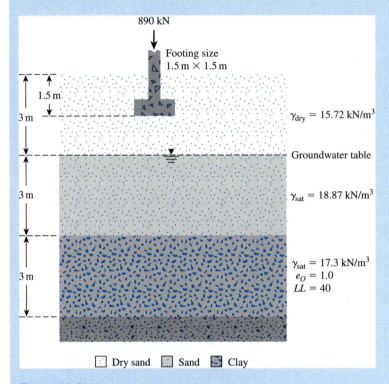

Figure 11.30

Solution
For normally consolidated clay, from Eq. (11.31),

$$S_e = \frac{C_c H}{1 + e_o} \log \frac{\sigma'_o + \Delta \sigma'_{av}}{\sigma'_o}$$

where

$$C_c = 0.009(LL - 10) = 0.009(40 - 10) = 0.27$$

$$H = 3 \times 1000 = 3000 \text{ mm}$$

$$e_o = 1.0$$

$$\sigma'_O = 3 \text{ m} \times \gamma_{dry(sand)} + 3 \text{ m}[\gamma_{sat(sand)} - 9.81] + \frac{3}{2}[\gamma_{sat(clay)} - 9.81]$$

$$= 3 \times 15.72 + 3(18.87 - 9.81) + 1.5(17.3 - 9.81)$$

$$= 85.58 \text{ kN/m}^2$$

From Eq. (11.70),

$$\Delta\sigma'_{av} = \frac{\Delta\sigma'_t + 4\Delta\sigma'_m + \Delta\sigma'_b}{6}$$

$\Delta\sigma'_t$, $\Delta\sigma'_m$, and $\Delta\sigma'_b$ below the center of the footing can be obtained from Eq. (10.34).
Now we can prepare the following table (*Note: L/B* = 1.5/1.5 = 1):

m_1	z (m)	$b = B/2$ (m)	$n_1 = z/b$	q (kN/m²)	I_4	$\Delta\sigma' = qI_4$ (kN/m²)
1	4.5	0.75	6	$\frac{890}{1.5 \times 1.5} = 395.6$	0.051	$20.18 = \Delta\sigma'_t$
1	6.0	0.75	8	359.6	0.029	$11.47 = \Delta\sigma'_m$
1	7.5	0.75	10	359.6	0.019	$7.52 = \Delta\sigma'_b$

So,

$$\Delta\sigma'_{av} = \frac{20.18 + (4)(11.47) + 7.52}{6} = 12.26 \text{ kN/m}^2$$

Hence,

$$S_c = \frac{(0.27)(3000)}{1 + 1} \log\frac{85.58 + 12.26}{85.58} \approx \textbf{23.6 mm} \qquad \blacksquare$$

11.15 A Case History—Settlement Due to a Preload Fill for Construction of Tampa VA Hospital

Wheeless and Sowers (1972) have presented the field measurements of settlement due to a preload fill used for the construction of Tampa Veterans Administration Hospital. Figure 11.31 shows the simplified general subsoil conditions at the building site. In general, the subsoil consisted of 4.6 to 6 m subangular quartz sand at the top followed by clayey soil of varying thicknesses. The void ratio of the clayey soil varied from 0.7 to 1.4. The silt and clay content of the clayey soil varied from 5 to 75%. The Tampa limestone underlying the clay layer is a complex assortment of chalky, poorly consolidated

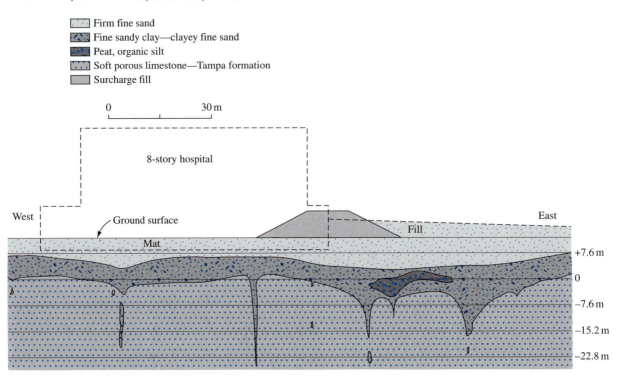

□ Firm fine sand
□ Fine sandy clay—clayey fine sand
□ Peat, organic silt
□ Soft porous limestone—Tampa formation
□ Surcharge fill

0   30 m

8-story hospital

West

Ground surface

Mat

East

Fill

+7.6 m

0

−7.6 m

−15.2 m

−22.8 m

Figure 11.31 Simplified general subsoil conditions at the site of Tampa VA Hospital (*After Wheeless and Sowers, 1972. With permission from ASCE.*)

calcareous deposits. The groundwater table was located at a depth of about 4.57 m below the ground surface (elevation +7.6 m). Figure 11.32 shows the consolidation curves obtained in the laboratory for clayey sand and sandy clay samples collected from various depths at the site.

The plan of the hospital building is shown in Figure 11.33 (broken lines). Figure 11.31 also shows the cross section of the building. For a number of reasons, it was decided that the hospital should be built with a mat foundation. As can be seen from Figure 11.31, some soil had to be excavated to build the mat. As reported by Wheeless and Sowers, preliminary calculations indicated that the average building load in the eight-story area would be equal to the weight of the soil to be excavated for the construction of the mat. In that case, the consolidation settlement of the clay layer under the building would be rather small. However, the grading plan required a permanent fill of 4.88 m over the original ground surface to provide access to the main floor on the east side. This is also shown in Figure 11.31. Preliminary calculations indicated that the weight of this fill could be expected to produce a soil settlement of about 101.6 mm near the east side of the building. This settlement would produce undue bending and overstressing of the mat foundation. For that reason, it was decided to build a temporary fill that was 7.93 m high and limited to the front area of the proposed building. The fill area is shown in Figures 11.31 and 11.33. This temporary fill was built because the vertical stress that it induced in the clay layer would be greater than the stress induced by the permanent fill of 4.88 m as required

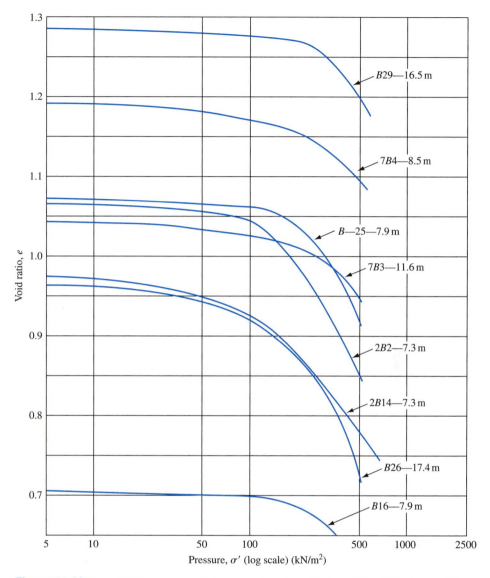

Figure 11.32 Consolidation curves of clayey sands and sandy clays (*After Wheeless and Sowers, 1972. With permission from ASCE.*)

by the grading plan. This would produce faster consolidation settlement. In a period of about four months, the settlement would be approximately 101.6 mm which is the magnitude of maximum settlement expected from the required permanent fill of 4.88 m. At that time, if the excess fill material is removed and the building constructed, the settlement of the mat on the east side will be negligible. This technique of achieving the probable settlement of soil before construction is referred to as *preloading*.

Figure 11.33 shows the locations of eight settlement plates placed on the original ground surface before the construction of the temporary fill. Figure 11.34 shows the

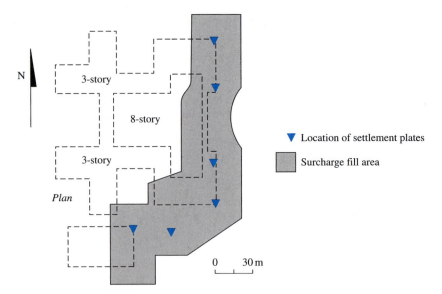

Figure 11.33 Plan of the Tampa VA Hospital (*After Wheeless and Sowers, 1972. With permission from ASCE.*)

time-settlement records beneath the surcharge fill area as observed from the settlement plates. Following is a comparison of total estimated and observed consolidation settlement due to preloading.

Settlement plate location	Observed settlement (mm)	Estimated consolidation settlement (mm)
3	66.0	73.7
4	63.5	73.7
6	73.7	76.2
7	86.4	96.5

From the preceding comparisons of observed and estimated settlements given by Wheeless and Sowers and Figure 11.34, the following conclusions can be drawn:

1. In all cases, the estimated settlement slightly exceeded the observed settlement.
2. Most of the settlement was complete in a period of about 90 days.
3. The difference between the estimated and observed settlement varied from 3 to 16%, with an average of 13%.
4. Two-thirds to four-fifths of the total observed settlement was completed during the period of fill construction. The rate of consolidation was much faster than anticipated.

Wheeless and Sowers have suggested that the accelerated rate of consolidation may be due primarily to irregular sandy seams within the clay stratum. In Section 11.12, it was shown that the average degree of consolidation is related to the time factor, T_v. Also

$$t = \frac{T_v H_{dr}^2}{c_v}$$

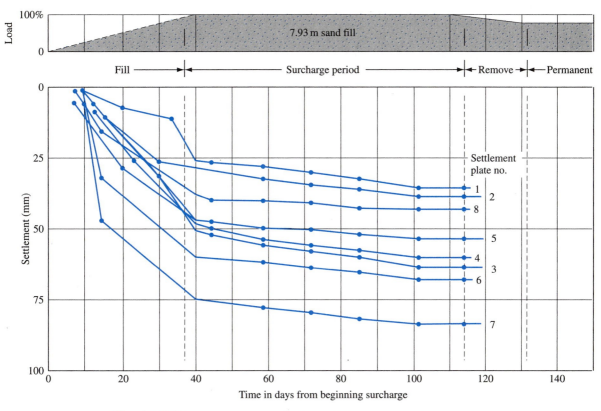

Figure 11.34 Settlement-time curves beneath the surcharge fill area for the construction of Tampa VA Hospital (*After Wheeless and Sowers, 1972. With permission from ASCE.*)

For similar values of T_v (or average degree of consolidation) and c_v, the time t will be less if the maximum length of the drainage path (H_{dr}) is less. The presence of irregular sandy seams in the clay layer tends to reduce the magnitude of H_{dr}. This is the reason why a faster rate of consolidation was attained in this area.

The structural load of the VA hospital was completed in the early part of 1970. No noticeable foundation movement has occurred.

11.16 *Methods for Accelerating Consolidation Settlement*

In many instances, *sand drains* and *prefabricated vertical drains* are used in the field to accelerate consolidation settlement in soft, normally consolidated clay layers and to achieve precompression before the construction of a desired foundation. Sand drains are constructed by drilling holes through the clay layer(s) in the field at regular intervals. The holes then are backfilled with sand. This can be achieved by several means, such as (a) rotary drilling and then backfilling with sand; (b) drilling by continuous flight auger with a hollow stem and backfilling with sand (through the

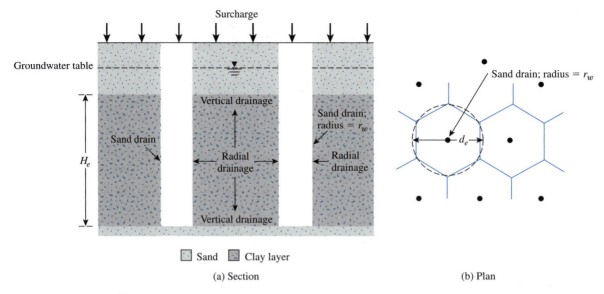

Figure 11.35 Sand drains

hollow stem); and (c) driving hollow steel piles. The soil inside the pile is then jetted out, and the hole is backfilled with sand. Figure 11.35 shows a schematic diagram of sand drains. After backfilling the drill holes with sand, a surcharge is applied at the ground surface. This surcharge will increase the pore water pressure in the clay. The excess pore water pressure in the clay will be dissipated by drainage—both vertically and radially to the sand drains—which accelerates settlement of the clay layer. In Figure 11.35a, note that the radius of the sand drains is r_w. Figure 11.35b shows the plan of the layout of the sand drains. The effective zone from which the radial drainage will be directed toward a given sand drain is approximately cylindrical, with a diameter of d_e. The surcharge that needs to be applied at the ground surface and the length of time it has to be maintained to achieve the desired degree of consolidation will be a function of r_w, d_e, and other soil parameters. Figure 11.36 shows a sand drain installation in progress.

Prefabricated vertical drains (PVDs), which also are referred to as *wick* or *strip drains,* originally were developed as a substitute for the commonly used sand drain. With the advent of materials science, these drains are manufactured from synthetic polymers such as polypropylene and high-density polyethylene. PVDs normally are manufactured with a corrugated or channeled synthetic core enclosed by a geotextile filter, as shown schematically in Figure 11.37. Installation rates reported in the literature are on the order of 0.1 to 0.3 m/s, excluding equipment mobilization and setup time. PVDs have been used extensively in the past for expedient consolidation of low permeability soils under surface surcharge. The main advantage of PVDs over sand drains is that they do not require drilling and, thus, installation is much faster. Figure 11.38 shows the installation of PVDs in the field.

Figure 11.36 Sand drain installation in progress (*Courtesy of E. C. Shin, University of Incheon, South Korea*)

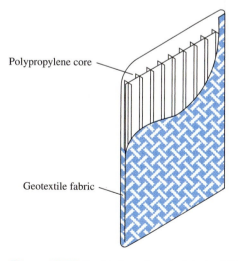

Polypropylene core

Geotextile fabric

Figure 11.37 Prefabricated vertical drain (PVD)

Figure 11.38
Installation of PVDs in progress
(*Courtesy of E. C. Shin, University of Incheon, South Korea*)

11.17   *Precompression*

When highly compressible, normally consolidated clayey soil layers lie at a limited depth and large consolidation settlements are expected as a result of the construction of large buildings, highway embankments, or earth dams, precompression of soil may be used to minimize postconstruction settlement. The principles of precompression are best explained by referring to Figure 11.39. Here, the proposed structural load per unit area is $\Delta\sigma_{(p)}$ and the thickness of the clay layer undergoing consolidation is H. The maximum primary consolidation settlement caused by the structural load, $S_c = S_{(p)}$, then is

$$S_c = S_{(p)} = \frac{C_c H}{1 + e_o} \log \frac{\sigma'_o + \Delta\sigma_{(p)}}{\sigma'_o} \qquad (11.71)$$

Note that at the end of consolidation, $\Delta\sigma' = \Delta\sigma_{(p)}$.

The settlement–time relationship under the structural load will be like that shown in Figure 11.39b. However, if a surcharge of $\Delta\sigma_{(p)} + \Delta\sigma_{(f)}$ is placed on the ground, then the primary consolidation settlement will be

$$S_c = S_{(p+f)} = \frac{C_c H}{1 + e_o} \log \frac{\sigma'_o + [\Delta\sigma_{(p)} + \Delta\sigma_{(f)}]}{\sigma'_o} \qquad (11.72)$$

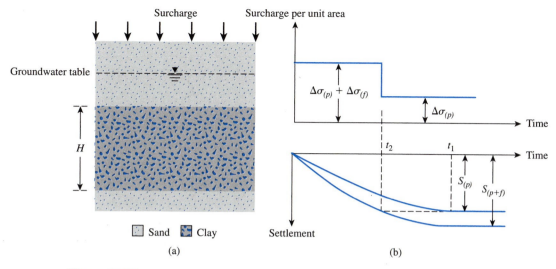

Figure 11.39 Principles of precompression

Note that at the end of consolidation,

$$\Delta\sigma' = \Delta\sigma_{(p)} + \Delta\sigma_{(f)}$$

The settlement–time relationship under a surcharge of $\Delta\sigma_{(p)} + \Delta\sigma_{(f)}$ is also shown in Figure 11.39b. Note that a total settlement of $S_{(p)}$ would occur at a time t_2, which is much shorter than t_1. So, if a temporary total surcharge of $\Delta\sigma_{(f)} + \Delta\sigma_{(p)}$ is applied on the ground surface for time t_2, the settlement will equal $S_{(p)}$. At that time, if the surcharge is removed and a structure with a permanent load per unit area of $\Delta\sigma_{(p)}$ is built, no appreciable settlement will occur. The procedure just described is *precompression*. The total surcharge, $\Delta\sigma_{(p)} + \Delta\sigma_{(f)}$, can be applied by using temporary fills.

Derivation of Equations to Obtain $\Delta\sigma_{(f)}$ and t_2

Figure 11.39b shows that, under a surcharge of $\Delta\sigma_{(p)} + \Delta\sigma_{(f)}$, the degree of consolidation at time t_2 after load application is

$$U = \frac{S_{(p)}}{S_{(p+f)}} \tag{11.73}$$

Substitution of Eqs. (11.71) and (11.72) into Eq. (11.73) yields

$$U = \frac{\log\left[\dfrac{\sigma'_o + \Delta\sigma_{(p)}}{\sigma'_o}\right]}{\log\left[\dfrac{\sigma'_o + \Delta\sigma_{(p)} + \Delta\sigma_{(f)}}{\sigma'_o}\right]} = \frac{\log\left[1 + \dfrac{\Delta\sigma_{(p)}}{\sigma'_o}\right]}{\log\left\{1 + \dfrac{\Delta\sigma_{(p)}}{\sigma'_o}\left[1 + \dfrac{\Delta\sigma_{(f)}}{\Delta\sigma_{(p)}}\right]\right\}} \tag{11.74}$$

From Eq. (11.61), we know that

$$U = f(T_v) \tag{11.75}$$

where T_v = time factor = $c_v t_2 / H_{dr}^2$

c_v = coefficient of consolidation

t_2 = time

H_{dr} = maximum drainage path ($H/2$ for two-way drainage and H for one-way drainage)

The variation of U with T_v is shown in Table 11.8.

Procedure for Obtaining Precompression Parameters

Engineers may encounter two problems during precompression work in the field:

1. The value of $\Delta\sigma_{(f)}$ is known, but t_2 must be obtained. In such case, obtain σ_o' and $\Delta\sigma_{(p)}$ and solve for U using Eq. (11.74). For this value of U, obtain T_v from Table 11.8. Then,

$$t_2 = \frac{T_v H_{dr}^2}{c_v} \tag{11.76}$$

2. For a specified value of t_2, $\Delta\sigma_{(f)}$ must be obtained. In such case, calculate T_v. Then refer to Table 11.8 to obtain the degree of consolidation, U. With the estimated value of U, go to Eq. (11.74) to find the required $\Delta\sigma_{(f)}/\Delta\sigma_{(p)}$ and then calculate $\Delta\sigma_{(f)}$.

The case history given in Section 11.15 is an example of precompression.

Example 11.13

Refer to Figure 11.39. During the construction of a highway bridge, the average permanent load on the clay layer is expected to increase by about 115 kN/m². The average effective overburden pressure at the middle of the clay layer is 210 kN/m². Here, H = 6 m, C_c = 0.28, e_o = 0.9, and c_v = 0.36 m²/mo. The clay is normally consolidated.

a. Determine the total primary consolidation settlement of the bridge without precompression.
b. What is the surcharge, $\Delta\sigma_{(f)}$, needed to eliminate by precompression the entire primary consolidation settlement in nine months?

Solution
Part a
The total primary consolidation settlement may be calculated from Eq. (11.71):

$$S_c = S_{(p)} = \frac{C_c H}{1 + e_o} \log\left[\frac{\sigma_o' + \Delta\sigma_{(p)}}{\sigma_o'}\right] = \frac{(0.28)(6)}{1 + 0.9} \log\left[\frac{210 + 115}{210}\right]$$

$$= 0.1677 \text{ m} = \mathbf{167.7 \text{ mm}}$$

Part b

$$T_v = \frac{c_v t_2}{H_{dr}^2}$$

$$c_v = 0.36 \text{ m}^2/\text{mo.}$$

$$H_{dr} = 3 \text{ m (two-way drainage)}$$

$$t_2 = 9 \text{ mo.}$$

Hence,

$$T_v = \frac{(0.36)(9)}{3^2} = 0.36$$

According to Table 11.8, for $T_v = 0.36$, the value of U is 67%. Now

$$\Delta\sigma_{(p)} = 115 \text{ kN/m}^2$$

$$\sigma_o' = 210 \text{ kN/m}^2$$

So,

$$\frac{\Delta\sigma_{(p)}}{\sigma_o'} = \frac{115}{210} = 0.548$$

Thus, from Eq. (11.74),

$$U = 0.67 = \frac{\log(1 + 0.548)}{\log\left\{1 + 0.548\left[1 + \frac{\Delta\sigma_{(f)}}{\Delta\sigma_{(p)}}\right]\right\}}$$

$$\frac{\Delta\sigma_{(f)}}{\Delta\sigma_{(p)}} = 0.677$$

So,

$$\Delta\sigma_{(f)} = (0.677)(115) = \textbf{78 kN/m}^2 \qquad \blacksquare$$

11.18   Summary and General Comments

In this chapter, we discussed the fundamental concepts and theories for estimating elastic and consolidation (primary and secondary) settlement. Elastic settlement of a foundation is primarily a function of the size and rigidity of the foundation, the modulus of elasticity, the Poisson's ratio of the soil, and the intensity of load on the foundation.

Consolidation is a time-dependent process of settlement of saturated clay layers located below the ground water table by extrusion of excess water pressure generated by application of load on the foundation. Total consolidation settlement of a clay foundation is a function of compression index (C_c), swell index (C_s), initial void ratio, (e_O) and the average stress increase in the clay layer. The degree of consolidation for a given soil layer at a certain time after the load application depends on its coefficient of consolidation (c_v) and also on the length of the minimum drainage path. Installation of sand drains and wick drains helps reduce the time for accomplishing the desired degree of consolidation for a given construction project.

There are several case histories in the literature for which the fundamental principles of soil compressibility have been used to predict and compare the actual total settlement and the time rate of settlement of soil profiles under superimposed loading. In some cases, the actual and predicted maximum settlements agree remarkably; in many others, the predicted settlements deviate to a large extent from the actual settlements observed. The disagreement in the latter cases may have several causes:

1. Improper evaluation of soil properties
2. Nonhomogeneity and irregularity of soil profiles
3. Error in the evaluation of the net stress increase with depth, which induces settlement

The variation between the predicted and observed time rate of settlement may also be due to

- Improper evaluation of c_v (see Section 11.13)
- Presence of irregular sandy seams within the clay layer, which reduces the length of the maximum drainage path, H_{dr}

Problems

11.1 Refer to Figure 11.3 for the rigid foundation. Given: $B = 1$ m; $L = 2$ m; $D_f = 1$ m; $H = 5$ m; $E_s = 14{,}000$ kN/m²; $\mu_s = 0.4$; and net increase of pressure on the foundation, $\Delta\sigma = 200$ kN/m². Estimate the elastic settlement.

11.2 Refer to Figure 11.3 for a square rigid foundation measuring 3 m × 3 m in plan supported by a layer of sand. Given that $D_f = 1.5$ m, $E_s = 16{,}000$ kN/m², $\mu_s = 0.3$; $H = 20$ m, and $\Delta\sigma = 100$ kN/m², calculate the elastic settlement.

11.3 The following are the results of a consolidation test.

e	Pressure, σ' (kN/m²)
1.1	25
1.085	50
1.055	100
1.01	200
0.94	400
0.79	800
0.63	1600

a. Plot the e-log σ' curve
b. Using Casagrande's method, determine the preconsolidation pressure
c. Calculate the compression index, C_c, from the laboratory e-log σ' curve

11.4 Repeat Problem 11.3 using the following values.

e	Pressure, σ'' (kN/m^2)
1.21	25
1.195	50
1.15	100
1.06	200
0.98	400
0.925	500

11.5 The results of a laboratory consolidation test on a clay specimen are the following.

Pressure, σ' (kN/m^2)	Total height of specimen at end of consolidation (mm)
25	17.65
50	17.40
100	17.03
200	16.56
400	16.15
800	15.88

Given the initial height of specimen = 19 mm., G_s = 2.68, mass of dry specimen = 95.2 g, and area of specimen = 31.68 cm^2:
a. Plot the e-log σ' curve
b. Determine the preconsolidation pressure
c. Calculate the compression index, C_c

11.6 Figure 11.40 shows a soil profile. The uniformly distributed load on the ground surface is $\Delta\sigma$. Given: $\Delta\sigma$ = 50 kN/m^2, H_1 = 2.44 m, H_2 = 4.57 m, and H_3 = 5.18 m. Also,
- Sand: γ_{dry} = 17.29 kN/m^3, γ_{sat} = 18.08 kN/m^3
- Clay: γ_{sat} = 18.87 kN/m^3, LL = 50, e = 0.9
Estimate the primary consolidation settlement if
a. The clay is normally consolidated
b. The preconsolidation pressure is 130 kN/m^2 ($C_s \approx \frac{1}{6}C_c$)

11.7 Refer to Figure 11.40. Given: H_1 = 2.5 m, H_2 = 2.5 m, H_3 = 3 m, and $\Delta\sigma$ = 100 kN/m^2. Also,
- Sand: e = 0.64, G_s = 2.65
- Clay: e = 0.9, G_s = 2.75, LL = 55
Estimate the primary consolidation settlement of the clay layer assuming that it is normally consolidated.

11.8 Refer to Figure 11.40. Given: H_1 = 1.5 m, H_2 = 2.1 m, H_3 = 2 m, and $\Delta\sigma$ = 150 kN/m^2. Also,
- Clay: e = 1.1, G_s = 2.72, LL = 45
- Sand: e = 0.58, G_s = 2.65

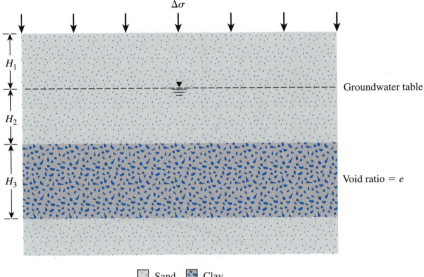

$\Delta\sigma$

H_1

Groundwater table

H_2

H_3

Void ratio $= e$

☐ Sand   ▨ Clay

Figure 11.40

Estimate the primary consolidation settlement if the preconsolidation pressure is 175 kN/m². Assume $C_S \approx \frac{1}{5}C_c$.

11.9 The coordinates of two points on a virgin compression curve are as follows:

- $e_1 = 0.82$ • $\sigma'_1 = 125$ kN/m²
- $e_2 = 0.70$ • $\sigma'_2 = 200$ kN/m²

Determine the void ratio that corresponds to a pressure of 300 kN/m².

11.10 The laboratory consolidation data for an undisturbed clay specimen are as follows:

- $e_1 = 1.1$ • $\sigma'_1 = 100$ kN/m²
- $e_2 = 0.9$ • $\sigma'_2 = 300$ kN/m²

What is the void ratio for a pressure of 350 kN/m²? (*Note:* $\sigma'_c = 80$ kN/m².)

11.11 Refer to Problem 11.7. Given: $c_v = 2.8 \times 10^{-6}$ m²/min.

How long will it take for 60% consolidation to occur?

11.12 Following are the relationships of e and σ' for a clay soil:

e	σ' (kN/m²)
1.0	20
0.97	50
0.85	180
0.75	320

For this clay soil in the field, the following values are given: $H = 1.37$ m, $\sigma'_o = 70$ kN/m², and $\sigma'_o + \Delta\sigma' = 200$ kN/m². Calculate the expected settlement caused by primary consolidation.

11.13 The coordinates of two points on a virgin compression curve are as follows:
- $e_1 = 1.7$ • $\sigma_1' = 150$ kN/m^2
- $e_2 = 1.48$ • $\sigma_2' = 400$ kN/m^2

a. Determine the coefficient of volume compressibility for the pressure range stated.
b. Given that $c_v = 0.002$ cm^2/sec, determine k in cm/sec corresponding to the average void ratio.

11.14 The time for 50% consolidation of a 25 mm thick clay layer (drained at top and bottom) in the laboratory is 2 min, 20 sec. How long (in days) will it take for a 2.44 mm thick layer of the same clay in the field (under the same pressure increment) to reach 30% consolidation? In the field, there is a rock layer at the bottom of the clay.

11.15 For a normally consolidated clay, the following are given:
- $\sigma_o' = 200$ kN/m^2 • $e = e_o = 1.21$
- $\sigma_o' + \Delta\sigma' = 400$ kN/m^2 • $e = 0.96$

The hydraulic conductivity k of the clay for the preceding loading range is 54.9×10^{-4} cm/day.

a. How long (in days) will it take for a 2.7 m thick clay layer (drained on both sides) in the field to reach 60% consolidation?
b. What is the settlement at that time (that is, at 60% consolidation)?

11.16 For a laboratory consolidation test on a clay specimen (drained on both sides), the following were obtained:
- Thickness of the clay layer = 25 mm
- $\sigma_1' = 200$ kN/m^2 • $e_1 = 0.73$
- $\sigma_2' = 400$ kN/m^2 • $e_2 = 0.61$
- Time for 50% consolidation (t_{50}) = 2.8 min

Determine the hydraulic conductivity of the clay for the loading range.

11.17 The time for 50% consolidation of a 25-mm thick clay layer (drained at top and bottom) in the laboratory is 225 sec. How long (in days) will it take for a 2-m thick layer of the same clay in the field (under the same pressure increment) to reach 50% consolidation? There is a rock layer at the bottom of the clay in the field.

11.18 A 3-m thick layer of saturated clay under a surcharge loading underwent 90% primary consolidation in 100 days. The laboratory test's specimen will have two-way drainage.

a. Find the coefficient of consolidation of clay for the pressure range.
b. For a 25 mm. thick undisturbed clay specimen, how long will it take to undergo 80% consolidation in the laboratory for a similar pressure range?

11.19 A normally consolidated clay layer is 3 m thick (one-way drainage). From the application of a given pressure, the total anticipated primary consolidation settlement will be 80 mm.

a. What is the average degree of consolidation for the clay layer when the settlement is 25 mm?
b. If the average value of c_v for the pressure range is 0.002 cm^2/sec, how long will it take for 50% settlement to occur?
c. How long will it take for 50% consolidation to occur if the clay layer is drained at both top and bottom?

11.20 Refer to Figure 11.41. Given that $B = 1$ m, $L = 3$ m, and $Q = 110$ kN, calculate the primary consolidation settlement of the foundation.

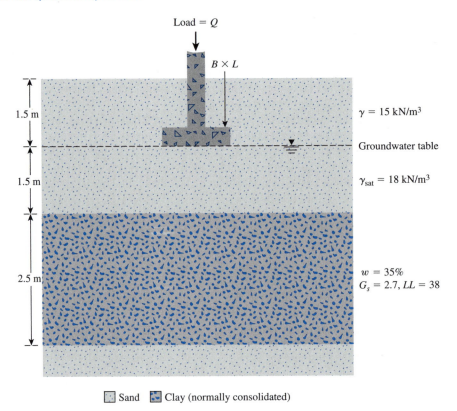

Figure 11.41

References

ADACHI, K., and TODO, H. (1979). "A Case Study on Settlement of Soil Clay in Penang," *Proceedings,* 6th Asian Regional Conference on Soil Mechanics and Foundation Engineering, Vol. 1, 117–120.

BOWLES, J. E. (1987). "Elastic Foundation Settlement on Sand Deposits," *Journal of Geotechnical Engineering*, ASCE, Vol. 113, No. 8, 846–860.

CASAGRANDE, A. (1936). "Determination of the Preconsolidation Load and Its Practical Significance," *Proceedings,* 1st International Conference on Soil Mechanics and Foundation Engineering, Cambridge, Mass., Vol. 3, 60–64.

CASAGRANDE, A., and FADUM, R. E. (1940). "Notes on Soil Testing for Engineering Purposes," Harvard University Graduate School of Engineering Publication No. 8.

FOX, E. N. (1948). "The Mean Elastic Settlement of a Uniformly Loaded Area at a Depth Below the Ground Surface," *Proceedings*, 2nd International Conference on Soil Mechanics and Foundation Engineering, Rotterdam, Vol. 1, pp. 129–132.

HANSBO, S. (1957). "A New Approach to the Determination of Shear Strength of Clay by the Fall Cone Test," *Report 14*, Swedish Geotechnical Institute, Stockholm.

HOLTZ, R. D., and BROMS, B. (1972). "Long Term Loading Tests on Ska-Edeby, Sweden," *Proceedings,* Conference on Performance of Earth and Earth-Supported Structures, ASCE, Vol. 1.1, 435–464.

HOUGH, B. K. (1957). *Basic Soils Engineering,* Ronald Press, New York.

KABBAJ, M. (1985). "Aspects Rhéologiques des Argiles Naturelles en Consolidation." Ph.D. Thesis, Université Laval, Québec, Canada.

KULHAWY, F. H., AND MAYNE, P. W. (1990). *Manual of Estimating Soil Properties for Foundation Design*, Final Report (*EL-6800*) submitted to Electric Power Research Institute (EPRI), Palo Alto, CA.

LEROUEIL, S. (1988). "Tenth Canadian Geotechnical Colloquium: Recent Developments in Consolidation of Natural Clays," *Canadian Geotechnical Journal*, Vol. 25, No. 1, 85–107.

LEROUEIL, S., MAGNAN, J. P., and TAVENAS, F. (1985). *Remblais sur Argiles Molles*. Technique et Documentation, Paris, France.

LEWIS, W. A., MURRAY, R. T., and SYMONS, I. F. (1975). "Settlement and Stability of Embankments Constructed on Soft Alluvial Soils," *Proceedings*, The Institute of Civil Engineers, Vol. 59, 571–593.

MAGNAN, J. P., MIEUSSENS, C., and QUEYROI, D. (1983). Research Report LPC 127, *Étude d'un Remblai sur Sols Compressibles: le Remblai B du Site Expérimental de Cubzac-les-Ponts*. Laboratorie Central des Ponts et Chaussées, Paris.

MESRI, G. (1973). "Coefficient of Secondary Compression," *Journal of the Soil Mechanics and Foundations Division*, ASCE, Vol. 99, No. SM1, 122–137.

MESRI, G., AND GODLEWSKI, P. M. (1977). "Time and Stress—Compressiblity Interrelationship," *Journal of the Geotechnical Engineering Division*, ASCE, Vol. 103, No. GT5, 417–430.

NAGARAJ, T., and MURTY, B. R. S. (1985). "Prediction of the Preconsolidation Pressure and Recompression Index of Soils," *Geotechnical Testing Journal*, Vol. 8, No. 4, 199–202.

NISHIDA, Y. (1956). "A Brief Note on Compression Index of Soils," *Journal of the Soil Mechanics and Foundations Division*, ASCE, Vol. 82, No. SM3, 1027-1–1027-14.

PARK, J. H., and KOUMOTO, T. (2004). "New Compression Index Equation," *Journal of Geotechnical and Geoenvironmental Engineering*, ASCE, Vol. 130, No. 2, 223–226.

RENDON-HERRERO, O. (1983). "Universal Compression Index Equation," *Discussion, Journal of Geotechnical Engineering*, ASCE, Vol. 109, No. 10, 1349.

RENDON-HERRERO, O. (1980). "Universal Compression Index Equation," *Journal of the Geotechnical Engineering Division*, ASCE, Vol. 106, No. GT11, 1179–1200.

ROBINSON, R. G., and ALLAM, M. M. (1996). "Determination of Coefficient of Consolidation from Early Stage of log *t* Plot," *Geotechnical Testing Journal*, ASTM, Vol. 19, No. 3, 316–320.

SCHMERTMANN, J. H. (1953). "Undisturbed Consolidation Behavior of Clay," *Transactions*, ASCE, Vol. 120, 1201.

SIVARAM, B., and SWAMEE, A. (1977). "A Computational Method for Consolidation Coefficient," *Soils and Foundations*, Vol. 17, No. 2, 48–52.

SKEMPTON, A. W. (1944). "Notes on the Compressibility of Clays," *Quarterly Journal of the Geological Society of London*, Vol. 100, 119–135.

SRIDHARAN, A., and PRAKASH, K. (1985). "Improved Rectangular Hyperbola Method for the Determination of Coefficient of Consolidation," *Geotechnical Testing Journal*, ASTM, Vol. 8, No. 1, 37–40.

STAS, C.V., AND KULHAWY, F. H. (1984). "Critical Evaluation of Design Methods for Foundations under Axial Uplift and Compression Loading," *Report EL-3771*, Electric Power Research Institute (EPRI), Palo Alto, CA.

STEINBRENNER, W. (1934). "Tafeln zur Setzungsberechnung," *Die Strasse*, Vol. 1, 121–124.

TAYLOR, D. W. (1942). "Research on Consolidation of Clays," *Serial No. 82*, Department of Civil and Sanitary Engineering, Massachusetts Institute of Technology, Cambridge, Mass.

TAYLOR, D. W. (1948). *Fundamentals of Soil Mechanics*, Wiley, New York.

TERZAGHI, K. (1925). *Erdbaumechanik auf Bodenphysikalischer Grundlager*, Deuticke, Vienna.

TERZAGHI, K., and PECK, R. B. (1967). *Soil Mechanics in Engineering Practice*, 2nd ed., Wiley, New York.

WALKER, L. K. and MORGAN, J. R. (1977). "Field Performance of a Firm Silty Clay," *Proceedings,* 9th International Conference on Soil Mechanics and Foundation Engineering, Tokyo, Vol. 1, 341–346.

WHEELESS, L. D., AND SOWERS, G. F. (1972). "Mat Foundation and Preload Fill, VA Hospital, Tampa," *Proceedings, Specialty Conference on Performance of Earth and Earth-Supported Structures*, ASCE, Vol. 1, Part 2, 939–951.

WROTH, C. P., AND WOOD, D. M. (1978). "The Correlation of Index Properties with Some Basic Engineering Properties of Soils," *Canadian Geotechnical Journal*, Vol. 15, No. 2, 137–145.

12 Shear Strength of Soil

The *shear strength* of a soil mass is the internal resistance per unit area that the soil mass can offer to resist failure and sliding along any plane inside it. One must understand the nature of shearing resistance in order to analyze soil stability problems, such as bearing capacity, slope stability, and lateral pressure on earth-retaining structures.

12.1 Mohr–Coulomb Failure Criterion

Mohr (1900) presented a theory for rupture in materials that contended that a material fails because of a critical combination of normal stress and shearing stress and not from either maximum normal or shear stress alone. Thus, the functional relationship between normal stress and shear stress on a failure plane can be expressed in the following form:

$$\tau_f = f(\sigma) \tag{12.1}$$

The failure envelope defined by Eq. (12.1) is a curved line. For most soil mechanics problems, it is sufficient to approximate the shear stress on the failure plane as a linear function of the normal stress (Coulomb, 1776). This linear function can be written as

$$\tau_f = c + \sigma \tan \phi \tag{12.2}$$

where c = cohesion
ϕ = angle of internal friction
σ = normal stress on the failure plane
τ_f = shear strength

The preceding equation is called the *Mohr–Coulomb failure criterion.*

In saturated soil, the total normal stress at a point is the sum of the effective stress (σ') and pore water pressure (u), or

$$\sigma = \sigma' + u$$

The effective stress σ' is carried by the soil solids. The Mohr–Coulomb failure criterion, expressed in terms of effective stress, will be of the form

$$\tau_f = c' + \sigma' \tan \phi' \qquad (12.3)$$

where c' = cohesion and ϕ' = friction angle, based on effective stress.

Thus, Eqs. (12.2) and (12.3) are expressions of shear strength based on total stress and effective stress. The value of c' for sand and inorganic silt is 0. For normally consolidated clays, c' can be approximated at 0. Overconsolidated clays have values of c' that are greater than 0. The angle of friction, ϕ', is sometimes referred to as the *drained angle of friction*. Typical values of ϕ' for some granular soils are given in Table 12.1.

The significance of Eq. (12.3) can be explained by referring to Fig. 12.1, which shows an elemental soil mass. Let the effective normal stress and the shear stress on the

Table 12.1 Typical Values of Drained Angle of Friction for Sands and Silts

Soil type	ϕ' (deg)
Sand: Rounded grains	
Loose	27–30
Medium	30–35
Dense	35–38
Sand: Angular grains	
Loose	30–35
Medium	35–40
Dense	40–45
Gravel with some sand	34–48
Silts	26–35

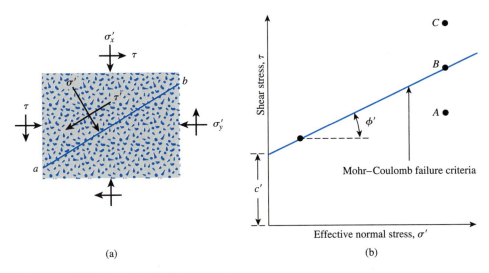

(a)   (b)

Figure 12.1 Mohr–Coulomb failure criterion

plane ab be σ' and τ, respectively. Figure 12.1b shows the plot of the failure envelope defined by Eq. (12.3). If the magnitudes of σ' and τ on plane ab are such that they plot as point A in Figure 12.1b, shear failure will not occur along the plane. If the effective normal stress and the shear stress on plane ab plot as point B (which falls on the failure envelope), shear failure will occur along that plane. A state of stress on a plane represented by point C cannot exist, because it plots above the failure envelope, and shear failure in a soil would have occurred already.

12.2 *Inclination of the Plane of Failure Caused by Shear*

As stated by the Mohr–Coulomb failure criterion, failure from shear will occur when the shear stress on a plane reaches a value given by Eq. (12.3). To determine the inclination of the failure plane with the major principal plane, refer to Figure 12.2, where σ_1' and σ_3' are, respectively, the major and minor effective principal stresses. The failure plane EF makes an angle θ with the major principal plane. To determine the angle θ and the relationship between σ_1' and σ_3', refer to Figure 12.3, which is a plot of the Mohr's circle for the state of stress shown in Figure 12.2 (see Chapter 10). In Figure 12.3, fgh is the failure envelope defined by the relationship $\tau_f = c' + \sigma' \tan \phi'$. The radial line ab defines the major principal plane (CD in Figure 12.2), and the radial line ad defines the failure plane (EF in Figure 12.2). It can be shown that $\angle bad = 2\theta = 90 + \phi'$, or

$$\theta = 45 + \frac{\phi'}{2} \tag{12.4}$$

Again, from Figure 12.3,

$$\frac{\overline{ad}}{\overline{fa}} = \sin \phi' \tag{12.5}$$

$$\overline{fa} = \overline{fO} + \overline{Oa} = c' \cot \phi' + \frac{\sigma_1' + \sigma_3'}{2} \tag{12.6a}$$

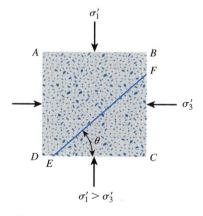

Figure 12.2 Inclination of failure plane in soil with major principal plane

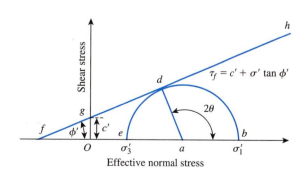

Figure 12.3 Mohr's circle and failure envelope

Also,

$$\overline{ad} = \frac{\sigma_1' - \sigma_3'}{2} \tag{12.6b}$$

Substituting Eqs. (12.6a) and (12.6b) into Eq. (12.5), we obtain

$$\sin \phi' = \frac{\dfrac{\sigma_1' - \sigma_3'}{2}}{c' \cot \phi' + \dfrac{\sigma_1' + \sigma_3'}{2}}$$

or

$$\sigma_1' = \sigma_3'\left(\frac{1 + \sin \phi'}{1 - \sin \phi'}\right) + 2c\left(\frac{\cos \phi'}{1 - \sin \phi'}\right) \tag{12.7}$$

However,

$$\frac{1 + \sin \phi'}{1 - \sin \phi'} = \tan^2\left(45 + \frac{\phi'}{2}\right)$$

and

$$\frac{\cos \phi'}{1 - \sin \phi'} = \tan\left(45 + \frac{\phi'}{2}\right)$$

Thus,

$$\sigma_1' = \sigma_3' \tan^2\left(45 + \frac{\phi'}{2}\right) + 2c' \tan\left(45 + \frac{\phi'}{2}\right) \tag{12.8}$$

An expression similar to Eq. (12.8) could also be derived using Eq. (12.2) (that is, total stress parameters c and ϕ), or

$$\sigma_1 = \sigma_3 \tan^2\left(45 + \frac{\phi}{2}\right) + 2c \tan\left(45 + \frac{\phi}{2}\right) \tag{12.9}$$

12.3 *Laboratory Test for Determination of Shear Strength Parameters*

There are several laboratory methods now available to determine the shear strength parameters (i.e., c, ϕ, c', ϕ') of various soil specimens in the laboratory. They are as follows:

- Direct shear test
- Triaxial test
- Direct simple shear test
- Plane strain triaxial test
- Torsional ring shear test

The direct shear test and the triaxial test are the two commonly used techniques for determining the shear strength parameters. These two tests will be described in detail in the sections that follow.

12.4 *Direct Shear Test*

The direct shear test is the oldest and simplest form of shear test arrangement. A diagram of the direct shear test apparatus is shown in Figure 12.4. The test equipment consists of a metal shear box in which the soil specimen is placed. The soil specimens may be square or circular in plan. The size of the specimens generally used is about 51 mm × 51 mm or 102 mm × 102 mm across and about 25 mm high. The box is split horizontally into halves. Normal force on the specimen is applied from the top of the shear box. The normal stress on the specimens can be as great as 1050 kN/m². Shear force is applied by moving one-half of the box relative to the other to cause failure in the soil specimen.

Depending on the equipment, the shear test can be either stress controlled or strain controlled. In stress-controlled tests, the shear force is applied in equal increments until the specimen fails. The failure occurs along the plane of split of the shear box. After the application of each incremental load, the shear displacement of the top half of the box is measured by a horizontal dial gauge. The change in the height of the specimen (and thus the volume change of the specimen) during the test can be obtained from the readings of a dial gauge that measures the vertical movement of the upper loading plate.

In strain-controlled tests, a constant rate of shear displacement is applied to one-half of the box by a motor that acts through gears. The constant rate of shear displacement is measured by a horizontal dial gauge. The resisting shear force of the soil corresponding to any shear displacement can be measured by a horizontal proving ring or load cell. The volume change of the specimen during the test is obtained in a manner

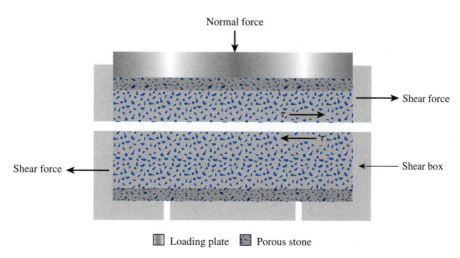

Normal force

Shear force

Shear force

Shear box

■ Loading plate ▨ Porous stone

Figure 12.4 Diagram of direct shear test arrangement

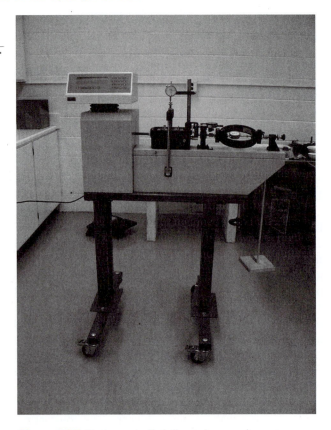

Figure 12.5 Strain-controlled direct shear equipment
(*Courtesy of Braja M. Das, Henderson, Nevada*)

similar to that in the stress-controlled tests. Figure 12.5 shows a photograph of strain-controlled direct shear test equipment. Figure 12.6 shows a photograph taken from the top of the direct shear test equipment with the dial gages and proving ring in place.

The advantage of the strain-controlled tests is that in the case of dense sand, peak shear resistance (that is, at failure) as well as lesser shear resistance (that is, at a point after failure called *ultimate strength*) can be observed and plotted. In stress-controlled tests, only the peak shear resistance can be observed and plotted. Note that the peak shear resistance in stress-controlled tests can be only approximated because failure occurs at a stress level somewhere between the prefailure load increment and the failure load increment. Nevertheless, compared with strain-controlled tests, stress-controlled tests probably model real field situations better.

For a given test, the normal stress can be calculated as

$$\sigma = \text{Normal stress} = \frac{\text{Normal force}}{\text{Cross-sectional area of the specimen}} \qquad (12.10)$$

The resisting shear stress for any shear displacement can be calculated as

$$\tau = \text{Shear stress} = \frac{\text{Resisting shear force}}{\text{Cross-sectional area of the specimen}} \qquad (12.11)$$

Figure 12.6 A photograph showing the dial gauges and proving ring in place (*Courtesy of Braja M. Das, Henderson, Nevada*)

Figure 12.7 shows a typical plot of shear stress and change in the height of the specimen against shear displacement for dry loose and dense sands. These observations were obtained from a strain-controlled test. The following generalizations can be developed from Figure 12.7 regarding the variation of resisting shear stress with shear displacement:

1. In loose sand, the resisting shear stress increases with shear displacement until a failure shear stress of τ_f is reached. After that, the shear resistance remains approximately constant for any further increase in the shear displacement.
2. In dense sand, the resisting shear stress increases with shear displacement until it reaches a failure stress of τ_f. This τ_f is called the *peak shear strength*. After failure stress is attained, the resisting shear stress gradually decreases as shear displacement increases until it finally reaches a constant value called the *ultimate shear strength*.

Since the height of the specimen changes during the application of the shear force (as shown in Figure 12.7), it is obvious that the void ratio of the sand changes (at least in the vicinity of the split of the shear box). Figure 12.8 shows the nature of variation of the void ratio for loose and dense sands with shear displacement. At large shear displacement, the void ratios of loose and dense sands become practically the same, and this is termed the *critical void ratio*.

It is important to note that, in dry sand,

$$\sigma = \sigma'$$

and

$$c' = 0$$

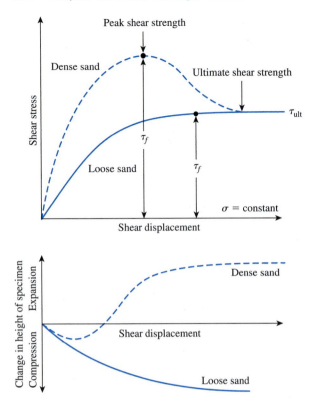

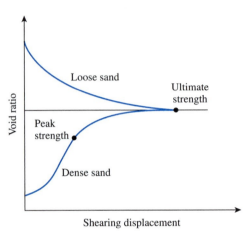

Figure 12.7 Plot of shear stress and change in height of specimen against shear displacement for loose and dense dry sand (direct shear test)

Figure 12.8 Nature of variation of void ratio with shearing displacement

Direct shear tests are repeated on similar specimens at various normal stresses. The normal stresses and the corresponding values of τ_f obtained from a number of tests are plotted on a graph from which the shear strength parameters are determined. Figure 12.9 shows such a plot for tests on a dry sand. The equation for the average line obtained from experimental results is

$$\tau_f = \sigma' \tan \phi' \tag{12.12}$$

So, the friction angle can be determined as follows:

$$\phi' = \tan^{-1}\left(\frac{\tau_f}{\sigma'}\right) \tag{12.13}$$

It is important to note that *in situ* cemented sands may show a c' intercept.

If the variation of the ultimate shear strength (τ_{ult}) with normal stress is known, it can be plotted as shown in Figure 12.9. The average plot can be expressed as

$$\tau_{ult} = \sigma' \tan \phi'_{ult} \tag{12.14}$$

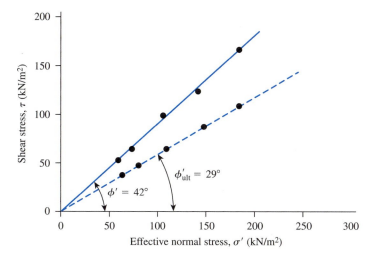

Figure 12.9 Determination of shear strength parameters for a dry sand using the results of direct shear tests

or

$$\phi'_{ult} = \tan^{-1}\left(\frac{\tau_{ult}}{\sigma'}\right) \qquad (12.15)$$

12.5 Drained Direct Shear Test on Saturated Sand and Clay

In the direct shear test arrangement, the shear box that contains the soil specimen is generally kept inside a container that can be filled with water to saturate the specimen. A *drained test* is made on a saturated soil specimen by keeping the rate of loading slow enough so that the excess pore water pressure generated in the soil is dissipated completely by drainage. Pore water from the specimen is drained through two porous stones. (See Figure 12.4.)

Because the hydraulic conductivity of sand is high, the excess pore water pressure generated due to loading (normal and shear) is dissipated quickly. Hence, for an ordinary loading rate, essentially full drainage conditions exist. The friction angle, ϕ', obtained from a drained direct shear test of saturated sand will be the same as that for a similar specimen of dry sand.

The hydraulic conductivity of clay is very small compared with that of sand. When a normal load is applied to a clay soil specimen, a sufficient length of time must elapse for full consolidation—that is, for dissipation of excess pore water pressure. For this reason, the shearing load must be applied very slowly. The test may last from two to five days. Figure 12.10 shows the results of a drained direct shear test on an overconsolidated clay. Figure 12.11 shows the plot of τ_f against σ' obtained from a number of drained direct shear tests on a normally consolidated clay and an overconsolidated clay. Note that the value of $c' \approx 0$ for a normally consolidated clay.

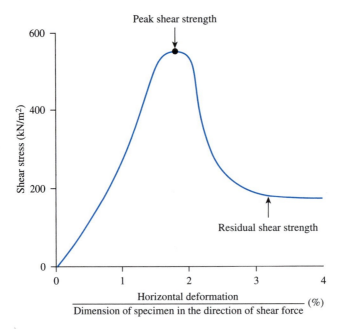

Figure 12.10 Results of a drained direct shear test on an overconsolidated clay [*Note:* Residual shear strength in clay is similar to ultimate shear strength in sand (see Figure 12.7)]

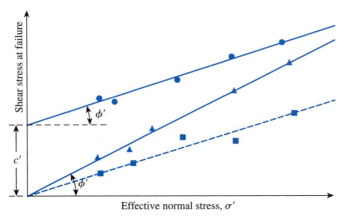

● Overconsolidated clay $\tau_f = c' + \sigma' \tan \phi'$ $(c' \neq 0)$
▲ Normally consolidated clay $\tau_f = \sigma' \tan \phi'$ $(c' \approx 0)$
■ Residual strength plot $\tau_r = \sigma' \tan \phi_r'$

Figure 12.11
Failure envelope for clay obtained from drained direct shear tests

Similar to the ultimate shear strength in the case of sand (Figure 12.8), at large shearing displacements, we can obtain the *residual shear strength* of clay (τ_r) in a drained test. This is shown in Figure 12.10. Figure 12.11 shows the plot of τ_r versus σ'. The average plot *will pass through the origin* and can be expressed as

$$\tau_r = \sigma' \tan \phi_r'$$

or

$$\phi'_r = \tan^{-1}\left(\frac{\tau_r}{\sigma'}\right) \tag{12.16}$$

The drained angle of friction, ϕ', of normally consolidated clays generally decreases with the plasticity index of soil. This fact is illustrated in Figure 12.12 for a number of clays from data reported by Kenney (1959). Although the data are scattered considerably, the general pattern seems to hold.

Skempton (1964) provided the results of the variation of the residual angle of friction, ϕ'_r, of a number of clayey soils with the clay-size fraction ($\leq 2\ \mu$m) present. The following table shows a summary of these results.

Soil	Clay-size fraction (%)	Residual friction angle, ϕ'_r (deg)
Selset	17.7	29.8
Wiener Tegel	22.8	25.1
Jackfield	35.4	19.1
Oxford clay	41.9	16.3
Jari	46.5	18.6
London clay	54.9	16.3
Walton's Wood	67	13.2
Weser-Elbe	63.2	9.3
Little Belt	77.2	11.2
Biotite	100	7.5

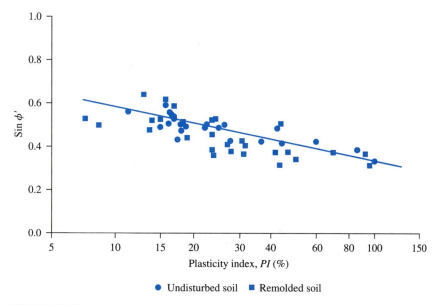

● Undisturbed soil ■ Remolded soil

Figure 12.12 Variation of sin ϕ' with plasticity index for a number of soils (*After Kenney, 1959. With permission from ASCE.*)

12.6   *General Comments on Direct Shear Test*

The direct shear test is simple to perform, but it has some inherent shortcomings. The reliability of the results may be questioned because the soil is not allowed to fail along the weakest plane but is forced to fail along the plane of split of the shear box. Also, the shear stress distribution over the shear surface of the specimen is not uniform. Despite these shortcomings, the direct shear test is the simplest and most economical for a dry or saturated sandy soil.

In many foundation design problems, one must determine the angle of friction between the soil and the material in which the foundation is constructed (Figure 12.13). The foundation material may be concrete, steel, or wood. The shear strength along the surface of contact of the soil and the foundation can be given as

$$\tau_f = c_a' + \sigma' \tan \delta' \tag{12.17}$$

where c_a' = adhesion
      δ' = effective angle of friction between the soil and the foundation material

Note that the preceding equation is similar in form to Eq. (12.3). The shear strength parameters between a soil and a foundation material can be conveniently determined by a direct shear test. This is a great advantage of the direct shear test. The foundation material can be placed in the bottom part of the direct shear test box and then the soil can be placed above it (that is, in the top part of the box), as shown in Figure 12.14, and the test can be conducted in the usual manner.

Figure 12.15 shows the results of direct shear tests conducted in this manner with a quartz sand and concrete, wood, and steel as foundation materials, with $\sigma' = 100$ kN/m².

It was mentioned briefly in Section 12.1 [related to Eq. (12.1)] that Mohr's failure envelope is curvilinear in nature, and Eq. (12.2) is only an approximation. This fact should be kept in mind when considering problems at higher confining pressures. Figure 12.16 shows the decrease of ϕ' and δ' with the increase of normal stress (σ') for the same materials discussed in Figure 12.15. This can be explained by referring to Figure 12.17, which shows a curved Mohr's failure envelope. If a direct shear test is conducted with $\sigma' = \sigma_{(1)}'$, the shear strength will be $\tau_{f(1)}$. So,

$$\delta_1' = \tan^{-1}\left[\frac{\tau_{f(1)}}{\sigma_{(1)}'}\right]$$

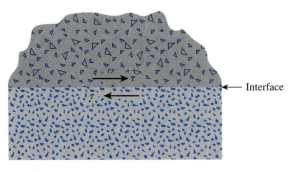

← Interface

Figure 12.13
Interface of a foundation material and soil

☐ Foundation material   ☐ Soil

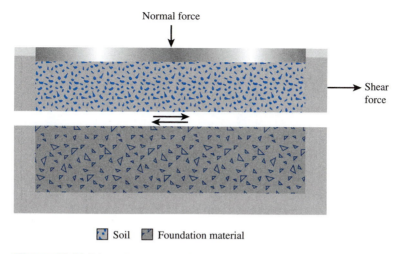

Figure 12.14 Direct shear test to determine interface friction angle

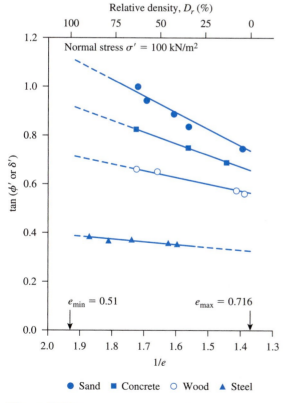

Figure 12.15 Variation of tan ϕ' and tan δ' with $1/e$ [*Note: e* = void ratio, σ' = 100 kN/m², quartz sand] (*After Acar, Durgunoglu, and Tumay, 1982. With permission from ASCE.*)

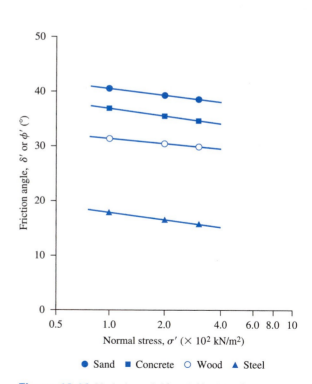

Figure 12.16 Variation of ϕ' and δ' with σ' (*Note:* Relative density = 45%; quartz sand) (*After Acar, Durgunoglu, and Tumay, 1982. With permission from ASCE.*)

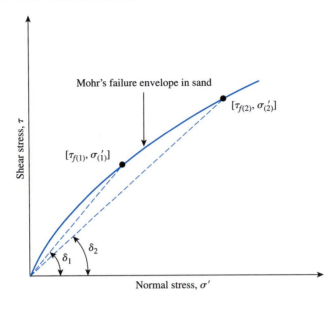

Figure 12.17
Curvilinear nature of Mohr's failure envelope in sand

This is shown in Figure 12.17. In a similar manner, if the test is conducted with $\sigma' = \sigma'_{(2)}$, then

$$\delta' = \delta'_2 = \tan^{-1}\left[\frac{\tau_{f(2)}}{\sigma'_{(2)}}\right]$$

As can be seen from Figure 12.17, $\delta'_2 < \delta'_1$ since $\sigma'_2 > \sigma'_{(1)}$. Keeping this in mind, it must be realized that the values of ϕ' given in Table 12.1 are only the average values.

Example 12.1

Following are the results of four drained direct shear tests on an *overconsolidated clay:*

- Diameter of specimen = 50 mm
- Height of specimen = 25 mm

Test no.	Normal force, N (N)	Shear force at failure, S_{peak} (N)	Residual shear force, $S_{residual}$ (N)
1	150	157.5	44.2
2	250	199.9	56.6
3	350	257.6	102.9
4	550	363.4	144.5

Determine the relationships for *peak shear strength* (τ_f) and *residual shear strength* (τ_r).

Solution

Area of the specimen $(A) = (\pi/4)\left(\dfrac{50}{1000}\right)^2 = 0.0019634$ m². Now the following table can be prepared.

Test no.	Normal force, N (N)	Normal stress, σ' (kN/m²)	Peak shear force, S_{peak} (N)	$\tau_f = \dfrac{S_{peak}}{A}$ (kN/m²)	Residual shear force, $S_{residual}$ (N)	$\tau_r = \dfrac{S_{residual}}{A}$ (kN/m²)
1	150	76.4	157.5	80.2	44.2	22.5
2	250	127.3	199.9	101.8	56.6	28.8
3	350	178.3	257.6	131.2	102.9	52.4
4	550	280.1	363.4	185.1	144.5	73.6

The variations of τ_f and τ_r with σ' are plotted in Figure 12.18. From the plots, we find that

Peak strength: $\quad \tau_f(\text{kN/m}^2) = \mathbf{40 + \sigma' \tan 27}$

Residual strength: $\tau_r(\text{kN/m}^2) = \boldsymbol{\sigma' \tan 14.6}$

(*Note:* For all *overconsolidated clays,* the residual shear strength can be expressed as

$$\tau_r = \sigma' \tan \phi_r'$$

where $\phi_r' = $ effective residual friction angle.)

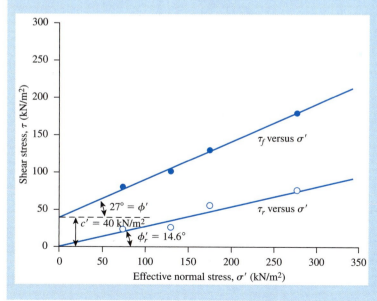

Figure 12.18 Variations of τ_f and τ_r with σ' ∎

<table>
<tr><td>**12.7**</td></tr>
</table>

12.7 *Triaxial Shear Test (General)*

The triaxial shear test is one of the most reliable methods available for determining shear strength parameters. It is used widely for research and conventional testing. A diagram of the triaxial test layout is shown in Figure 12.19.

In this test, a soil specimen about 36 mm in diameter and 76 mm long generally is used. The specimen is encased by a thin rubber membrane and placed inside a plastic cylindrical chamber that usually is filled with water or glycerine. The specimen is subjected to a confining pressure by compression of the fluid in the chamber. (*Note:* Air is sometimes used as a compression medium.) To cause shear failure in the specimen, one must apply axial stress through a vertical loading ram (sometimes called *deviator stress*). This stress can be applied in one of two ways:

1. Application of dead weights or hydraulic pressure in equal increments until the specimen fails. (Axial deformation of the specimen resulting from the load applied through the ram is measured by a dial gauge.)
2. Application of axial deformation at a constant rate by means of a geared or hydraulic loading press. This is a strain-controlled test.

The axial load applied by the loading ram corresponding to a given axial deformation is measured by a proving ring or load cell attached to the ram.

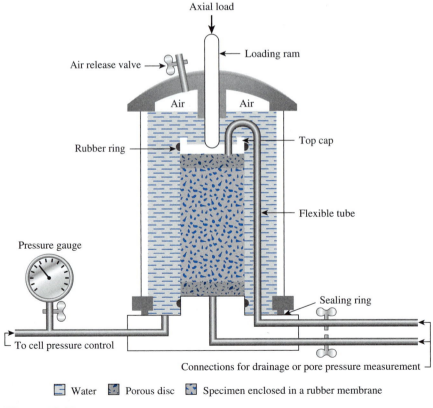

Water   Porous disc   Specimen enclosed in a rubber membrane

Figure 12.19 Diagram of triaxial test equipment (*After Bishop and Bjerrum, 1960. With permission from ASCE.*)

Connections to measure drainage into or out of the specimen, or to measure pressure in the pore water (as per the test conditions), also are provided. The following three standard types of triaxial tests generally are conducted:

1. Consolidated-drained test or drained test (CD test)
2. Consolidated-undrained test (CU test)
3. Unconsolidated-undrained test or undrained test (UU test)

The general procedures and implications for each of the tests in *saturated soils* are described in the following sections.

12.8   *Consolidated-Drained Triaxial Test*

In the CD test, the saturated specimen first is subjected to an all around confining pressure, σ_3, by compression of the chamber fluid (Figure 12.20a). As confining pressure is applied, the pore water pressure of the specimen increases by u_c (if drainage is prevented). This increase in the pore water pressure can be expressed as a nondimensional parameter in the form

$$B = \frac{u_c}{\sigma_3} \tag{12.18}$$

where B = Skempton's pore pressure parameter (Skempton, 1954).

For saturated soft soils, B is approximately equal to 1; however, for saturated stiff soils, the magnitude of B can be less than 1. Black and Lee (1973) gave the theoretical values of B for various soils at complete saturation. These values are listed in Table 12.2.

Now, if the connection to drainage is opened, dissipation of the excess pore water pressure, and thus consolidation, will occur. With time, u_c will become equal to 0. In saturated soil, the change in the volume of the specimen (ΔV_c) that takes place during consolidation can be obtained from the volume of pore water drained (Figure 12.21a). Next, the deviator stress, $\Delta\sigma_d$, on the specimen is increased very slowly (Figure 12.20b).

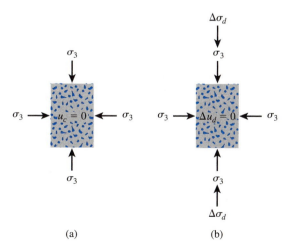

(a)   (b)

Figure 12.20
Consolidated-drained triaxial test: (a) specimen under chamber-confining pressure; (b) deviator stress application

Table 12.2 Theoretical Values of B at Complete Saturation

Type of soil	Theoretical value
Normally consolidated soft clay	0.9998
Lightly overconsolidated soft clays and silts	0.9988
Overconsolidated stiff clays and sands	0.9877
Very dense sands and very stiff clays at high confining pressures	0.9130

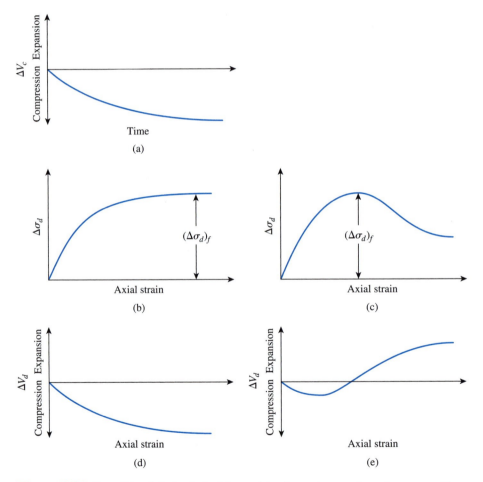

Figure 12.21 Consolidated-drained triaxial test: (a) volume change of specimen caused by chamber-confining pressure; (b) plot of deviator stress against strain in the vertical direction for loose sand and normally consolidated clay; (c) plot of deviator stress against strain in the vertical direction for dense sand and overconsolidated clay; (d) volume change in loose sand and normally consolidated clay during deviator stress application; (e) volume change in dense sand and overconsolidated clay during deviator stress application

The drainage connection is kept open, and the slow rate of deviator stress application allows complete dissipation of any pore water pressure that developed as a result ($\Delta u_d = 0$).

A typical plot of the variation of deviator stress against strain in loose sand and normally consolidated clay is shown in Figure 12.21b. Figure 12.21c shows a similar plot for dense sand and overconsolidated clay. The volume change, ΔV_d, of specimens that occurs because of the application of deviator stress in various soils is also shown in Figures 12.21d and 12.21e.

Because the pore water pressure developed during the test is completely dissipated, we have

$$\text{Total and effective confining stress} = \sigma_3 = \sigma_3'$$

and

$$\text{Total and effective axial stress at failure} = \sigma_3 + (\Delta\sigma_d)_f = \sigma_1 = \sigma_1'$$

In a triaxial test, σ_1' is the major principal effective stress at failure and σ_3' is the minor principal effective stress at failure.

Several tests on similar specimens can be conducted by varying the confining pressure. With the major and minor principal stresses at failure for each test the Mohr's circles can be drawn and the failure envelopes can be obtained. Figure 12.22 shows the type of effective stress failure envelope obtained for tests on sand and normally consolidated clay. The coordinates of the point of tangency of the failure envelope with a Mohr's circle (that is, point A) give the stresses (normal and shear) on the failure plane of that test specimen.

For normally consolidated clay, referring to Figure 12.22

$$\sin \phi' = \frac{AO'}{OO'}$$

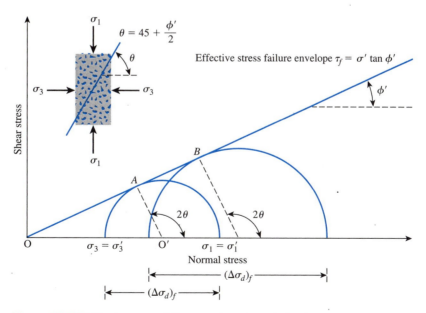

Figure 12.22 Effective stress failure envelope from drained tests on sand and normally consolidated clay

or

$$\sin \phi' = \frac{\left(\dfrac{\sigma'_1 - \sigma'_3}{2}\right)}{\left(\dfrac{\sigma'_1 + \sigma'_3}{2}\right)}$$

$$\phi' = \sin^{-1}\left(\frac{\sigma'_1 - \sigma'_3}{\sigma'_1 + \sigma'_3}\right) \tag{12.19}$$

Also, the failure plane will be inclined at an angle of $\theta = 45 + \phi'/2$ to the major principal plane, as shown in Figure 12.22.

Overconsolidation results when a clay initially is consolidated under an all-around chamber pressure of $\sigma_c \ (= \sigma'_c)$ and is allowed to swell by reducing the chamber pressure to $\sigma_3 \ (= \sigma'_3)$. The failure envelope obtained from drained triaxial tests of such overconsolidated clay specimens shows two distinct branches (*ab* and *bc* in Figure 12.23). The portion *ab* has a flatter slope with a cohesion intercept, and the shear strength equation for this branch can be written as

$$\tau_f = c' + \sigma' \tan \phi'_1 \tag{12.20}$$

The portion *bc* of the failure envelope represents a normally consolidated stage of soil and follows the equation $\tau_f = \sigma' \tan \phi'$.

If the triaxial test results of two overconsolidated soil specimens are known, the magnitudes of ϕ'_1 and c' can be determined as follows. From Eq. (12.8), for Specimen 1:

$$\sigma'_{1(1)} = \sigma'_{3(1)} \tan^2 (45 + \phi'_1/2) + 2c' \tan(45 + \phi'_1/2) \tag{12.21}$$

And, for Specimen 2:

$$\sigma'_{1(2)} = \sigma'_{3(2)} \tan^2 (45 + \phi'_1/2) + 2c' \tan(45 + \phi'_1/2) \tag{12.22}$$

or

$$\sigma'_{1(1)} - \sigma'_{1(2)} = [\sigma'_{3(1)} - \sigma'_{3(2)}] \tan^2 (45 + \phi'_1/2)$$

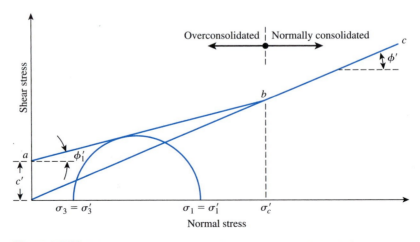

Figure 12.23 Effective stress failure envelope for overconsolidated clay

Hence,

$$\phi_1' = 2 \left\{ \tan^{-1} \left[\frac{\sigma_{1(1)}' - \sigma_{1(2)}'}{\sigma_{3(1)}' - \sigma_{3(2)}'} \right]^{0.5} - 45° \right\} \qquad (12.23)$$

Once the value of ϕ_1' is known, we can obtain c' as

$$c' = \frac{\sigma_{1(1)}' - \sigma_{3(1)}' \tan^2 \left(45 + \dfrac{\phi_1'}{2} \right)}{2 \tan \left(45 + \dfrac{\phi_1'}{2} \right)} \qquad (12.24)$$

A consolidated-drained triaxial test on a clayey soil may take several days to complete. This amount of time is required because deviator stress must be applied very slowly to ensure full drainage from the soil specimen. For this reason, the CD type of triaxial test is uncommon.

Example 12.2

A consolidated-drained triaxial test was conducted on a normally consolidated clay. The results are as follows:

- $\sigma_3 = 110.4 \text{ kN/m}^2$
- $(\Delta\sigma_d)_f = 172.5 \text{ kN/m}^2$

Determine

a. Angle of friction, ϕ'
b. Angle θ that the failure plane makes with the major principal plane

Solution

For normally consolidated soil, the failure envelope equation is

$$\tau_f = \sigma' \tan \phi' \qquad (\text{because } c' = 0)$$

For the triaxial test, the effective major and minor principal stresses at failure are as follows:

$$\sigma_1' = \sigma_1 = \sigma_3 + (\Delta\sigma_d)_f = 110.4 + 172.5 = 282.9 \text{ kN/m}^2$$

and

$$\sigma_3' = \sigma_3 = 110.4 \text{ kN/m}^2$$

Part a

The Mohr's circle and the failure envelope are shown in Figure 12.24. From Eq. (12.19),

$$\sin \phi' = \frac{\sigma_1' - \sigma_3'}{\sigma_1' + \sigma_3'} = \frac{282.9 - 110.4}{282.9 + 110.4} = 0.438$$

or

$$\phi' = \mathbf{26°}$$

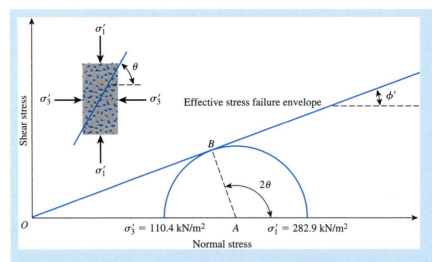

Figure 12.24 Mohr's circle and failure envelope for a normally consolidated clay

Part b
From Eq. (12.4),

$$\theta = 45 + \frac{\phi'}{2} = 45° + \frac{26°}{2} = \mathbf{58°}$$

■

Example 12.3

Refer to Example 12.2.

    a. Find the normal stress σ' and the shear stress τ_f on the failure plane.
    b. Determine the effective normal stress on the plane of maximum shear stress.

Solution
Part a
From Eqs. (10.8) and (10.9),

$$\sigma'\,(\text{on the failure plane}) = \frac{\sigma_1' + \sigma_3'}{2} + \frac{\sigma_1' - \sigma_3'}{2}\cos 2\theta$$

and

$$\tau_f = \frac{\sigma_1' - \sigma_3'}{2}\sin 2\theta$$

Substituting the values of $\sigma_1' = 282.9\ \text{kN/m}^2$, $\sigma_3' = 110.4\ \text{kN/m}^2$, and $\theta = 58°$ into the preceding equations, we get

$$\sigma' = \frac{282.9 + 110.4}{2} + \frac{282.9 - 110.4}{2}\cos(2 \times 58) = \mathbf{158.84\ kN/m^2}$$

and

$$\tau_f = \frac{282.9 - 110.4}{2} \sin (2 \times 58) = \textbf{77.52 kN/m}^2$$

Part b

From Eq. (10.9), it can be seen that the maximum shear stress will occur on the plane with $\theta = 45°$. From Eq. (10.8),

$$\sigma' = \frac{\sigma_1' + \sigma_3'}{2} + \frac{\sigma_1' - \sigma_3'}{2} \cos 2\theta$$

Substituting $\theta = 45°$ into the preceding equation gives

$$\sigma' = \frac{282.9 + 110.4}{2} + \frac{282.9 - 110.4}{2} \cos 90 = \textbf{196.65 kN/m}^2 \qquad \blacksquare$$

Example 12.4

The equation of the effective stress failure envelope for normally consolidated clayey soil is $\tau_f = \sigma' \tan 30°$. A drained triaxial test was conducted with the same soil at a chamber-confining pressure of 69 kN/m^2. Calculate the deviator stress at failure.

Solution

For normally consolidated clay, $c' = 0$. Thus, from Eq. (12.8),

$$\sigma_1' = \sigma_3' \tan^2 \left(45 + \frac{\phi'}{2} \right)$$

$$\phi' = 30°$$

$$\sigma_1' = 69 \tan^2 \left(45 + \frac{30}{2} \right) = 207 \text{ kN/m}^2$$

So,

$$(\Delta\sigma_d)_f = \sigma_1' - \sigma_3' = 207 - 69 = \textbf{138 kN/m}^2 \qquad \blacksquare$$

Example 12.5

The results of two drained triaxial tests on a saturated clay follow:

Specimen I:

$$\sigma_3 = 70 \text{ kN/m}^2$$

$$(\Delta\sigma_d)_f = 130 \text{ kN/m}^2$$

Specimen II:

$$\sigma_3 = 160 \text{ kN/m}^2$$

$$(\Delta\sigma_d)_f = 223.5 \text{ kN/m}^2$$

Determine the shear strength parameters.

Solution

Refer to Figure 12.25. For Specimen I, the principal stresses at failure are

$$\sigma_3' = \sigma_3 = 70 \text{ kN/m}^2$$

and

$$\sigma_1' = \sigma_1 = \sigma_3 + (\Delta\sigma_d)_f = 70 + 130 = 200 \text{ kN/m}^2$$

Similarly, the principal stresses at failure for Specimen II are

$$\sigma_3' = \sigma_3 = 160 \text{ kN/m}^2$$

and

$$\sigma_1' = \sigma_1 = \sigma_3 + (\Delta\sigma_d)_f = 160 + 223.5 = 383.5 \text{ kN/m}^2$$

Now, from Eq. (12.23),

$$\phi_1' = 2\left\{\tan^{-1}\left[\frac{\sigma_{1(\text{I})}' - \sigma_{1(\text{II})}'}{\sigma_{3(\text{I})}' - \sigma_{3(\text{II})}'}\right]^{0.5} - 45°\right\} = 2\left\{\tan^{-1}\left[\frac{200 - 383.5}{70 - 160}\right]^{0.5} - 45°\right\} = \mathbf{20°}$$

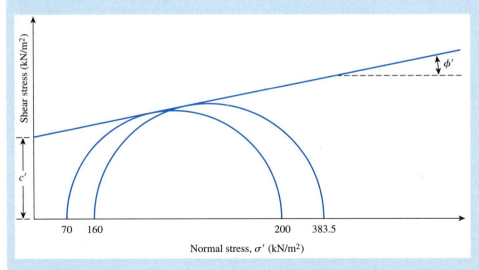

Figure 12.25 Effective stress failure envelope and Mohr's circles for Specimens I and II

Again, from Eq. (12.24),

$$c' = \frac{\sigma'_{1(\mathrm{I})} - \sigma'_{3(\mathrm{I})} \tan^2\left(45 + \dfrac{\phi'_1}{2}\right)}{2 \tan\left(45 + \dfrac{\phi'_1}{2}\right)} = \frac{200 - 70 \tan^2\left(45 + \dfrac{20}{2}\right)}{2 \tan (45 + 20)} = 20 \text{ kN/m}^2 \qquad \blacksquare$$

12.9 Consolidated-Undrained Triaxial Test

The consolidated-undrained test is the most common type of triaxial test. In this test, the saturated soil specimen is first consolidated by an all-around chamber fluid pressure, σ_3, that results in drainage (Figures 12.26a and 12.26b). After the pore water pressure generated by the application of confining pressure is dissipated, the deviator stress, $\Delta\sigma_d$, on the specimen is increased to cause shear failure (Figure 12.26c). During this phase of the test, the drainage line from the specimen is kept closed. Because drainage is not permitted, the pore water pressure, Δu_d, will increase. During the test, simultaneous measurements of $\Delta\sigma_d$ and Δu_d are made. The increase in the pore water pressure, Δu_d, can be expressed in a nondimensional form as

$$\overline{A} = \frac{\Delta u_d}{\Delta\sigma_d} \qquad (12.25)$$

where $\overline{A}$ = Skempton's pore pressure parameter (Skempton, 1954).

The general patterns of variation of $\Delta\sigma_d$ and Δu_d with axial strain for sand and clay soils are shown in Figures 12.26d through 12.26g. In loose sand and normally consolidated clay, the pore water pressure increases with strain. In dense sand and overconsolidated clay, the pore water pressure increases with strain to a certain limit, beyond which it decreases and becomes negative (with respect to the atmospheric pressure). This decrease is because of a tendency of the soil to dilate.

Unlike the consolidated-drained test, the total and effective principal stresses are not the same in the consolidated-undrained test. Because the pore water pressure at failure is measured in this test, the principal stresses may be analyzed as follows:

- Major principal stress at failure (total): $\sigma_3 + (\Delta\sigma_d)_f = \sigma_1$

- Major principal stress at failure (effective): $\sigma_1 - (\Delta u_d)_f = \sigma'_1$

- Minor principal stress at failure (total): σ_3

- Minor principal stress at failure (effective): $\sigma_3 - (\Delta u_d)_f = \sigma'_3$

In these equations, $(\Delta u_d)_f$ = pore water pressure at failure. The preceding derivations show that

$$\sigma_1 - \sigma_3 = \sigma'_1 - \sigma'_3$$

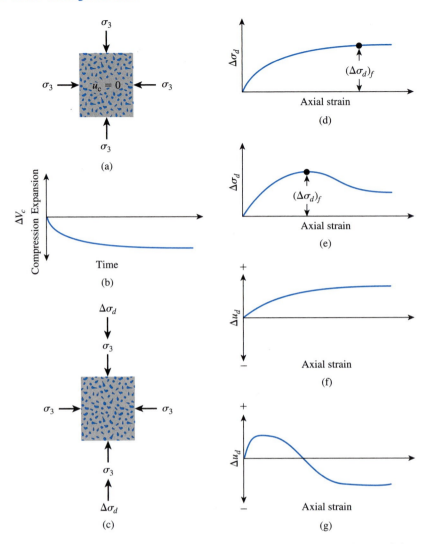

Figure 12.26 Consolidated undrained test: (a) specimen under chamber confining pressure; (b) volume change in specimen caused by confining pressure; (c) deviator stress application; (d) deviator stress against axial strain for loose sand and normally consolidated clay; (e) deviator stress against axial strain for dense sand and overconsolidated clay; (f) variation of pore water pressure with axial strain for loose sand and normally consolidated clay; (g) variation of pore water pressure with axial strain for dense sand and overconsolidated clay

Tests on several similar specimens with varying confining pressures may be conducted to determine the shear strength parameters. Figure 12.27 shows the total and effective stress Mohr's circles at failure obtained from consolidated-undrained triaxial tests in sand and normally consolidated clay. Note that A and B are two total stress Mohr's circles obtained from two tests. C and D are the effective stress Mohr's circles corresponding to total stress circles A and B, respectively. The diameters of circles A and C are the same; similarly, the diameters of circles B and D are the same.

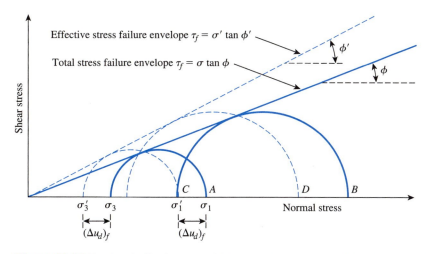

Figure 12.27 Total and effective stress failure envelopes for consolidated undrained triaxial tests. (*Note:* The figure assumes that no back pressure is applied.)

In Figure 12.27, the total stress failure envelope can be obtained by drawing a line that touches all the total stress Mohr's circles. For sand and normally consolidated clays, this will be approximately a straight line passing through the origin and may be expressed by the equation

$$\tau_f = \sigma \tan \phi \qquad (12.26)$$

where σ = total stress

ϕ = the angle that the total stress failure envelope makes with the normal stress axis, also known as the *consolidated-undrained angle of shearing resistance*

Equation (12.26) is seldom used for practical considerations. Similar to Eq. (12.19), for sand and normally consolidated clay, we can write

$$\phi = \sin^{-1}\left(\frac{\sigma_1 - \sigma_3}{\sigma_1 + \sigma_3}\right) \qquad (12.27)$$

and

$$\begin{aligned}
\phi' &= \sin^{-1}\left(\frac{\sigma_1' - \sigma_3'}{\sigma_1' + \sigma_3'}\right) \\
&= \sin^{-1}\left\{\frac{[\sigma_1 - (\Delta u_d)_f] - [\sigma_3 - (\Delta u_d)_f]}{[\sigma_1 - (\Delta u_d)_f] + [\sigma_3 - (\Delta u_d)_f]}\right\} \\
&= \sin^{-1}\left[\frac{\sigma_1 - \sigma_3}{\sigma_1 + \sigma_3 - 2(\Delta u_d)_f}\right] \qquad (12.28)
\end{aligned}$$

Again referring to Figure 12.27, we see that the failure envelope that is tangent to all the effective stress Mohr's circles can be represented by the equation $\tau_f = \sigma' \tan \phi'$, which is the same as that obtained from consolidated-drained tests (see Figure 12.22).

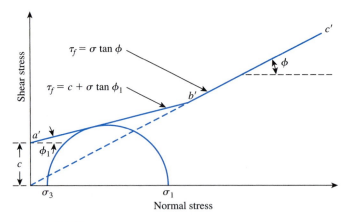

Figure 12.28 Total stress failure envelope obtained from consolidated-undrained tests in over-consolidated clay

In overconsolidated clays, the total stress failure envelope obtained from consolidated-undrained tests will take the shape shown in Figure 12.28. The straight line $a'b'$ is represented by the equation

$$\tau_f = c + \sigma \tan \phi_1 \tag{12.29}$$

and the straight line $b'c'$ follows the relationship given by Eq. (12.26). The effective stress failure envelope drawn from the effective stress Mohr's circles will be similar to that shown in Figure 12.28.

Consolidated-drained tests on clay soils take considerable time. For this reason, consolidated-undrained tests can be conducted on such soils with pore pressure measurements to obtain the drained shear strength parameters. Because drainage is not allowed in these tests during the application of deviator stress, they can be performed quickly.

Skempton's pore water pressure parameter $\overline{A}$ was defined in Eq. (12.25). At failure, the parameter $\overline{A}$ can be written as

$$\overline{A} = \overline{A}_f = \frac{(\Delta u_d)_f}{(\Delta \sigma_d)_f} \tag{12.30}$$

The general range of $\overline{A}_f$ values in most clay soils is as follows:

- Normally consolidated clays: 0.5 to 1
- Overconsolidated clays: −0.5 to 0

Table 12.3 gives the values of $\overline{A}_f$ for some normally consolidated clays as obtained by the Norwegian Geotechnical Institute.

Laboratory triaxial tests of Simons (1960) on Oslo clay, Weald clay, and London clay showed that $\overline{A}_f$ becomes approximately zero at an overconsolidation value of about 3 or 4 (Figure 12.29).

Table 12.3 Triaxial Test Results for Some Normally Consolidated Clays Obtained by the Norwegian Geotechnical Institute[*]

Location	Liquid limit	Plastic limit	Liquidity index	Sensitivity[a]	Drained friction angle, ϕ' (deg)	$\overline{A}_f$
Seven Sisters, Canada	127	35	0.28		19	0.72
Sarpborg	69	28	0.68	5	25.5	1.03
Lilla Edet, Sweden	68	30	1.32	50	26	1.10
Fredrikstad	59	22	0.58	5	28.5	0.87
Fredrikstad	57	22	0.63	6	27	1.00
Lilla Edet, Sweden	63	30	1.58	50	23	1.02
Göta River, Sweden	60	27	1.30	12	28.5	1.05
Göta River, Sweden	60	30	1.50	40	24	1.05
Oslo	48	25	0.87	4	31.5	1.00
Trondheim	36	20	0.50	2	34	0.75
Drammen	33	18	1.08	8	28	1.18

[*]After Bjerrum and Simons, 1960. With permission from ASCE.
[a]See Section 12.13 for the definition of sensitivity

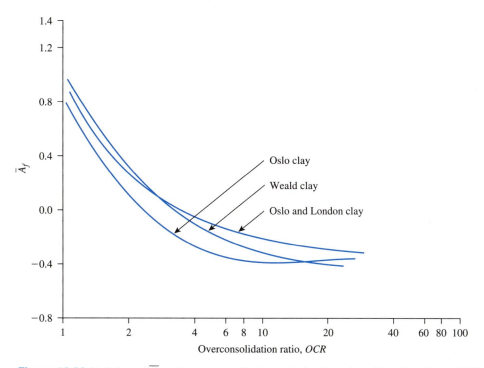

Figure 12.29 Variation of $\overline{A}_f$ with overconsolidation ratio for three clays (*Based on Simon, 1960*)

Example 12.6

A specimen of saturated sand was consolidated under an all-around pressure of 82.8 kN/m^2. The axial stress was then increased and drainage was prevented. The specimen failed when the axial deviator stress reached 62.8 kN/m^2. The pore water pressure at failure was 46.9 kN/m^2. Determine

    a. Consolidated-undrained angle of shearing resistance, ϕ
    b. Drained friction angle, ϕ'

Solution

Part a

For this case, $\sigma_3 = 82.8 \text{ kN/m}^2$, $\sigma_1 = 82.8 + 62.8 = 145.6 \text{ kN/m}^2$, and $(\Delta u_d)_f = 46.9 \text{ kN/m}^2$. The total and effective stress failure envelopes are shown in Figure 12.30. From Eq. (12.27),

$$\phi = \sin^{-1}\left(\frac{\sigma_1 - \sigma_3}{\sigma_1 + \sigma_3}\right) = \sin^{-1}\left(\frac{145.6 - 82.8}{145.6 + 82.8}\right) \approx \mathbf{16}°$$

From Eq. (12.28),

Part b

$$\phi' = \sin^{-1}\left[\frac{\sigma_1 - \sigma_3}{\sigma_1 + \sigma_3 - 2(\Delta u_d)_f}\right] = \sin^{-1}\left[\frac{145.6 - 82.8}{145.6 + 82.8 - (2)(46.9)}\right] = \mathbf{27.8}°$$

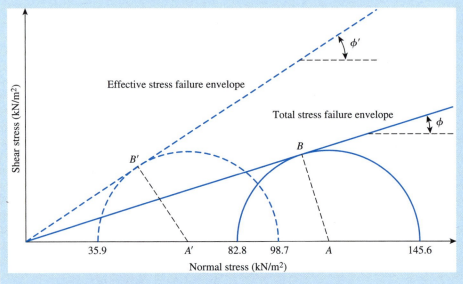

Figure 12.30 Failure envelopes and Mohr's circles for a saturated sand

Example 12.7

Refer to the soil specimen described in Example 12.6. What would be the deviator stress at failure, $(\Delta\sigma_d)_f$, if a drained test was conducted with the same chamber all-around pressure (82.8 kN/m^2)?

Solution
From Eq. (12.8) (with $c' = 0$),

$$\sigma'_1 = \sigma'_3 \tan^2\left(45 + \frac{\phi'}{2}\right)$$

$\sigma'_3 = 82.8$ kN/m^2 and $\phi' = 27.8°$ (from Example 12.6). So,

$$\sigma'_1 = 82.8 \tan^2\left(45 + \frac{27.8}{2}\right) \approx 227.5 \text{ kN/m}^2$$

$$(\Delta\sigma_d)_f = \sigma'_1 - \sigma'_3 = 227.5 - 82.8 = \textbf{144.7 kN/m}^2$$ ■

12.10 Unconsolidated-Undrained Triaxial Test

In unconsolidated-undrained tests, drainage from the soil specimen is not permitted during the application of chamber pressure σ_3. The test specimen is sheared to failure by the application of deviator stress, $\Delta\sigma_d$, and drainage is prevented. Because drainage is not allowed at any stage, the test can be performed quickly. Because of the application of chamber confining pressure σ_3, the pore water pressure in the soil specimen will increase by u_c. A further increase in the pore water pressure (Δu_d) will occur because of the deviator stress application. Hence, the total pore water pressure u in the specimen at any stage of deviator stress application can be given as

$$u = u_c + \Delta u_d \tag{12.31}$$

From Eqs. (12.18) and (12.25), $u_c = B\sigma_3$ and $\Delta u_d = \overline{A}\Delta\sigma_d$, so

$$u = B\sigma_3 + \overline{A}\Delta\sigma_d = B\sigma_3 + \overline{A}(\sigma_1 - \sigma_3) \tag{12.32}$$

This test usually is conducted on clay specimens and depends on a very important strength concept for cohesive soils if the soil is fully saturated. The added axial stress at failure $(\Delta\sigma_d)_f$ is practically the same regardless of the chamber confining pressure. This property is shown in Figure 12.31. The failure envelope for the total stress Mohr's circles becomes a horizontal line and hence is called a $\phi = 0$ condition. From Eq. (12.9) with $\phi = 0$, we get

$$\tau_f = c = c_u \tag{12.33}$$

where c_u is the undrained shear strength and is equal to the radius of the Mohr's circles. Note that the $\phi = 0$ concept is applicable to only saturated clays and silts.

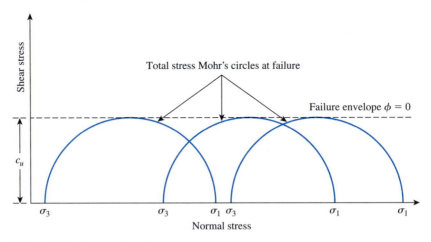

Figure 12.31 Total stress Mohr's circles and failure envelope ($\phi = 0$) obtained from unconsolidated-undrained triaxial tests on fully saturated cohesive soil

The reason for obtaining the same added axial stress $(\Delta\sigma_d)_f$ regardless of the confining pressure can be explained as follows. If a clay specimen (No. I) is consolidated at a chamber pressure σ_3 and then sheared to failure without drainage, the total stress conditions at failure can be represented by the Mohr's circle P in Figure 12.32. The pore pressure developed in the specimen at failure is equal to $(\Delta u_d)_f$. Thus, the major and minor principal effective stresses at failure are, respectively,

$$\sigma_1' = [\sigma_3 + (\Delta\sigma_d)_f] - (\Delta u_d)_f = \sigma_1 - (\Delta u_d)_f$$

and

$$\sigma_3' = \sigma_3 - (\Delta u_d)_f$$

Q is the effective stress Mohr's circle drawn with the preceding principal stresses. Note that the diameters of circles P and Q are the same.

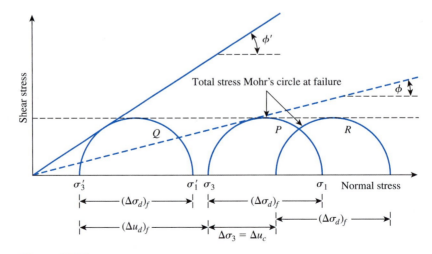

Figure 12.32 The $\phi = 0$ concept

Now let us consider another similar clay specimen (No. II) that has been consolidated under a chamber pressure σ_3 with initial pore pressure equal to zero. If the chamber pressure is increased by $\Delta\sigma_3$ without drainage, the pore water pressure will increase by an amount Δu_c. For saturated soils under isotropic stresses, the pore water pressure increase is equal to the total stress increase, so $\Delta u_c = \Delta\sigma_3$ ($B = 1$). At this time, the effective confining pressure is equal to $\sigma_3 + \Delta\sigma_3 - \Delta u_c = \sigma_3 + \Delta\sigma_3 - \Delta\sigma_3 = \sigma_3$. This is the same as the effective confining pressure of Specimen I before the application of deviator stress. Hence, if Specimen II is sheared to failure by increasing the axial stress, it should fail at the same deviator stress $(\Delta\sigma_d)_f$ that was obtained for Specimen I. The total stress Mohr's circle at failure will be R (see Figure 12.32). The added pore pressure increase caused by the application of $(\Delta\sigma_d)_f$ will be $(\Delta u_d)_f$.

At failure, the minor principal effective stress is

$$[(\sigma_3 + \Delta\sigma_3)] - [\Delta u_c + (\Delta u_d)_f] = \sigma_3 - (\Delta u_d)_f = \sigma_3'$$

and the major principal effective stress is

$$[\sigma_3 + \Delta\sigma_3 + (\Delta\sigma_d)_f] - [\Delta u_c + (\Delta u_d)_f] = [\sigma_3 + (\Delta\sigma_d)_f] - (\Delta u_d)_f$$
$$= \sigma_1 - (\Delta u_d)_f = \sigma_1'$$

Thus, the effective stress Mohr's circle will still be Q because strength is a function of effective stress. Note that the diameters of circles P, Q, and R are all the same.

Any value of $\Delta\sigma_3$ could have been chosen for testing Specimen II. In any case, the deviator stress $(\Delta\sigma_d)_f$ to cause failure would have been the same as long as the soil was fully saturated and fully undrained during both stages of the test.

12.11 Unconfined Compression Test on Saturated Clay

The unconfined compression test is a special type of unconsolidated-undrained test that is commonly used for clay specimens. In this test, the confining pressure σ_3 is 0. An axial load is rapidly applied to the specimen to cause failure. At failure, the total minor principal stress is zero and the total major principal stress is σ_1 (Figure 12.33). Because the undrained shear strength is independent of the confining pressure as long as the soil is fully saturated and fully undrained, we have

$$\tau_f = \frac{\sigma_1}{2} = \frac{q_u}{2} = c_u \tag{12.34}$$

where q_u is the *unconfined compression strength*. Table 12.4 gives the approximate consistencies of clays on the basis of their unconfined compression strength. A photograph of

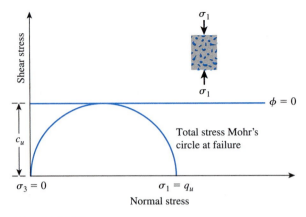

Figure 12.33 Unconfined compression test

Table 12.4 General Relationship of Consistency and Unconfined Compression Strength of Clays

	q_u
Consistency	**kN/m²**
Very soft	0–25
Soft	25–50
Medium	50–100
Stiff	100–200
Very stiff	200–400
Hard	>400

unconfined compression test equipment is shown in Figure 12.34. Figures 12.35 and 12.36 show the failure in two specimens—one by shear and one by bulging—at the end of unconfined compression tests.

Theoretically, for similar saturated clay specimens, the unconfined compression tests and the unconsolidated-undrained triaxial tests should yield the same values of c_u. In practice, however, unconfined compression tests on saturated clays yield slightly lower values of c_u than those obtained from unconsolidated-undrained tests.

12.12 *Empirical Relationships Between Undrained Cohesion (c_u) and Effective Overburden Pressure (σ'_0)*

Several empirical relationships have been proposed between c_u and the effective overburden pressure σ'_o. The most commonly cited relationship is that given by Skempton (1957) which can be expressed as

$$\frac{c_{u(\text{VST})}}{\sigma'_o} = 0.11 + 0.0037(PI) \quad \text{(for normally consolidated clay)} \qquad (12.35)$$

Figure 12.34 Unconfined compression test equipment (*Courtesy of ELE International*)

Figure 12.35 Failure by shear of an unconfined compression test specimen (*Courtesy of Braja M. Das, Henderson, Nevada*)

where $c_{u(\text{VST})}$ = undrained shear strength from vane shear test (see Section 12.15)
$\quad PI$ = plasticity index (%)

Chandler (1988) suggested that the preceding relationship will hold good for overconsolidated soil with an accuracy of $\pm 25\%$. This does not include sensitive and fissured clays. Ladd, *et al.* (1977) proposed that

$$\frac{\left(\dfrac{c_u}{\sigma'_o}\right)_{\text{overconsolidated}}}{\left(\dfrac{c_u}{\sigma'_o}\right)_{\text{normally consolidated}}} = (\text{OCR})^{0.8} \tag{12.36}$$

where OCR = overconsolidation ration.

Figure 12.36
Failure by bulging of an unconfined
compression test specimen (*Courtesy
of Braja M. Das, Henderson, Nevada*)

Example 12.8

An overconsolidated clay deposit located below the groundwater table has the
following:

- Average present effective overburden pressure = 160 kN/m^2
- Overconsolidation ratio = 3.2
- Plasticity index = 28

Estimate the average undrained shear strength of the clay [that is, $c_{u(\text{VST})}$].

Solution
From Eq. (12.35),

$$\left[\frac{c_{u(\text{VST})}}{\sigma'_o}\right]_{\text{normally consolidated}} = 0.11 + 0.0037(PI) = 0.11 + (0.0037)(28) = 0.2136$$

From Eq. (12.36),

$$\left[\frac{c_{u(\text{VST})}}{\sigma'_o}\right]_{\text{overconsolidated}} = (\text{OCR})^{0.8}\left[\frac{c_{u(\text{VST})}}{\sigma'_o}\right]_{\text{normally consolidated}} = (3.2)^{0.8}(0.2136) = 0.542$$

Thus,

$$c_{u(\text{VST})} = 0.542\sigma'_o = (0.542)(160) = \mathbf{86.7\ kN/m^2}$$

12.13 *Sensitivity and Thixotropy of Clay*

For many naturally deposited clay soils, the unconfined compression strength is reduced greatly when the soils are tested after remolding without any change in the moisture content, as shown in Figure 12.37. This property of clay soils is called *sensitivity*. The degree of sensitivity may be defined as the ratio of the unconfined compression strength in an undisturbed state to that in a remolded state, or

$$S_t = \frac{q_{u(\text{undisturbed})}}{q_{u(\text{remolded})}} \tag{12.37}$$

The sensitivity ratio of most clays ranges from about 1 to 8; however, highly flocculent marine clay deposits may have sensitivity ratios ranging from about 10 to 80. Some clays turn to viscous fluids upon remolding. These clays are found mostly in the previously glaciated areas of North America and Scandinavia. Such clays are referred to as *quick* clays. Rosenqvist (1953) classified clays on the basis of their sensitivity. This general classification is shown in Figure 12.38.

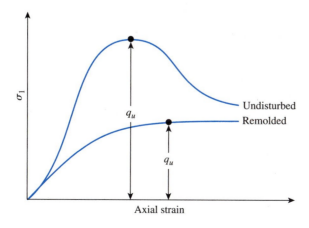

Figure 12.37
Unconfined compression strength for undisturbed and remolded clay

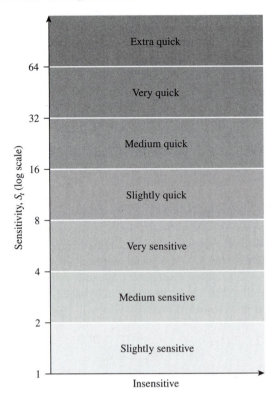

Figure 12.38 Classification of clays based on sensitivity

The loss of strength of clay soils from remolding is caused primarily by the destruction of the clay particle structure that was developed during the original process of sedimentation.

If, however, after remolding, a soil specimen is kept in an undisturbed state (that is, without any change in the moisture content), it will continue to gain strength with time. This phenomenon is referred to as *thixotropy*. Thixotropy is a time-dependent, reversible process in which materials under constant composition and volume soften when remolded. This loss of strength is gradually regained with time when the materials are allowed to rest. This phenomenon is illustrated in Figure 12.39a.

Most soils, however, are partially thixotropic—that is, part of the strength loss caused by remolding is never regained with time. The nature of the strength-time variation for partially thixotropic materials is shown in Figure 12.39b. For soils, the difference between the undisturbed strength and the strength after thixotropic hardening can be attributed to the destruction of the clay-particle structure that was developed during the original process of sedimentation.

Seed and Chan (1959) conducted several tests on three compacted clays with a water content near or below the plastic limit to study the thixotropic strength regain

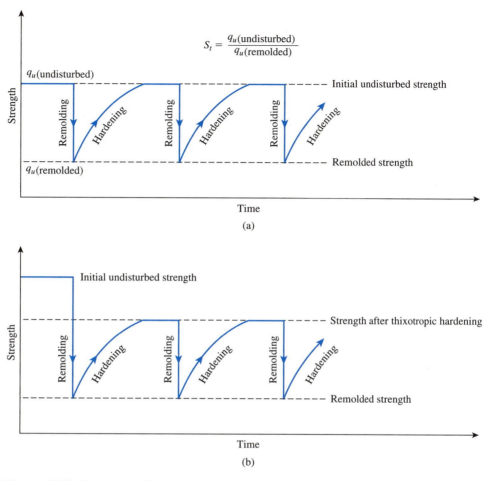

$$S_t = \frac{q_u(\text{undisturbed})}{q_u(\text{remolded})}$$

Figure 12.39 Behavior of (a) thixotropic material; (b) partially thixotropic material

characteristics of the clays. The results of these tests are shown in Figure 12.40. Note that in Figure 12.40,

$$\text{Thixotropic strength ratio} = \frac{c_{u(\text{at time } t \text{ after compaction})}}{c_{u(\text{at time } t = 0 \text{ after compaction})}} \qquad (12.38)$$

12.14 Strength Anisotropy in Clay

The unconsolidated-undrained shear strength of some saturated clays can vary, depending on the direction of load application; this variation is referred to as *anisotropy with respect to strength*. Anisotropy is caused primarily by the nature of the deposition of the cohesive soils, and subsequent consolidation makes the clay particles orient perpendicular to the

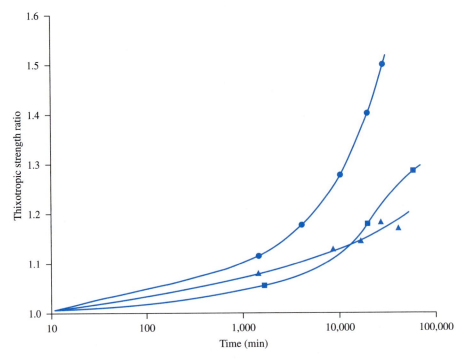

● Vicksburg silty clay $PL = 23; w = 19.5\%$
■ Pittsburgh sandy clay $PL = 20; w = 17.4\%$
▲ Friant-Kern clay $PL = 35; w = 22\%$

Figure 12.40 Thixotropic strength increase with time for three clays (*Based on Seed and Chan, 1959*)

direction of the major principal stress. Parallel orientation of the clay particles can cause the strength of clay to vary with direction. Figure 12.41 shows an element of saturated clay in a deposit with the major principal stress making an angle α with respect to the horizontal. For anisotropic clays, the magnitude of c_u is a function of α.

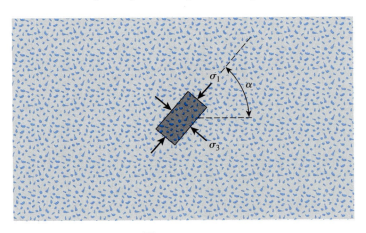

Saturated clay

Figure 12.41 Strength anisotropy in clay

As an example, the variation of c_u with α for undisturbed specimens of Winnipeg Upper Brown clay (Loh and Holt, 1974) is shown in Figure 12.42. Based on several laboratory test results, Casagrande and Carrillo (1944) proposed the following relationship for the directional variation of undrained shear strength:

$$c_{u(\alpha)} = c_{u(\alpha=0°)} + \left[c_{u(\alpha=90°)} - c_{u(\alpha=0°)}\right]\sin^2 \alpha \qquad (12.39)$$

For normally consolidated clays, $c_{u(\alpha=90°)} > c_{u(\alpha=0°)}$; for overconsolidated clays, $c_{u(\alpha=90°)} < c_{u(\alpha=0°)}$. Figure 12.43 shows the directional variation for $c_{u(\alpha)}$ based on Eq. (12.39). The anisotropy with respect to strength for clays can have an important effect on various stability calculations.

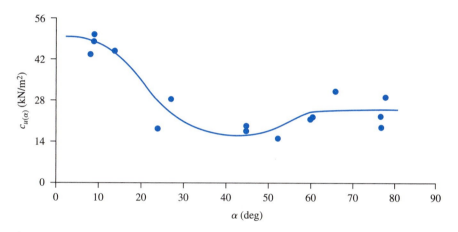

Figure 12.42 Directional variation of c_u for undisturbed Winnipeg Upper Brown clay (*Based on Loh and Holt, 1974*)

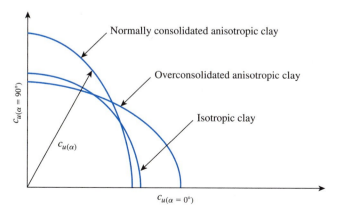

Figure 12.43 Graphical representation of Eq. (12.39)

12.15 *Vane Shear Test*

Fairly reliable results for the undrained shear strength, c_u ($\phi = 0$ concept), of very soft to medium cohesive soils may be obtained directly from vane shear tests. The shear vane usually consists of four thin, equal-sized steel plates welded to a steel torque rod (Figure 12.44). First, the vane is pushed into the soil. Then torque is applied at the top of the torque rod to rotate the vane at a uniform speed. A cylinder of soil of height h and diameter d will resist the torque until the soil fails. The undrained shear strength of the soil can be calculated as follows.

If T is the maximum torque applied at the head of the torque rod to cause failure, it should be equal to the sum of the resisting moment of the shear force along the side surface of the soil cylinder (M_s) and the resisting moment of the shear force at each end (M_e) (Figure 12.45):

$$T = M_s + \underbrace{M_e + M_e}_{\text{Two ends}} \tag{12.40}$$

The resisting moment can be given as

$$M_s = \underbrace{(\pi d h)c_u}_{\substack{\text{Surface} \\ \text{area}}} \underbrace{(d/2)}_{\substack{\text{Moment} \\ \text{arm}}} \tag{12.41}$$

where d = diameter of the shear vane
$\qquad h$ = height of the shear vane

For the calculation of M_e, investigators have assumed several types of distribution of shear strength mobilization at the ends of the soil cylinder:

1. *Triangular.* Shear strength mobilization is c_u at the periphery of the soil cylinder and decreases linearly to zero at the center.
2. *Uniform.* Shear strength mobilization is constant (that is, c_u) from the periphery to the center of the soil cylinder.
3. *Parabolic.* Shear strength mobilization is c_u at the periphery of the soil cylinder and decreases parabolically to zero at the center.

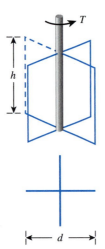

Figure 12.44
Diagram of vane shear test equipment

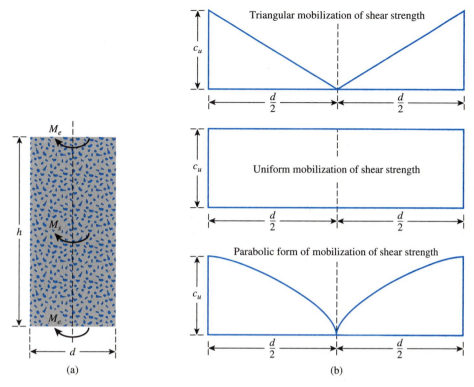

Figure 12.45 Derivation of Eq. (12.43): (a) resisting moment of shear force; (b) variations in shear strength-mobilization

These variations in shear strength mobilization are shown in Figure 12.45b. In general, the torque, T, at failure can be expressed as

$$T = \pi c_u \left[\frac{d^2 h}{2} + \beta \frac{d^3}{4} \right]$$

(12.42)

or

$$c_u = \frac{T}{\pi \left[\frac{d^2 h}{2} + \beta \frac{d^3}{4} \right]}$$

(12.43)

where $\beta = \frac{1}{2}$ for triangular mobilization of undrained shear strength

$\beta = \frac{2}{3}$ for uniform mobilization of undrained shear strength

$\beta = \frac{3}{5}$ for parabolic mobilization of undrained shear strength

Note that Eq. (12.43) usually is referred to as *Calding's equation.*

Vane shear tests can be conducted in the laboratory and in the field during soil exploration. The laboratory shear vane has dimensions of about 13 mm in diameter and 25 mm in height. Figure 12.46 shows a photograph of laboratory vane shear test equipment. Figure 12.47 shows the field vanes recommended by ASTM (2004). Table 12.5 gives the ASTM recommended dimensions of field vanes.

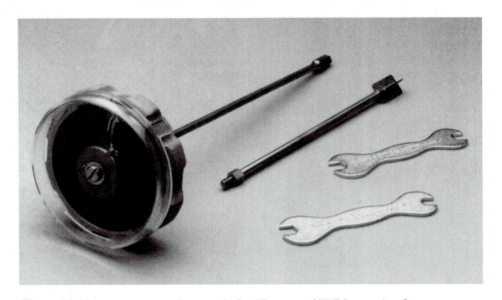

Figure 12.46 Laboratory vane shear test device (*Courtesy of ELE International*)

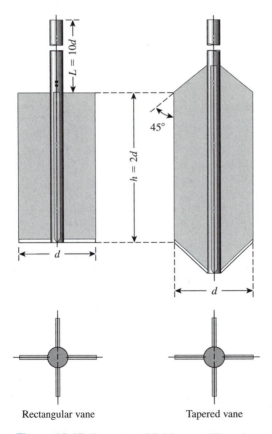

Rectangular vane   Tapered vane

Figure 12.47 Geometry of field vanes (*From Annual Book of ASTM Standards, 04.08, p. 273. Copyright 2004 ASTM INTERNATIONAL. Reprinted with permission.*)

Table 12.5 Recommended Dimensions of Field Vanes[*][a]

Casing size	Diameter, (mm)	Height, (mm)	Thickness of blade, (mm)	Diameter of rod, (mm)
AX	38.1	76.2	1.6	12.7
BX	50.8	101.6	1.6	12.7
NX	63.5	127.0	3.2	12.7
101.6[b] mm	92.1	184.1	3.2	12.7

[*]*After ASTM, 2004. Copyright ASTM INTERNATIONAL. Reprinted with permission.*
[a]Selection of vane size is directly related to the consistency of the soil being tested; that is, the softer the soil, the larger the vane diameter should be
[b]Inside diameter

According to ASTM (2004), if $h/d = 2$, then

$$c_u\,(\text{kN/m}^2) = \frac{T\,(\text{N}\cdot\text{m})}{(366 \times 10^{-8})\underset{\uparrow}{d^3}}$$
$$\text{(cm)}$$

(12.44)

or

$$c_u\,(\text{kN/m}^2) = 273 \times 10^3\, \frac{T\,(\text{N}\cdot\text{m})}{(d\ \text{cm})^3}$$

(12.45)

In the field, where considerable variation in the undrained shear strength can be found with depth, vane shear tests are extremely useful. In a short period, one can establish a reasonable pattern of the change of c_u with depth. However, if the clay deposit at a given site is more or less uniform, a few unconsolidated-undrained triaxial tests on undisturbed specimens will allow a reasonable estimation of soil parameters for design work. Vane shear tests also are limited by the strength of soils in which they can be used. The undrained shear strength obtained from a vane shear test also depends on the rate of application of torque T.

Figure 12.48 shows a comparison of the variation of c_u with the depth obtained from field vane shear tests, unconfined compression tests, and unconsolidated-undrained triaxial tests for Morgan City recent alluvium (Arman, *et al.*, 1975). It can be seen that the vane shear test values are higher compared to the others.

Bjerrum (1974) also showed that, as the plasticity of soils increases, c_u obtained from vane shear tests may give results that are unsafe for foundation design. For this reason, he suggested the correction

$$c_{u(\text{design})} = \lambda c_{u(\text{vane shear})}$$

(12.46)

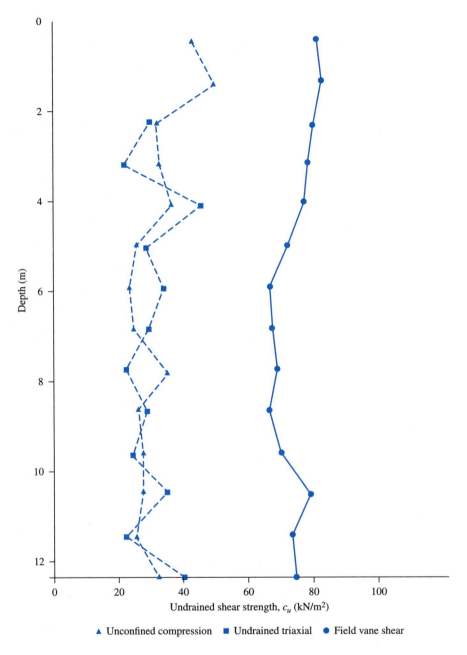

Figure 12.48 Variation of c_u with depth obtained from various tests for Morgan City recent alluvium (*Drawn from the test results of Arman, et al., 1975*)

where

$$\lambda = \text{correction factor} = 1.7 - 0.54 \log(PI) \qquad (12.47)$$

$$PI = \text{plasticity index}$$

12.16 *Other Methods for Determining Undrained Shear Strength*

A modified form of the vane shear test apparatus is the *Torvane* (Figure 12.49), which is a handheld device with a calibrated spring. This instrument can be used for determining c_u for tube specimens collected from the field during soil exploration, and it can be used in the field. The Torvane is pushed into the soil and then rotated until the soil fails. The undrained shear strength can be read at the top of the calibrated dial.

Figure 12.50 shows a *pocket penetrometer,* which is pushed directly into the soil. The unconfined compression strength (q_u) is measured by a calibrated spring. This device can be used both in the laboratory and in the field.

Figure 12.49
Torvane (*Courtesy of ELE International*)

Figure 12.50
Pocket penetrometer
(*Courtesy of ELE International*)

12.17 *Shear Strength of Unsaturated Cohesive Soils*

The equation relating total stress, effective stress, and pore water pressure for unsaturated soils, can be expressed as

$$\sigma' = \sigma - u_a + \chi(u_a - u_w) \tag{12.48}$$

where σ' = effective stress
σ = total stress
u_a = pore air pressure
u_w = pore water pressure

When the expression for σ' is substituted into the shear strength equation [Eq. (12.3)], which is based on effective stress parameters, we get

$$\tau_f = c' + [\sigma - u_a + \chi(u_a - u_w)]\tan\phi' \tag{12.49}$$

The values of χ depend primarily on the degree of saturation. With ordinary triaxial equipment used for laboratory testing, it is not possible to determine accurately the effective stresses in unsaturated soil specimens, so the common practice is to conduct undrained triaxial tests on unsaturated specimens and measure only the total stress. Figure 12.51 shows a total stress failure envelope obtained from a number of undrained triaxial tests conducted with a given initial degree of saturation. The failure envelope is generally curved. Higher confining pressure causes higher compression of the air in void spaces; thus, the solubility of void air in void water is increased. For design purposes, the curved envelope is sometimes approximated as a straight line, as shown in Figure 12.51, with an equation as follows:

$$\tau_f = c + \sigma \tan\phi \tag{12.50}$$

(*Note: c* and ϕ in the preceding equation are empirical constants.)

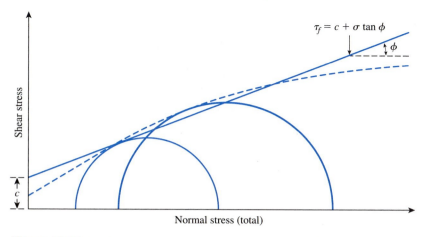

Figure 12.51 Total stress failure envelope for unsaturated cohesive soils

Figure 12.52 shows the variation of the total stress envelopes with change of the initial degree of saturation obtained from undrained tests on an inorganic clay. Note that for these tests the specimens were prepared with approximately the same initial dry unit weight of about 16.7 kN/m^3. For a given total normal stress, the shear stress needed to cause failure decreases as the degree of saturation increases. When the degree of saturation reaches 100%, the total stress failure envelope becomes a horizontal line that is the same as with the $\phi = 0$ concept.

In practical cases where a cohesive soil deposit may become saturated because of rainfall or a rise in the groundwater table, the strength of partially saturated clay should not be used for design considerations. Instead, the unsaturated soil specimens collected from the field must be saturated in the laboratory and the undrained strength determined.

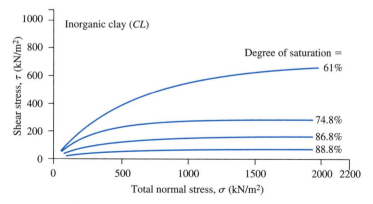

Figure 12.52 Variation of the total stress failure envelope with change of initial degree of saturation obtained from undrained tests of an inorganic clay (*After Casagrande and Hirschfeld, 1960. With permission from ASCE.*)

12.18	**Stress Path**

Results of triaxial tests can be represented by diagrams called *stress paths*. A stress path is a line that connects a series of points, each of which represents a successive stress state experienced by a soil specimen during the progress of a test. There are several ways in which a stress path can be drawn. This section covers one of them.

Lambe (1964) suggested a type of stress path representation that plots q' against p' (where p' and q' are the coordinates of the top of the Mohr's circle). Thus, relationships for p' and q' are as follows:

$$p' = \frac{\sigma_1' + \sigma_3'}{2} \tag{12.51}$$

$$q' = \frac{\sigma_1' - \sigma_3'}{2} \tag{12.52}$$

This type of stress path plot can be explained with the aid of Figure 12.53. Let us consider a normally consolidated clay specimen subjected to an isotropically consolidated-drained triaxial test. At the beginning of the application of deviator stress, $\sigma_1' = \sigma_3' = \sigma_3$, so

$$p' = \frac{\sigma_3' + \sigma_3'}{2} = \sigma_3' = \sigma_3 \tag{12.53}$$

and

$$q' = \frac{\sigma_3' - \sigma_3'}{2} = 0 \tag{12.54}$$

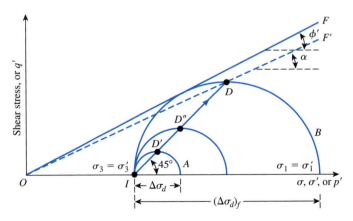

Figure 12.53 Stress path—plot of q' against p' for a consolidated-drained triaxial test on a normally consolidated clay

For this condition, p' and q' will plot as a point (that is, I in Figure 12.53). At some other time during deviator stress application, $\sigma_1' = \sigma_3' + \Delta\sigma_d = \sigma_3 + \Delta\sigma_d; \sigma_3' = \sigma_3$. The Mohr's circle marked A in Figure 12.53 corresponds to this state of stress on the soil specimen. The values of p' and q' for this stress condition are

$$p' = \frac{\sigma_1' + \sigma_3'}{2} = \frac{(\sigma_3' + \Delta\sigma_d) + \sigma_3'}{2} = \sigma_3' + \frac{\Delta\sigma_d}{2} = \sigma_3 + \frac{\Delta\sigma_d}{2} \qquad (12.55)$$

and

$$q' = \frac{(\sigma_3' + \Delta\sigma_d) - \sigma_3'}{2} = \frac{\Delta\sigma_d}{2} \qquad (12.56)$$

If these values of p' and q' were plotted in Figure 12.53, they would be represented by point D' at the top of the Mohr's circle. So, if the values of p' and q' at various stages of the deviator stress application are plotted and these points are joined, a straight line like ID will result. The straight line ID is referred to as the *stress path* in a q'-p' plot for a con-solidated-drained triaxial test. Note that the line ID makes an angle of 45° with the horizontal. Point D represents the failure condition of the soil specimen in the test. Also, we can see that Mohr's circle B represents the failure stress condition.

For normally consolidated clays, the failure envelope can be given by $\tau_f = \sigma' \tan \phi'$. This is the line OF in Figure 12.53. (See also Figure 12.22.) A modified failure envelope now can be defined by line OF'. This modified line commonly is called the K_f line. The equation of the K_f line can be expressed as

$$q' = p' \tan \alpha \qquad (12.57)$$

where $\alpha =$ the angle that the modified failure envelope makes with the horizontal.

The relationship between the angles ϕ' and α can be determined by referring to Figure 12.54, in which, for clarity, the Mohr's circle at failure (that is, circle B) and lines

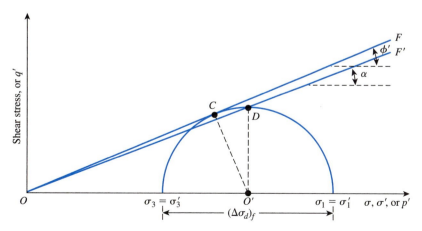

Figure 12.54 Relationship between ϕ' and α

OF and OF', as shown in Figure 12.53, have been redrawn. Note that O' is the center of the Mohr's circle at failure. Now,

$$\frac{DO'}{OO'} = \tan \alpha$$

and thus, we obtain

$$\tan \alpha = \frac{\dfrac{\sigma_1' - \sigma_3'}{2}}{\dfrac{\sigma_1' + \sigma_3'}{2}} = \frac{\sigma_1' - \sigma_3'}{\sigma_1' + \sigma_3'} \qquad (12.58)$$

Again,

$$\frac{CO'}{OO'} = \sin \phi'$$

or

$$\sin \phi' = \frac{\dfrac{\sigma_1' - \sigma_3'}{2}}{\dfrac{\sigma_1' + \sigma_3'}{2}} = \frac{\sigma_1' - \sigma_3'}{\sigma_1' + \sigma_3'} \qquad (12.59)$$

Comparing Eqs. (12.58) and (12.59), we see that

$$\sin \phi' = \tan \alpha \qquad (12.60)$$

or

$$\phi' = \sin^{-1}(\tan \alpha) \qquad (12.61)$$

Figure 12.55 shows a q'–p' plot for a normally consolidated clay specimen subjected to an isotropically consolidated-undrained triaxial test. At the beginning of the application of deviator stress, $\sigma_1' = \sigma_3' = \sigma_3$. Hence, $p' = \sigma_3'$ and $q' = 0$. This relationship is represented by point I. At some other stage of the deviator stress application,

$$\sigma_1' = \sigma_3 + \Delta\sigma_d - \Delta u_d$$

and

$$\sigma_3' = \sigma_3 - \Delta u_d$$

So,

$$p' = \frac{\sigma_1' + \sigma_3'}{2} = \sigma_3 + \frac{\Delta\sigma_d}{2} - \Delta u_d \qquad (12.62)$$

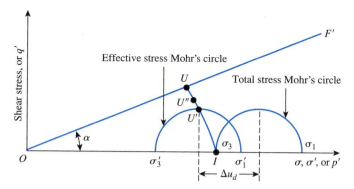

Figure 12.55 Stress path—plot of q' against p' for a consolidated-undrained triaxial test on a normally consolidated clay

and

$$q' = \frac{\sigma_1' - \sigma_3'}{2} = \frac{\Delta\sigma_d}{2} \qquad (12.63)$$

The preceding values of p' and q' will plot as point U' in Figure 12.55. Points such as U'' represent values of p' and q' as the test progresses. At failure of the soil specimen,

$$p' = \sigma_3 + \frac{(\Delta\sigma_d)_f}{2} - (\Delta u_d)_f \qquad (12.64)$$

and

$$q' = \frac{(\Delta\sigma_d)_f}{2} \qquad (12.65)$$

The values of p' and q' given by Eqs. (12.64) and (12.65) will plot as point U. Hence, the effective stress path for a consolidated-undrained test can be given by the curve $IU'U$. Note that point U will fall on the modified failure envelope, OF' (see Figure 12.54), which is inclined at an angle α to the horizontal. Lambe (1964) proposed a technique to evaluate the elastic and consolidation settlements of foundations on clay soils by using the stress paths determined in this manner.

Example 12.9

For a normally consolidated clay, the failure envelope is given by the equation $\tau_f = \sigma'$ $\tan \phi'$. The corresponding modified failure envelope (q'-p' plot) is given by Eq. (12.57) as $q' = p' \tan \alpha$. In a similar manner, if the failure envelope is $\tau_f = c' + \sigma' \tan \phi'$, the corresponding modified failure envelope is a q'-p' plot that can be expressed as $q' = m + p' \tan \alpha$. Express α as a function of ϕ', and give m as a function of c' and ϕ'.

Solution
From Figure 12.56,

$$\sin \phi' = \frac{AB}{AC} = \frac{AB}{CO + OA} = \frac{\left(\dfrac{\sigma_1' - \sigma_3'}{2}\right)}{c' \cot \phi' + \left(\dfrac{\sigma_1' + \sigma_3'}{2}\right)}$$

So,

$$\frac{\sigma_1' - \sigma_3'}{2} = c' \cos \phi' + \left(\frac{\sigma_1' + \sigma_3'}{2}\right) \sin \phi' \qquad \text{(a)}$$

or

$$q' = m + p' \tan \alpha \qquad \text{(b)}$$

Comparing Eqs. (a) and (b), we find that

$$m = c' \cos \phi'$$

and

$$\tan \alpha = \sin \phi'$$

or

$$\alpha = \tan^{-1}(\sin \phi')$$

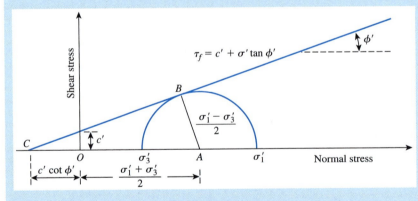

Figure 12.56 Derivation of α as a function of ϕ' and m as a function of c' and ϕ' ∎

12.19   *Summary and General Comments*

In this chapter, the shear strengths of granular and cohesive soils were examined. Laboratory procedures for determining the shear strength parameters were described.

In textbooks, determination of the shear strength parameters of cohesive soils appears to be fairly simple. However, in practice, the proper choice of these parameters for design and stability checks of various earth, earth-retaining, and earth-supported structures is very difficult and requires experience and an appropriate theoretical background in geotechnical engineering. In this chapter, three types of strength parameters (*consolidated-drained, consolidated-undrained,* and *unconsolidated-undrained*) were introduced. Their use depends on drainage conditions.

Consolidated-drained strength parameters can be used to determine the long-term stability of structures such as earth embankments and cut slopes. Consolidated-undrained shear strength parameters can be used to study stability problems relating to cases where the soil initially is fully consolidated and then there is rapid loading. An excellent example of this is the stability of slopes of earth dams after rapid drawdown. The unconsolidated-undrained shear strength of clays can be used to evaluate the end-of-construction stability of saturated cohesive soils with the assumption that the load caused by construction has been applied rapidly and there has been little time for drainage to take place. The bearing capacity of foundations on soft saturated clays and the stability of the base of embankments on soft clays are examples of this condition.

The unconsolidated-undrained shear strength of some saturated clays can vary depending on the direction of load application; this is referred to as *anisotropy with respect to strength.* Anisotropy is caused primarily by the nature of the deposition of the cohesive soils, and subsequent consolidation makes the clay particles orient perpendicular to the direction of the major principal stress. Parallel orientation of the clay particles can cause the strength of clay to vary with direction. The anisotropy with respect to strength for clays can have an important effect on the load-bearing capacity of foundations and the stability of earth embankments because the direction of the major principal stress along the potential failure surfaces changes.

The sensitivity of clays was discussed in Section 12.13. It is imperative that sensitive clay deposits are properly identified. For instance, when machine foundations (which are subjected to vibratory loading) are constructed over sensitive clays, the clay may lose its load-bearing capacity substantially, and failure may occur.

Problems

12.1 For a direct shear test on a dry sand, the following are given:
- Specimen size: 75 mm × 75 mm × 30 mm (height)
- Normal stress: 200 kN/m^2
- Shear stress at failure: 175 kN/m^2
  a. Determine the angle of friction, ϕ'
  b. For a normal stress of 150 kN/m^2, what shear force is required to cause failure in the specimen?

12.2 For a dry sand specimen in a direct shear test box, the following are given:
- Angle of friction: 38°
- Size of specimen: 50 mm × 50 mm × 30.5 mm (height)
- Normal stress: 138 kN/m^2

Determine the shear force required to cause failure.

12.3 The following are the results of four drained, direct shear tests on a normally consolidated clay. Given:
- Size of specimen = 60 mm × 60 mm
- Height of specimen = 30 mm

Test no.	Normal force (N)	Shear force at failure (N)
1	200	155
2	300	230
3	400	310
4	500	385

Draw a graph for the shear stress at failure against the normal stress, and determine the drained angle of friction from the graph.

12.4   Repeat Problem 12.3 with the following data. Given specimen size:
- Diameter = 50 mm
- Height = 25 mm

Test no.	Normal force (N)	Shear force at failure (N)
1	267	166.8
2	400	244.6
3	489	311.4
4	556	355.8

12.5   The equation of the effective stress failure envelope for a loose, sandy soil was obtained from a direct shear test at $\tau_f = \sigma' \tan 30°$. A drained triaxial test was conducted with the same soil at a chamber confining pressure of 69 kN/m^2. Calculate the deviator stress at failure.

12.6   For the triaxial test described in Problem 12.5:
  a. Estimate the angle that the failure plane makes with the major principal plane.
  b. Determine the normal stress and shear stress (when the specimen failed) on a plane that makes an angle of 30° with the major principal plane. Also, explain why the specimen did not fail along the plane during the test.

12.7   The relationship between the relative density, D_r, and the angle of friction, ϕ', of a sand can be given as $\phi' = 25 + 0.18D_r$ (D_r is in %). A drained triaxial test on the same sand was conducted with a chamber-confining pressure of 124 kN/m^2. The relative density of compaction was 60%. Calculate the major principal stress at failure.

12.8   For a normally consolidated clay, the results of a drained triaxial test are as follows.
- Chamber confining pressure: 103.5 kN/m^2
- Deviator stress at failure: 234.6 kN/m^2

Determine the soil friction angle, ϕ'.

12.9   For a normally consolidated clay, $\phi' = 24°$. In a drained triaxial test, the specimen failed at a deviator stress of 175 kN/m^2. What was the chamber confining pressure, σ'_3?

12.10   For a normally consolidated clay, $\phi' = 28°$. In a drained triaxial test, the specimen failed at a deviator stress of 207 kN/m^2. What was the chamber confining pressure, σ'_3?

12.11 A consolidated-drained triaxial test was conducted on a normally consolidated clay. The results were as follows:

$$\sigma_3 = 250 \text{ kN/m}^2$$
$$(\Delta\sigma_d)_f = 275 \text{ kN/m}^2$$

Determine:

a. Angle of friction, ϕ'
b. Angle θ that the failure plane makes with the major principal plane
c. Normal stress, σ', and shear stress, τ_f, on the failure plane

12.12 The results of two drained triaxial tests on a saturated clay are given next:

Specimen I: Chamber confining pressure = 103.5 kN/m²
                Deviator stress at failure = 216.7 kN/m²
Specimen II: Chamber-confining pressure = 172.5 kN/m²
                Deviator stress at failure = 324.3 kN/m²

Calculate the shear strength parameters of the soil.

12.13 If the clay specimen described in Problem 12.12 is tested in a triaxial apparatus with a chamber-confining pressure of 172.5 kN/m², what is the major principal stress at failure?

12.14 A sandy soil has a drained angle of friction of 38°. In a drained triaxial test on the same soil, the deviator stress at failure is 175 kN/m². What is the chamber-confining pressure?

12.15 A consolidated-undrained test on a normally consolidated clay yielded the following results:

- $\sigma_3 = 103.5 \text{ kN/m}^2$
- Deviator stress: $(\Delta\sigma_d)_f = 75.9 \text{ kN/m}^2$
  Pore pressure: $(\Delta u_d)_f = 49.7 \text{ kN/m}^2$

Calculate the consolidated-undrained friction angle and the drained friction angle.

12.16 Repeat Problem 12.15 with the following:

$$\sigma_3 = 140 \text{ kN/m}^2$$
$$(\Delta\sigma_d)_f = 125 \text{ kN/m}^2$$
$$(\Delta u_d)_f = 75 \text{ kN/m}^2$$

12.17 The shear strength of a normally consolidated clay can be given by the equation $\tau_f = \sigma' \tan 31°$. A consolidated-undrained triaxial test was conducted on the clay. Following are the results of the test:

- Chamber confining pressure = 112 kN/m²
- Deviator stress at failure = 100 kN/m²

Determine:

a. Consolidated-undrained friction angle
b. Pore water pressure developed in the clay specimen at failure

12.18 For the clay specimen described in Problem 12.17, what would have been the deviator stress at failure if a drained test had been conducted with the same chamber-confining pressure (that is, $\sigma_3 = 112 \text{ kN/m}^2$)?

12.19 For a normally consolidated clay soil, $\phi' = 32°$ and $\phi = 22°$. A consolidated-undrained triaxial test was conducted on this clay soil with a chamber-confining pressure of 103.5 kN/m². Determine the deviator stress and the pore water pressure at failure.

12.20 The friction angle, ϕ', of a normally consolidated clay specimen collected during field exploration was determined from drained triaxial tests to be 25°. The unconfined compression strength, q_u, of a similar specimen was found to be 100 kN/m². Determine the pore water pressure at failure for the unconfined compression test.

12.21 Repeat Problem 12.20 using the following values.

$\phi' = 23°$

$q_u = 120$ kN/m²

12.22 The results of two consolidated-drained triaxial tests on a clayey soil are as follows.

Test no.	σ'_3 (kN/m²)	$\sigma'_{1(failure)}$ (kN/m²)
1	186.3	503.7
2	82.8	331.2

Use the failure envelope equation given in Example 12.9—that is, $q' = m + p'$ tan α. (Do not plot the graph.)

a. Find m and α

b. Find c' and ϕ'

12.23 A 15-m thick normally consolidated clay layer is shown in Figure 12.57. The plasticity index of the clay is 18. Estimate the undrained cohesion as would be determined from a vane shear test at a depth of 8 m below the ground surface. Use Eq. (12.35).

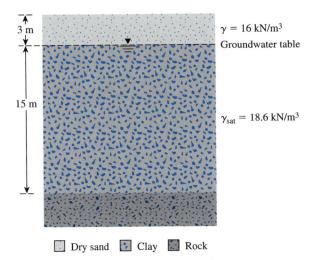

3 m

15 m

$\gamma = 16$ kN/m³

Groundwater table

$\gamma_{sat} = 18.6$ kN/m³

☐ Dry sand   ☐ Clay   ☐ Rock

Figure 12.57

References

ACAR, Y. B., DURGUNOGLU, H. T., and TUMAY, M. T. (1982). "Interface Properties of Sand," *Journal of the Geotechnical Engineering Division*, ASCE, Vol. 108, No. GT4, 648–654.

ARMAN, A., POPLIN, J. K., and AHMAD, N. (1975). "Study of Vane Shear," *Proceedings*, Conference on *In Situ* Measurement and Soil Properties, ASCE, Vol. 1, 93–120.

AMERICAN SOCIETY FOR TESTING AND MATERIALS (2004). *Annual Book of ASTM Standards*, Vol. 04.08, Philadelphia, Pa.

BISHOP, A. W., and BJERRUM, L. (1960). "The Relevance of the Triaxial Test to the Solution of Stability Problems," *Proceedings,* Research Conference on Shear Strength of Cohesive Soils, ASCE, 437–501.

BJERRUM, L. (1974). "Problems of Soil Mechanics and Construction on Soft Clays," Norwegian Geotechnical Institute, *Publication No. 110,* Oslo.

BLACK, D. K., and LEE, K. L. (1973). "Saturating Laboratory Samples by Back Pressure," *Journal of the Soil Mechanics and Foundations Division,* ASCE, Vol. 99, No. SM1, 75–93.

CASAGRANDE, A., and CARRILLO, N. (1944). "Shear Failure of Anisotropic Materials," in *Contribution to Soil Mechanics 1941–1953,* Boston Society of Civil Engineers, Boston.

CASAGRANDE, A., and HIRSCHFELD, R. C. (1960). "Stress Deformation and Strength Characteristics of a Clay Compacted to a Constant Dry Unit Weight," *Proceedings,* Research Conference on Shear Strength of Cohesive Soils, ASCE, 359–417.

CHANDLER, R. J. (1988). "The *in situ* Measurement of the Undrained Shear Strength of Clays Using the Field Vane," *STP 1014, Vane Shear Strength Testing in Soils: Field and Laboratory Studies,* ASTM, 13–44.

COULOMB, C. A. (1776). "Essai sur une application des regles de Maximums et Minimis á quelques Problèmes de Statique, relatifs á l'Architecture," *Memoires de Mathematique et de Physique,* Présentés, á l'Academie Royale des Sciences, Paris, Vol. 3, 38.

KENNEY, T. C. (1959). "Discussion," *Proceedings,* ASCE, Vol. 85, No. SM3, 67–79.

LADD, C. C., FOOTE, R., ISHIHARA, K., SCHLOSSER, F., and POULOS, H. G. (1977). "Stress Deformation and Strength Characteristics," *Proceedings,* 9th International Conference on Soil Mechanics and Foundation Engineering, Tokyo, Vol. 2, 421–494.

LAMBE, T. W. (1964). "Methods of Estimating Settlement," *Journal of the Soil Mechanics and Foundations Division,* ASCE, Vol. 90, No. SM5, 47–74.

LOH, A. K., and HOLT, R. T. (1974). "Directional Variation in Undrained Shear Strength and Fabric of Winnipeg Upper Brown Clay," *Canadian Geotechnical Journal,* Vol. 11, No. 3, 430–437.

MOHR, O. (1900). "Welche Umstände Bedingen die Elastizitätsgrenze und den Bruch eines Materiales?" *Zeitschrift des Vereines Deutscher Ingenieure,* Vol. 44, 1524–1530, 1572–1577.

ROSENQVIST, I. TH. (1953). "Considerations on the Sensitivity of Norwegian Quick Clays," *Geotechnique,* Vol. 3, No. 5, 195–200.

SEED, H. B., and CHAN, C. K. (1959). "Thixotropic Characteristics of Compacted Clays," *Transactions,* ASCE, Vol. 124, 894–916.

SIMONS, N. E. (1960). "The Effect of Overconsolidation on the Shear Strength Characteristics of an Undisturbed Oslo Clay," *Proceedings,* Research Conference on Shear Strength of Cohesive Soils, ASCE, 747–763.

SKEMPTON, A. W. (1954). "The Pore Water Coefficients *A* and *B*," *Geotechnique,* Vol. 4, 143–147.

SKEMPTON, A. W. (1957). "Discussion: The Planning and Design of New Hong Kong Airport," *Proceedings,* Institute of Civil Engineers, London, Vol. 7, 305–307.

SKEMPTON, A. W. (1964). "Long-Term Stability of Clay Slopes," *Geotechnique,* Vol. 14, 77.

13

Lateral Earth Pressure: At-Rest, Rankine, and Coulomb

Retaining structures such as retaining walls, basement walls, and bulkheads commonly are encountered in foundation engineering as they support slopes of earth masses. Proper design and construction of these structures require a thorough knowledge of the lateral forces that act between the retaining structures and the soil masses being retained. These lateral forces are caused by lateral earth pressure. This chapter is devoted to the study of the various earth pressure theories.

13.1 At-Rest, Active, and Passive Pressures

Consider a mass of soil shown in Figure. 13.1a. The mass is bounded by a *frictionless wall* of height AB. A soil element located at a depth z is subjected to a vertical effective pressure, σ'_o, and a horizontal effective pressure, σ'_h. There are no shear stresses on the vertical and horizontal planes of the soil element. Let us define the ratio of σ'_h to σ'_o as a nondimensional quantity K, or

$$K = \frac{\sigma'_h}{\sigma'_o} \tag{13.1}$$

Now, three possible cases may arise concerning the retaining wall: and they are described

Case 1 If the wall AB is static—that is, if it does not move either to the right or to the left of its initial position—the soil mass will be in a state of *static equilibrium*. In that case, σ'_h is referred to as the *at-rest earth pressure*, or

$$K = K_o = \frac{\sigma'_h}{\sigma'_o} \tag{13.2}$$

where K_o = at-rest earth pressure coefficient.

Case 2 If the frictionless wall rotates sufficiently about its bottom to a position of $A'B$ (Figure 13.1b), then a triangular soil mass ABC' adjacent to the wall will reach a state of

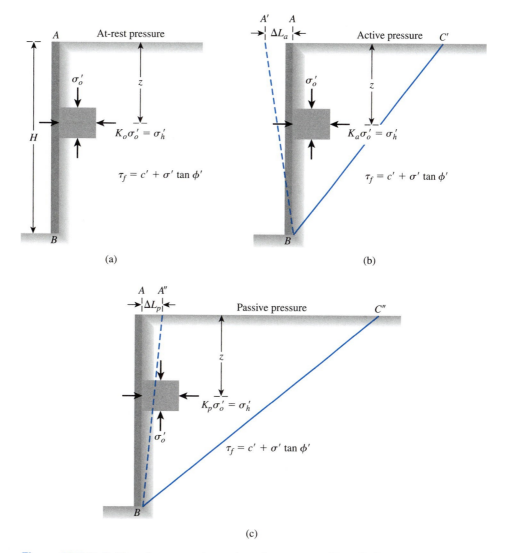

Figure 13.1 Definition of at-rest, active, and passive pressures (*Note:* Wall AB is frictionless)

plastic equilibrium and will fail sliding down the plane BC'. At this time, the horizontal effective stress, $\sigma'_h = \sigma'_a$, will be referred to as *active pressure*. Now,

$$K = K_a = \frac{\sigma'_h}{\sigma'_o} = \frac{\sigma'_a}{\sigma'_o} \qquad (13.3)$$

where K_a = active earth pressure coefficient.

Case 3 If the frictionless wall rotates sufficiently about its bottom to a position $A''B$ (Figure 13.1c), then a triangular soil mass ABC'' will reach a state of *plastic*

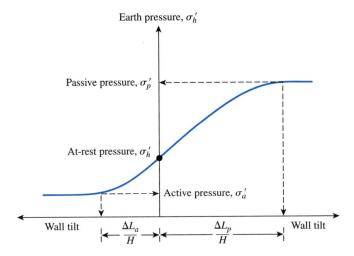

Figure 13.2 Variation of the magnitude of lateral earth pressure with wall tilt

Table 13.1 Typical Values of $\Delta L_a/H$ and $\Delta L_p/H$

Soil type	$\Delta L_a/H$	$\Delta L_p/H$
Loose sand	0.001–0.002	0.01
Dense sand	0.0005–0.001	0.005
Soft clay	0.02	0.04
Stiff clay	0.01	0.02

equilibrium and will fail sliding upward along the plane BC''. The horizontal effective stress at this time will be $\sigma'_h = \sigma'_p$, the so-called *passive pressure*. In this case,

$$K = K_p = \frac{\sigma'_h}{\sigma'_o} = \frac{\sigma'_p}{\sigma'_o} \tag{13.4}$$

where K_p = passive earth pressure coefficient

Figure 13.2 shows the nature of variation of lateral earth pressure with the wall tilt. Typical values of $\Delta L_a/H$ ($\Delta L_a = A'A$ in Figure 13.1b) and $\Delta L_p/H$ ($\Delta L_p = A''A$ in Figure 13.1c) for attaining the active and passive states in various soils are given in Table 13.1.

AT-REST LATERAL EARTH PRESSURE

13.2 *Earth Pressure At-Rest*

The fundamental concept of earth pressure at rest was discussed in the preceding section. In order to define the earth pressure coefficient K_o at rest, we refer to Figure 13.3,

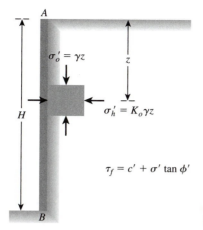

Figure 13.3
Earth pressure at rest

which shows a wall AB retaining a dry soil with a unit weight of γ. The wall is static. At a depth z,

$$\text{Vertical effective stress} = \sigma'_o = \gamma z$$
$$\text{Horizontal effective stress} = \sigma'_h = K_o \gamma z$$

So,

$$K_o = \frac{\sigma'_h}{\sigma'_o} = \text{at-rest earth pressure coefficient}$$

For coarse-grained soils, the coefficient of earth pressure at rest can be estimated by using the empirical relationship (Jaky, 1944)

$$K_o = 1 - \sin \phi' \tag{13.5}$$

where $\phi' = $ drained friction angle.

While designing a wall that may be subjected to lateral earth pressure at rest, one must take care in evaluating the value of K_o. Sherif, Fang, and Sherif (1984), on the basis of their laboratory tests, showed that Jaky's equation for K_o [Eq. (13.5)] gives good results when the backfill is loose sand. However, for a dense, compacted sand backfill, Eq. (13.5) may grossly underestimate the lateral earth pressure at rest. This underestimation results because of the process of compaction of backfill. For this reason, they recommended the design relationship

$$K_o = (1 - \sin \phi) + \left[\frac{\gamma_d}{\gamma_{d(\min)}} - 1 \right] 5.5 \tag{13.6}$$

where $\gamma_d = $ actual compacted dry unit weight of the sand behind the wall
$\gamma_{d(\min)} = $ dry unit weight of the sand in the loosest state (Chapter 3)

The increase of K_o observed from Eq. (13.6) compared to Eq. (13.5) is due to over-consolidation. For that reason, Mayne and Kulhawy (1982), after evaluating 171 soils, recommended a modification to Eq. (13.5). Or

$$K_o = (1 - \sin \phi')(OCR)^{\sin\phi'} \qquad (13.7)$$

where

$$OCR = \text{overconsolidation ratio}$$
$$= \frac{\text{preconsolidation pressure, } \sigma'_c}{\text{present effective overburden pressure, } \sigma'_o}$$

Equation (13.7) is valid for soils ranging from clay to gravel.

For fine-grained, normally consolidated soils, Massarsch (1979) suggested the following equation for K_o:

$$K_o = 0.44 + 0.42\left[\frac{PI\,(\%)}{100}\right] \qquad (13.8)$$

For overconsolidated clays, the coefficient of earth pressure at rest can be approximated as

$$K_{o(\text{overconsolidated})} = K_{o(\text{normally consolidated})}\sqrt{OCR} \qquad (13.9)$$

Figure 13.4 shows the distribution of lateral earth pressure at rest on a wall of height H retaining a dry soil having a unit weight of γ. The total force per unit length of the wall, P_o, is equal to the area of the pressure diagram, so

$$P_o = \tfrac{1}{2}K_o\gamma H^2 \qquad (13.10)$$

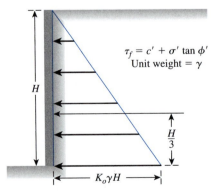

Figure 13.4 Distribution of lateral earth pressure at-rest on a wall

13.3 Earth Pressure At-Rest for Partially Submerged Soil

Figure 13.5a shows a wall of height H. The groundwater table is located at a depth H_1 below the ground surface, and there is no compensating water on the other side of the wall. For $z \le H_1$, the lateral earth pressure at rest can be given as $\sigma'_h = K_o \gamma z$. The variation of σ'_h with depth is shown by triangle ACE in Figure 13.5a. However, for $z \ge H_1$ (i.e., below the groundwater table), the pressure on the wall is found from the effective stress and pore water pressure components via the equation

$$\text{Effective vertical pressure} = \sigma'_o = \gamma H_1 + \gamma'(z - H_1) \tag{13.11}$$

where $\gamma' = \gamma_{\text{sat}} - \gamma_w =$ the effective unit weight of soil. So, the effective lateral pressure at rest is

$$\sigma'_h = K_o \sigma'_o = K_o [\gamma H_1 + \gamma'(z - H_1)] \tag{13.12}$$

The variation of σ'_h with depth is shown by $CEGB$ in Figure 13.5a. Again, the lateral pressure from pore water is

$$u = \gamma_w(z - H_1) \tag{13.13}$$

The variation of u with depth is shown in Figure 13.5b.

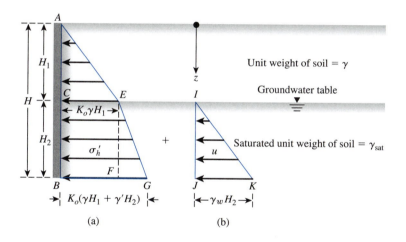

(a)   (b)

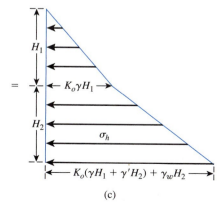

(c)

Figure 13.5
Distribution of earth pressure at rest for partially submerged soil

Hence, the total lateral pressure from earth and water at any depth $z \geq H_1$ is equal to

$$\sigma_h = \sigma_h' + u$$
$$= K_o[\gamma H_1 + \gamma'(z - H_1)] + \gamma_w(z - H_1) \qquad (13.14)$$

The force per unit length of the wall can be found from the sum of the areas of the pressure diagrams in Figures 13.5a and 13.5b and is equal to (Figure 13.5c)

$$P_o = \underbrace{\tfrac{1}{2}K_o\gamma H_1^2}_{\substack{\text{Area} \\ ACE}} + \underbrace{K_o\gamma H_1 H_2}_{\substack{\text{Area} \\ CEFB}} + \underbrace{\tfrac{1}{2}(K_o\gamma' + \gamma_w)H_2^2}_{\substack{\text{Areas} \\ EFG \text{ and } IJK}} \qquad (13.15)$$

Example 13.1

Figure 13.6a shows a 4.57 m-high retaining wall. The wall is restrained from yielding. Calculate the lateral force P_o per unit length of the wall. Also, determine the location of the resultant force. Assume that for sand $OCR = 2$.

Solution

$$K_o = (1 - \sin\phi')(OCR)^{\sin\phi'}$$
$$= (1 - \sin 30)(2)^{\sin 30} = 0.707$$

At $z = 0$: $\sigma_o' = 0$; $\sigma_h' = 0$; $u = 0$

At $z = 3.05$ m: $\sigma_o' = (3.05)(15.72) = 47.95 \text{ kN/m}^2$

$$\sigma_h' = K_o\sigma_o' = (0.707)(47.95) = 33.9 \text{ kN/m}^2$$

$$u = 0$$

At $z = 4.57$ m: $\sigma_o' = (3.05)(15.72) + (1.52)(19.24 - 9.81) = 62.28 \text{ kN/m}^2$

$$\sigma_h' = K_o\sigma_o' = (0.707)(62.28) = 44.03 \text{ kN/m}^2$$

$$u = (1.52)(\gamma_w) = (1.52)(9.81) = 14.91 \text{ kN/m}^2$$

The variations of σ_h' and u with depth are shown in Figures 13.6b and 13.6c.

Lateral force P_o = Area 1 + Area 2 + Area 3 + Area 4

or

$$P_o = \left(\frac{1}{2}\right)(3.05)(33.9) + (1.52)(33.9) + \left(\frac{1}{2}\right)(1.52)(10.13) + \left(\frac{1}{2}\right)(1.52)(14.91)$$

$$= 51.7 + 51.53 + 7.7 + 11.33 = 122.26 \text{ kN/m}$$

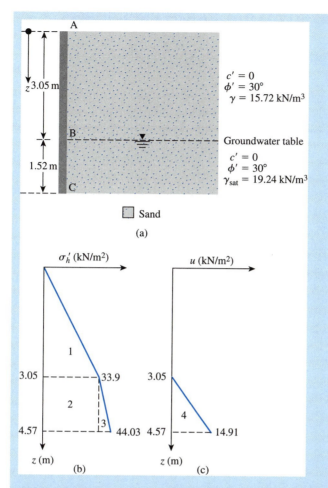

Sand

(a)

(b)

(c)

Figure 13.6

The location of the resultant, measured from the bottom of the wall, is

$$\bar{z} = \frac{\Sigma \text{ moment of pressure diagram about } C}{P_o}$$

or

$$\bar{z} = \frac{(51.7)\left(1.52 + \dfrac{3.05}{3}\right) + (51.53)\left(\dfrac{1.52}{2}\right) + (7.7)\left(\dfrac{1.52}{3}\right) + (11.33)\left(\dfrac{1.52}{3}\right)}{122.26}$$

$$= \textbf{1.47 m}$$

RANKINE'S LATERAL EARTH PRESSURE

Rankine's Theory of Active Pressure

The phrase *plastic equilibrium in soil* refers to the condition where every point in a soil mass is on the verge of failure. Rankine (1857) investigated the stress conditions in soil at a state of plastic equilibrium. In this section and in Section 13.5, we deal with Rankine's theory of earth pressure.

Figure 13.7a shows a soil mass that is bounded by a frictionless wall, *AB,* that extends to an infinite depth. The vertical and horizontal effective principal stresses on a soil element at a depth z are σ'_o and σ'_h, respectively. As we saw in Section 13.2, if the wall *AB* is not allowed to move, then $\sigma'_h = K_o\sigma'_o$. The stress condition in the soil element can be

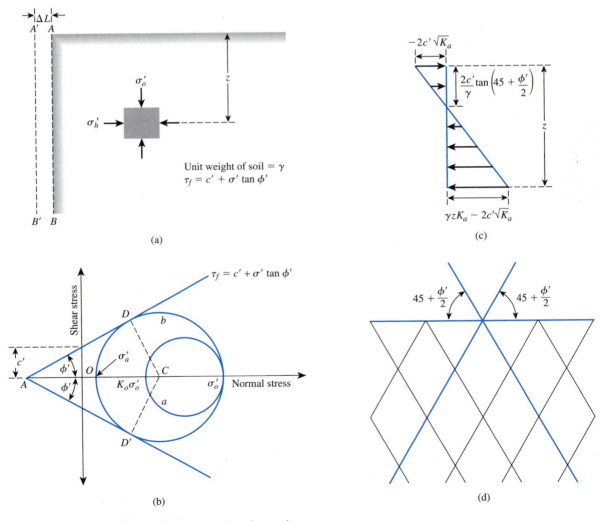

Figure 13.7 Rankine's active earth pressure

represented by the Mohr's circle a in Figure 13.7b. However, if the wall AB is allowed to *move away from the soil mass* gradually, the horizontal principal stress will decrease. Ultimately a state will be reached when the stress condition in the soil element can be represented by the Mohr's circle b, the state of plastic equilibrium and failure of the soil will occur. This situation represents *Rankine's active state*, and the effective pressure σ'_a on the vertical plane (which is a principal plane) is Rankine's *active earth pressure*. We next derive σ'_a in terms of γ, z, c', and ϕ' from Figure 13.7b

$$\sin \phi' = \frac{CD}{AC} = \frac{CD}{AO + OC}$$

But

$$CD = \text{radius of the failure circle} = \frac{\sigma'_o - \sigma'_a}{2}$$

$$AO = c' \cot \phi'$$

and

$$OC = \frac{\sigma'_o + \sigma'_a}{2}$$

So,

$$\sin \phi' = \frac{\dfrac{\sigma'_o - \sigma'_a}{2}}{c' \cot \phi' + \dfrac{\sigma'_o + \sigma'_a}{2}}$$

or

$$c' \cos \phi' + \frac{\sigma'_o + \sigma'_a}{2} \sin \phi' = \frac{\sigma'_o - \sigma'_a}{2}$$

or

$$\sigma'_a = \sigma'_o \frac{1 - \sin \phi'}{1 + \sin \phi'} - 2c' \frac{\cos \phi'}{1 + \sin \phi'} \tag{13.16}$$

But

$$\sigma'_o = \text{vertical effective overburden pressure} = \gamma z$$

$$\frac{1 - \sin \phi'}{1 + \sin \phi'} = \tan^2 \left(45 - \frac{\phi'}{2} \right)$$

and

$$\frac{\cos \phi'}{1 + \sin \phi'} = \tan \left(45 - \frac{\phi'}{2} \right)$$

Substituting the preceding values into Eq. (13.16), we get

$$\sigma_a' = \gamma z \, \tan^2\left(45 - \frac{\phi'}{2}\right) - 2c' \tan\left(45 - \frac{\phi'}{2}\right) \qquad (13.17)$$

The variation of σ_a' with depth is shown in Figure 13.7c. For cohesionless soils, $c' = 0$ and

$$\sigma_a' = \sigma_o' \, \tan^2\left(45 - \frac{\phi'}{2}\right) \qquad (13.18)$$

The ratio of σ_a' to σ_o' is called the *coefficient of Rankine's active earth pressure* and is given by

$$K_a = \frac{\sigma_a'}{\sigma_o'} = \tan^2\left(45 - \frac{\phi'}{2}\right) \qquad (13.19)$$

Again, from Figure 13.7b we can see that the failure planes in the soil make $\pm(45 + \phi'/2)$-degree angles with the direction of the major principal plane—that is, the horizontal. These are called potential *slip planes* and are shown in Figure 13.7d.

It is important to realize that a similar equation for σ_a could be derived based on the total stress shear strength parameters—that is, $\tau_f = c + \sigma \tan \phi$. For this case,

$$\sigma_a = \gamma z \, \tan^2\left(45 - \frac{\phi}{2}\right) - 2c \tan\left(45 - \frac{\phi}{2}\right) \qquad (13.20)$$

13.5   *Theory of Rankine's Passive Pressure*

Rankine's passive state can be explained with the aid of Figure 13.8. *AB* is a frictionless wall that extends to an infinite depth (Figure 13.8a). The initial stress condition on a soil element is represented by the Mohr's circle *a* in Figure 13.8b. If the wall gradually is *pushed into the soil mass,* the effective principal stress σ_h' will increase. Ultimately, the wall will reach a situation where the stress condition for the soil element can be expressed by the Mohr's circle *b*. At this time, failure of the soil will occur. This situation is referred to as *Rankine's passive state.* The lateral earth pressure σ_p', which is the major principal stress, is called *Rankine's passive earth pressure.* From Figure 13.8b, it can be shown that

$$\sigma_p' = \sigma_o' \, \tan^2\left(45 + \frac{\phi'}{2}\right) + 2c' \tan\left(45 + \frac{\phi'}{2}\right)$$

$$= \gamma z \, \tan^2\left(45 + \frac{\phi'}{2}\right) + 2c' \tan\left(45 + \frac{\phi'}{2}\right) \qquad (13.21)$$

The derivation is similar to that for Rankine's active state.

Figure 13.8c shows the variation of passive pressure with depth. For cohesionless soils ($c' = 0$),

$$\sigma'_p = \sigma'_o \tan^2\left(45 + \frac{\phi'}{2}\right)$$

or

$$\frac{\sigma'_p}{\sigma'_o} = K_p = \tan^2\left(45 + \frac{\phi'}{2}\right) \tag{13.22}$$

K_p (the ratio of effective stresses) in the preceding equation is referred to as the *coefficient of Rankine's passive earth pressure.*

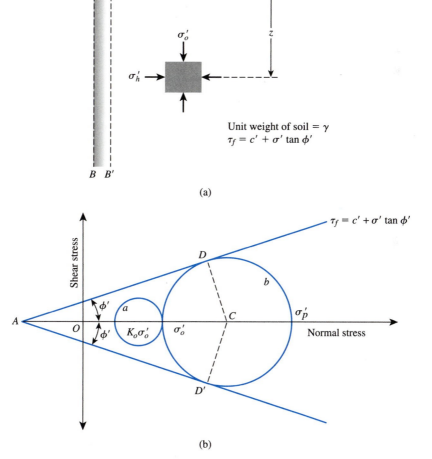

Figure 13.8 Rankine's passive earth pressure

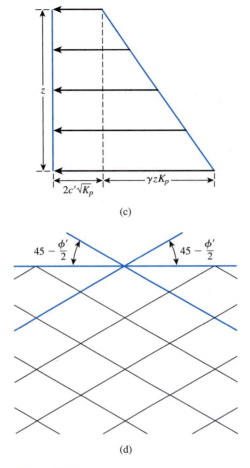

(c)

(d)

Figure 13.8 (*continued*)

The points D and D' on the failure circle (see Figure 13.8b) correspond to the slip planes in the soil. For Rankine's passive state, the slip planes make $\pm(45 - \phi'/2)$-degree angles with the direction of the minor principal plane—that is, in the horizontal direction. Figure 13.8d shows the distribution of slip planes in the soil mass.

13.6   *Yielding of Wall of Limited Height*

We learned in the preceding discussion that sufficient movement of a frictionless wall extending to an infinite depth is necessary to achieve a state of plastic equilibrium. However, the distribution of lateral pressure against a wall of limited height is influenced very much by the manner in which the wall actually yields. In most retaining walls of limited height, movement may occur by simple translation or, more frequently, by rotation about the bottom.

For preliminary theoretical analysis, let us consider a frictionless retaining wall represented by a plane AB as shown in Figure 13.9a. If the wall AB rotates sufficiently about its bottom to a position $A'B$, then a triangular soil mass ABC' adjacent to the wall will reach Rankine's active state. Because the slip planes in Rankine's active state make angles of $\pm(45 + \phi'/2)$ degrees with the major principal plane, the soil mass in the state of plastic equilibrium is bounded by the plane BC', which makes an angle of $(45 + \phi'/2)$ degrees with the horizontal. The soil inside the zone ABC' undergoes the same unit deformation in the horizontal direction everywhere, which is equal to $\Delta L_a/L_a$. The lateral earth pressure on the wall at any depth z from the ground surface can be calculated by using Eq. (13.17).

In a similar manner, if the frictionless wall AB (Figure 13.9b) rotates sufficiently into the soil mass to a position $A''B$, then the triangular mass of soil ABC'' will reach Rankine's

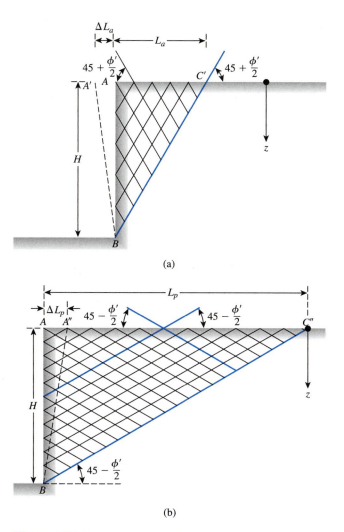

(a)

(b)

Figure 13.9 Rotation of frictionless wall about the bottom

passive state. The slip plane BC'' bounding the soil wedge that is at a state of plastic equilibrium will make an angle of $(45 - \phi'/2)$ degrees with the horizontal. Every point of the soil in the triangular zone ABC'' will undergo the same unit deformation in the horizontal direction, which is equal to $\Delta L_p/L_p$. The passive pressure on the wall at any depth z can be evaluated by using Eq. (13.21).

13.7 A Generalized Case for Rankine Active and Passive Pressure—Granular Backfill

In Sections 13.4, 13.5, and 13.6, we discussed the Rankine active and passive pressure cases for a frictionless wall with a vertical back and a horizontal backfill of granular soil. This can be extended to general cases of frictionless wall with inclined backfill (granular soil) as shown in Figure 13.10 (Chu, 1991).

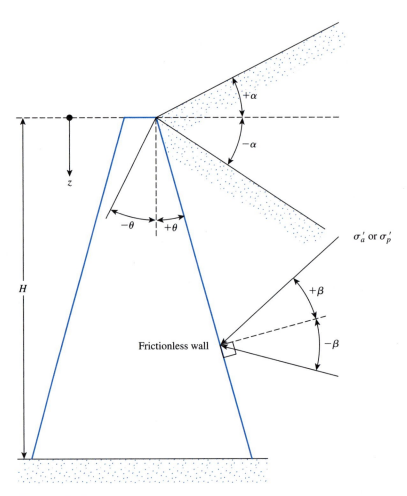

Figure 13.10 General case for Rankine active and passive pressures

Rankine Active Case

For the Rankine active case, the lateral earth pressure (σ'_a) at a depth z can be given as

$$\sigma'_a = \frac{\gamma z \cos \alpha \sqrt{1 + \sin^2 \phi' - 2 \sin \phi' \cos \psi_a}}{\cos \alpha + \sqrt{\sin^2 \phi' - \sin^2 \alpha}} \qquad (13.23)$$

where

$$\psi_a = \sin^{-1}\left(\frac{\sin \alpha}{\sin \phi'}\right) - \alpha + 2\theta \qquad (13.24)$$

The pressure σ'_a will be inclined at an angle β with the plane drawn at right angle to the backface of the wall, and

$$\beta = \tan^{-1}\left(\frac{\sin \phi' \sin \psi_a}{1 - \sin \phi' \cos \psi_a}\right) \qquad (13.25)$$

The active force P_a for unit length of the wall can then be calculated as

$$P_a = \frac{1}{2}\gamma H^2 K_{a(R)} \qquad (13.26)$$

where

$$K_{a(R)} = \frac{\cos(\alpha - \theta)\sqrt{1 + \sin^2 \phi' - 2 \sin \phi' \cos \psi_a}}{\cos^2 \theta(\cos \alpha + \sqrt{\sin^2 \phi' - \sin^2 \alpha})}$$
$$= \text{Rankine active earth-pressure coefficient}$$
$$\text{for generalized case} \qquad (13.27)$$

The location and direction of the resultant force P_a is shown in Figure 13.11a. Also shown in this figure is the failure wedge, ABC. Note that BC will be inclined at an angle η. Or

$$\eta = \frac{\pi}{4} + \frac{\phi'}{2} + \frac{\alpha}{2} - \frac{1}{2}\sin^{-1}\left(\frac{\sin \alpha}{\sin \phi'}\right) \qquad (13.28)$$

As a special case, for a vertical backface of the wall (that is, $\theta = 0$) as shown in Figure 13.12, Eqs. (13.26) and (13.27) simplify to the following

$$P_a = \frac{1}{2}K_{a(R)}\gamma H^2$$

where

$$K_{a(R)} = \cos \alpha \frac{\cos \alpha - \sqrt{\cos^2 \alpha - \cos^2 \phi'}}{\cos \alpha + \sqrt{\cos^2 \alpha - \cos^2 \phi'}} \qquad (13.29)$$

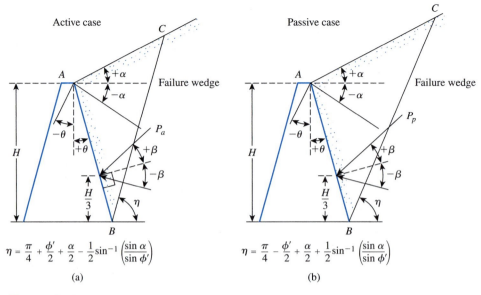

$$\eta = \frac{\pi}{4} + \frac{\phi'}{2} + \frac{\alpha}{2} - \frac{1}{2}\sin^{-1}\left(\frac{\sin\alpha}{\sin\phi'}\right)$$

(a)

$$\eta = \frac{\pi}{4} - \frac{\phi'}{2} + \frac{\alpha}{2} + \frac{1}{2}\sin^{-1}\left(\frac{\sin\alpha}{\sin\phi'}\right)$$

(b)

Figure 13.11 Location and direction of resultant Rankine force

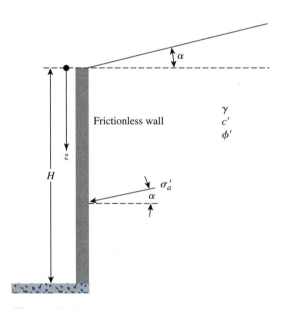

Figure 13.12 Frictionless vertical retaining wall with sloping backfill

Also note that, for this case, the angle β will be equal to α. The variation of $K_{a(R)}$ given in Eq. (13.29) with α and ϕ' is given in Table 13.2.

Rankine Passive Case

Similar to the active case, for the Rankine passive case, we can obtain the following relationships.

Table 13.2 Values of $K_{a(R)}$ [Eq. (13.29)]

	ϕ' (deg) $\rightarrow$						
$\downarrow \alpha$ (deg)	28	30	32	34	36	38	40
0	0.361	0.333	0.307	0.283	0.260	0.238	0.217
5	0.366	0.337	0.311	0.286	0.262	0.240	0.219
10	0.380	0.350	0.321	0.294	0.270	0.246	0.225
15	0.409	0.373	0.341	0.311	0.283	0.258	0.235
20	0.461	0.414	0.374	0.338	0.306	0.277	0.250
25	0.573	0.494	0.434	0.385	0.343	0.307	0.275

$$\sigma'_p = \frac{\gamma z \cos \alpha \sqrt{1 + \sin^2 \phi' + 2 \sin \phi' \cos \psi_p}}{\cos \alpha - \sqrt{\sin^2 \phi' - \sin^2 \alpha}} \tag{13.30}$$

where

$$\psi_p = \sin^{-1}\left(\frac{\sin \alpha}{\sin \phi'}\right) + \alpha - 2\theta \tag{13.31}$$

The inclination β of σ'_p, as shown in Figure 13.10, is

$$\beta = \tan^{-1}\left(\frac{\sin \phi' \sin \psi_p}{1 + \sin \phi' \cos \psi_p}\right) \tag{13.32}$$

The passive force per unit length of the wall is

$$P_p = \frac{1}{2}\gamma H^2 K_{p(R)}$$

where

$$K_{p(R)} = \frac{\cos(\alpha - \theta)\sqrt{1 + \sin^2 \phi' + 2\sin \phi' \cos \psi_p}}{\cos^2 \theta(\cos \alpha - \sqrt{\sin^2 \phi' - \sin^2 \alpha})} \tag{13.33}$$

The location and direction of P_a along with the failure wedge is shown in Figure 13.11b. For walls with vertical backface, $\theta = 0$,

$$P_p = \frac{1}{2}K_{p(R)}\gamma H^2$$

where

$$K_{p(R)} = \cos \alpha \frac{\cos \alpha + \sqrt{\cos^2 \alpha - \cos^2 \phi'}}{\cos \alpha - \sqrt{\cos^2 \alpha - \cos^2 \phi'}} \tag{13.34}$$

Table 13.3 Passive Earth Pressure Coefficient, $K_{p(R)}$ [Eq. (13.34)]

	ϕ' (deg) →						
↓ α (deg)	28	30	32	34	36	38	40
0	2.770	3.000	3.255	3.537	3.852	4.204	4.599
5	2.715	2.943	3.196	3.476	3.788	4.136	4.527
10	2.551	2.775	3.022	3.295	3.598	3.937	4.316
15	2.284	2.502	2.740	3.003	3.293	3.615	3.977
20	1.918	2.132	2.362	2.612	2.886	3.189	3.526
25	1.434	1.664	1.894	2.135	2.394	2.676	2.987

The variation of $K_{p(R)}$ with α and ϕ', as expressed by Eq. (13.34), is given in Table 13.3. Again, for this special case, the angle of P_p with the normal drawn to the back of the wall will be equal to α (that is, $\beta = \alpha$).

13.8 Diagrams for Lateral Earth-Pressure Distribution Against Retaining Walls

Backfill—Cohesionless Soil with Horizontal Ground Surface

Active Case Figure 13.13a shows a retaining wall with cohensionless soil backfill that has a horizontal ground surface. The unit weight and the angle of friction of the soil are γ and ϕ', respectively.

For Rankine's active state, the earth pressure at any depth against the retaining wall can be given by Eq. (13.17):

$$\sigma'_a = K_a \gamma z \qquad (Note: c' = 0)$$

Note that σ'_a increases linearly with depth, and at the bottom of the wall, it is

$$\sigma'_a = K_a \gamma H \qquad (13.35)$$

The total force per unit length of the wall is equal to the area of the pressure diagram, so

$$P_a = \tfrac{1}{2} K_a \gamma H^2 \qquad (13.36)$$

Passive Case The lateral pressure distribution against a retaining wall of height H for Rankine's passive state is shown in Figure 13.13b. The lateral earth pressure at any depth z [Eq. (13.22), $c' = 0$] is

$$\sigma'_p = K_p \gamma H \qquad (13.37)$$

The total force per unit length of the wall is

$$P_p = \tfrac{1}{2} K_p \gamma H^2 \qquad (13.38)$$

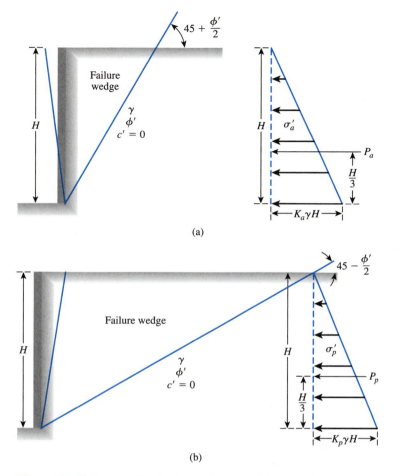

Figure 13.13 Pressure distribution against a retaining wall for cohensionless soil backfill with horizontal ground surface: (a) Rankine's active state; (b) Rankine's passive state

Backfill—Partially Submerged Cohensionless Soil Supporting a Surcharge

Active Case Figure 13.14a shows a frictionless retaining wall of height H and a backfill of cohensionless soil. The groundwater table is located at a depth of H_1 below the ground surface, and the backfill is supporting a surcharge pressure of q per unit area. From Eq. (13.19), the effective active earth pressure at any depth can be given by

$$\sigma'_a = K_a \sigma'_o \qquad (13.39)$$

where σ'_o and σ'_a = the effective vertical pressure and lateral pressure, respectively. At $z = 0$,

$$\sigma_o = \sigma'_o = q \qquad (13.40)$$

and

$$\sigma'_a = K_a q \qquad (13.41)$$

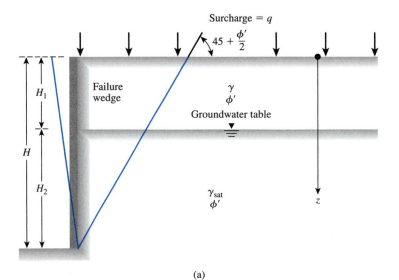

(a)

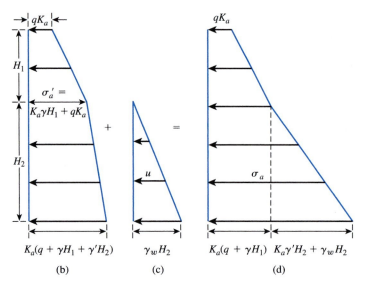

(b)   (c)   (d)

Figure 13.14
Rankine's active earth-pressure distribution against a retaining wall with partially submerged cohesionless soil backfill supporting a surcharge

At depth $z = H_1$,

$$\sigma'_o = (q + \gamma H_1) \tag{13.42}$$

and

$$\sigma'_a = K_a(q + \gamma H_1) \tag{13.43}$$

At depth $z = H$,

$$\sigma'_o = (q + \gamma H_1 + \gamma' H_2) \tag{13.44}$$

and

$$\sigma'_a = K_a(q + \gamma H_1 + \gamma' H_2) \tag{13.45}$$

where $\gamma' = \gamma_{\text{sat}} - \gamma_w$. The variation of σ'_a with depth is shown in Figure 13.14b.

The lateral pressure on the wall from the pore water between $z = 0$ and H_1 is 0, and for $z > H_1$, it increases linearly with depth (Figure 13.14c). At $z = H$,

$$u = \gamma_w H_2$$

The total lateral-pressure diagram (Figure 13.14d) is the sum of the pressure diagrams shown in Figures 13.14b and 13.14c. The total active force per unit length of the wall is the area of the total pressure diagram. Thus,

$$P_a = K_a q H + \tfrac{1}{2} K_a \gamma H_1^2 + K_a \gamma H_1 H_2 + \tfrac{1}{2}(K_a \gamma' + \gamma_w) H_2^2 \qquad (13.46)$$

Passive Case Figure 13.15a shows the same retaining wall as was shown in Figure 13.14a. Rankine's passive pressure at any depth against the wall can be given by Eq. (13.22):

$$\sigma'_p = K_p \sigma'_o$$

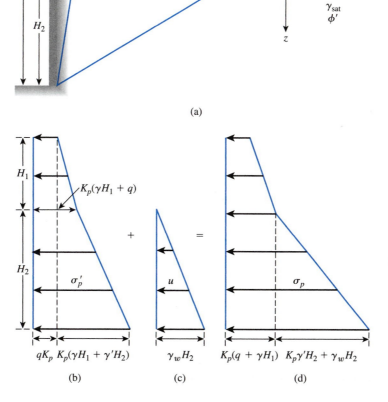

Figure 13.15
Rankine's passive earth-pressure distribution against a retaining wall with partially submerged cohesionless soil backfill supporting a surcharge

Using the preceding equation, we can determine the variation of σ'_p with depth, as shown in Figure 13.15b. The variation of the pressure on the wall from water with depth is shown in Figure 13.15c. Figure 13.15d shows the distribution of the total pressure σ_p with depth. The total lateral passive force per unit length of the wall is the area of the diagram given in Figure 13.15d, or

$$P_p = K_p q H + \tfrac{1}{2} K_p \gamma H_1^2 + K_p \gamma H_1 H_2 + \tfrac{1}{2}(K_p \gamma' + \gamma_w) H_2^2 \qquad (13.47)$$

Backfill—Cohesive Soil with Horizontal Backfill

Active Case Figure 13.16a shows a frictionless retaining wall with a cohesive soil backfill. The active pressure against the wall at any depth below the ground surface can be expressed as [Eq. (13.17)]

$$\sigma'_a = K_a \gamma z - 2\sqrt{K_a}\, c'$$

The variation of $K_a \gamma z$ with depth is shown in Figure 13.16b, and the variation of $2\sqrt{K_a}\, c'$ with depth is shown in Figure 13.16c. Note that $2\sqrt{K_a}\, c'$ is not a function of z; hence, Figure 13.16c is a rectangle. The variation of the net value of σ'_a with depth is plotted in Figure 13.16d. Also note that, because of the effect of cohesion, σ'_a is negative in the upper part of the retaining wall. The depth z_o at which the active pressure becomes equal to 0 can be found from Eq. (13.17) as

$$K_a \gamma z_o - 2\sqrt{K_a}\, c' = 0$$

or

$$z_o = \frac{2c'}{\gamma\sqrt{K_a}} \qquad (13.48)$$

For the undrained condition—that is, $\phi = 0$, $K_a = \tan^2 45 = 1$, and $c = c_u$ (undrained cohesion)—from Eq. (13.20),

$$z_o = \frac{2c_u}{\gamma} \qquad (13.49)$$

So, with time, tensile cracks at the soil–wall interface will develop up to a depth z_o.

The total active force per unit length of the wall can be found from the area of the total pressure diagram (Figure 13.16d), or

$$P_a = \tfrac{1}{2} K_a \gamma H^2 - 2\sqrt{K_a}\, c' H \qquad (13.50)$$

For the $\phi = 0$ condition,

$$P_a = \tfrac{1}{2} \gamma H^2 - 2c_u H \qquad (13.51)$$

For calculation of the total active force, common practice is to take the tensile cracks into account. Because no contact exists between the soil and the wall up to a depth of z_o after the development of tensile cracks, only the active pressure distribution

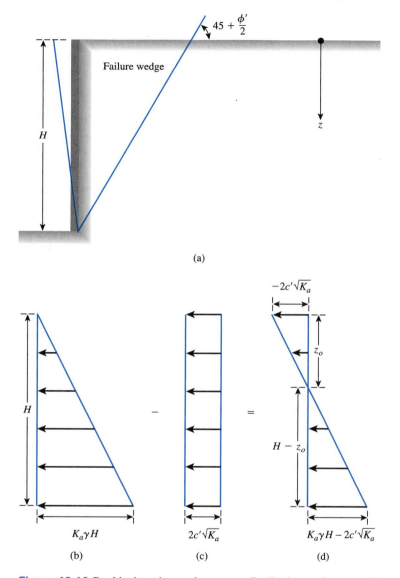

Figure 13.16 Rankine's active earth-pressure distribution against a retaining wall with cohesive soil backfill

against the wall between $z = 2c'/(\gamma\sqrt{K_a})$ and H (Figure 13.16d) is considered. In this case,

$$P_a = \tfrac{1}{2}(K_a\gamma H - 2\sqrt{K_a}c')\left(H - \frac{2c'}{\gamma\sqrt{K_a}}\right)$$

$$= \tfrac{1}{2}K_a\gamma H^2 - 2\sqrt{K_a}c'H + 2\frac{c'^2}{\gamma} \qquad (13.52)$$

For the $\phi = 0$ condition,

$$P_a = \tfrac{1}{2}\gamma H^2 - 2c_u H + 2\frac{c_u^2}{\gamma} \tag{13.53}$$

Passive Case Figure 13.17a shows the same retaining wall with backfill similar to that considered in Figure 13.16a. Rankine's passive pressure against the wall at depth z can be given by [Eq. (13.21)]

$$\sigma'_p = K_p \gamma z + 2\sqrt{K_p}\,c'$$

At $z = 0$,

$$\sigma'_p = 2\sqrt{K_p}\,c' \tag{13.54}$$

and at $z = H$,

$$\sigma'_p = K_p \gamma H + 2\sqrt{K_p}\,c' \tag{13.55}$$

The variation of σ'_p with depth is shown in Figure 13.17b. The passive force per unit length of the wall can be found from the area of the pressure diagrams as

$$P_p = \tfrac{1}{2}K_p \gamma H^2 + 2\sqrt{K_p}\,c'H \tag{13.56}$$

For the $\phi = 0$ condition, $K_p = 1$ and

$$P_p = \tfrac{1}{2}\gamma H^2 + 2c_u H \tag{13.57}$$

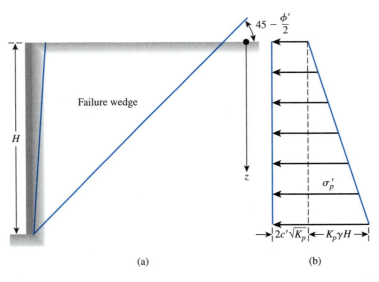

(a)   (b)

Figure 13.17 Rankine's passive earth-pressure distribution against a retaining wall with cohesive soil backfill

Example 13.2

An 6-m high retaining wall is shown in Figure 13.18a. Determine

    a. Rankine active force per unit length of the wall and the location of the resultant

    b. Rankine passive force per unit length of the wall and the location of the resultant

Solution
Part a

Because $c' = 0$, to determine the active force we can use from Eq. (13.20).

$$\sigma'_a = K_a\sigma'_o = K_a\gamma z$$

$$K_a = \frac{1 - \sin \phi'}{1 + \sin \phi'} = \frac{1 - \sin 36}{1 + \sin 36} = 0.26$$

At $z = 0$, $\sigma'_a = 0$; at $z = 6$ m,

$$\sigma'_a = (0.26)(16)(6) = 24.96 \text{ kN/m}^2$$

The pressure-distribution diagram is shown in Figure 13.18b. The active force per unit length of the wall is as follows:

$$P_a = \tfrac{1}{2}(6)(24.96) = \mathbf{74.88 \text{ kN/m}}$$

Also,

$$\bar{z} = \mathbf{2 \text{ m}}$$

Part b

To determine the passive force, we are given that $c' = 0$. So, from Eq. (13.22),

$$\sigma'_p = K_p\sigma'_o = K_p\gamma z$$

$$K_p = \frac{1 + \sin \phi'}{1 - \sin \phi'} = \frac{1 + \sin 36}{1 - \sin 36} = 3.85$$

At $z = 0$, $\sigma'_p = 0$; at $z = 6$ m,

$$\sigma'_p = (3.85)(16)(6) = 369.6 \text{ kN/m}^2$$

The pressure-distribution diagram is shown in Figure 13.18d. The passive force per unit length of the wall is as follows.

$$P_p = \tfrac{1}{2}(6)(369.6) = \mathbf{1108.8 \text{ kN/m}^2}$$

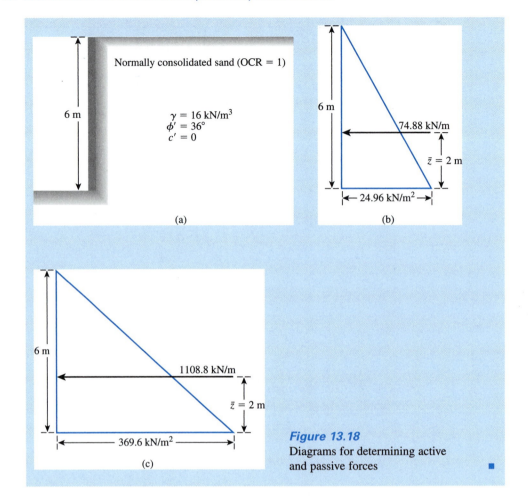

Normally consolidated sand (OCR = 1)

6 m

$\gamma = 16 \ kN/m^3$
$\phi' = 36°$
$c' = 0$

(a)

6 m

74.88 kN/m

$\bar{z} = 2$ m

← 24.96 kN/m² →

(b)

6 m

1108.8 kN/m

$\bar{z} = 2$ m

369.6 kN/m²

(c)

Figure 13.18
Diagrams for determining active
and passive forces

Example 13.3

Refer to Figure 13.10. Given: $H = 3.66$ m, $\alpha = +20°$, and $\theta = +20°$. For the granular backfill, it is given that $\gamma = 18.08 \ kN/m^3$ and $\phi' = 30°$. Determine the active force P_a per unit length of the wall as well as the location and direction of the resultant.

Solution
From Eq. (13.24),

$$\psi_a = \sin^{-1}\left(\frac{\sin \alpha}{\sin \phi'}\right) - \alpha + 2\theta = \sin^{-1}\left(\frac{\sin 20}{\sin 30}\right) - 20° + (2)(20°)$$

$$= 43.16° - 20° + 40° = 63.16°$$

From Eq. (13.27),

$$K_{a(R)} = \frac{\cos(\alpha - \theta)\sqrt{1 + \sin^2\phi' - 2\sin\phi' \cdot \cos\psi_a}}{\cos^2\theta\,(\cos\alpha + \sqrt{\sin^2\phi' - \sin^2\alpha})}$$

$$= \frac{\cos(20 - 20)\sqrt{1 + \sin^2 30 - (2)(\sin 30)(\cos 63.16)}}{\cos^2 20\,(\cos 20 + \sqrt{\sin^2 30 - \sin^2 20})} = 0.776$$

From Eq. (13.26),

$$P_a = \tfrac{1}{2}\gamma H^2 K_{a(R)} = \tfrac{1}{2}(18.08)(3.66)^2(0.776) = \textbf{93.97 kN/m}$$

From Eq. (13.25),

$$\beta = \tan^{-1}\left(\frac{\sin\phi'\sin\psi_a}{1 - \sin\phi'\cos\psi_a}\right)$$

$$= \tan^{-1}\left[\frac{(\sin 30)(\sin 63.16)}{1 - (\sin 30)(\cos 63.16)}\right] = 29.95° \approx 30°$$

The resultant will act a distance of $\dfrac{3.66}{3} = \textbf{1.22 m}$ above the bottom of the wall with $\beta = \textbf{30°}$.

■

Example 13.4

For the retaining wall shown in Figure 13.19a, determine the force per unit length of the wall for Rankine's active state. Also find the location of the resultant.

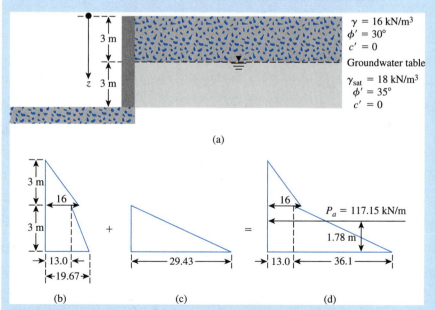

(a)

(b)

(c)

(d)

Figure 13.19 Retaining wall and pressure diagrams for determining Rankine's active earth pressure. (*Note:* The units of pressure in (b), (c), and (d) are kN/m²)

Solution

Given that $c' = 0$, we known that $\sigma'_a = K_a\sigma'_o$. For the upper layer of the soil, Rankine's active earth-pressure coefficient is

$$K_a = K_{a(1)} = \frac{1 - \sin 30°}{1 + \sin 30°} = \frac{1}{3}$$

For the lower layer,

$$K_a = K_{a(2)} = \frac{1 - \sin 35°}{1 + \sin 35°} = 0.271$$

At $z = 0$, $\sigma'_o = 0$. At $z = 3$ m (just inside the bottom of the upper layer), $\sigma'_o = 3 \times 16 = 48$ kN/m^2. So

$$\sigma'_a = K_{a(1)}\sigma'_o = \tfrac{1}{3} \times 48 = 16 \text{ kN/m}^2$$

Again, at $z = 3$ m (in the lower layer), $\sigma'_o = 3 \times 16 = 48$ kN/m^2, and

$$\sigma'_a = K_{a(2)}\sigma'_o = (0.271) \times (48) = 13.0 \text{ kN/m}^2$$

At $z = 6$ m,

$$\sigma'_o = 3 \times 16 + 3(18 - 9.81) = 72.57 \text{ kN/m}^2$$
$$\uparrow$$
$$\gamma_w$$

and

$$\sigma'_a = K_{a(2)}\sigma'_o = (0.271) \times (72.57) = 19.67 \text{ kN/m}^2$$

The variation of σ'_a with depth is shown in Figure 13.19b.
 The lateral pressures due to the pore water are as follows.

$$\text{At } z = 0: \quad u = 0$$

$$\text{At } z = 3 \text{ m: } \quad u = 0$$

$$\text{At } z = 6 \text{ m: } \quad u = 3 \times \gamma_w = 3 \times 9.81 = 29.43 \text{ kN/m}^2$$

The variation of u with depth is shown in Figure 13.19c, and that for σ_a (total active pressure) is shown in Figure 13.19d. Thus,

$$P_a = (\tfrac{1}{2})(3)(16) + 3(13.0) + (\tfrac{1}{2})(3)(36.1) = 24 + 39.0 + 54.15 = \textbf{117.15 kN/m}$$

The location of the resultant can be found by taking the moment about the bottom of the wall:

$$\bar{z} = \frac{24\left(3 + \dfrac{3}{3}\right) + 39.0\left(\dfrac{3}{2}\right) + 54.15\left(\dfrac{3}{3}\right)}{117.15}$$

$$= \textbf{1.78 m}$$

Example 13.5

A retaining wall that has a soft, saturated clay backfill is shown in Figure 13.20a. For the undrained condition ($\phi = 0$) of the backfill, determine

a. Maximum depth of the tensile crack
b. P_a before the tensile crack occurs
c. P_a after the tensile crack occurs

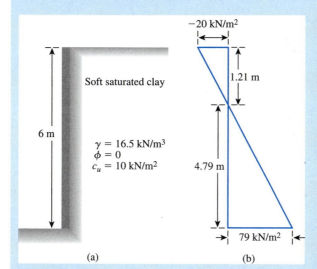

Figure 13.20
Rankine active pressure due to a soft, saturated clay backfill

(a) (b)

Solution

For $\phi = 0$, $K_a = \tan^2 45 = 1$ and $c = c_u$. From Eq. (13.20),

$$\sigma_a = \gamma z - 2c_u$$

At $z = 0$,

$$\sigma_a = -2c_u = -(2)(10) = -20 \ \text{kN/m}^2$$

At $z = 6$ m,

$$\sigma_a = (16.5)(6) - (2)(10) = 79 \ \text{kN/m}^2$$

The variation of σ_a with depth is shown in Figure 13.20b.

Part a
From Eq. (13.49), the depth of the tensile crack equals

$$z_o = \frac{2c_u}{\gamma} = \frac{(2)(10)}{16.5} = \textbf{1.21 m}$$

Part b

Before the tensile crack occurs [Eq. (13.51)],

$$P_a = \tfrac{1}{2}\gamma H^2 - 2c_u H$$

or

$$P_a = \tfrac{1}{2}(16.5)(6)^2 - 2(10)(6) = \textbf{177 kN/m}$$

Part c

After the tensile crack occurs,

$$P_a = \tfrac{1}{2}(6 - 1.21)(79) = \textbf{189.2 kN/m}$$

[*Note:* The preceding P_a can also be obtained by substituting the proper values into Eq. (13.53).] ∎

13.9 Rankine Pressure for c'–ϕ' Soil—Inclined Backfill

Figure 13.21 shows a retaining wall with a vertical back with an inclined backfill which is a c'–ϕ' soil. The analysis for determining the active and passive Rankine earth pressure for this condition has been provided by Mazindrani and Ganjali (1997). According to this analysis,

$$\text{Active pressure: } \sigma_a' = \gamma z K_{a(R)} = \gamma z K_{a(R)}' \cos \alpha \qquad (13.58)$$

where $K_{a(R)}$ = Rankline active earth-pressure coefficient

$$K_{a(R)}' = \frac{K_{a(R)}}{\cos \alpha} \qquad (13.59)$$

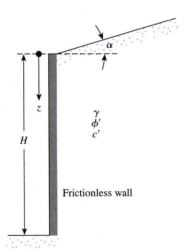

Frictionless wall

Figure 13.21
Rankine active and passive pressures
with an inclined backfill of c'–ϕ' soil

Similarly,

$$\text{Passive pressure: } \sigma'_p = \gamma z K_{p(R)} = \gamma z K'_{p(R)} \cos \alpha \qquad (13.60)$$

where $K_{p(R)} = $ Rankine passive earth-pressure coefficient

$$K'_{p(R)} = \frac{K_{p(R)}}{\cos \alpha} \qquad (13.61)$$

Also,

$$K'_{a(R)}, K'_{p(R)} = \frac{1}{\cos^2 \phi'}$$

$$\times \left\{ \begin{array}{l} 2 \cos^2 \alpha + 2 \left(\dfrac{c'}{\gamma z} \right) \cos \phi' \sin \phi' \\[2mm] \pm \sqrt{\left[4 \cos^2 \alpha (\cos^2 \alpha - \cos^2 \phi') + 4 \left(\dfrac{c'}{\gamma z} \right)^2 \cos^2 \phi' + 8 \left(\dfrac{c'}{\gamma z} \right) \cos^2 \alpha \sin \phi' \cos \phi' \right]} \end{array} \right\} - 1$$

$$(13.62)$$

Tables 13.4 and 13.5 give the variation of $K'_{a(R)}$ and $K'_{p(R)}$ with α, $c'/\gamma z$, and ϕ'.

Table 13.4 Variation of $K'_{a(R)}$ with α, $c'/\gamma z$, and ϕ'

α (deg)	ϕ' (deg)	c'/γz			
		0	0.025	0.050	0.100
0	15	0.589	0.550	0.512	0.435
	20	0.490	0.455	0.420	0.351
	25	0.406	0.374	0.342	0.278
	30	0.333	0.305	0.276	0.218
	35	0.271	0.245	0.219	0.167
	40	0.217	0.194	0.171	0.124
5	15	0.607	0.566	0.525	0.445
	20	0.502	0.465	0.429	0.357
	25	0.413	0.381	0.348	0.283
	30	0.339	0.309	0.280	0.221
	35	0.275	0.248	0.222	0.169
	40	0.220	0.196	0.173	0.126
10	15	0.674	0.621	0.571	0.477
	20	0.539	0.497	0.456	0.377
	25	0.438	0.402	0.366	0.296
	30	0.355	0.323	0.292	0.230
	35	0.286	0.258	0.230	0.175
	40	0.228	0.203	0.179	0.130
15	15	1.000	0.776	0.683	0.546
	20	0.624	0.567	0.514	0.417
	25	0.486	0.443	0.401	0.321
	30	0.386	0.350	0.315	0.246
	35	0.307	0.276	0.246	0.186
	40	0.243	0.216	0.190	0.337

Table 13.5 Variation of $K'_{p(R)}$ with α, $c'/\gamma z$, and ϕ'

		$c'/\gamma z$			
α (deg)	ϕ' (deg)	0	0.025	0.050	0.100
0	15	1.698	1.764	1.829	1.959
	20	2.040	2.111	2.182	2.325
	25	2.464	2.542	2.621	2.778
	30	3.000	3.087	3.173	3.346
	35	3.690	3.786	3.882	4.074
	40	4.599	4.706	4.813	5.028
5	15	1.674	1.716	1.783	1.916
	20	1.994	2.067	2.140	2.285
	25	2.420	2.499	2.578	2.737
	30	2.954	3.042	3.129	3.303
	35	3.641	3.738	3.834	4.027
	40	5.545	4.652	4.760	4.975
10	15	1.484	1.564	1.641	1.788
	20	1.854	1.932	2.010	2.162
	25	2.285	2.368	2.450	2.614
	30	2.818	2.907	2.996	3.174
	35	3.495	3.593	3.691	3.887
	40	4.383	4.491	4.600	4.817
15	15	1.000	1.251	1.370	1.561
	20	1.602	1.696	1.786	1.956
	25	2.058	2.147	2.236	2.409
	30	2.500	2.684	2.777	2.961
	35	3.255	3.356	3.456	3.656
	40	4.117	4.228	4.338	4.558

For the *active* case, the depth of the tensile crack can be given as

$$z_o = \frac{2c'}{\gamma} \sqrt{\frac{1 + \sin \phi'}{1 - \sin \phi'}} \tag{13.63}$$

COULOMB'S EARTH PRESSURE THEORY

More than 200 years ago, Coulomb (1776) presented a theory for active and passive earth pressures against retaining walls. In this theory, Coulomb assumed that the failure

---

Example 13.6

Refer to Figure 13.21. Given: $H = 4.57$ m, $\alpha = 10°$, $\gamma = 18.55$ kN/m³, $\phi' = 20°$, and $c' = 12$ kN/m². Determine the Rankine active force P_a on the retaining wall after the tensile crack occurs.

Solution

From Eq. (13.63), the depth of tensile crack is

$$z_o = \frac{2c'}{\gamma}\sqrt{\frac{1 + \sin\phi'}{1 - \sin\phi'}} = \frac{(2)(12)}{18.55}\sqrt{\frac{1 + \sin 20}{1 - \sin 20}} = 1.85 \text{ m}$$

So

$$\text{At } z = 0: \qquad \sigma'_a = 0$$

$$\text{At } z = 4.57 \text{ m}: \quad \sigma'_a = \gamma z K'_{a(R)} \cos\alpha$$

$$\frac{c'}{\gamma z} = \frac{12}{(18.55)(4.57)} \approx 0.14$$

From Table 13.4, for $\alpha = 10°$ and $c'/\gamma z = 0.14$, the magnitude of $K'_{a(R)} \approx 0.3$. So

$$\sigma'_a = (18.55)(4.57)(0.3)(\cos 10°) = 25 \text{ kN/m}^2$$

Hence,

$$P_a = \frac{1}{2}(H - z_o)(25) = \frac{1}{2}(4.57 - 1.85)(25) = \textbf{34 kN/m} \qquad \blacksquare$$

surface is a plane. The *wall friction* was taken into consideration. The following sections discuss the general principles of the derivation of Coulomb's earth-pressure theory for a cohesionless backfill (shear strength defined by the equation $\tau_f = \sigma' \tan\phi'$).

13.10 *Coulomb's Active Pressure*

Let AB (Figure 13.22a) be the back face of a retaining wall supporting a granular soil; the surface of which is constantly sloping at an angle α with the horizontal. BC is a trial failure surface. In the stability consideration of the probable failure wedge ABC, the following forces are involved (per unit length of the wall):

1. W—the weight of the soil wedge.
2. F—the resultant of the shear and normal forces on the surface of failure, BC. This is inclined at an angle of ϕ' to the normal drawn to the plane BC.
3. P_a—the active force per unit length of the wall. The direction of P_a is inclined at an angle δ' to the normal drawn to the face of the wall that supports the soil. δ' is the angle of friction between the soil and the wall.

The force triangle for the wedge is shown in Figure 13.22b. From the law of sines, we have

$$\frac{W}{\sin(90 + \theta + \delta' - \beta + \phi')} = \frac{P_a}{\sin(\beta - \phi')} \qquad (13.64)$$

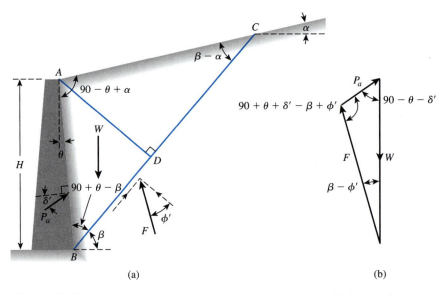

Figure 13.22 Coulomb's active pressure: (a) trial failure wedge; (b) force polygon

or

$$P_a = \frac{\sin(\beta - \phi')}{\sin(90 + \theta + \delta' - \beta + \phi')} W \tag{13.65}$$

The preceding equation can be written in the form

$$P_a = \frac{1}{2} \gamma H^2 \left[\frac{\cos(\theta - \beta)\cos(\theta - \alpha)\sin(\beta - \phi')}{\cos^2 \theta \sin(\beta - \alpha)\sin(90 + \theta + \delta' - \beta + \phi')} \right] \tag{13.66}$$

where γ = unit weight of the backfill. The values of γ, H, θ, α, ϕ', and δ' are constants, and β is the only variable. To determine the critical value of β for maximum P_a, we have

$$\frac{dP_a}{d\beta} = 0 \tag{13.67}$$

After solving Eq. (13.67), when the relationship of β is substituted into Eq. (13.66), we obtain Coulomb's active earth pressure as

$$P_a = \tfrac{1}{2} K_a \gamma H^2 \tag{13.68}$$

where K_a is Coulomb's active earth-pressure coefficient and is given by

$$K_a = \frac{\cos^2(\phi' - \theta)}{\cos^2 \theta \cos(\delta' + \theta) \left[1 + \sqrt{\dfrac{\sin(\delta' + \phi')\sin(\phi' - \alpha)}{\cos(\delta' + \theta)\cos(\theta - \alpha)}} \right]^2} \tag{13.69}$$

Table 13.6 Values of K_a [Eq. (13.69)] for $\theta = 0°$, $\alpha = 0°$

↓ϕ' (deg)	δ' (deg) →					
	0	5	10	15	20	25
28	0.3610	0.3448	0.3330	0.3251	0.3203	0.3186
30	0.3333	0.3189	0.3085	0.3014	0.2973	0.2956
32	0.3073	0.2945	0.2853	0.2791	0.2755	0.2745
34	0.2827	0.2714	0.2633	0.2579	0.2549	0.2542
36	0.2596	0.2497	0.2426	0.2379	0.2354	0.2350
38	0.2379	0.2292	0.2230	0.2190	0.2169	0.2167
40	0.2174	0.2089	0.2045	0.2011	0.1994	0.1995
42	0.1982	0.1916	0.1870	0.1841	0.1828	0.1831

Note that when $\alpha = 0°$, $\theta = 0°$, and $\delta' = 0°$, Coulomb's active earth-pressure coefficient becomes equal to $(1 - \sin \phi')/(1 + \sin \phi')$, which is the same as Rankine's earth-pressure coefficient given earlier in this chapter.

The variation of the values of K_a for retaining walls with a vertical back ($\theta = 0°$) and horizontal backfill ($\alpha = 0°$) is given in Table 13.6. From this table, note that for a given value of ϕ', the effect of wall friction is to reduce somewhat the active earth-pressure coefficient.

Tables 13.7 and 13.8 also give the variation of K_a [Eq. (13.69)] for various values of α, ϕ', θ, and δ' ($\delta' = \frac{2}{3}\phi'$ in Table 13.7 and $\delta' = \frac{1}{2}\phi'$ in Table 13.8).

Example 13.7

Refer to Figure 13.22. Given: $\alpha = 10°$; $\theta = 5°$; $H = 4$ m; unit weight of soil, $\gamma = 15$ kN/m³; soil friction angle, $\phi' = 30°$; and $\delta' = 15°$. Estimate the active force, P_a, per unit length of the wall. Also, state the direction and location of the resultant force, P_a.

Solution
From Eq. (13.68),

$$P_a = \frac{1}{2}\gamma H^2 K_a$$

For $\phi' = 30°$; $\delta' = 15°$—that is, $\dfrac{\delta'}{\phi'} = \dfrac{15}{30} = \dfrac{1}{2}$; $\alpha = 10°$; and $\theta = 5°$, the magnitude of K_a is 0.3872 (Table 13.8). So,

$$P_a = \frac{1}{2}(15)(4)^2(0.3872) = \textbf{46.46 kN/m}$$

The resultant will act at a vertical distance equal to $H/3 = 4/3 = 1.33$ m above the bottom of the wall and will be inclined at an angle of $15°$ $(= \delta')$ to the back face of the wall. ∎

Table 13.7 Values of K_a [Eq. (13.69)] (*Note:* $\delta' = \frac{2}{3}\phi'$)

α (deg)	ϕ' (deg)	θ (deg)					
		0	5	10	15	20	25
0	28	0.3213	0.3588	0.4007	0.4481	0.5026	0.5662
	29	0.3091	0.3467	0.3886	0.4362	0.4908	0.5547
	30	0.2973	0.3349	0.3769	0.4245	0.4794	0.5435
	31	0.2860	0.3235	0.3655	0.4133	0.4682	0.5326
	32	0.2750	0.3125	0.3545	0.4023	0.4574	0.5220
	33	0.2645	0.3019	0.3439	0.3917	0.4469	0.5117
	34	0.2543	0.2916	0.3335	0.3813	0.4367	0.5017
	35	0.2444	0.2816	0.3235	0.3713	0.4267	0.4919
	36	0.2349	0.2719	0.3137	0.3615	0.4170	0.4824
	37	0.2257	0.2626	0.3042	0.3520	0.4075	0.4732
	38	0.2168	0.2535	0.2950	0.3427	0.3983	0.4641
	39	0.2082	0.2447	0.2861	0.3337	0.3894	0.4553
	40	0.1998	0.2361	0.2774	0.3249	0.3806	0.4468
	41	0.1918	0.2278	0.2689	0.3164	0.3721	0.4384
	42	0.1840	0.2197	0.2606	0.3080	0.3637	0.4302
5	28	0.3431	0.3845	0.4311	0.4843	0.5461	0.6190
	29	0.3295	0.3709	0.4175	0.4707	0.5325	0.6056
	30	0.3165	0.3578	0.4043	0.4575	0.5194	0.5926
	31	0.3039	0.3451	0.3916	0.4447	0.5067	0.5800
	32	0.2919	0.3329	0.3792	0.4324	0.4943	0.5677
	33	0.2803	0.3211	0.3673	0.4204	0.4823	0.5558
	34	0.2691	0.3097	0.3558	0.4088	0.4707	0.5443
	35	0.2583	0.2987	0.3446	0.3975	0.4594	0.5330
	36	0.2479	0.2881	0.3338	0.3866	0.4484	0.5221
	37	0.2379	0.2778	0.3233	0.3759	0.4377	0.5115
	38	0.2282	0.2679	0.3131	0.3656	0.4273	0.5012
	39	0.2188	0.2582	0.3033	0.3556	0.4172	0.4911
	40	0.2098	0.2489	0.2937	0.3458	0.4074	0.4813
	41	0.2011	0.2398	0.2844	0.3363	0.3978	0.4718
	42	0.1927	0.2311	0.2753	0.3271	0.3884	0.4625
10	28	0.3702	0.4164	0.4686	0.5287	0.5992	0.6834
	29	0.3548	0.4007	0.4528	0.5128	0.5831	0.6672
	30	0.3400	0.3857	0.4376	0.4974	0.5676	0.6516
	31	0.3259	0.3713	0.4230	0.4826	0.5526	0.6365
	32	0.3123	0.3575	0.4089	0.4683	0.5382	0.6219
	33	0.2993	0.3442	0.3953	0.4545	0.5242	0.6078
	34	0.2868	0.3314	0.3822	0.4412	0.5107	0.5942
	35	0.2748	0.3190	0.3696	0.4283	0.4976	0.5810
	36	0.2633	0.3072	0.3574	0.4158	0.4849	0.5682
	37	0.2522	0.2957	0.3456	0.4037	0.4726	0.5558
	38	0.2415	0.2846	0.3342	0.3920	0.4607	0.5437
	39	0.2313	0.2740	0.3231	0.3807	0.4491	0.5321
	40	0.2214	0.2636	0.3125	0.3697	0.4379	0.5207
	41	0.2119	0.2537	0.3021	0.3590	0.4270	0.5097
	42	0.2027	0.2441	0.2921	0.3487	0.4164	0.4990

Table 13.7 (*continued*)

α (deg)	ϕ' (deg)	θ (deg)					
		0	5	10	15	20	25
15	28	0.4065	0.4585	0.5179	0.5868	0.6685	0.7670
	29	0.3881	0.4397	0.4987	0.5672	0.6483	0.7463
	30	0.3707	0.4219	0.4804	0.5484	0.6291	0.7265
	31	0.3541	0.4049	0.4629	0.5305	0.6106	0.7076
	32	0.3384	0.3887	0.4462	0.5133	0.5930	0.6895
	33	0.3234	0.3732	0.4303	0.4969	0.5761	0.6721
	34	0.3091	0.3583	0.4150	0.4811	0.5598	0.6554
	35	0.2954	0.3442	0.4003	0.4659	0.5442	0.6393
	36	0.2823	0.3306	0.3862	0.4513	0.5291	0.6238
	37	0.2698	0.3175	0.3726	0.4373	0.5146	0.6089
	38	0.2578	0.3050	0.3595	0.4237	0.5006	0.5945
	39	0.2463	0.2929	0.3470	0.4106	0.4871	0.5805
	40	0.2353	0.2813	0.3348	0.3980	0.4740	0.5671
	41	0.2247	0.2702	0.3231	0.3858	0.4613	0.5541
	42	0.2146	0.2594	0.3118	0.3740	0.4491	0.5415
20	28	0.4602	0.5205	0.5900	0.6714	0.7689	0.8880
	29	0.4364	0.4958	0.5642	0.6445	0.7406	0.8581
	30	0.4142	0.4728	0.5403	0.6195	0.7144	0.8303
	31	0.3935	0.4513	0.5179	0.5961	0.6898	0.8043
	32	0.3742	0.4311	0.4968	0.5741	0.6666	0.7799
	33	0.3559	0.4121	0.4769	0.5532	0.6448	0.7569
	34	0.3388	0.3941	0.4581	0.5335	0.6241	0.7351
	35	0.3225	0.3771	0.4402	0.5148	0.6044	0.7144
	36	0.3071	0.3609	0.4233	0.4969	0.5856	0.6947
	37	0.2925	0.3455	0.4071	0.4799	0.5677	0.6759
	38	0.2787	0.3308	0.3916	0.4636	0.5506	0.6579
	39	0.2654	0.3168	0.3768	0.4480	0.5342	0.6407
	40	0.2529	0.3034	0.3626	0.4331	0.5185	0.6242
	41	0.2408	0.2906	0.3490	0.4187	0.5033	0.6083
	42	0.2294	0.2784	0.3360	0.4049	0.4888	0.5930

13.11 Graphic Solution for Coulomb's Active Earth Pressure

An expedient method for creating a graphic solution of Coulomb's earth-pressure theory was given by Culmann (1875). Culmann's solution can be used for any wall friction, regardless of irregularity of backfill and surcharges. Hence, it provides a powerful technique for estimating lateral earth pressure. The steps in Culmann's solution of active pressure with granular backfill ($c' = 0$) are described next, with reference to Figure 13.23a:

Step 1: Draw the features of the retaining wall and the backfill to a convenient scale.

Step 2: Determine the value of ψ (degrees) $= 90 - \theta - \delta'$, where $\theta =$ the inclination of the back face of the retaining wall with the vertical, and $\delta' =$ angle of wall friction.

Table 13.8 Values of K_a [Eq. (13.69)] (*Note:* $\delta' = \phi'/2$)

α (deg)	ϕ' (deg)	θ (deg)					
		0	**5**	**10**	**15**	**20**	**25**
0	28	0.3264	0.3629	0.4034	0.4490	0.5011	0.5616
	29	0.3137	0.3502	0.3907	0.4363	0.4886	0.5492
	30	0.3014	0.3379	0.3784	0.4241	0.4764	0.5371
	31	0.2896	0.3260	0.3665	0.4121	0.4645	0.5253
	32	0.2782	0.3145	0.3549	0.4005	0.4529	0.5137
	33	0.2671	0.3033	0.3436	0.3892	0.4415	0.5025
	34	0.2564	0.2925	0.3327	0.3782	0.4305	0.4915
	35	0.2461	0.2820	0.3221	0.3675	0.4197	0.4807
	36	0.2362	0.2718	0.3118	0.3571	0.4092	0.4702
	37	0.2265	0.2620	0.3017	0.3469	0.3990	0.4599
	38	0.2172	0.2524	0.2920	0.3370	0.3890	0.4498
	39	0.2081	0.2431	0.2825	0.3273	0.3792	0.4400
	40	0.1994	0.2341	0.2732	0.3179	0.3696	0.4304
	41	0.1909	0.2253	0.2642	0.3087	0.3602	0.4209
	42	0.1828	0.2168	0.2554	0.2997	0.3511	0.4117
5	28	0.3477	0.3879	0.4327	0.4837	0.5425	0.6115
	29	0.3337	0.3737	0.4185	0.4694	0.5282	0.5972
	30	0.3202	0.3601	0.4048	0.4556	0.5144	0.5833
	31	0.3072	0.3470	0.3915	0.4422	0.5009	0.5698
	32	0.2946	0.3342	0.3787	0.4292	0.4878	0.5566
	33	0.2825	0.3219	0.3662	0.4166	0.4750	0.5437
	34	0.2709	0.3101	0.3541	0.4043	0.4626	0.5312
	35	0.2596	0.2986	0.3424	0.3924	0.4505	0.5190
	36	0.2488	0.2874	0.3310	0.3808	0.4387	0.5070
	37	0.2383	0.2767	0.3199	0.3695	0.4272	0.4954
	38	0.2282	0.2662	0.3092	0.3585	0.4160	0.4840
	39	0.2185	0.2561	0.2988	0.3478	0.4050	0.4729
	40	0.2090	0.2463	0.2887	0.3374	0.3944	0.4620
	41	0.1999	0.2368	0.2788	0.3273	0.3840	0.4514
	42	0.1911	0.2276	0.2693	0.3174	0.3738	0.4410
10	28	0.3743	0.4187	0.4688	0.5261	0.5928	0.6719
	29	0.3584	0.4026	0.4525	0.5096	0.5761	0.6549
	30	0.3432	0.3872	0.4368	0.4936	0.5599	0.6385
	31	0.3286	0.3723	0.4217	0.4782	0.5442	0.6225
	32	0.3145	0.3580	0.4071	0.4633	0.5290	0.6071
	33	0.3011	0.3442	0.3930	0.4489	0.5143	0.5920
	34	0.2881	0.3309	0.3793	0.4350	0.5000	0.5775
	35	0.2757	0.3181	0.3662	0.4215	0.4862	0.5633
	36	0.2637	0.3058	0.3534	0.4084	0.4727	0.5495
	37	0.2522	0.2938	0.3411	0.3957	0.4597	0.5361
	38	0.2412	0.2823	0.3292	0.3833	0.4470	0.5230
	39	0.2305	0.2712	0.3176	0.3714	0.4346	0.5103
	40	0.2202	0.2604	0.3064	0.3597	0.4226	0.4979
	41	0.2103	0.2500	0.2956	0.3484	0.4109	0.4858
	42	0.2007	0.2400	0.2850	0.3375	0.3995	0.4740

Table 13.8 (*continued*)

α (deg)	φ' (deg)	θ (deg) 0	5	10	15	20	25
15	28	0.4095	0.4594	0.5159	0.5812	0.6579	0.7498
	29	0.3908	0.4402	0.4964	0.5611	0.6373	0.7284
	30	0.3730	0.4220	0.4777	0.5419	0.6175	0.7080
	31	0.3560	0.4046	0.4598	0.5235	0.5985	0.6884
	32	0.3398	0.3880	0.4427	0.5059	0.5803	0.6695
	33	0.3244	0.3721	0.4262	0.4889	0.5627	0.6513
	34	0.3097	0.3568	0.4105	0.4726	0.5458	0.6338
	35	0.2956	0.3422	0.3953	0.4569	0.5295	0.6168
	36	0.2821	0.3282	0.3807	0.4417	0.5138	0.6004
	37	0.2692	0.3147	0.3667	0.4271	0.4985	0.5846
	38	0.2569	0.3017	0.3531	0.4130	0.4838	0.5692
	39	0.2450	0.2893	0.3401	0.3993	0.4695	0.5543
	40	0.2336	0.2773	0.3275	0.3861	0.4557	0.5399
	41	0.2227	0.2657	0.3153	0.3733	0.4423	0.5258
	42	0.2122	0.2546	0.3035	0.3609	0.4293	0.5122
20	28	0.4614	0.5188	0.5844	0.6608	0.7514	0.8613
	29	0.4374	0.4940	0.5586	0.6339	0.7232	0.8313
	30	0.4150	0.4708	0.5345	0.6087	0.6968	0.8034
	31	0.3941	0.4491	0.5119	0.5851	0.6720	0.7772
	32	0.3744	0.4286	0.4906	0.5628	0.6486	0.7524
	33	0.3559	0.4093	0.4704	0.5417	0.6264	0.7289
	34	0.3384	0.3910	0.4513	0.5216	0.6052	0.7066
	35	0.3218	0.3736	0.4331	0.5025	0.5851	0.6853
	36	0.3061	0.3571	0.4157	0.4842	0.5658	0.6649
	37	0.2911	0.3413	0.3991	0.4668	0.5474	0.6453
	38	0.2769	0.3263	0.3833	0.4500	0.5297	0.6266
	39	0.2633	0.3120	0.3681	0.4340	0.5127	0.6085
	40	0.2504	0.2982	0.3535	0.4185	0.4963	0.5912
	41	0.2381	0.2851	0.3395	0.4037	0.4805	0.5744
	42	0.2263	0.2725	0.3261	0.3894	0.4653	0.5582

Step 3: Draw a line BD that makes an angle ϕ' with the horizontal.
Step 4: Draw a line BE that makes an angle ψ with line BD.
Step 5: To consider some trial failure wedges, draw lines $BC_1, BC_2, BC_3, \ldots, BC_n$.
Step 6: Find the areas of $ABC_1, ABC_2, ABC_3, \ldots, ABC_n$.
Step 7: Determine the weight of soil, W, per unit length of the retaining wall in each of the trial failure wedges as follows:

$$W_1 = (\text{Area of } ABC_1) \times (\gamma) \times (1)$$
$$W_2 = (\text{Area of } ABC_2) \times (\gamma) \times (1)$$
$$W_3 = (\text{Area of } ABC_3) \times (\gamma) \times (1)$$
$$\vdots$$
$$W_n = (\text{Area of } ABC_n) \times (\gamma) \times (1)$$

Step 8: Adopt a convenient load scale and plot the weights W_1, W_2, W_3, . . . , W_n determined from step 7 on line BD. (*Note*: $Bc_1 = W_1$, $Bc_2 = W_2$, $Bc_3 = W_3$, . . . , $Bc_n = W_n$.)

Step 9: Draw c_1c_1', c_2c_2', c_3c_3', . . . , c_nc_n' parallel to the line BE. (*Note*: c_1', c_2', c_3', . . . , c_n' are located on lines BC_1, BC_2, BC_3, . . . , BC_n, respectively.)

Step 10: Draw a smooth curve through points c_1', c_2', c_3', . . . , c_n'. This curve is called the *Culmann line*.

Step 11: Draw a tangent $B'D'$ to the smooth curve drawn in Step 10. $B'D'$ is parallel to line BD. Let c_a' be the point of tangency.

Step 12: Draw a line c_ac_a' parallel to the line BE.

Step 13: Determine the active force per unit length of wall as
$$P_a = (\text{Length of } c_ac_a') \times (\text{Load scale})$$

Step 14: Draw a line $Bc_a'C_a$. ABC_a is the desired failure wedge.

Note that the construction procedure entails, in essence, drawing a number of force polygons for a number of trial wedges and finding the maximum value of the active force that the wall can be subjected to. For example, Figure 13.23b shows the force polygon for the failure wedge ABC_a (similar to that in Figure 13.22b), in which

W = weight of the failure wedge of soil ABC_a
P_a = active force on the wall
F = the resultant of the shear and normal forces acting along BC_a
$\beta = \angle C_aBF$ (the angle that the failure wedge makes with the horizontal)

The force triangle (Figure 13.23b) is simply rotated in Figure 13.23a and is represented by the triangle Bc_ac_a'. Similarly, the force triangles Bc_1c_1', Bc_2c_2', Bc_3c_3', . . . , Bc_nc_n' correspond to the trial wedges ABC_1, ABC_2, ABC_3, . . . , ABC_n.

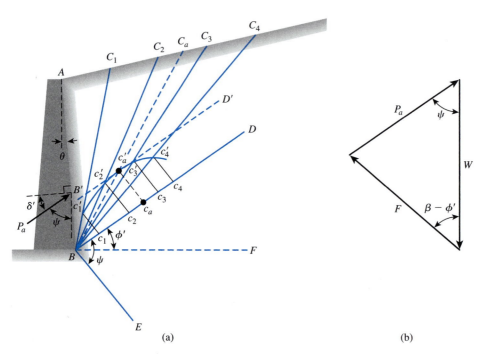

Figure 13.23
Culmann's solution for active earth pressure

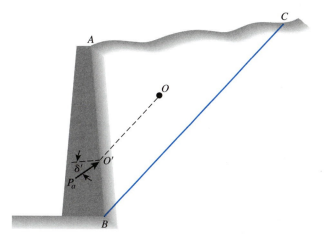

Figure 13.24
Approximate method for finding the point of application of the resultant active force

The preceding graphic procedure is given in a step-by-step manner only to facilitate basic understanding. These problems can be easily and effectively solved by the use of computer programs.

The Culmann solution provides us with only the magnitude of the active force per unit length of the retaining wall—not with the point of application of the resultant. The analytic procedure used to find the point of application of the resultant can be tedious. For this reason, an approximate method, which does not sacrifice much accuracy, can be used. This method is demonstrated in Figure 13.24, in which ABC is the failure wedge determined by Culmann's method. O is the center of gravity of the wedge ABC. If a line OO' is drawn parallel to the surface of sliding, BC, the point of intersection of this line with the back face of the wall will give the point of application of P_a. Thus, P_a acts at O' inclined at angle δ' with the normal drawn to the back face of the wall.

Example 13.8

A 4.57-m-high retaining wall with a granular soil backfill is shown in Figure 13.25. Given that $\gamma = 15.72$ kN/m^3, $\phi' = 35°$, $\theta = 5°$, and $\delta' = 10°$, determine the active thrust per foot length of the wall.

Solution
For this problem, $\psi = 90 - \theta - \delta' = 90° - 5° - 10° = 75°$. The graphic construction is shown in Figure 13.25. The weights of the wedges considered are as follows.

Wedge	Weight (kN/m)
ABC_1	$\frac{1}{2}(1.33)(5.45)(15.72) = 56.97$ kN/m
ABC_2	$56.97 + [\frac{1}{2}(0.72)(5.66)](15.72) = 89.10$ kN/m
ABC_3	$89.10 + [\frac{1}{2}(0.68)(5.96)](15.72) = 120.86$ kN/m
ABC_4	$120.86 + [\frac{1}{2}(0.64)(6.33)](15.72) = 152.70$ kN/m
ABC_5	$152.70 + [\frac{1}{2}(0.6)(6.77)](15.72) = 184.63$ kN/m

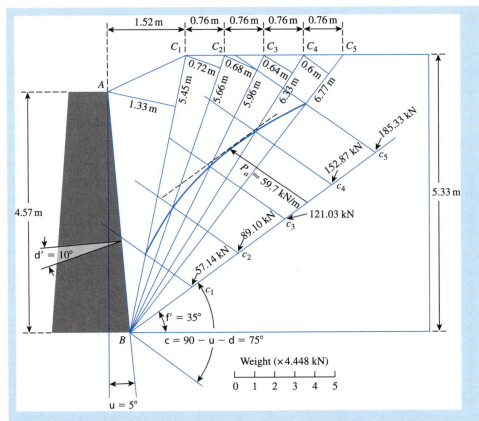

Figure 13.25 Culmann's solution for determining active thrust per unit length of wall

In Figure 13.25,

$$\overline{Bc_1} = 56.97 \text{ kN}$$

$$\overline{Bc_2} = 89.10 \text{ kN}$$

$$\overline{Bc_3} = 120.86 \text{ kN}$$

$$\overline{Bc_4} = 152.70 \text{ kN}$$

$$\overline{Bc_5} = 184.63 \text{ kN}$$

The active thrust per unit length of the wall is **59.7 kN/m.** ∎

13.12   *Coulomb's Passive Pressure*

Figure 13.26a shows a retaining wall with a sloping cohensionless backfill similar to that considered in Figure 13.22a. The force polygon for equilibrium of the wedge *ABC* for the passive state is shown in Figure 13.26b. P_p is the notation for the passive force. Other

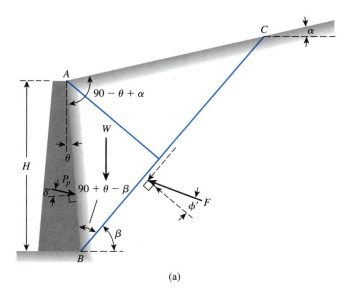

(a)

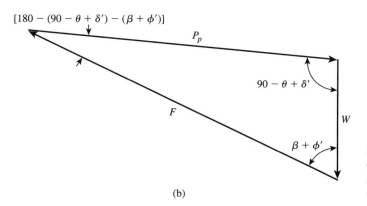

(b)

Figure 13.26
Coulomb's passive pressure: (a) trial failure wedge; (b) force polygon

notations used are the same as those for the active case (Section 13.10). In a procedure similar to the one that we followed in the active case [Eq. (13.68)], we get

$$P_p = \tfrac{1}{2} K_p \gamma H^2 \tag{13.70}$$

where K_p = Coulomb's passive earth-pressure coefficient, or

$$K_p = \frac{\cos^2(\phi' + \theta)}{\cos^2 \theta \cos(\delta' - \theta)\left[1 - \sqrt{\dfrac{\sin(\phi' + \delta')\sin(\phi' + \alpha)}{\cos(\delta' - \theta)\cos(\alpha - \theta)}}\right]^2} \tag{13.71}$$

Table 13.9 Values of K_p [Eq. 13.71] for $\theta = 0°$, $\alpha = 0°$

↓ ϕ' (deg)	δ' (deg) →				
	0	5	10	15	20
15	1.698	1.900	2.130	2.405	2.735
20	2.040	2.313	2.636	3.030	3.525
25	2.464	2.830	3.286	3.855	4.597
30	3.000	3.506	4.143	4.977	6.105
35	3.690	4.390	5.310	6.854	8.324
40	4.600	5.590	6.946	8.870	11.772

For a frictionless wall with the vertical back face supporting granular soil backfill with a horizontal surface (that is, $\theta = 0°$, $\alpha = 0°$, and $\delta' = 0°$), Eq. (13.71) yields

$$K_p = \frac{1 + \sin \phi'}{1 - \sin \phi'} = \tan^2 \left(45 + \frac{\phi'}{2} \right)$$

This relationship is the same as that obtained for the passive earth-pressure coefficient in Rankine's case, given by Eq. (13.22).

The variation of K_p with ϕ' and δ' (for $\theta = 0°$ and $\alpha = 0°$) is given in Table 13.9. We can see from this table that for given value of ϕ', the value of K_p increases with the wall friction.

13.13 Active Force on Retaining Walls with Earthquake Forces

Active Case (Granular Backfill)

Coulomb's analysis for active force on retaining walls discussed in Section 13.10 can be conveniently extended to include earthquake forces. To do so, let us consider a retaining wall of height H with a sloping *granular backfill,* as shown in Figure 13.27a. Let the unit weight and the friction angle of the granular soil retained by the wall be equal to γ and ϕ', respectively. Also, let δ' be the angle of friction between the soil and the wall. *ABC* is a trial failure wedge. The forces acting on the wedge are as follows:

1. Weight of the soil in the wedge, W
2. Resultant of the shear and normal forces on the failure surface *BC, F*
3. Active force per unit length of the wall, P_{ae}
4. Horizontal inertial force, $k_h W$
5. Vertical inertial force, $k_v W$

Note that

$$k_h = \frac{\text{Horizontal component of earthquake acceleration}}{g} \tag{13.72}$$

$$k_v = \frac{\text{Vertical component of earthquake acceleration}}{g} \tag{13.73}$$

where g = acceleration due to gravity.

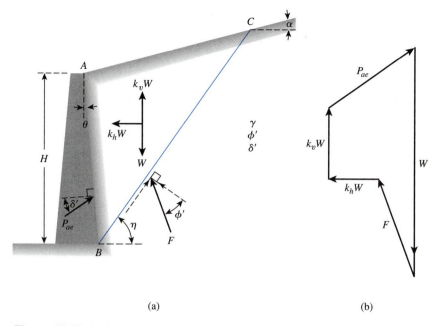

Figure 13.27 Active force on a retaining wall with earthquake forces

The force polygon demonstrating these forces is shown in Figure 13.27b. The dynamic active force on the wall is the maximum value of P_{ae} exerted by any wedge. This value can be expressed as

$$P_{ae} = \tfrac{1}{2}\gamma H^2(1 - k_v)K_a''$$ (13.74)

where

$$K_a'' = \frac{\cos^2(\phi' - \theta - \bar{\beta})}{\cos^2\theta\cos\bar{\beta}\cos(\delta' + \theta + \bar{\beta})\left\{1 + \left[\dfrac{\sin(\delta' + \phi')\sin(\phi' - \alpha - \bar{\beta})}{\cos(\delta' + \theta + \bar{\beta})\cos(\theta - \alpha)}\right]^{1/2}\right\}^2}$$

(13.75)

and

$$\bar{\beta} = \tan^{-1}\left(\frac{k_h}{1 - k_v}\right)$$ (13.76)

Note that with no inertia forces from earthquakes, $\bar{\beta}$ is equal to 0. Hence, $K_a'' = K_a$ as given in Eq. (13.69). Equations (13.74) and (13.75) generally are referred to as the *Mononobe-Okabe equations* (Mononobe, 1929; Okabe, 1926). The variation of K_a'' with $\theta = 0°$ and $k_v = 0$ is given in Table 13.10.

Table 13.10 Values of K_a'' [Eq. (13.75)] with $\theta = 0°$ and $k_v = 0$

k_h	δ' (deg)	α (deg)	28	30	35	40	45
					ϕ' (deg)		
0.1	0	0	0.427	0.397	0.328	0.268	0.217
0.2			0.508	0.473	0.396	0.382	0.270
0.3			0.611	0.569	0.478	0.400	0.334
0.4			0.753	0.697	0.581	0.488	0.409
0.5			1.005	0.890	0.716	0.596	0.500
0.1	0	5	0.457	0.423	0.347	0.282	0.227
0.2			0.554	0.514	0.424	0.349	0.285
0.3			0.690	0.635	0.522	0.431	0.356
0.4			0.942	0.825	0.653	0.535	0.442
0.5			—	—	0.855	0.673	0.551
0.1	0	10	0.497	0.457	0.371	0.299	0.238
0.2			0.623	0.570	0.461	0.375	0.303
0.3			0.856	0.748	0.585	0.472	0.383
0.4			—	—	0.780	0.604	0.486
0.5			—	—	—	0.809	0.624
0.1	$\phi'/2$	0	0.396	0.368	0.306	0.253	0.207
0.2			0.485	0.452	0.380	0.319	0.267
0.3			0.604	0.563	0.474	0.402	0.340
0.4			0.778	0.718	0.599	0.508	0.433
0.5			1.115	0.972	0.774	0.648	0.552
0.1	$\phi'/2$	5	0.428	0.396	0.326	0.268	0.218
0.2			0.537	0.497	0.412	0.342	0.283
0.3			0.699	0.640	0.526	0.438	0.367
0.4			1.025	0.881	0.690	0.568	0.475
0.5			—	—	0.962	0.752	0.620
0.1	$\phi'/2$	10	0.472	0.433	0.352	0.285	0.230
0.2			0.616	0.562	0.454	0.371	0.303
0.3			0.908	0.780	0.602	0.487	0.400
0.4			—	—	0.857	0.656	0.531
0.5			—	—	—	0.944	0.722
0.1	$\frac{2}{3}\phi'$	0	0.393	0.366	0.306	0.256	0.212
0.2			0.486	0.454	0.384	0.326	0.276
0.3			0.612	0.572	0.486	0.416	0.357
0.4			0.801	0.740	0.622	0.533	0.462
0.5			1.177	1.023	0.819	0.693	0.600
0.1	$\frac{2}{3}\phi'$	5	0.427	0.395	0.327	0.271	0.224
0.2			0.541	0.501	0.418	0.350	0.294
0.3			0.714	0.655	0.541	0.455	0.386
0.4			1.073	0.921	0.722	0.600	0.509
0.5			—	—	1.034	0.812	0.679
0.1	$\frac{2}{3}\phi'$	10	0.472	0.434	0.354	0.290	0.237
0.2			0.625	0.570	0.463	0.381	0.317
0.3			0.942	0.807	0.624	0.509	0.423
0.4			—	—	0.909	0.699	0.573
0.5			—	—	—	1.037	0.800

Seed and Whitman (1970) provided a simple procedure to obtain the value of K_a'' from the standard charts of K_a [see Eq. (13.69)]. This procedure is explained next. Referring to Eqs. (13.68) and (13.69), we can write

$$P_a = \frac{1}{2}\gamma H^2 K_a = \left(\frac{1}{2}\gamma H^2\right)\left(\frac{1}{\cos^2\theta}\right)(A_c) \tag{13.77}$$

where

$$A_c = K_a \cos^2\theta = \frac{\cos^2(\phi' - \theta)}{\cos^2(\delta' + \theta)\left[1 + \sqrt{\dfrac{\sin(\delta' + \phi')\sin(\phi' - \alpha)}{\cos(\delta' + \theta)\cos(\theta - \alpha)}}\right]^2} \tag{13.78}$$

Now, referring to Eqs. (13.74) and (13.75), we can write

$$P_{ae} = \frac{1}{2}\gamma H^2(1 - k_v)K_a'' = \left(\frac{1}{2}\gamma H^2\right)(1 - k_v)\left(\frac{1}{\cos^2\theta\cos\bar{\beta}}\right)(A_m) \tag{13.79}$$

where

$$
\begin{aligned}
A_m &= K_a''\cos^2\theta\cos\bar{\beta} \\
&= \frac{\cos^2(\phi' - \theta - \bar{\beta})}{\cos(\delta' + \theta + \bar{\beta})\left[1 + \sqrt{\dfrac{\sin(\delta' + \phi')\sin(\phi' - \alpha - \bar{\beta})}{\cos(\delta' + \theta + \bar{\beta})\cos(\theta - \alpha)}}\right]^2}
\end{aligned}
\tag{13.80}
$$

Now, let us define

$$\theta^* = \theta + \bar{\beta} \tag{13.81}$$

and

$$\alpha^* = \alpha + \bar{\beta} \tag{13.82}$$

Substituting Eqs. (13.81) and (13.82) into the relationship for A_m [that is, Eq. (13.80)], we obtain

$$A_m = \frac{\cos^2(\phi' - \phi^*)}{\cos(\delta' + \theta^*)\left[1 + \sqrt{\dfrac{\sin(\delta' + \phi')\sin(\phi' - \alpha^*)}{\cos(\delta' + \theta^*)\cos(\theta^* - \alpha^*)}}\right]^2} \tag{13.83}$$

Comparing Eqs. (13.78) and (13.83), we have

$$A_m = A_c(\theta^*, \alpha^*) = K_a(\theta^*, \alpha^*)\cos^2\theta^* \tag{13.84}$$

Hence,

$$P_{ae} = \frac{1}{2}\gamma H^2 (1 - k_v) \frac{1}{\cos^2\theta \cos\bar{\beta}} K_a(\theta^*, \alpha^*) \cos^2\theta^*$$

or

$$P_{ae} = [P_a(\theta^*, \alpha^*)](1 - k_v)\left(\frac{\cos^2\theta^*}{\cos^2\theta \cos\bar{\beta}}\right)$$

$$= [P_a(\theta^*, \alpha^*)](1 - k_v)\left[\frac{\cos^2(\theta + \bar{\beta})}{\cos^2\theta \cos\bar{\beta}}\right] \tag{13.84}$$

The term $P_a(\theta^*, \alpha^*)$ in Eq. (13.85) is the active earth pressure on an imaginary retaining wall with a wall inclination of θ^* and backfill slope inclination of α^*. The value of K_a can be obtained from standard charts or tables, such as Tables 13.7 and 13.8.

Considering the active force relation given by Eqs. (13.74) through (13.76), we find that the term $\sin(\phi' - \alpha - \bar{\beta})$ in Eq. (13.75) has two important implications. First, if $\phi' - \alpha - \bar{\beta} < 0$ (i.e., negative), no real solution of K_a'' is possible. Physically, this implies that an *equilibrium condition will not exist*. Hence, for stability, the limiting slope of the backfill may be given as

$$\alpha \le \phi' - \bar{\beta} \tag{13.86}$$

For no earthquake condition, $\bar{\beta} = 0°$; for stability, Eq. (13.86) gives the familiar relation

$$\alpha \le \phi' \tag{13.87}$$

Second, for horizontal backfill, $\alpha = 0°$; for stability,

$$\bar{\beta} \le \phi' \tag{13.88}$$

Because $\bar{\beta} = \tan^{-1}[k_h/(1 - k_v)]$, for stability, combining Eqs. (13.76) and (13.88) results in

$$k_h \le (1 - k_v)\tan\phi' \tag{13.89}$$

Hence, the critical value of the horizontal acceleration can be defined as

$$k_{h(cr)} = (1 - k_v)\tan\phi' \tag{13.90}$$

where $k_{h(cr)}$ = critical of horizontal acceleration (Figure 13.28).

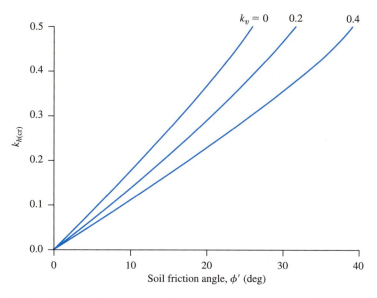

Figure 13.28 Critical values of horizontal acceleration (Eq. 13.90)

Location of Line of Action of Resultant Force, P_{ae}

Seed and Whitman (1970) proposed a simple procedure to determine the location of the line of action of the resultant, P_{ae}. Their method is as follows:

1. Let

$$P_{ae} = P_a + \Delta P_{ae} \tag{13.91}$$

   where P_a = Coulomb's active force as determined from Eq. (13.68)
   ΔP_{ae} = additional active force caused by the earthquake effect

2. Calculate P_a [Eq. (13.68)].
3. Calculate P_{ae} [Eq. (13.74)].
4. Calculate $\Delta P_{ae} = P_{ae} - P_a$.
5. According to Figure 13.29, P_a will act at a distance of $H/3$ from the base of the wall. Also, ΔP_{ae} will act at a distance of $0.6H$ from the base of the wall.
6. Calculate the location of P_{ae} as

$$\bar{z} = \frac{P_a\left(\dfrac{H}{3}\right) + \Delta P_{ae}(0.6H)}{P_{ae}} \tag{13.92}$$

   where $\bar{z}$ = distance of the line of action of P_{ae} from the base of the wall.

Note that the line of action of P_{ae} will be inclined at an angle of δ' to the normal drawn to the back face of the retaining wall. It is very important to realize that this method of determining P_{ae} is approximate and does not actually model the soil dynamics.

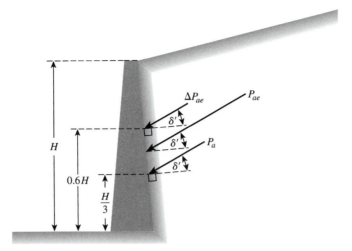

Figure 13.29
Location of the line of
action of P_{ae}

Active Case (c'–φ' Backfill)

The Mononobe–Okabe equation for estimating P_{ae} for cohesionless backfill also can be extended to $c'–\phi'$ soil (Prakash and Saran, 1966; Saran and Prakash, 1968). Figure 13.30 shows a retaining wall of height H with a horizontal $c'–\phi'$ backfill. The depth of tensile crack that may develop in a $c'–\phi'$ soil was given in Eq. (13.48) as

$$z_o = \frac{2c'}{\gamma\sqrt{K_a}}$$

where $K_a = \tan^2(45 - \phi'/2)$.

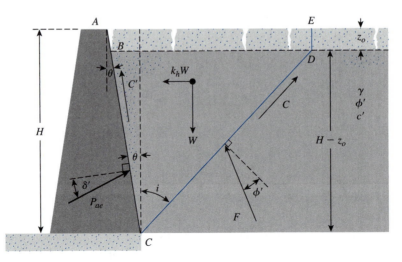

Figure 13.30 Trial failure wedge behind a retaining wall with a $c'–\phi'$ backfill

Referring to Figure 13.30, the forces acting on the soil wedge (per unit length of the wall) are as follows:

- The weight of the wedge *ABCDE, W*
- Resultant of the shear and normal forces on the failure surface *CD, F*
- Active force, P_{ae}
- Horizontal inertia force, $k_h W$
- Cohesive force along *CD, $C = c'(\overline{CD})$*
- Adhesive force along *BC, $C' = c'(\overline{BC})$*

It is important to realize that the following two assumptions have been made:

1. The vertical inertia force ($k_v W$) has been taken to be zero.
2. The *unit adhesion* along the soil–wall interface (*BC*) has been taken to be equal to the cohesion (c') of the soil.

Considering these forces, we can show that

$$P_{ae} = \gamma(H - z_o)^2 N'_{a\gamma} - c'(H - z_o)N'_{ac} \tag{13.93}$$

where

$$N'_{ac} = \frac{\cos \eta' \sec \theta + \cos \phi' \sec i}{\sin(\eta' + \delta')} \tag{13.94}$$

$$N'_{a\gamma} = \frac{[(n + 0.5)(\tan \theta + \tan i) + n^2 \tan \theta][\cos(i + \phi') + k_h \sin(i + \phi')]}{\sin(\eta' + \delta')} \tag{13.95}$$

in which

$$\eta' = \theta + i + \phi' \tag{13.96}$$

$$n = \frac{z_o}{H - z_o} \tag{13.97}$$

The values of N'_{ac} and $N'_{a\gamma}$ can be determined by optimizing each coefficient separately. Thus, Eq. (13.93) gives the upper bound of P_{ae}.

For the static condition, $k_h = 0$. Thus,

$$P_{ae} = \gamma(H - z_o)^2 N_{a\gamma} - c'(H - z_o)N_{ac} \tag{13.98}$$

The relationships for N_{ac} and $N_{a\gamma}$ can be determined by substituting $k_h = 0$ into Eqs. (13.94) and (13.95). Hence,

$$N_{ac} = N'_{ac} = \frac{\cos \eta' \sec \theta + \cos \phi' \sec i}{\sin(\eta' + \delta')} \tag{13.99}$$

$$N_{a\gamma} = \frac{N'_{a\gamma}}{\lambda} = \frac{[(n + 0.5)(\tan \theta + \tan i) + n^2 \tan \theta]\cos(i + \phi')}{\sin(\eta' + \delta')} \tag{13.100}$$

The variations of N_{ac}, $N_{a\gamma}$, and λ with ϕ' and θ are shown in Figures 13.31 through 13.34.

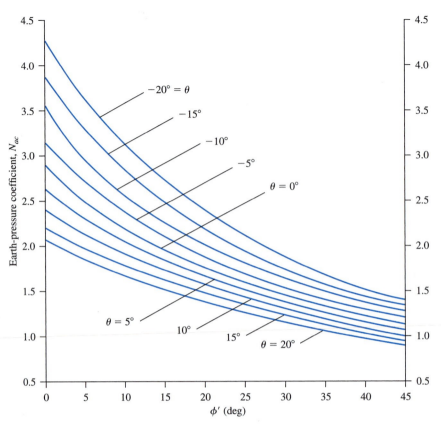

Figure 13.31 Variation of $N_{ac} = N'_{ac}$ with ϕ' and θ (*Based on Prakash and Saran, 1966, and Saran and Prakash, 1968*)

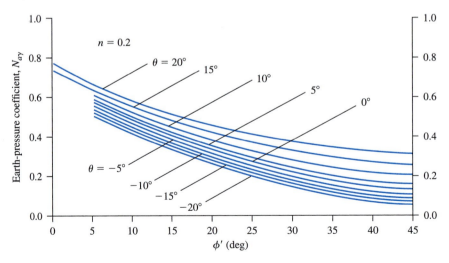

Figure 13.32 Variation of $N_{a\gamma}$ with ϕ' and θ ($n = 0.2$) (*Based on Prakash and Saran, 1966, and Saran and Prakash, 1968*)

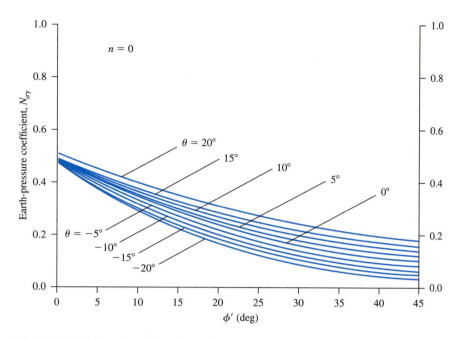

Figure 13.33 Variation of $N_{a\gamma}$ with ϕ' and θ ($n = 0$) (*Based on Prakash and Saran, 1966, and Saran and Prakash, 1968*)

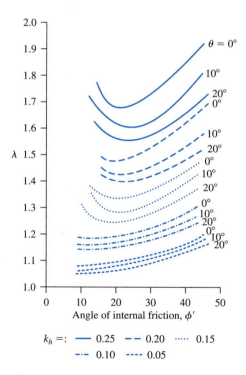

$k_h =:$ —— 0.25 — — 0.20 ⋯⋯ 0.15 —·— 0.10 - - - - 0.05

Figure 13.34 Variation of λ with k_h, ϕ', and θ (*Based on Prakash and Saran, 1966, and Saran and Prakash, 1968*)

Example 13.9

For a retaining wall with a cohesionless soil backfill, $\gamma = 15.5 \text{ kN/m}^3$, $\phi' = 30°$, $\delta' = 15°$, $\theta = 0°$, $\alpha = 0°$, $H = 4$ m, $k_v = 0$, and $k_h = 0.2$. Determine P_{ae}. Also determine the location of the resultant line of action of P_{ae}—that is, $\bar{z}$.

Solution

From Eqs. (13.81) and (13.82),

$$\theta^* = \theta + \bar{\beta}$$

$$\alpha^* = \alpha + \bar{\beta}$$

$$\bar{\beta} = \tan^{-1}\left(\frac{k_h}{1 - k_v}\right) = \tan^{-1}\left(\frac{0.2}{1 - 0}\right) = 11.3°$$

So,

$$\theta^* = 0 + 11.3° = 11.3°$$

$$\alpha^* = 0 + 11.3° = 11.3°$$

From Eq. (13.85),

$$P_{ae} = [P_a(\theta^*, \alpha^*)](1 - k_v)\left(\frac{\cos^2 \theta^*}{\cos^2 \theta \cos^2 \bar{\beta}}\right)$$

$$= [P_a(\theta^*, \alpha^*)](1 - k_v)\left[\frac{\cos^2(\theta + \bar{\beta})}{\cos \theta \cos^2 \bar{\beta}}\right]$$

$$P_a(\theta^*, \alpha) = \tfrac{1}{2}\gamma H^2 K_a$$

where K_a is a function of θ^* and α^*.

Since $\delta'/\phi' = 15/30 = 0.5$, we will use Table 13.8. For $\theta^* = \alpha^* = 11.3°$, the value of $K_a \approx 0.452$. Thus,

$$P_{ae} = \tfrac{1}{2}(15.5)(4)^2(0.452)(1 - 0)\left[\frac{\cos^2(0 + 11.3)}{(\cos 0)(\cos^2 0)}\right] = \textbf{56.05 kN/m}$$

[*Note:* We can also get the same values of P_{ae} using Eq. (13.75) and K'_{ae} from Table 13.10.]
We now locate the resultant line of action. From Eq. (13.68),

$$P_a = \tfrac{1}{2}K_a\gamma H^2$$

For $\phi' = 30°$ and $\delta' = 15°$, $K_a = 0.3014$ (Table 13.6), so

$$P_a = \tfrac{1}{2}(0.3014)(15.5)(4)^2 = 37.37 \text{ kN/m}$$

Hence, $\Delta P_{ae} = 56.05 - 37.37 = 18.68 \text{ kN/m}$. From Eq. (13.92),

$$\bar{z} = \frac{P_a\left(\dfrac{H}{3}\right) + \Delta P_{ae}(0.6H)}{P_{ae}} = \frac{(37.37)\left(\dfrac{4}{3}\right) + (18.68)(2.4)}{56.05} = \textbf{1.69 m} \quad \blacksquare$$

Example 13.10

For a retaining wall, the following are given.

- $H = 8.54$ m
- $c' = 10.06$ kN/m^2
- $\theta = +10°$
- $\gamma = 18.55$ kN/m^3
- $\phi' = 20°$
- $k_h = 0.1$

Determine the magnitude of the active force, P_{ae}.

Solution

From Eq. (13.48),

$$z_o = \frac{2c'}{\gamma\sqrt{K_a}} = \frac{2c}{\gamma \tan\left(45 - \dfrac{\phi'}{2}\right)} = \frac{(2)(10.06)}{(18.55)\tan\left(45 - \dfrac{20}{2}\right)} = 1.55 \text{ m}$$

From Eq. (13.97),

$$n = \frac{z_o}{H - z_o} = \frac{1.55}{8.54 - 1.55} = 0.22 \approx 0.2$$

From Eqs. (13.93), (13.99), and (13.100),

$$P_{ae} = \gamma(H - z_o)^2(\lambda N_{a\gamma}) - c'(H - z_o)N_{ac}$$

For $\theta = 10°$, $\phi' = 20°$, $k_h = 0.1$, and $n \approx 0.2$,

$$N_{ac} = 1.60 \quad \text{(Figure 13.31)}$$

$$N_{a\gamma} = 0.375 \quad \text{(Figure 13.32)}$$

$$\lambda = 1.17 \quad \text{(Figure 13.34)}$$

Thus,

$$P_{ae} = (18.55)(8.54 - 1.55)^2(1.17 \times 0.375) - (10.06)(8.54 - 1.55)(1.60)$$
$$= \mathbf{285.15 \text{ kN/m}}$$

■

13.14 *Common Types of Retaining Walls in the Field*

The preceding sections present the theoretical concepts for estimating the lateral earth pressure for retaining walls. In practice, the common types of retaining walls constructed can be divided into two major categories: rigid retaining walls and mechanically stabilized earth (MSE) walls. The following is a brief overview of the various types of retaining walls constructed in the field.

Rigid Retaining Walls

Under this category, the wall may be subdivided to four categories. They are:

1. Gravity retaining walls
2. Semigravity retaining walls

3. Cantilever retaining walls
4. Counterfort retaining walls

Gravity retaining walls (Figure 13.35a) are constructed with plain concrete or stone masonry. They depend on their own weight and any soil resting on the masonry for stability. This type of construction is not economical for high walls.

In many cases, a small amount of steel may be used for the construction of gravity walls, thereby minimizing the size of wall sections. Such walls generally are referred to as *semigravity walls* (Figure 13.35b).

Cantilever retaining walls (Figure 13.35c) are made of reinforced concrete that consists of a thin stem and a base slab. This type of wall is economical to a height of about 8 m.

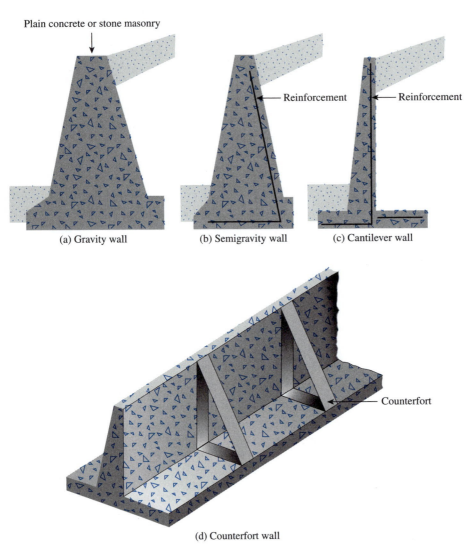

(a) Gravity wall (b) Semigravity wall (c) Cantilever wall

(d) Counterfort wall

Figure 13.35
Types of retaining wall

Counterfort retaining walls (Figure 13.35d) are similar to cantilever walls. At regular intervals, however, they have thin, vertical concrete slabs known as *counterforts* that tie the wall and the base slab together. The purpose of the counterforts is to reduce the shear and the bending moments.

Mechanically Stabilized Earth (MSE) Walls

Mechanically stabilized earth walls are flexible walls, and they are becoming more common nowadays. The main components of these types of walls are

- *Backfill*—which is granular soil
- *Reinforcement* in the backfill
- A *cover* (or *skin*) on the front face

The reinforcement can be thin galvanized steel strips, geogrid, or geotextile (see Chapter 17 for descriptions of geogrid and geotextile). In most cases, precast concrete slabs are used as skin. The slabs are grooved to fit into each other so that soil cannot flow between the joints. Thin galvanized steel also can be used as skin when the reinforcements are metallic strips. When metal skins are used, they are bolted together, and reinforcing strips are placed between the skins.

Figure 13.36 shows an MSE wall with metallic strips as reinforcement along with a metal skin. Figure 13.37 shows some typical MSE walls with geogrid reinforcement in the backfill.

The retaining walls are designed using various earth-pressure theories described in this chapter. For actual wall design, refer to any foundation engineering book (for instance, Das, 2007).

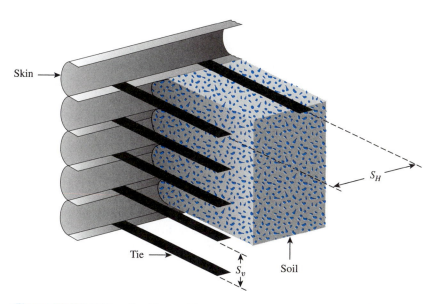

Figure 13.36 *MSE wall with metallic strip reinforcement and metallic skin*

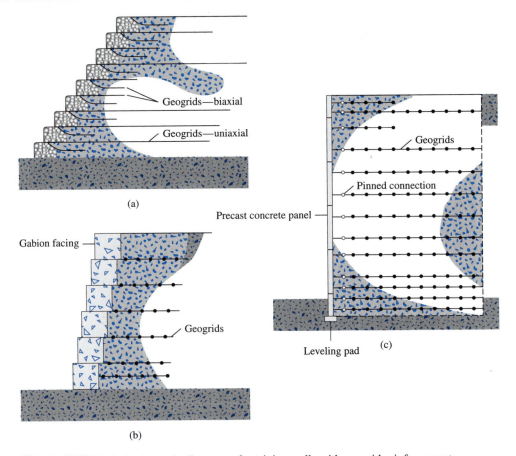

Figure 13.37 Typical schematic diagrams of retaining walls with geogrid reinforcement: (a) geogrid wraparound wall; (b) wall with gabion facing; (c) concrete-panel-faced wall

13.15 Summary and General Comments

This chapter covers the general topics of lateral earth pressure, including the following:

1. At-rest earth pressure
2. Active earth pressure—Rankine's and Coulomb's
3. Passive earth pressure—Rankine's and Coulomb's
4. Active earth pressure, which includes earthquake forces. This is an extension of Coulomb's theory

For design, it is important to realize that the lateral active pressure on a retaining wall can be calculated using Rankine's theory only when the wall moves *sufficiently* outward by rotation about the toe of the footing or by deflection of the wall. If sufficient wall movement cannot occur (or is not allowed to occur), then the lateral earth pressure will be greater than the Rankine active pressure and sometimes may be closer to the at-rest earth pressure. Hence, proper selection of the lateral earth-pressure coefficient is crucial for safe and proper design. It is a general practice to assume a value for the soil friction angle (ϕ') of the backfill in order to calculate the Rankine active pressure distribution, ignoring the contribution of the cohesion (c'). The general range of ϕ' used for the design of retaining walls is given in the following table.

Soil type	Soil friction angle, ϕ' (deg)
Soft clay	0–15
Compacted clay	20–30
Dry sand and gravel	30–40
Silty sand	20–30

In Section 13.8, we saw that the lateral earth pressure on a retaining wall is increased greatly in the presence of a water table above the base of the wall. Most retaining walls are not designed to withstand full hydrostatic pressure; hence, it is important that adequate drainage facilities are provided to ensure that the backfill soil does not become fully saturated. This can be achieved by providing weepholes at regular intervals along the length of the wall.

Problems

13.1 through 13.4   Figure 13.38 shows a retaining wall that is restrained from yielding. For each problem, determine the magnitude of the lateral earth force per unit length of the wall. Also, state the location of the resultant, $\bar{z}$, measured from the bottom of the wall.

Problem	H	ϕ' (deg)	γ	Over-consolidation ratio, OCR
13.1	7 m	38	17 kN/m³	2.5
13.2	4.57 m	32	17.29 kN/m³	2
13.3	6 m	35	18 kN/m³	2
13.4	6.1 m	30	16.51 kN/m³	1

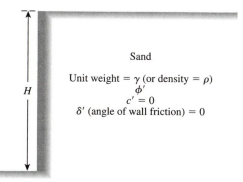

Figure 13.38

13.5 through 13.8   Assume that the retaining wall shown in Figure 13.38 is frictionless. For each problem, determine the Rankine active force per unit length of the wall, the variation of active earth pressure with depth, and the location of the resultant.

Problem	H	ϕ' (deg)	γ
13.5	4.57 m	30	16.51 kN/m³
13.6	5.49 m	32	15.72 kN/m³
13.7	4 m	36	18 kN/m³
13.8	5 m	40	17 kN/m³

13.9 through 13.12   Assume that the retaining wall shown in Figure 13.38 is frictionless. For each problem, determine the Rankine passive force per unit length of the wall, the variation of lateral earth pressure with depth, and the location of the resultant.

Problem	H	ϕ' (deg)	γ
13.9	2.44 m	34	17.29 kN/m³
13.10	3.05 m	36	16.51 kN/m³
13.11	5 m	35	14 kN/m³
13.12	4 m	30	15 kN/m³

13.13 through 13.15   A retaining wall is shown in Figure 13.39. For each problem, determine the Rankine active force, P_a, per unit length of the wall and the location of the resultant.

Problem	H	H_1	γ_1	γ_2	ϕ'_1 (deg)	ϕ'_2 (deg)	q
13.13	3.05 m	1.52 m	16.51 kN/m³	19.18 kN/m³	30	30	0
13.14	6.1 m	1.83 m	17.29 kN/m³	19.81 kN/m³	34	34	14.38 kN/m³
13.15	6 m	3 m	15.5 kN/m³	19.0 kN/m³	30	36	15 kN/m³

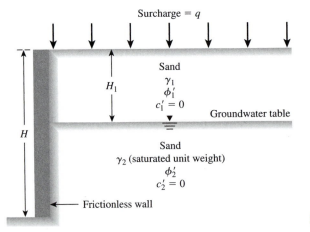

Figure 13.39

13.16 Refer to the frictionless wall shown in Figure 13.10. Given: $H = 5$ m, $\theta = 10°$, and $\alpha = 15°$. For the granular backfill, $\phi' = 35°$ and $\gamma = 15$ kN/m³.
  a. Determine the magnitude of active earth pressure, σ'_a, at the bottom of the wall. Also, state the direction of application of σ'_a.
  b. Determine P_a per meter length of the wall and its location and direction.

13.17 Refer to the retaining wall shown in Figure 13.40. Given: $\gamma = 16.5$ kN/m³, $H = 4$ m, and $\delta' = 0$. Determine the magnitude of the passive force per unit length of the wall, its location, and its direction of the line of action.

13.18 For the retaining wall described in Problem 13.13, determine the Rankine passive force per unit length of the wall and the location of the resultant.

13.19 A 4.57-m-high retaining wall with a vertical back face retains a homogeneous saturated soft clay. The saturated unit weight of the clay is 19.18 kN/m³. Laboratory tests showed that the undrained shear strength, c_u, of the clay is equal to 16.77 kN/m².
  a. Make the necessary calculations and draw the variation of Rankine's active pressure on the wall with depth.
  b. Find the depth up to which a tensile crack can occur.
  c. Determine the total active force per unit length of the wall before the tensile crack occurs.
  d. Determine the total active force per unit length of the wall after the tensile crack occurs. Also, find the location of the resultant.

13.20 Redo Problem 13.19 assuming that the backfill is supported by a surcharge of 9.6 kN/m².

13.21 A 5-m-high retaining wall with a vertical back face has a $c' - \phi'$ soil for backfill. For the backfill, $\gamma = 19$ kN/m³, $c' = 26$ kN/m², and $\phi' = 16°$. Considering the existence of the tensile crack, determine the active force, P_a, on the wall for Rankine's active state.

13.22 Consider a 5-m-high frictionless retaining wall with a vertical back and inclined backfill. The inclination of the backfill with the horizontal, α, is 10°. For the backfill, given: $\gamma = 18$ kN/m³, $\phi' = 25°$, and $c' = 5$ kN/m³. Determine the Rankine active force, P_a, per unit length of the wall after the occurrence of the tensile crack.

13.23 Consider the retaining wall shown in Figure 13.40. The height of the wall is 5 m, and the unit weight of the sand backfill is 18 kN/m³. Using Coulomb's equation, calculate the active force, P_a, on the wall for the following values of the angle of wall friction.
  a. $\delta' = 15°$
  b. $\delta' = 20°$
  Comment on the direction and location of the resultant.

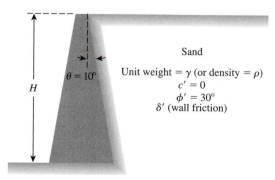

Sand

Unit weight $= \gamma$ (or density $= \rho$)
$c' = 0$
$\phi' = 30°$
δ' (wall friction)

$\theta = 10°$

H

Figure 13.40

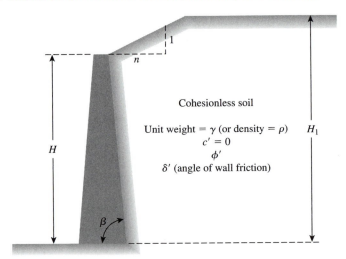

Figure 13.41

13.24 Referring to Figure 13.41, determine Coulomb's active force, P_a, per unit length of the wall for the following cases.
  a. $H = 4.57$ m, $\beta = 85°$, $n = 1$, $H_1 = 6.1$ m, $\gamma = 20.12$ kN/m^3, $\phi' = 38°$, $\delta' = 20°$
  b. $H = 5.5$ m, $\beta = 80°$, $n = 1$, $H_1 = 6.5$ m, $\rho = 1680$ kg/m^3, $\phi' = 30°$, $\delta' = 30°$
  Use Culmann's graphic construction procedure.

13.25 Refer to Figure 13.27. Given: $H = 4$ m, $\theta = 10°$, $\alpha = 10°$, $\gamma = 15$ kN/m^3, $\phi' = 35°$, $\delta' = \frac{2}{3}\phi'$, $k_h = 0.15$, and $k_v = 0$. Determine the active force, P_{ae}, per unit length of the retaining wall.

13.26 Refer to Figure 13.30. Given: $H = 6$ m, $\theta = 10°$, $\phi = 15°$, $c' = 20$ kN/m^2, $\gamma = 19$ kN/m^3, and $k_h = 0.15$. Using the method described in Section 13.13, determine P_{ae}. Assume that the depth of tensile crack is zero.

References

CHU, S. C. (1991). "Rankine Analysis of Active and Passive Pressures on Dry Sand," *Soils and Foundations,* Vol. 31, No. 4, 115–120.

COULOMB, C. A. (1776). "Essai sur une Application des Règles de Maximis et Minimis a quelques Problèmes de Statique, relatifs a l'Architecture," *Mem. Roy. des Sciences,* Paris, Vol. 3, 38.

CULMANN, C. (1875). *Die graphische Statik,* Meyer and Zeller, Zurich.

JAKY, J. (1944). "The Coefficient of Earth Pressure at Rest," *Journal of the Society of Hungarian Architects and Engineers,* Vol. 7, 355–358.

MASSARSCH, K. R. (1979). "Lateral Earth Pressure in Normally Consolidated Clay," *Proceedings of the Seventh European Conference on Soil Mechanics and Foundation Engineering,* Brighton, England, Vol. 2, 245–250.

MAYNE, P. W., and KULHAWY, F. H. (1982). "K_o—OCR Relationships in Soil," *Journal of the Geotechnical Division,* ASCE, Vol. 108, No. 6, 851–872.

MAZINDRANI, Z. H., and GANJALI, M. H. (1997). "Lateral Earth Pressure Problem of Cohesive Backfill with Inclined Surface," *Journal of Geotechnical and Geoenvironmental Engineering,* ASCE, Vol. 123, No. 2, 110–112.

OKABE, S. (1926). "General Theory of Earth Pressure," *Journal of the Japanese Society of Civil Engineers,* Tokyo, Vol. 12, No. 1.

PRAKASH, S., and SARAN, S. (1966). "Static and Dynamic Earth Pressure Behind Retaining Walls," *Proceedings,* 3rd Symposium on Earthquake Engineering, Roorkee, India, Vol. 1, 277–288.

RANKINE, W. M. J. (1857). "On Stability on Loose Earth," *Philosophic Transactions of Royal Society,* London, Part I, 9–27.

SARAN, S., and PRAKASH, S. (1968). "Dimensionless Parameters for Static and Dynamic Earth Pressure for Retaining Walls," *Indian Geotechnical Journal,* Vol. 7, No. 3, 295–310.

SEED, H. B., and WHITMAN, R. V. (1970). "Design of Earth Retaining Structures for Dynamic Loads," *Proceedings,* Specialty Conference on Lateral Stresses in the Ground and Design of Earth Retaining Structures, ASCE, 103–147.

SHERIF, M. A., FANG, Y. S., and SHERIF, R. I. (1984). "K_A and K_O Behind Rotating and Non-Yielding Walls," *Journal of Geotechnical Engineering,* ASCE, Vol. 110, No. GT1, 41–56.

14 Lateral Earth Pressure: Curved Failure Surface

In Chapter 13, we considered Coulomb's earth pressure theory, in which the retaining wall was considered to be rough. The potential failure surfaces in the backfill were considered to be planes. In reality, most failure surfaces in soil are curved. There are several instances where the assumption of plane failure surfaces in soil may provide unsafe results. Examples of these cases are the estimation of passive pressure and braced cuts. This chapter describes procedures by which passive earth pressure and lateral earth pressure on braced cuts can be estimated using curved failure surfaces in the soil.

14.1 Retaining Walls with Friction

In reality, retaining walls are rough, and shear forces develop between the face of the wall and the backfill. To understand the effect of wall friction on the failure surface, let us consider a rough retaining wall AB with a horizontal granular backfill as shown in Figure 14.1.

In the active case (Figure 14.1a), when the wall AB moves to a position $A'B$, the soil mass in the active zone will be stretched outward. This will cause a downward motion of the soil relative to the wall. This motion causes a downward shear on the wall (Figure 14.1b), and it is called a *positive wall friction in the active case*. If δ' is the angle of friction between the wall and the backfill, then the resultant active force P_a will be inclined at an angle δ' to the normal drawn to the back face of the retaining wall. Advanced studies show that the failure surface in the backfill can be represented by BCD, as shown in Figure 14.1a. The portion BC is curved, and the portion CD of the failure surface is a straight line. Rankine's active state exists in the zone ACD.

Under certain conditions, if the wall shown in Figure 14.1a is forced downward with reference to the backfill, the direction of the active force, P_a, will change as shown in Figure 14.1c. This is a situation of negative wall friction $(-\delta')$ in the active case. Figure 14.1c also shows the nature of the failure surface in the backfill.

The effect of wall friction for the passive state is shown in Figures 14.1d and e. When the wall AB is pushed to a position $A'B$ (Figure 14.1d), the soil in the passive zone will be

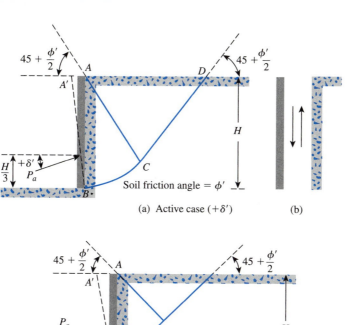

$45 + \dfrac{\phi'}{2}$ A D $45 + \dfrac{\phi'}{2}$

A'

H

$\dfrac{H}{3}$ $+\delta'$

P_a

C

B

Soil friction angle $= \phi'$

(a) Active case $(+\delta')$ (b)

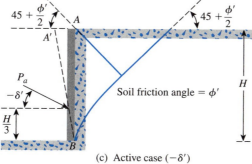

$45 + \dfrac{\phi'}{2}$ A $45 + \dfrac{\phi'}{2}$

A'

P_a

$-\delta'$

$\dfrac{H}{3}$

B

H

Soil friction angle $= \phi'$

(c) Active case $(-\delta')$

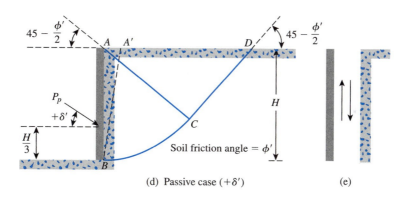

$45 - \dfrac{\phi'}{2}$ A A' D $45 - \dfrac{\phi'}{2}$

P_p

$+\delta'$

$\dfrac{H}{3}$

B

C

H

Soil friction angle $= \phi'$

(d) Passive case $(+\delta')$ (e)

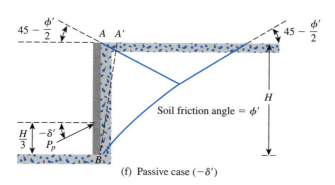

$45 - \dfrac{\phi'}{2}$ A A' $45 - \dfrac{\phi'}{2}$

$\dfrac{H}{3}$ $-\delta'$

P_p

B

H

Soil friction angle $= \phi'$

(f) Passive case $(-\delta')$

Figure 14.1 Effect of wall friction on failure surface

489

compressed. The result is an upward motion relative to the wall. The upward motion of the soil will cause an upward shear on the retaining wall (Figure 14.1e). This is referred to as *positive wall friction in the passive case*. The resultant passive force, P_p, will be inclined at an angle δ' to the normal drawn to the back face of the wall. The failure surface in the soil has a curved lower portion *BC* and a straight upper portion *CD*. Rankine's passive state exists in the zone *ACD*.

If the wall shown in Figure 14.1d is forced upward relative to the backfill by a force, then the direction of the passive force P_p will change as shown in Figure 14.1f. This is *negative wall friction in the passive case* $(-\delta')$. Figure 14.1f also shows the nature of the failure surface in the backfill under such a condition.

For practical considerations, in the case of loose granular backfill, the angle of wall friction δ' is taken to be equal to the angle of friction of soil, ϕ'. For dense granular backfills, δ' is smaller than ϕ' and is in the range of $\phi'/2 \leq \delta' \leq (2/3)\phi'$.

The assumption of plane failure surface gives reasonably good results while calculating active earth pressure. However, the assumption that the failure surface is a plane in Coulomb's theory grossly overestimates the passive resistance of walls, particularly for $\delta' > \phi'/2$.

14.2   *Properties of a Logarithmic Spiral*

The case of passive pressure shown in Figure 14.1d (case of $+\delta'$) is the most common one encountered in design and construction. Also, the curved failure surface represented by *BC* in Figure 14.1d is assumed most commonly to be the arc of a logarithmic spiral. In a similar manner, the failure surface in soil in the case of braced cuts (Sections 14.8 to 14.9) also is assumed to be the arc of a logarithmic spiral. Hence, some useful ideas concerning the properties of a logarithmic spiral are described in this section.

The equation of the logarithmic spiral generally used in solving problems in soil mechanics is of the form

$$r = r_o e^{\theta \tan \phi'} \tag{14.1}$$

where r = radius of the spiral
r_o = starting radius at $\theta = 0$
ϕ' = angle of friction of soil
θ = angle between r and r_o

The basic parameters of a logarithmic spiral are shown in Figure 14.2, in which O is the center of the spiral. The area of the sector *OAB* is given by

$$A = \int_0^\theta \frac{1}{2}r(r\,d\theta) \tag{14.2}$$

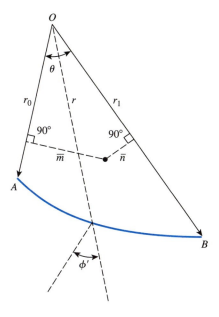

Figure 14.2
General parameters of a logarithmic spiral

Substituting the values of r from Eq. (14.1) into Eq. (14.2), we get

$$A = \int_0^{\theta_1} \frac{1}{2} r_o^2 e^{2\theta \tan \phi'} \, d\theta$$

$$= \frac{r_1^2 - r_o^2}{4 \tan \phi'} \tag{14.3}$$

The location of the centroid can be defined by the distances $\overline{m}$ and $\overline{n}$ (Figure 14.2), measured from OA and OB, respectively, and can be given by the following equations (Hijab, 1956):

$$\overline{m} = \frac{4}{3} r_o \frac{\tan \phi'}{(9 \tan^2 \phi' + 1)} \left[\frac{\left(\dfrac{r_1}{r_o}\right)^3 (3 \tan \phi' \sin \theta - \cos \theta) + 1}{\left(\dfrac{r_1}{r_o}\right)^2 - 1} \right] \tag{14.4}$$

$$\overline{n} = \frac{4}{3} r_o \frac{\tan \phi'}{(9 \tan^2 \phi' + 1)} \left[\frac{\left(\dfrac{r_1}{r_o}\right)^3 - 3 \tan \phi' \sin \theta - \cos \theta}{\left(\dfrac{r_1}{r_o}\right)^2 - 1} \right] \tag{14.5}$$

Another important property of the logarithmic spiral defined by Eq. (14.1) is that any radial line makes an angle ϕ' with the normal to the curve drawn at the point where the radial line and the spiral intersect. This basic property is useful particularly in solving problems related to lateral earth pressure.

PASSIVE EARTH PRESSURE

Procedure for Determination of Passive Earth Pressure (P_p)—Cohesionless Backfill

Figure 14.1d shows the curved failure surface in the granular backfill of a retaining wall of height H. The shear strength of the granular backfill is expressed as

$$\tau_f = \sigma' \tan \phi' \tag{14.6}$$

The curved lower portion BC of the failure wedge is an arc of a logarithmic spiral defined by Eq. (14.1). The center of the log spiral lies on the line CA (not necessarily within the limits of points C and A). The upper portion CD is a straight line that makes an angle of $(45 - \phi'/2)$ degrees with the horizontal. The soil in the zone ACD is in *Rankine's passive state*.

Figure 14.3 shows the procedure for evaluating the passive resistance by trial wedges (Terzaghi and Peck, 1967). The retaining wall is first drawn to scale as shown in Figure 14.3a. The line C_1A is drawn in such a way that it makes an angle of $(45 - \phi'/2)$ degrees with the surface of the backfill. BC_1D_1 is a trial wedge in which BC_1 is the arc of a logarithmic spiral. According to the equation $r_1 = r_o e^{\theta \tan \phi'}$, O_1 is the center of the spiral. (*Note:* $\overline{O_1B} = r_o$ and $\overline{O_1C_1} = r_1$ and $\angle BO_1C_1 = \theta_1$; refer to Figure 14.2.)

Now let us consider the stability of the soil mass ABC_1C_1' (Figure 14.3b). For equilibrium, the following forces per unit length of the wall are to be considered:

1. Weight of the soil in zone $ABC_1C_1' = W_1 = (\gamma)(\text{Area of } ABC_1C_1')(1)$.
2. The vertical face, C_1C_1', is in the zone of Rankine's passive state; hence, the force acting on this face is

$$P_{d(1)} = \frac{1}{2}\gamma(d_1)^2 \tan^2\left(45 + \frac{\phi'}{2}\right) \tag{14.7}$$

where $d_1 = \overline{C_1C_1'}$. $P_{d(1)}$ acts horizontally at a distance of $d_1/3$ measured vertically upward from C_1.
3. F_1 is the resultant of the shear and normal forces that act along the surface of sliding, BC_1. At any point on the curve, according to the property of the logarithmic spiral, a radial line makes an angle ϕ' with the normal. Because the resultant, F_1, makes an angle ϕ' with the normal to the spiral at its point of application, its line of application will coincide with a radial line and will pass through the point O_1.
4. P_1 is the passive force per unit length of the wall. It acts at a distance of $H/3$ measured vertically from the bottom of the wall. The direction of the force P_1 is inclined at an angle δ' with the normal drawn to the back face of the wall.

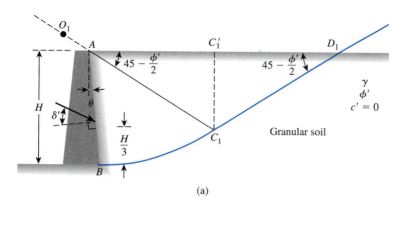

(a)

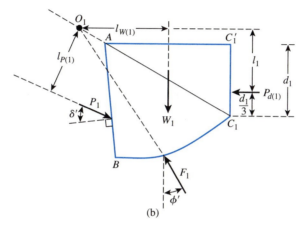

(b)

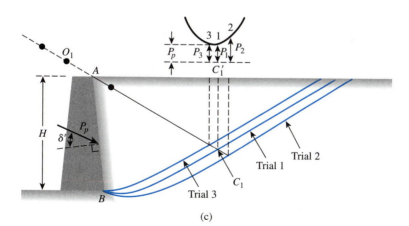

(c)

Figure 14.3 Passive earth pressure against retaining wall with curved failure surface

Now, taking the moments of W_1, $P_{d(1)}$, F_1, and P_1 about the point O_1, for equilibrium, we have

$$W_1[l_{W(1)}] + P_{d(1)}[l_1] + F_1[0] = P_1[l_{P(1)}] \qquad (14.8)$$

or

$$P_1 = \frac{1}{l_{P(1)}}[W_1 l_{W(1)} + P_{d(1)} l_1] \qquad (14.9)$$

where $l_{W(1)}$, l_1, and $l_{P(1)}$ are moment arms for the forces W_1, $P_{d(1)}$, and P_1, respectively.

The preceding procedure for finding the trial passive force per unit length of the wall is repeated for several trial wedges such as those shown in Figure 14.3c. Let P_1, P_2, $P_3, \ldots, P_n$ be the forces that correspond to trial wedges 1, 2, 3, $\ldots$, n, respectively. The forces are plotted to some scale as shown in the upper part of the figure. A smooth curve is plotted through the points 1, 2, 3, $\ldots$, n. The lowest point of the smooth curve defines the actual passive force, P_p, per unit length of the wall.

14.4 Coefficient of Passive Earth Pressure (K_p)

Referring to the retaining wall with a *granular backfill* ($c' = 0$) shown in Fig. 14.3, the passive earth pressure K_p can be expressed as

$$P_p = \frac{1}{2}K_p\gamma H^2 \qquad (14.10)$$

or

$$K_p = \frac{P_p}{0.5\gamma H^2} \qquad (14.11)$$

Following is a summary of results obtained by several investigators.

Procedure of Terzaghi and Peck

Using the procedure of Terzaghi and Peck (1967) described in Section 14.3, the passive earth-pressure coefficient can be evaluated for various combinations of θ, δ', and ϕ'. Figure 14.4 shows the variation of K_p for $\phi' = 30°$ and $40°$ (for $\theta = 0$) with δ'.

Solution by the Method of Slices

Shields and Tolunay (1973) improved the trail wedge solutions described in Section 14.3 using the *method of slices* to consider the stability of the trial soil wedge such as $ABC_1 C'_1$ in Fig. 14.3a. The details of the analysis are beyond the scope of this text. However the values of K_p (passive earth-pressure coefficient) obtained by this method are given in Fig. 14.5. Note that the values of K_p shown in Fig. 14.5 are for retaining walls with a vertical back (that is, $\theta = 0$ in Fig. 14.3) supporting a granular backfill with a horizontal ground surface.

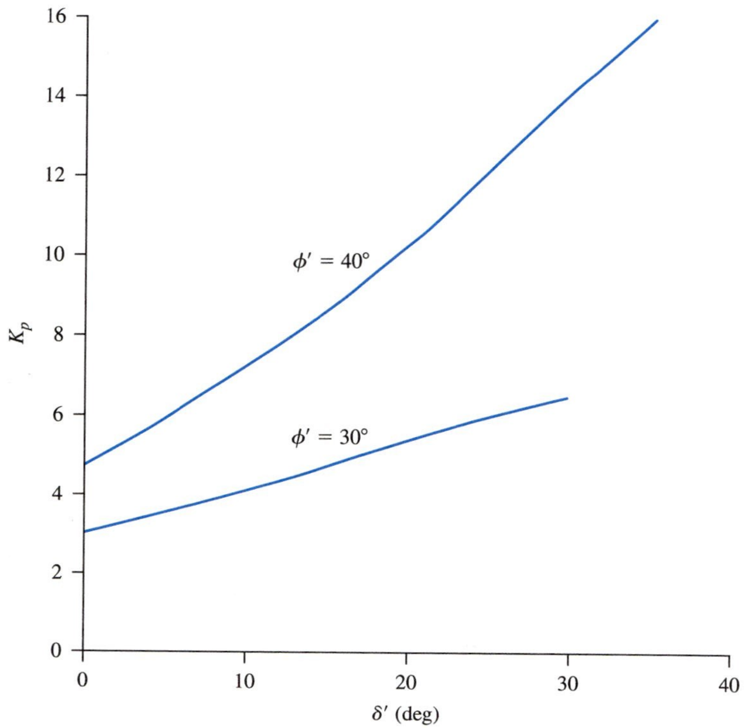

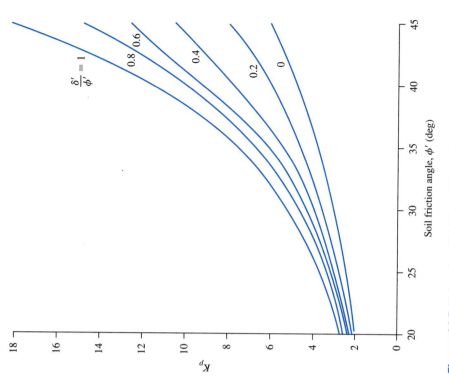

Figure 14.5 K_p based on Shields and Tolunay's analysis (*Note* $\theta = 0$) (*Based on Shields and Tolunay's analysis*)

Figure 14.4 Variation of K_p with ϕ' and δ' based on the procedure of Terzaghi and Peck (1967) (*Note:* $\theta = 0$) (*Based on Terzaghi and Peck, 1967*)

Solution by the Method of Triangular Slices

Zhu and Qian (2000) used the method of triangular slices (such as in the zone of ABC_1 in Fig. 14.3a) to obtain the variation of K_p. According to this analysis

$$K_p = K_{p(\delta' = 0)}R \qquad (14.12)$$

where K_p = passive earth pressure coefficient for a given value of θ, δ', and ϕ'
$K_{p(\delta' = 0)} = K_p$ for a given value of θ, ϕ', with $\delta' = 0$
R = modification factor which is a function of ϕ', θ, δ'/ϕ'

The variations of $K_{p(\delta' = 0)}$ are given in Table 14.1. The interpolated values of R are given in Table 14.2.

Table 14.1 Variation of $K_{p(\delta' = 0)}$ [see Eq. (14.12) and Figure 14.3a]*

ϕ' (deg)	θ (deg)						
	30	25	20	15	10	5	0
20	1.70	1.69	1.72	1.77	1.83	1.92	2.04
21	1.74	1.73	1.76	1.81	1.89	1.99	2.12
22	1.77	1.77	1.80	1.87	1.95	2.06	2.20
23	1.81	1.81	1.85	1.92	2.01	2.13	2.28
24	1.84	1.85	1.90	1.97	2.07	2.21	2.37
25	1.88	1.89	1.95	2.03	2.14	2.28	2.46
26	1.91	1.93	1.99	2.09	2.21	2.36	2.56
27	1.95	1.98	2.05	2.15	2.28	2.45	2.66
28	1.99	2.02	2.10	2.21	2.35	2.54	2.77
29	2.03	2.07	2.15	2.27	2.43	2.63	2.88
30	2.07	2.11	2.21	2.34	2.51	2.73	3.00
31	2.11	2.16	2.27	2.41	2.60	2.83	3.12
32	2.15	2.21	2.33	2.48	2.68	2.93	3.25
33	2.20	2.26	2.39	2.56	2.77	3.04	3.39
34	2.24	2.32	2.45	2.64	2.87	3.16	3.53
35	2.29	2.37	2.52	2.72	2.97	3.28	3.68
36	2.33	2.43	2.59	2.80	3.07	3.41	3.84
37	2.38	2.49	2.66	2.89	3.18	3.55	4.01
38	2.43	2.55	2.73	2.98	3.29	3.69	4.19
39	2.48	2.61	2.81	3.07	3.41	3.84	4.38
40	2.53	2.67	2.89	3.17	3.53	4.00	4.59
41	2.59	2.74	2.97	3.27	3.66	4.16	4.80
42	2.64	2.80	3.05	3.38	3.80	4.34	5.03
43	2.70	2.88	3.14	3.49	3.94	4.52	5.27
44	2.76	2.94	3.23	3.61	4.09	4.72	5.53
45	2.82	3.02	3.32	3.73	4.25	4.92	5.80

*Based on Zhu and Qian, 2000

Table 14.2 Variation of R [Eq. (14.12)]

θ (deg)	δ'/ϕ'	R for ϕ' (deg) 30	35	40	45
0	0.2	1.2	1.28	1.35	1.45
	0.4	1.4	1.6	1.8	2.2
	0.6	1.65	1.95	2.4	3.2
	0.8	1.95	2.4	3.15	4.45
	1.0	2.2	2.85	3.95	6.1
5	0.2	1.2	1.25	1.32	1.4
	0.4	1.4	1.6	1.8	2.1
	0.6	1.6	1.9	2.35	3.0
	0.8	1.9	2.35	3.05	4.3
	1.0	2.15	2.8	3.8	5.7
10	0.2	1.15	1.2	1.3	1.4
	0.4	1.35	1.5	1.7	2.0
	0.6	1.6	1.85	2.25	2.9
	0.8	1.8	2.25	2.9	4.0
	1.0	2.05	2.65	3.6	5.3
15	0.2	1.15	1.2	1.3	1.35
	0.4	1.35	1.5	1.65	1.95
	0.6	1.55	1.8	2.2	2.7
	0.8	1.8	2.2	2.8	3.8
	1.0	2.0	2.6	3.4	4.95
20	0.2	1.15	1.2	1.3	1.35
	0.4	1.35	1.45	1.65	1.9
	0.6	1.5	1.8	2.1	2.6
	0.8	1.8	2.1	2.6	3.55
	1.0	1.9	2.4	3.2	4.8

Example 14.1

Consider a 3-m-high (H) retaining wall with a vertical back ($\theta = 0°$) and a horizontal granular backfill. Given: $\gamma = 15.7$ kN/m³, $\delta' = 15°$, and $\phi' = 30°$. Estimate the passive force, P_p, by using

    a. Coulomb's theory
    b. Terzaghi and Peck's wedge theory
    c. Shields and Tolunay's solution (method of slices)
    d. Zhu and Qian's solution (method of triangular slices)

Solution
Part a
From Eq. (13.70),

$$P_p = \tfrac{1}{2}K_p\gamma H^2$$

From Table 13.9, for $\phi' = 30°$ and $\delta' = 15°$, the value of K_p is 4.977. Thus,

$$P = (\tfrac{1}{2})(4.977)(15.7)(3)^2 = \textbf{351.6 kN/m}$$

Part b
From Fig. 14.4, for $\phi' = 30°$ and $\delta' = 15°$, the value of K_p is about 4.53. Thus,

$$P_p = (\tfrac{1}{2})(4.53)(15.7)(3)^2 = \textbf{320 kN/m}$$

Part c

$$P_p = \frac{1}{2} K_p \gamma H^2$$

From Figure 14.5, for $\phi' = 30°$ and $\delta' = 15°$ (i.e $\dfrac{\delta'}{\phi'} = 0.5$) the value of K_p is 4.13. Hence,

$$P_p = \left(\frac{1}{2}\right)(4.13)(15.7)(3)^2 \approx \textbf{292 kN/m}$$

Part d
From Eq. (14.12),

$$K_p = K_{p(\delta' = 0)} R$$

For $\phi' = 30°$ and $\theta = 0$, $K_{p(\delta' = 0)}$ is equal to 3.0 (Table 14.1). Again, for $\theta = 0$ and $\delta'/\phi' = 0.5$, the value of R is about 1.52. Thus, $K_p = (3)(1.52) = 4.56$.

$$P_p = (\tfrac{1}{2})(4.56(15.7)(3)^2 = \textbf{322 kN/m}$$ ∎

14.5   *Passive Force on Walls with Earthquake Forces*

The relationship for passive earth pressure on a retaining wall with a *granular horizontal backfill* and vertical back face under earthquake conditions was evaluated by Subba Rao and Choudhury (2005) using the pseudo-static approach to the method of limited equilibrium. The failure surface in soil assumed in the analysis was similar to that shown in Fig. 14.3 (with $\theta = 0$; that is, vertical back face). The notations used in the analysis were as follow:

$$H = \text{height of retaining wall}$$

$$P_{pe} = \text{passive force per unit length of the wall}$$

$$\phi' = \text{soil friction angle}$$

$$\delta' = \text{angle of wall friction}$$

$$k_h = \frac{\text{horizontal component of earthquake acceleration}}{\text{acceleration due to gravity}, g}$$

$$k_v = \frac{\text{vertical component of earthquake acceleration}}{\text{acceleration due to gravity}, g}$$

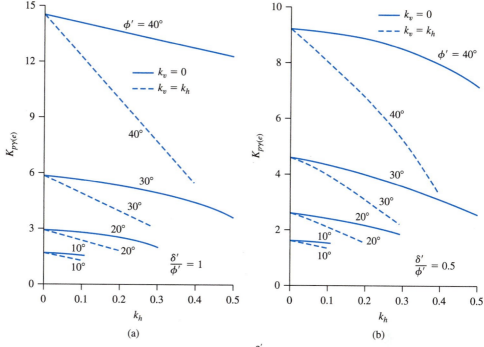

Figure 14.6 Variation of $K_{p\gamma(e)}$: (a) $\delta'/\phi' = 1$; (b) $\dfrac{\delta'}{\phi} = 0.5$

Based on this analysis, the passive force P_{pe} can be expressed as

$$P_{pe} = \left[\frac{1}{2}\gamma H^2 K_{p\gamma(e)} \right] \frac{1}{\cos \delta'} \tag{14.13}$$

where $K_{p\gamma(e)}$ = passive earth-pressure coefficient in the normal direction to the wall.

$K_{p\gamma(e)}$ is a function of k_h and k_v. The variations of $K_{p\gamma(e)}$ for $\delta'/\phi' = 0.5$ and 1 are shown in Figure 14.6. The passive pressure P_{pe} will be inclined at an angle δ' to the back face of the wall and will act at a distance of $H/3$ above the bottom of the wall.

BRACED CUTS

14.6 *Braced Cuts—General*

Frequently during the construction of foundations or utilities (such as sewers), open trenches with vertical soil slopes are excavated. Although most of these trenches are temporary, the sides of the cuts must be supported by proper bracing systems. Figure 14.7 shows one of several bracing systems commonly adopted in construction practice. The bracing consists of sheet piles, wales, and struts.

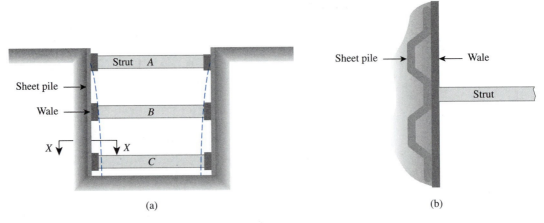

Figure 14.7 Braced cut: (a) cross section; (b) plan (section at X–X)

Proper design of these elements requires a knowledge of the lateral earth pressure exerted on the braced walls. The magnitude of the lateral earth pressure at various depths of the cut is very much influenced by the deformation condition of the sheeting. To understand the nature of the deformation of the braced walls, one needs to follow the sequence of construction. Construction of the unit begins with driving the sheetings. The top row of the wales and struts (marked A in Figure 14.7a) is emplaced immediately after a small cut is made. This emplacement must be done immediately so that the soil mass outside the cut has no time to deform and cause the sheetings to yield. As the sequence of driving the sheetings, excavating the soil, and placing rows of wales and struts (see B and C in Figure 14.7) continues, the sheetings move inward at greater depths. This action is caused by greater earth pressure exerted by the soil outside the cut. The deformation of the braced walls is shown by the broken lines in Figure 14.7a. Essentially, the problem models a condition where the walls are rotating about the level of the top row of struts. A photograph of braced cuts made for subway construction in Chicago is shown in Figure 14.8a. Figures 14.8b and 14.8c are photographs of two braced cuts—one in Seoul, South Korea, and the other in Taiwan.

The deformation of a braced wall differs from the deformation condition of a retaining wall in that, in a braced wall, the rotation is about the top. For this reason, neither Coulomb's nor Rankine's theory will give the actual earth-pressure distribution. This fact is illustrated in Figure 14.9, in which AB is a frictionless wall with a granular soil backfill. When the wall deforms to position AB', failure surface BC develops. Because the upper portion of the soil mass in the zone ABC does not undergo sufficient deformation, it does not pass into Rankine's active state. The sliding surface BC intersects the ground surface almost at $90°$. The corresponding earth pressure will be somewhat parabolic, like acb shown in Figure 14.9b. With this type of pressure distribution, the point of application of the resultant active thrust, P_a, will be at a height of $n_a H$ measured from the bottom of the wall, with $n_a > \frac{1}{3}$ (for triangular pressure distribution $n_a = \frac{1}{3}$). Theoretical evaluation and field measurements have shown that n_a could be as high as 0.55.

(a)

(b)

Figure 14.8 Braced cuts: (a) Chicago subway construction (*Courtesy of Ralph B. Peck*);
(b) in Seoul, South Korea (*Courtesy of E. C. Shin, University of Incheon, South Korea*);

(c)

Figure 14.8 (*continued*) (c) Braced cut in Taiwan (*Courtesy of Richard Tsai, C&M Hi-Tech Engineering Co., Ltd., Taipei, Taiwan*)

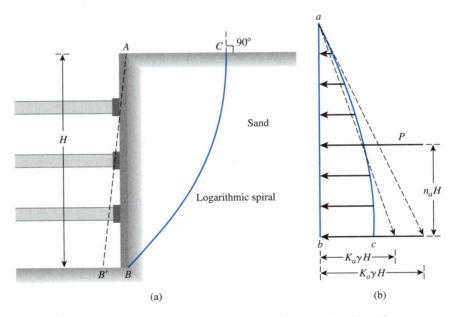

Figure 14.9 Earth-pressure distribution against a wall with rotation about the top

Determination of Active Thrust on Bracing Systems of Open Cuts—Granular Soil

The active thrust on the bracing system of open cuts can be estimated theoretically by using trial wedges and Terzaghi's general wedge theory (1941). The basic procedure for determination of the active thrust are described in this section.

Figure 14.10a shows a braced wall AB of height H that deforms by rotating about its top. The wall is assumed to be rough, with the angle of wall friction equal to δ'. The point of application of the active thrust (that is, $n_a H$) is assumed to be known. The curve of sliding is

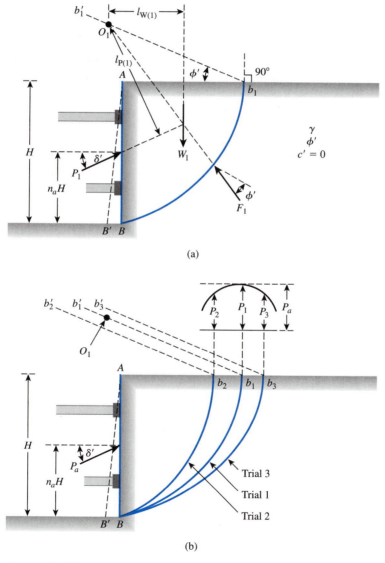

(a)

(b)

Figure 14.10 Determination of active force on bracing system of open cut in cohesionless soil

assumed to be an arc of a logarithmic spiral. As we discussed in the preceding section, the curve of sliding intersects the horizontal ground surface at 90°. To proceed with the trial wedge solution, let us select a point b_1. From b_1, a line b_1b_1' that makes an angle ϕ' with the ground surface is drawn. (Note that ϕ' = effective angle of friction of the soil.) The arc of the logarithmic spiral, b_1B, which defines the curve of sliding for this trial, can now be drawn, with the center of the spiral (point O_1) located on the line b_1b_1'. Note that the equation for the logarithmic spiral is given by $r_1 = r_o e^{\theta_1 \tan \phi'}$ and, in this case, $\overline{O_1b_1} = r_o$ and $\overline{O_1B} = r_1$. Also, it is interesting to see that the horizontal line that represents the ground surface is the normal to the curve of sliding at the point b_1, and that O_1b_1 is a radial line. The angle between them is equal to ϕ', which agrees with the property of the spiral.

To look at the equilibrium of the failure wedge, let us consider the following forces per unit length of the braced wall:

- W_1 = the weight of the wedge ABb_1 = (Area of ABb_1) × (γ) × (1).
- P_1 = the active thrust acting at a point n_aH measured vertically upward from the bottom of the cut and inclined at an angle δ' with the horizontal.
- F_1 = the resultant of the shear and normal forces that act along the trial failure surface. The line of action of the force F_1 will pass through the point O_1.

Now, taking the moments of these forces about O_1, we have

$$W_1[l_{W(1)}] + F_1(0) - P_1[l_{P(1)}] = 0$$

or

$$P_1 = \frac{W_1 l_{W(1)}}{l_{P(1)}} \tag{14.14}$$

where $l_{W(1)}$ and $l_{P(1)}$ are the moment arms for the forces W_1 and P_1, respectively.

This procedure of finding the active thrust can now be repeated for several wedges such as ABb_2, ABb_3, . . . , ABb_n (Figure 14.10b). Note that the centers of the logarithmic-spiral arcs will lie on lines b_2b_2', b_3b_3', . . . , b_nb_n', respectively. The active thrusts P_1, P_2, P_3, . . . , P_n derived from the trial wedges are plotted to some scale in the upper portion of Figure 14.10b. The maximum point of the smooth curve drawn through these points will yield the desired maximum active thrust, P_a, on the braced wall.

Kim and Preber (1969) determined the values of $P_a/0.5\gamma H^2$ for braced excavations for various values of ϕ', δ', and n_a. These values are given in Table 14.3. In general, the average magnitude of P_a is about 10% greater when the wall rotation is about the top as compared with the value obtained by Coulomb's active earth-pressure theory.

14.8 ## Determination of Active Thrust on Bracing Systems for Cuts—Cohesive Soil

Using the principles of the general wedge theory, we also can determine the active thrust on bracing systems for cuts made in $c'-\phi'$ soil. Table 14.4 gives the variation of P_a in a nondimensional form for various values of ϕ', δ', n_a, and $c'/\gamma H$.

Table 14.3 $P_a/0.5\gamma H^2$ Against ϕ', δ', and n_a ($c' = 0$) for Braced Cuts[*]

ϕ' (deg)	δ' (deg)	$P_a/0.5\gamma H^2$				ϕ' (deg)	δ' (deg)	$P_a/0.5\gamma H^2$			
		$n_a = 0.3$	$n_a = 0.4$	$n_a = 0.5$	$n_a = 0.6$			$n_a = 0.3$	$n_a = 0.4$	$n_a = 0.5$	$n_a = 0.6$
10	0	0.653	0.734	0.840	0.983	35	0	0.247	0.267	0.290	0.318
	5	0.623	0.700	0.799	0.933		5	0.239	0.258	0.280	0.318
	10	0.610	0.685	0.783	0.916		10	0.234	0.252	0.273	0.300
15	0	0.542	0.602	0.679	0.778		15	0.231	0.249	0.270	0.296
	5	0.518	0.575	0.646	0.739		20	0.231	0.248	0.269	0.295
	10	0.505	0.559	0.629	0.719		25	0.232	0.250	0.271	0.297
	15	0.499	0.554	0.623	0.714		30	0.236	0.254	0.276	0.302
20	0	0.499	0.495	0.551	0.622		35	0.243	0.262	0.284	0.312
	5	0.430	0.473	0.526	0.593	40	0	0.198	0.213	0.230	0.252
	10	0.419	0.460	0.511	0.575		5	0.192	0.206	0.223	0.244
	15	0.413	0.454	0.504	0.568		10	0.189	0.202	0.219	0.238
	20	0.413	0.454	0.504	0.569		15	0.187	0.200	0.216	0.236
25	0	0.371	0.405	0.447	0.499		20	0.187	0.200	0.216	0.235
	5	0.356	0.389	0.428	0.477		25	0.188	0.202	0.218	0.237
	10	0.347	0.378	0.416	0.464		30	0.192	0.205	0.222	0.241
	15	0.342	0.373	0.410	0.457		35	0.197	0.211	0.228	0.248
	20	0.341	0.372	0.409	0.456		40	0.205	0.220	0.237	0.259
	25	0.344	0.375	0.413	0.461	45	0	0.156	0.167	0.180	0.196
30	0	0.304	0.330	0.361	0.400		5	0.152	0.163	0.175	0.190
	5	0.293	0.318	0.347	0.384		10	0.150	0.160	0.172	0.187
	10	0.286	0.310	0.339	0.374		15	0.148	0.159	0.171	0.185
	15	0.282	0.306	0.334	0.368		20	0.149	0.159	0.171	0.185
	20	0.281	0.305	0.332	0.367		25	0.150	0.160	0.173	0.187
	25	0.284	0.307	0.335	0.370		30	0.153	0.164	0.176	0.190
	30	0.289	0.313	0.341	0.377		35	0.158	0.168	0.181	0.196
							40	0.164	0.175	0.188	0.204
							45	0.173	0.184	0.198	0.213

[*]*After Kim and Preber, 1969. With permission from ASCE.*

14.9 Pressure Variation for Design of Sheetings, Struts, and Wales

The active thrust against sheeting in a braced cut, calculated by using the general wedge theory, does not explain the variation of the earth pressure with depth that is necessary for design work. An important difference between bracings in open cuts and retaining walls is that retaining walls fail as single units, whereas bracings in an open cut undergo progressive failure where one or more struts fail at one time.

Empirical lateral pressure diagrams against sheetings for the design of bracing systems have been given by Peck (1969). These pressure diagrams for cuts in sand, soft to medium clay, and stiff clay are given in Figure 14.11. Strut loads may be determined by assuming that the vertical members are hinged at each strut level except the topmost and bottommost ones (Figure 14.12). Example 14.2 illustrates the procedure for the calculation of strut loads.

Table 14.4 Values of $P_a/0.5\gamma H^2$ for Cuts in a c'-ϕ' Soil with the Assumption $c'_a = c'(\tan \delta'/\tan \phi')^*$

δ' (deg)	$n_a = 0.3$ and $c'/\gamma H = 0.1$	$n_a = 0.4$ and $c'/\gamma H = 0.1$	$n_a = 0.5$ and $c'/\gamma H = 0.1$
$\phi' = 15°$			
0	0.254	0.285	0.322
5	0.214	0.240	0.270
10	0.187	0.210	0.238
15	0.169	0.191	0.218
$\phi' = 20°$			
0	0.191	0.210	0.236
5	0.160	0.179	0.200
10	0.140	0.156	0.173
15	0.122	0.127	0.154
20	0.113	0.124	0.140
$\phi' = 25°$			
0	0.138	0.150	0.167
5	0.116	0.128	0.141
10	0.099	0.110	0.122
15	0.085	0.095	0.106
20	0.074	0.083	0.093
25	0.065	0.074	0.083
$\phi' = 30°$			
0	0.093	0.103	0.113
5	0.078	0.086	0.094
10	0.066	0.073	0.080
15	0.056	0.060	0.067
20	0.047	0.051	0.056
25	0.036	0.042	0.047
30	0.029	0.033	0.038

*After Kim and Preber, 1969. With permission from ASCE.)

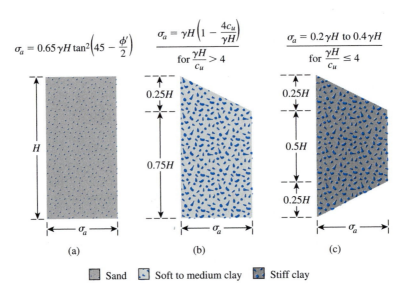

$$\sigma_a = 0.65\,\gamma H \tan^2\left(45 - \frac{\phi'}{2}\right)$$

$$\sigma_a = \gamma H\left(1 - \frac{4c_u}{\gamma H}\right)$$
for $\dfrac{\gamma H}{c_u} > 4$

$$\sigma_a = 0.2\gamma H \text{ to } 0.4\gamma H$$
for $\dfrac{\gamma H}{c_u} \leq 4$

0.25H

0.25H

0.5H

0.75H

0.25H

(a)   (b)   (c)

Sand   Soft to medium clay   Stiff clay

Figure 14.11 Peck's pressure diagrams for design of bracing systems

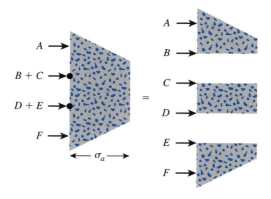

Figure 14.12
Determination of strut loads from empirical lateral pressure diagrams

Example 14.2

A 7-m-deep braced cut in sand is shown in Figure 14.13. In the plan, the struts are placed at $s = 2$ m center to center. Using Peck's empirical pressure diagram, calculate the design strut loads.

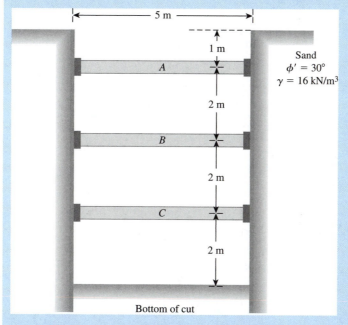

Figure 14.13 Braced cut in sand

Solution

Refer to Figure 14.11a. For the lateral earth pressure diagram,

$$\sigma_a = 0.65\gamma H \tan^2\left(45 - \frac{\phi'}{2}\right) = (0.65)(16)(7)\tan^2\left(45 - \frac{30}{2}\right) = 24.27 \, \text{kN/m}^2$$

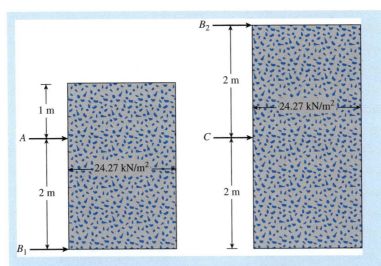

Figure 14.14 Calculation of strut loads from pressure envelope

Assume that the sheeting is hinged at strut level B. Now refer to the diagram in Figure 14.14 We need to find reactions at A, B_1, B_2, and C. Taking the moment about B_1, we have

$$2A = (24.27)(3)\left(\frac{3}{2}\right); \quad A = 54.61\,\text{kN/m}$$

Hence,

$$B_1 = (24.27)(3) - 54.61 = 18.2\ \text{kN/m}$$

Again, taking the moment about B_2, we have

$$2C = (24.27)(4)\left(\frac{4}{2}\right)$$

$$C = 97.08\,\text{kN/m}$$

So

$$B_2 = (24.27)(4) - 97.08 = 0$$

The strut loads are as follows:

$$\text{At level } A: \quad (A)(s) = (54.61)(2) = \textbf{109.22 kN}$$

$$\text{At level } B: \quad (B_1 + B_2)(s) = (18.2 + 0)(2) = \textbf{36.4 kN}$$

$$\text{At level } C: \quad (C)(s) = (97.08)(2) = \textbf{194.16 kN}$$

14.10   *Summary*

This chapter covers two major topics:

- Estimation of passive pressure using curved failure surface in soil
- Lateral earth pressure on braced cuts using the general wedge theory and pressure envelopes for design of struts, wales, and sheet piles.

   Passive pressure calculations using curved failure surface are essential for the case in which $\delta' > \phi'/2$, since plane-failure surface assumption provides results on the unsafe side for design.

   In the case of braced cuts, although the general wedge theory provides the force per unit length of the cut, it does not provide the nature of distribution of earth pressure with depth. For that reason, pressure envelopes are necessary for practical design.

Problems

14.1 Refer to the retaining wall shown in Figure 14.15. Given: $\theta = 10°$, $\alpha = 0$, $\gamma = 15.5$ kN/m³, $\phi' = 35°$, $\delta' = 21°$, and $H = 6$ m. Estimate the passive force, P_p, per unit length of the wall. Use Tables 14.1 and 14.2.

14.2 Refer to Figure 14.15. Given: $H = 4.75$ m, $\gamma = 15.72$ kN/m, $\phi' = 30°$, $\delta' = \frac{2}{3}\phi'$, $\alpha = 0$, and $\theta = 0$. Calculate the passive force per unit length of the wall using Figure 14.5.

14.3 Solve Problem 14.2 using Tables 14.1 and 14.2.

14.4 Refer to Figure 14.15. Given: $\theta = 0$, $\alpha = 0°$, $H = 2.5$ m, $\phi' = 30°$, $\delta' = 20°$, and $\gamma = 14.8$ kN/m³. Estimate the passive force per unit length of the wall using Figure 14.4.

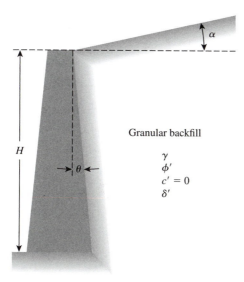

Granular backfill

γ
ϕ'
$c' = 0$
δ'

Figure 14.15

14.5 Refer to Figure 14.3a. Given: $H = 5$ m, $\gamma = 16$ kN/m³, $\phi' = 30°$, $\delta' = 15°$, $k_v = 0$, and $k_h = 0.3$. Calculate P_{pe} for the retaining wall (Section 14.5).

14.6 A braced wall is shown in Figure 14.16. Given: $H = 5$ m, $n_aH = 2$ m, $\phi' = 35°$, $\delta' = 20°$, $\gamma = 16$ kN/m³, and $c' = 0$. Determine the active thrust, P_a, on the wall using the general wedge theory.

14.7 Repeat Problem 14.6 with the following given: $H = 15.6$ m, $n_aH = 4.68$ m, $\gamma = 18$ kN/m³, $\phi' = 20°$, $\delta' = 15°$, and $c' = 28$ kN/m². Assume

$$\frac{c_a'}{c'} = \frac{\tan \delta'}{\tan \phi'}.$$

14.8 The elevation and plan of a bracing system for an open cut in sand are shown in Figure 14.17. Assuming $\gamma_{sand} = 16.51$ kN/m³ and $\phi' = 38°$, determine the strut loads.

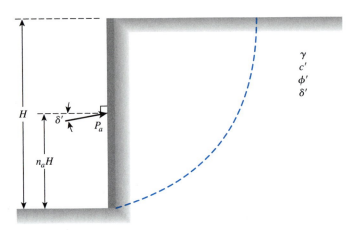

Figure 14.16

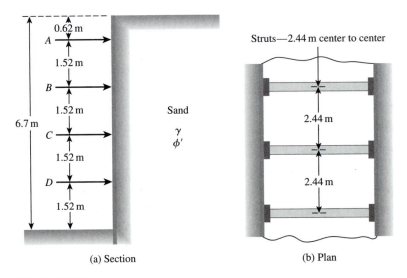

(a) Section   (b) Plan

Figure 14.17

References

HIJAB, W. (1956). "A Note on the Centroid of a Logarithmic Spiral Sector," *Geotechnique,* Vol. 4, No. 2, 96–99.

KIM, J. S., and PREBER, T. (1969). "Earth Pressure Against Braced Excavations," *Journal of the Soil Mechanics and Foundations Division,* ASCE, Vol. 95, No. SM6, 1581–1584.

PECK, R. B. (1969). "Deep Excavation and Tunneling in Soft Ground," *Proceedings,* 7th International Conference on Soil Mechanics and Foundation Engineering, Mexico City, State-of-the-Art Vol., 225–290.

SHIELDS, D. H., and TOLUNAY, A. Z. (1973). "Passive Pressure Coefficients by Method of Slices," *Journal of the Soil Mechanics and Foundations Division,* ASCE, Vol. 99, No. SM12, 1043–1053.

SUBBA RAO, K. S., AND CHOUDHURY, D. (2005). "Seismic Passive Earth Pressures in Soil," *Journal of Geotechnical and Geoenvironmental Engineering*, American Society of Civil Engineers, Vol. 131, No. 1, 131–135.

TERZAGHI, K. (1941). "General Wedge Theory of Earth Pressure," *Transactions,* ASCE, Vol. 106, 68–97.

TERZAGHI, K., and PECK, R. B. (1967). *Soil Mechanics in Engineering Practice,* 2nd ed., Wiley, New York.

ZHU, D. Y., AND QIAN, Q. (2000). "Determination of Passive Earth Pressure Coefficient by the Method of Triangular Slices," *Canadian Geotechnical Journal*, Vol. 37, No. 2, 485–491.

15 Slope Stability

15.1 Introduction—Modes of Slope Failure

An exposed ground surface that stands at an angle with the horizontal is called an unrestrained slope. The slope can be natural or man-made. It can fail in various modes. Cruden and Varnes (1996) classified the slope failures into the following five major categories. They are

1. **Fall.** This is the detachment of soil and/or rock fragments that fall down a slope (Figure 15.1). Figure 15.2 shows a fall in which a large amount of soil mass has slid down a slope.
2. **Topple.** This is a forward rotation of soil and/or rock mass about an axis below the center of gravity of mass being displaced (Figure 15.3).
3. **Slide.** This is the downward movement of a soil mass occurring on a surface of rupture (Figure 15.4).
4. **Spread.** This is a form of slide (Figure 15.5) by translation. It occurs by "sudden movement of water-bearing seams of sands or silts overlain by clays or loaded by fills" (Cruden and Varnes, 1996).
5. **Flow.** This is a downward movement of soil mass similar to a viscous fluid (Figure 15.6).

This chapter primarily relates to the quantitative analysis that fall under the category of *slide*.

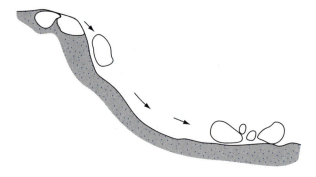

Figure 15.1 "Fall" type of landslide

Figure 15.2 Soil and rock "fall" in a slope (*Courtesy of E.C. Shin, University of Incheon, South Korea*)

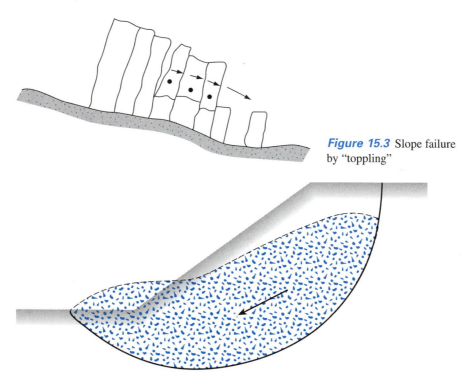

Figure 15.3 Slope failure by "toppling"

Figure 15.4 Slope failure by "sliding"

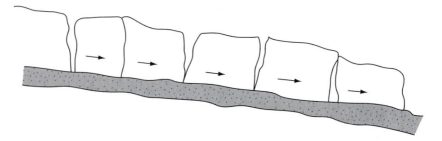

Figure 15.5 Slope failure by lateral "spreading"

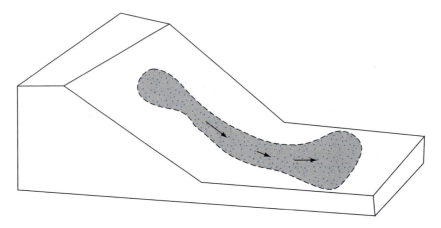

Figure 15.6 Slope failure by "flowing"

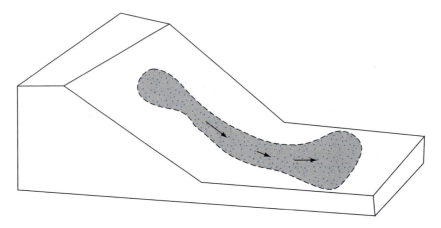

15.2 *Factor of Safety*

The task of the engineer charged with analyzing slope stability is to determine the factor of safety. Generally, the factor of safety is defined as

$$F_s = \frac{\tau_f}{\tau_d} \tag{15.1}$$

where F_s = factor of safety with respect to strength
τ_f = average shear strength of the soil
τ_d = average shear stress developed along the potential failure surface

The shear strength of a soil consists of two components, cohesion and friction, and may be written as

$$\tau_f = c' + \sigma' \tan \phi' \tag{15.2}$$

where c' = cohesion
ϕ' = angle of friction
σ' = normal stress on the potential failure surface

In a similar manner, we can write

$$\tau_d = c_d' + \sigma' \tan \phi_d' \tag{15.3}$$

where c_d' and ϕ_d' are, respectively, the cohesion and the angle of friction that develop along the potential failure surface. Substituting Eqs. (15.2) and (15.3) into Eq. (15.1), we get

$$F_s = \frac{c' + \sigma' \tan \phi'}{c_d' + \sigma' \tan \phi_d'} \tag{15.4}$$

Now we can introduce some other aspects of the factor of safety—that is, the factor of safety with respect to cohesion, $F_{c'}$, and the factor of safety with respect to friction, $F_{\phi'}$. They are defined as

$$F_{c'} = \frac{c'}{c_d'} \tag{15.5}$$

and

$$F_{\phi'} = \frac{\tan \phi'}{\tan \phi_d'} \tag{15.6}$$

When we compare Eqs. (15.4) through (15.6), we can see that when $F_{c'}$ becomes equal to $F_{\phi'}$, it gives the factor of safety with respect to strength. Or, if

$$\frac{c'}{c_d'} = \frac{\tan \phi'}{\tan \phi_d'}$$

then we can write

$$F_s = F_{c'} = F_{\phi'} \tag{15.7}$$

When F_s is equal to 1, the slope is in a state of impending failure. Generally, a value of 1.5 for the factor of safety with respect to strength is acceptable for the design of a stable slope.

15.3   *Stability of Infinite Slopes*

In considering the problem of slope stability, let us start with the case of an infinite slope as shown in Figure 15.7. The shear strength of the soil may be given by Eq. (15.2):

$$\tau_f = c' + \sigma' \tan \phi'$$

Assuming that the pore water pressure is zero, we will evaluate the factor of safety against a possible slope failure along a plane *AB* located at a depth *H* below the ground surface. The slope failure can occur by the movement of soil above the plane *AB* from right to left.

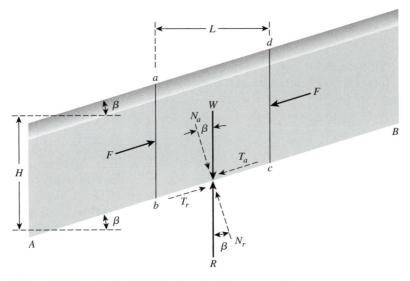

Figure 15.7 Analysis of infinite slope (without seepage)

Let us consider a slope element *abcd* that has a unit length perpendicular to the plane of the section shown. The forces, *F*, that act on the faces *ab* and *cd* are equal and opposite and may be ignored. The weight of the soil element is

$$W = (\text{Volume of soil element}) \times (\text{Unit weight of soil}) = \gamma L H \qquad (15.8)$$

The weight *W* can be resolved into two components:

1. Force perpendicular to the plane $AB = N_a = W \cos \beta = \gamma L H \cos \beta$.
2. Force parallel to the plane $AB = T_a = W \sin \beta = \gamma L H \sin \beta$. Note that this is the force that tends to cause the slip along the plane.

Thus, the effective normal stress and the shear stress at the base of the slope element can be given, respectively, as

$$\sigma' = \frac{N_a}{\text{Area of base}} = \frac{\gamma L H \cos \beta}{\left(\dfrac{L}{\cos \beta}\right)} = \gamma H \cos^2 \beta \qquad (15.9)$$

and

$$\tau = \frac{T_a}{\text{Area of base}} = \frac{\gamma L H \sin \beta}{\left(\dfrac{L}{\cos \beta}\right)} = \gamma H \cos \beta \sin \beta \qquad (15.10)$$

The reaction to the weight *W* is an equal and opposite force *R*. The normal and tangential components of *R* with respect to the plane *AB* are

$$N_r = R \cos \beta = W \cos \beta \qquad (15.11)$$

and

$$T_r = R \sin \beta = W \sin \beta \tag{15.12}$$

For equilibrium, the resistive shear stress that develops at the base of the element is equal to $(T_r)/(\text{Area of base}) = \gamma H \sin \beta \cos \beta$. The resistive shear stress also may be written in the same form as Eq. (15.3):

$$\tau_d = c_d' + \sigma' \tan \phi_d'$$

The value of the normal stress is given by Eq. (15.9). Substitution of Eq. (15.9) into Eq. (15.3) yields

$$\tau_d = c_d' + \gamma H \cos^2 \beta \tan \phi_d' \tag{15.13}$$

Thus,

$$\gamma H \sin \beta \cos \beta = c_d' + \gamma H \cos^2 \beta \tan \phi_d'$$

or

$$\frac{c_d'}{\gamma H} = \sin \beta \cos \beta - \cos^2 \beta \tan \phi_d'$$

$$= \cos^2 \beta (\tan \beta - \tan \phi_d') \tag{15.14}$$

The factor of safety with respect to strength has been defined in Eq. (15.7), from which we get

$$\tan \phi_d' = \frac{\tan \phi'}{F_s} \quad \text{and} \quad c_d' = \frac{c'}{F_s}$$

Substituting the preceding relationships into Eq. (15.14), we obtain

$$F_s = \frac{c'}{\gamma H \cos^2 \beta \tan \beta} + \frac{\tan \phi'}{\tan \beta} \tag{15.15}$$

For granular soils, $c' = 0$, and the factor of safety, F_s, becomes equal to $(\tan \phi')/(\tan \beta)$. This indicates that in an infinite slope in sand, the value of F_s is independent of the height H and the slope is stable as long as $\beta < \phi'$.

If a soil possesses cohesion and friction, the depth of the plane along which critical equilibrium occurs may be determined by substituting $F_s = 1$ and $H = H_{cr}$ into Eq. (15.15). Thus,

$$H_{cr} = \frac{c'}{\gamma} \frac{1}{\cos^2 \beta (\tan \beta - \tan \phi')} \tag{15.16}$$

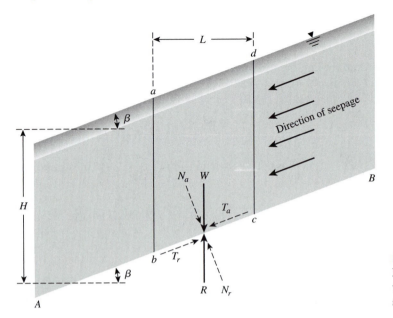

Figure 15.8
Infinite slope with steady-state seepage

If there is *steady state seepage through the soil* and the ground water table coincides with the ground surface, as shown in Figure 15.8, the factor of safety against sliding can be determined as

$$F_s = \frac{c'}{\gamma_{sat} H \cos^2 \beta \tan \beta} + \frac{\gamma' \tan \phi'}{\gamma_{sat} \tan \beta}$$ (15.17)

where γ_{sat} = saturated unit weight of soil
$\gamma' = \gamma_{sat} - \gamma_w$ = effective unit weight of soil

Example 15.1

For the infinite slope with a steady state seepage shown in Figure 15.9, determine:

a. The factor of safety against sliding along the soil-rock interface.
b. The height, H, that will give a factor of safety (F_s) of 2 against sliding along the soil-rock interface.

Solution
Part a
From Eq. (15.17),

$$F_s = \frac{c'}{\gamma_{sat} H \cos^2 \beta \tan \beta} + \frac{\gamma' \tan \phi'}{\gamma_{sat} \tan \beta}$$

$$\gamma_{sat} = 17.8 \text{ kN/m}^3$$

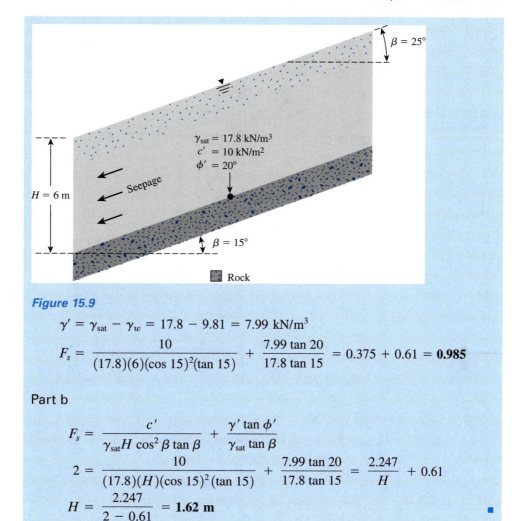

Figure 15.9

$$\gamma' = \gamma_{\text{sat}} - \gamma_w = 17.8 - 9.81 = 7.99 \text{ kN/m}^3$$

$$F_s = \frac{10}{(17.8)(6)(\cos 15)^2(\tan 15)} + \frac{7.99 \tan 20}{17.8 \tan 15} = 0.375 + 0.61 = \mathbf{0.985}$$

Part b

$$F_s = \frac{c'}{\gamma_{\text{sat}} H \cos^2 \beta \tan \beta} + \frac{\gamma' \tan \phi'}{\gamma_{\text{sat}} \tan \beta}$$

$$2 = \frac{10}{(17.8)(H)(\cos 15)^2(\tan 15)} + \frac{7.99 \tan 20}{17.8 \tan 15} = \frac{2.247}{H} + 0.61$$

$$H = \frac{2.247}{2 - 0.61} = \mathbf{1.62 \text{ m}}$$ ∎

15.4   *Finite Slopes—General*

When the value of H_{cr} approaches the height of the slope, the slope generally may be considered finite. For simplicity, when analyzing the stability of a finite slope in a homogeneous soil, we need to make an assumption about the general shape of the surface of potential failure. Although considerable evidence suggests that slope failures usually occur on curved failure surfaces, Culmann (1875) approximated the surface of potential failure as a plane. The factor of safety, F_s, calculated by using Culmann's approximation, gives fairly good results for near-vertical slopes only. After extensive investigation of slope failures in the 1920s, a Swedish geotechnical commission recommended that the actual surface of sliding may be approximated to be circularly cylindrical.

Since that time, most conventional stability analyses of slopes have been made by assuming that the curve of potential sliding is an arc of a circle. However, in many circumstances (for example, zoned dams and foundations on weak strata), stability analysis using plane failure of sliding is more appropriate and yields excellent results.

15.5 Analysis of Finite Slopes with Plane Failure Surfaces (Culmann's Method)

Culmann's analysis is based on the assumption that the failure of a slope occurs along a plane when the average shearing stress tending to cause the slip is more than the shear strength of the soil. Also, the most critical plane is the one that has a minimum ratio of the average shearing stress that tends to cause failure to the shear strength of soil.

Figure 15.10 shows a slope of height H. The slope rises at an angle β with the horizontal. AC is a trial failure plane. If we consider a unit length perpendicular to the section of the slope, we find that the weight of the wedge ABC is equal to

$$W = \tfrac{1}{2}(H)(\overline{BC})(1)(\gamma) = \tfrac{1}{2}H(H \cot \theta - H \cot \beta)\gamma$$

$$= \frac{1}{2}\gamma H^2 \left[\frac{\sin(\beta - \theta)}{\sin \beta \sin \theta} \right] \tag{15.18}$$

The normal and tangential components of W with respect to the plane AC are as follows.

$$N_a = \text{normal component} = W \cos \theta = \frac{1}{2}\gamma H^2 \left[\frac{\sin(\beta - \theta)}{\sin \beta \sin \theta} \right] \cos \theta \tag{15.19}$$

$$T_a = \text{tangential component} = W \sin \theta = \frac{1}{2}\gamma H^2 \left[\frac{\sin(\beta - \theta)}{\sin \beta \sin \theta} \right] \sin \theta \tag{15.20}$$

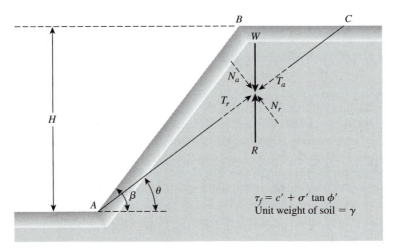

$$\tau_f = c' + \sigma' \tan \phi'$$
Unit weight of soil $= \gamma$

Figure 15.10
Finite slope analysis—Culmann's method

The average effective normal stress and the average shear stress on the plane AC are, respectively,

$$\sigma' = \frac{N_a}{(AC)(1)} = \frac{N_a}{\left(\dfrac{H}{\sin \theta}\right)}$$

$$= \frac{1}{2}\gamma H \left[\frac{\sin(\beta - \theta)}{\sin \beta \sin \theta}\right]\cos \theta \sin \theta \qquad (15.21)$$

and

$$\tau = \frac{T_a}{(AC)(1)} = \frac{T_a}{\left(\dfrac{H}{\sin \theta}\right)}$$

$$= \frac{1}{2}\gamma H \left[\frac{\sin(\beta - \theta)}{\sin \beta \sin \theta}\right]\sin^2 \theta \qquad (15.22)$$

The average resistive shearing stress developed along the plane AC also may be expressed as

$$\tau_d = c_d' + \sigma' \tan \phi_d'$$

$$= c_d' + \frac{1}{2}\gamma H \left[\frac{\sin(\beta - \theta)}{\sin \beta \sin \theta}\right]\cos \theta \sin \theta \tan \phi_d' \qquad (15.23)$$

Now, from Eqs. (15.22) and (15.23),

$$\frac{1}{2}\gamma H \left[\frac{\sin(\beta - \theta)}{\sin \beta \sin \theta}\right]\sin^2 \theta = c_d' + \frac{1}{2}\gamma H \left[\frac{\sin(\beta - \theta)}{\sin \beta \sin \theta}\right]\cos \theta \sin \theta \tan \phi_d' \qquad (15.24)$$

or

$$c_d = \frac{1}{2}\gamma H \left[\frac{\sin(\beta - \theta)(\sin \theta - \cos \theta \tan \phi_d')}{\sin \beta}\right] \qquad (15.25)$$

The expression in Eq. (15.25) is derived for the trial failure plane AC. In an effort to determine the critical failure plane, we must use the principle of maxima and minima (for a given value of ϕ_d') to find the angle θ where the developed cohesion would be maximum. Thus, the first derivative of c_d with respect to θ is set equal to zero, or

$$\frac{\partial c_d'}{\partial \theta} = 0 \qquad (15.26)$$

Because γ, H, and β are constants in Eq. (15.25), we have

$$\frac{\partial}{\partial \theta}\left[\sin(\beta - \theta)(\sin \theta - \cos \theta \tan \phi_d')\right] = 0 \qquad (15.27)$$

Solving Eq. (15.27) gives the critical value of θ, or

$$\theta_{cr} = \frac{\beta + \phi'_d}{2} \tag{15.28}$$

Substitution of the value of $\theta = \theta_{cr}$ into Eq. (15.25) yields

$$c'_d = \frac{\gamma H}{4}\left[\frac{1 - \cos(\beta - \phi'_d)}{\sin \beta \cos \phi'_d}\right] \tag{15.29}$$

The preceding equation also can be written as

$$\frac{c'_d}{\gamma H} = m = \frac{1 - \cos(\beta - \phi'_d)}{4 \sin \beta \cos \phi'_d} \tag{15.30}$$

where m = stability number.

The maximum height of the slope for which critical equilibrium occurs can be obtained by substituting $c'_d = c'$ and $\phi'_d = \phi'$ into Eq. (15.29). Thus,

$$H_{cr} = \frac{4c'}{\gamma}\left[\frac{\sin \beta \cos \phi'}{1 - \cos(\beta - \phi')}\right] \tag{15.31}$$

Example 15.2

A cut is to be made in a soil having $\gamma = 16.51$ kN/m³, $c' = 28.75$ kN/m², and $\phi' = 15°$. The side of the cut slope will make an angle of 45° with the horizontal. What should be the depth of the cut slope that will have a factor of safety (F_s) of 3?

Solution

Given: $\phi' = 15°$; $c' = 28.75$ kN/m². If $F_s = 3$, then $F_{c'}$ and $F_{\phi'}$ should both be equal to 3.

$$F_{c'} = \frac{c'}{c'_d}$$

or

$$c'_d = \frac{c'}{F_{c'}} = \frac{c'}{F_s} = \frac{28.75}{3} = 9.58 \text{ kN/m}^2$$

Similarly,

$$F_{\phi'} = \frac{\tan \phi'}{\tan \phi'_d}$$

$$\tan \phi'_d = \frac{\tan \phi'}{F_{\phi'}} = \frac{\tan \phi'}{F_s} = \frac{\tan 15}{3}$$

or

$$\phi'_d = \tan^{-1}\left[\frac{\tan 15}{3}\right] = 5.1°$$

Substituting the preceding values of c'_d and ϕ'_d in Eq. (15.29)

$$H = \frac{4c'_d}{\gamma}\left[\frac{\sin \beta \cdot \cos \phi'_d}{1 - \cos(\beta - \phi'_d)}\right]$$

$$= \frac{4 \times 9.58}{16.51}\left[\frac{\sin 45 \cdot \cos 5.1}{1 - \cos(45 - 5.1)}\right]$$

$$= \textbf{7.02 m}$$ ∎

15.6 Analysis of Finite Slopes with Circular Failure Surfaces—General

Modes of Failure

In general, finite slope failure occurs in one of the following modes (Figure 15.11):

1. When the failure occurs in such a way that the surface of sliding intersects the slope at or above its toe, it is called a *slope failure* (Figure 15.11a). The failure circle is referred to as a *toe circle* if it passes through the toe of the slope and as a *slope circle* if it passes above the toe of the slope. Under certain circumstances, a *shallow slope failure* can occur, as shown in Figure 15.11b.

2. When the failure occurs in such a way that the surface of sliding passes at some distance below the toe of the slope, it is called a *base failure* (Figure 15.11c). The failure circle in the case of base failure is called a *midpoint circle*.

Types of Stability Analysis Procedures

Various procedures of stability analysis may, in general, be divided into two major classes:

1. *Mass procedure:* In this case, the mass of the soil above the surface of sliding is taken as a unit. This procedure is useful when the soil that forms the slope is assumed to be homogeneous, although this is not the case in most natural slopes.

2. *Method of slices:* In this procedure, the soil above the surface of sliding is divided into a number of vertical parallel slices. The stability of each slice is calculated separately. This is a versatile technique in which the nonhomogeneity of the soils and pore water pressure can be taken into consideration. It also accounts for the variation of the normal stress along the potential failure surface.

The fundamentals of the analysis of slope stability by mass procedure and method of slices are given in the following sections.

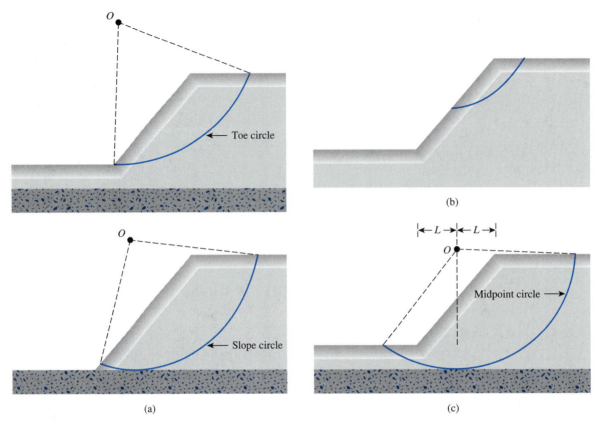

Figure 15.11 Modes of failure of finite slope: (a) slope failure; (b) shallow slope failure; (c) base failure

Mass Procedure—Slopes in Homogeneous Clay Soil with $\phi = 0$

Figure 15.12 shows a slope in a homogeneous soil. The undrained shear strength of the soil is assumed to be constant with depth and may be given by $\tau_f = c_u$. To perform the stability analysis, we choose a trial potential curve of sliding, *AED*, which is an arc of a circle that has a radius *r*. The center of the circle is located at *O*. Considering a unit length perpendicular to the section of the slope, we can give the weight of the soil above the curve *AED* as $W = W_1 + W_2$, where

$$W_1 = (\text{Area of } FCDEF)(\gamma)$$

and

$$W_2 = (\text{Area of } ABFEA)(\gamma)$$

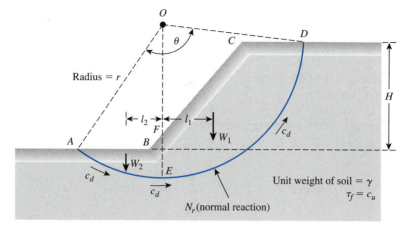

Figure 15.12 Stability analysis of slope in homogeneous saturated clay soil ($\phi = 0$)

Failure of the slope may occur by sliding of the soil mass. The moment of the driving force about O to cause slope instability is

$$M_d = W_1 l_1 - W_2 l_2 \tag{15.32}$$

where l_1 and l_2 are the moment arms.

The resistance to sliding is derived from the cohesion that acts along the potential surface of sliding. If c_d is the cohesion that needs to be developed, the moment of the resisting forces about O is

$$M_R = c_d(\widehat{AED})(1)(r) = c_d r^2 \theta \tag{15.33}$$

For equilibrium, $M_R = M_d$; thus,

$$c_d r^2 \theta = W_1 l_1 - W_2 l_2$$

or

$$c_d = \frac{W_1 l_1 - W_2 l_2}{r^2 \theta} \tag{15.34}$$

The factor of safety against sliding may now be found.

$$F_s = \frac{\tau_f}{c_d} = \frac{c_u}{c_d} \tag{15.35}$$

Note that the potential curve of sliding, *AED,* was chosen arbitrarily. The critical surface is that for which the ratio of c_u to c_d is a minimum. In other words, c_d is maximum. To find the critical surface for sliding, one must make a number of trials for different trial circles. The minimum value of the factor of safety thus obtained is the factor of safety against sliding for the slope, and the corresponding circle is the critical circle.

Stability problems of this type have been solved analytically by Fellenius (1927) and Taylor (1937). For the case of *critical circles,* the developed cohesion can be expressed by the relationship

$$c_d = \gamma H m$$

or

$$\frac{c_d}{\gamma H} = m \qquad (15.36)$$

Note that the term *m* on the right-hand side of the preceding equation is non-dimensional and is referred to as the *stability number.* The critical height (i.e., $F_s = 1$) of the slope can be evaluated by substituting $H = H_{cr}$ and $c_d = c_u$ (full mobilization of the undrained shear strength) into the preceding equation. Thus,

$$H_{cr} = \frac{c_u}{\gamma m} \qquad (15.37)$$

Values of the stability number, *m,* for various slope angles, β, are given in Figure 15.13. Terzaghi used the term $\gamma H/c_d$, the reciprocal of *m* and called it the *stability factor.* Readers should be careful in using Figure 15.13 and note that it is valid for slopes of saturated clay and is applicable to only undrained conditions ($\phi = 0$).

In reference to Figure 15.13, the following must be pointed out:

1. For a slope angle β greater than 53°, the critical circle is always a toe circle. The location of the center of the critical toe circle may be found with the aid of Figure 15.14.
2. For $\beta < 53°$, the critical circle may be a toe, slope, or midpoint circle, depending on the location of the firm base under the slope. This is called the *depth function,* which is defined as

$$D = \frac{\text{Vertical distance from top of slope to firm base}}{\text{Height of slope}} \qquad (15.38)$$

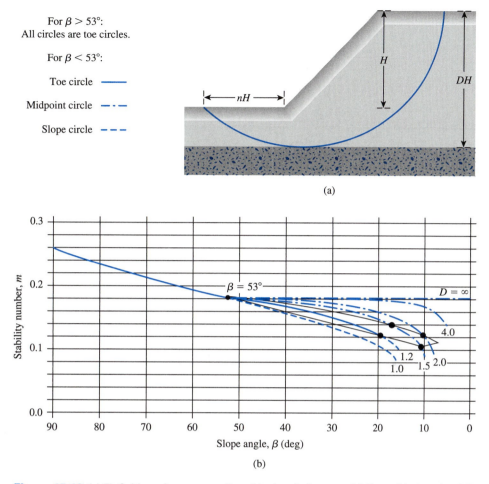

For $\beta > 53°$:
All circles are toe circles.

For $\beta < 53°$:

Toe circle ———

Midpoint circle —·—

Slope circle ———

(a)

(b)

Figure 15.13 (a) Definition of parameters for midpoint circle type of failure; (b) plot of stability number against slope angle (*Adapted from Terzaghi and Peck, 1967. With permission of John Wiley & Sons, Inc.*)

3. When the critical circle is a midpoint circle (i.e., the failure surface is tangent to the firm base), its position can be determined with the aid of Figure 15.15.
4. The maximum possible value of the stability number for failure as a midpoint circle is 0.181.

   Fellenius (1927) also investigated the case of critical toe circles for slopes with $\beta < 53°$. The location of these can be determined with the use of Figure 15.16 and Table 15.1. Note that these critical toe circles are not necessarily the most critical circles that exist.

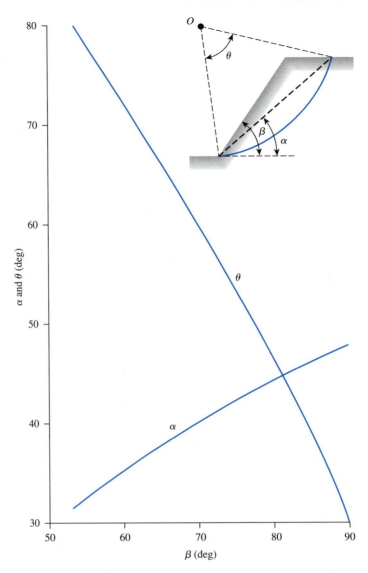

Figure 15.14 Location of the center of critical circles for $\beta > 53°$

Table 15.1 Location of the Center of Critical Toe Circles ($\beta < 53°$)

n'	β (deg)	α_1 (deg)	α_2 (deg)
1.0	45	28	37
1.5	33.68	26	35
2.0	26.57	25	35
3.0	18.43	25	35
5.0	11.32	25	37

Note: for notations of n', β, α_1, and α_2, see Figure 15.16

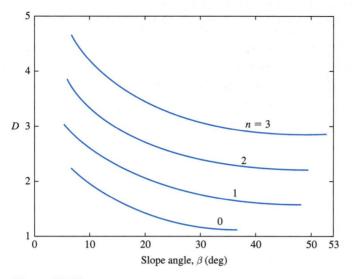

Figure 15.15 Location of midpoint circle (*Based on Fellenius, 1927; and Terzaghi and Peck, 1967*)

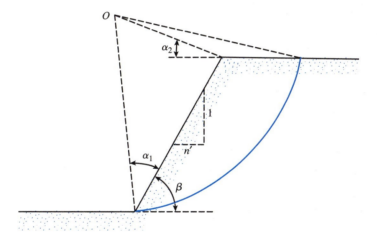

Figure 15.16 Location of the center of critical toe circles for $\beta < 53°$

Example 15.3

A cut slope is to be made in a soft saturated clay with its sides rising at an angle of 60°
to the horizontal (Figure 15.17).
    Given: $c_u = 40$ kN/m² and $\gamma = 17.5$ kN/m³.

  a. Determine the maximum depth up to which the excavation can be carried out.
  b. Find the radius, r, of the critical circle when the factor of safety is equal to 1
     (Part a).
  c. Find the distance $\overline{BC}$.

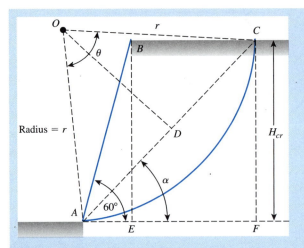

Figure 15.17

Solution

Part a

Since the slope angle $\beta = 60° > 53°$, the critical circle is a toe circle. From Figure 15.13, for $\beta = 60°$, the stability number = 0.195.

$$H_{cr} = \frac{c_u}{\gamma m} = \frac{40}{17.5 \times 0.195} = 11.72 \text{ m}$$

Part b

From Figure 15.17,

$$r = \frac{\overline{DC}}{\sin \dfrac{\theta}{2}}$$

But

$$\overline{DC} = \frac{\overline{AC}}{2} = \frac{\left(\dfrac{H_{cr}}{\sin \alpha}\right)}{2}$$

so,

$$r = \frac{H_{cr}}{2 \sin \alpha \ \sin \dfrac{\theta}{2}}$$

From Figure 15.14, for $\beta = 60°$, $\alpha = 35°$ and $\theta = 72.5°$. Substituting these values into the equation for r we get

$$r = \frac{H_{cr}}{2 \sin \alpha \ \sin \dfrac{\theta}{2}}$$

$$= \frac{11.72}{2(\sin 35)(\sin 36.25)} = 17.28 \text{ m}$$

Part c

$$\overline{BC} = \overline{EF} = \overline{AF} - \overline{AE}$$
$$= H_{cr}(\cot \alpha - \cot 75°)$$
$$= 11.72(\cot 35 - \cot 60) = \mathbf{9.97 \ m}$$ ∎

Example 15.4

A cut slope was excavated in a saturated clay. The slope made an angle of 40° with the horizontal. Slope failure occurred when the cut reached a depth of 6.1 m. Previous soil explorations showed that a rock layer was located at a depth of 9.15 m below the ground surface. Assuming an undrained condition and $\gamma_{sat} = 17.29$ kN/m^3, find the following.

  a. Determine the undrained cohesion of the clay (Figure 15.13).
  b. What was the nature of the critical circle?
  c. With reference to the toe of the slope, at what distance did the surface of sliding intersect the bottom of the excavation?

Solution
Part a
Referring to Figure 15.13,

$$D = \frac{9.15}{6.1} = 1.5$$

$$\gamma_{sat} = 17.29 \text{ kN/m}^3$$

$$H_{cr} = \frac{c_u}{\gamma m}$$

From Figure 15.13, for $\beta = 40°$ and $D = 1.5$, $m = 0.175$. So,

$$c_u = (H_{cr})(\gamma)(m) = (6.1)(17.29)(0.175) = \mathbf{18.5 \ kN/m^2}$$

Part b
Midpoint circle.

Part c
Again, from Figure 15.15, for $D = 1.5$, $\beta = 40°$; $n = 0.9$. So,

$$\text{Distance} = (n)(H_{cr}) = (0.9)(6.1) = \mathbf{5.49 \ m}$$ ∎

15.8 *Mass Procedure—Stability of Saturated Clay Slope (ϕ = 0 condition) with Earthquake Forces*

The stability of saturated clay slopes ($\phi = 0$ condition) with earthquake forces has been analyzed by Koppula (1984). Figure 15.18 shows a clay slope with a potential curve of sliding *AED*, which is an arc of a circle that has radius *r*. The center of the circle is located at *O*. Taking a unit length perpendicular to the slope, we consider these forces for stability analysis:

1. Weight of the soil wedge:

$$W = (\text{area of } ABCDEA)(\gamma)$$

2. Horizontal inertia force, $k_h W$:

$$k_h = \frac{\text{horizontal component of earthquake acceleration}}{g}$$

(g = acceleration from gravity)

3. Cohesive force along the surface of sliding, which will have a magnitude of $(\overset{\frown}{AED})c_u$.

   The moment of the driving forces about O can now be given as

$$M_d = Wl_1 + k_h Wl_2 \tag{15.39}$$

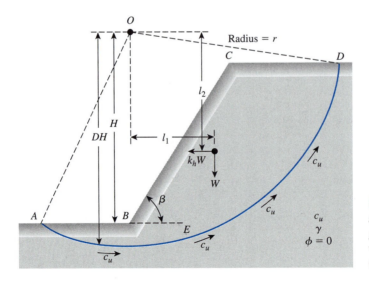

Figure 15.18 Stability analysis of slope in homogeneous saturated clay with earthquake forces ($\phi = 0$ condition)

Similarly, the moment of the resisting forces about O is

$$M_r = (\overset{\frown}{AED})(c_u)r \tag{15.40}$$

Thus, the factor of safety against sliding is

$$F_s = \frac{M_r}{M_d} = \frac{(\overset{\frown}{AED})(c_u)(r)}{Wl_1 + k_hWl_2} = \frac{c_u}{\gamma H}M \tag{15.41}$$

where M = stability factor.

The variations of the stability factor M with slope angle β and k_h based on Koppula's (1984) analysis are given in Figures 15.19 and 15.20.

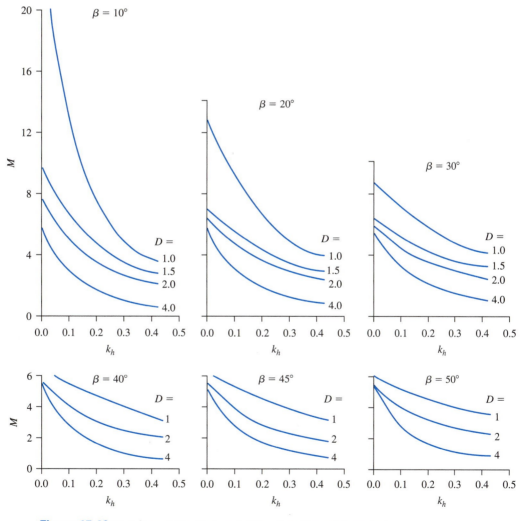

Figure 15.19 Variation of M with k_h and β based on Koppula's analysis

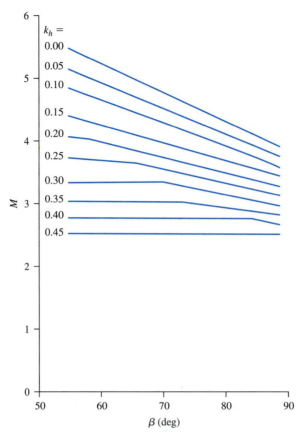

Figure 15.20 Variation of M with k_h based on Koppula's analysis (for $\beta \geq 55°$)

Example 15.5

A cut slope in saturated clay (Figure 15.21) makes an angle of 56° with the horizontal. Assume $k_h = 0.25$.

    a. Determine the maximum depth up to which the cut could be made.
    b. How deep should the cut be made if a factor of safety of 2 against sliding is required?

Solution
Part a
For the critical height of the slopes, $F_s = 1$. So, from Eq. (15.41),

$$H_{cr} = \frac{c_u M}{\gamma}$$

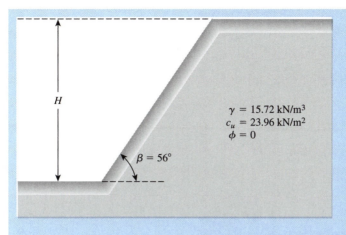

Figure 15.21

From Figure 15.20, for $\beta = 56°$ and $k_h = 0.25$, $M = 3.66$. Thus,

$$H_{cr} = \frac{(23.96)(3.66)}{15.72} = \textbf{5.58 m}$$

Part b
From Eq. (15.41),

$$H = \frac{c_u M}{\gamma F_s} = \frac{(23.96)(3.66)}{(15.72)(2)} = \textbf{2.79 m} \quad \blacksquare$$

15.9 *Mass Procedure—Slopes in Homogeneous c'-ϕ' Soil*

A slope in a homogeneous soil is shown in Figure 15.22a. The shear strength of the soil is given by

$$\tau_f = c' + \sigma' \tan \phi'$$

The pore water pressure is assumed to be zero. $\overset{\frown}{AC}$ is a trial circular arc that passes through the toe of the slope, and O is the center of the circle. Considering a unit length perpendicular to the section of the slope, we find

$$\text{Weight of soil wedge } ABC = W = (\text{Area of } ABC)(\gamma)$$

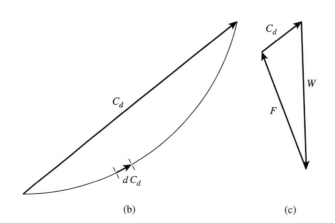

(a)

(b)   (c)

Figure 15.22 Stability analysis of slope in homogeneous c'–ϕ' soil

For equilibrium, the following other forces are acting on the wedge:

- C_d—resultant of the cohesive force that is equal to the cohesion per unit area developed times the length of the cord $\overline{AC}$. The magnitude of C_d is given by the following (Figure 15.22b).

$$C_d = c'_d(\overline{AC}) \tag{15.42}$$

C_d acts in a direction parallel to the cord $\overline{AC}$ (see Figure 15.22 b) and at a distance a from the center of the circle O such that

$$C_d(a) = c'_d(\widehat{AC})r$$

or

$$a = \frac{c'_d(\widehat{AC})r}{C_d} = \frac{\widehat{AC}}{AC}r \tag{15.43}$$

- *F*—the resultant of the normal and frictional forces along the surface of sliding. For equilibrium, the line of action of *F* will pass through the point of intersection of the line of action of *W* and C_d.

Now, if we assume that full friction is mobilized ($\phi'_d = \phi'$ or $F_{\phi'} = 1$), the line of action of *F* will make an angle of ϕ' with a normal to the arc and thus will be a tangent to a circle with its center at *O* and having a radius of $r \sin \phi'$. This circle is called the *friction circle*. Actually, the radius of the friction circle is a little larger than $r \sin \phi'$.

Because the directions of *W*, C_d, and *F* are known and the magnitude of *W* is known, a force polygon, as shown in Figure 15.22c, can be plotted. The magnitude of C_d can be determined from the force polygon. So the cohesion per unit area developed can be found.

$$c'_d = \frac{C_d}{AC}$$

Determination of the magnitude of c'_d described previously is based on a trial surface of sliding. Several trials must be made to obtain the most critical sliding surface, along which the developed cohesion is a maximum. Thus, we can express the maximum cohesion developed along the critical surface as

$$c'_d = \gamma H [f(\alpha, \beta, \theta, \phi')] \tag{15.44}$$

For critical equilibrium—that is, $F_{c'} = F_{\phi'} = F_s = 1$—we can substitute $H = H_{cr}$ and $c'_d = c'$ into Eq. (15.44) and write

$$c' = \gamma H_{cr}[f(\alpha, \beta, \theta, \phi')]$$

or

$$\frac{c'}{\gamma H_{cr}} = f(\alpha, \beta, \theta, \phi') = m \tag{15.45}$$

where m = stability number. The values of m for various values of ϕ' and β are given in Figure 15.23, which is based on Taylor (1937). This can be used to determine the factor of safety, F_s, of the homogeneous slope. The procedure to do the analysis is given as

Step 1: Determine c', ϕ', γ, β and *H*.
Step 2: Assume several values of ϕ'_d (Note: $\phi'_d \leq \phi'$, such as $\phi'_{d(1)}$, $\phi'_{d(2)}$ . . . .
(Column 1 of Table 15.2).

Table 15.2 Determination of F_s by Frication Circle Method

ϕ'_d	$F_{\phi'} = \dfrac{\tan \phi'}{\tan \phi'_d}$	m	c'_d	$F_{c'}$
(1)	(2)	(3)	(4)	(5)
$\phi'_{d(1)}$	$\dfrac{\tan \phi'}{\tan \phi'_{d(1)}}$	m_1	$m_1 \gamma H = c'_{d(1)}$	$\dfrac{c'}{c'_{d(1)}} = F_{c'(1)}$
$\phi'_{d(2)}$	$\dfrac{\tan \phi'}{\tan \phi'_{d(2)}}$	m_2	$m_2 \gamma H = c'_{d(2)}$	$\dfrac{c'}{c'_{d(2)}} = F_{c'(2)}$

Step 3: Determine $F_{\phi'}$, for each assumed value of ϕ'_d as (Column 2, Table 15.2)

$$FS_{\phi'(1)} = \frac{\tan \phi'}{\tan \phi'_{d(1)}}$$

$$FS_{\phi'(2)} = \frac{\tan \phi'}{\tan \phi'_{d(2)}}$$

Step 4: For each assumed value of ϕ'_d and β, determine m (that is, m_1, m_2, m_3, . . .) from Figure 15.23 (Column 3, Table 15.2).

Step 5: Determine the developed cohesion for each value of m as (Column 4, Table 15.2)

$$c'_{d(1)} = m_1 \gamma H$$
$$c'_{d(2)} = m_2 \gamma H$$

Step 6: Calculate $F_{c'}$ for each value of c'_d (Column 5, Table 15.2), or

$$F_{c'(1)} = \frac{c'}{c'_{d(1)}}$$

$$F_{c'(2)} = \frac{c'}{c'_{d(2)}}$$

Step 7: Plot a graph of $F_{\phi'}$ versus the corresponding $F_{c'}$ (Figure 15.24) and determine $F_s = F_{\phi'} = F_{c'}$.

An example of determining FS_s using the procedure just described is given in Example 15.6.

Using Taylor's friction circle method of slope stability (as shown in Example 15.6) Singh (1970) provided graphs of equal factors of safety, FS_s, for various slopes. This is shown in Figure 15.25.

Calculations have shown that for $\phi > \sim 3°$, the critical circles are all *toe circles*.

More recently, Michalowski (2002) made a stability analysis of simple slopes using the kinematic approach of limit analysis applied to a rigid rotational collapse mechanism. The failure surface in soil assumed in this study is an arc of a logarithmic spiral (Figure 15.26). The results of this study are summarized in Figure 15.27, from which F_s can be obtained directly.

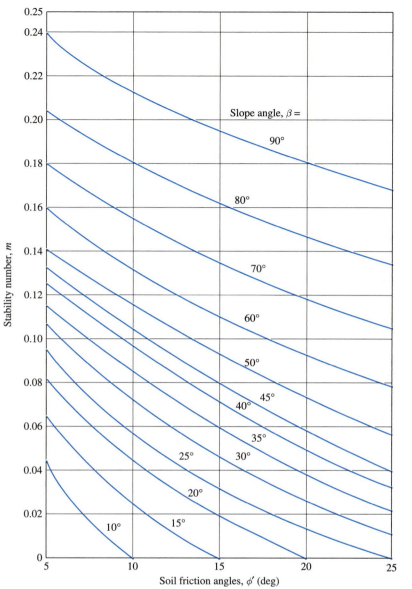

Figure 15.23 Taylor's stability number

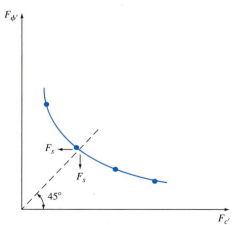

Figure 15.24 Plot of $F_{\phi'}$ versus $F_{c'}$ to determine F_s

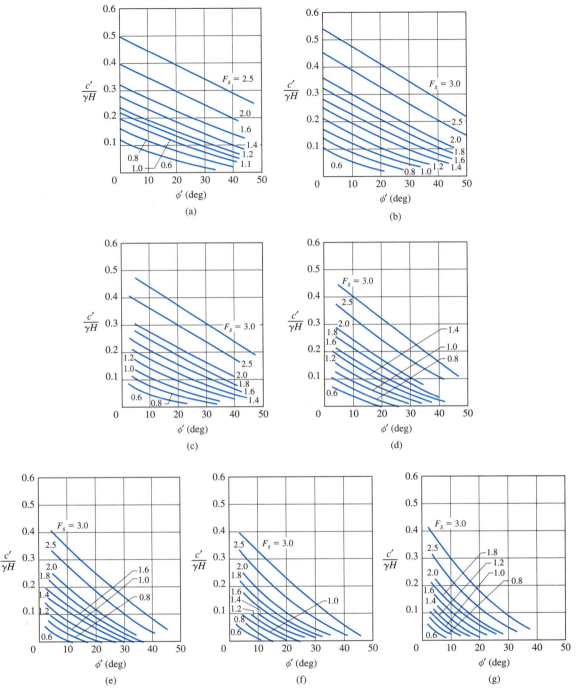

Figure 15.25 Contours of equal factors of safety: (a) slope – 1 vertical to 0.5 horizontal; (b) slope – 1 vertical to 0.75 horizontal; (c) slope – 1 vertical to 1 horizontal; (d) slope – 1 vertical to 1.5 horizontal; (e) slope – 1 vertical to 2 horizontal; (f) slope – 1 vertical to 2.5 horizontal; (g) slope – 1 vertical to 3 horizontal (*After Singh, 1970. With permission from ASCE.*)

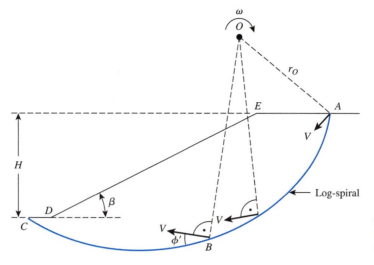

Figure 15.26 Stability analysis using rotational collapse mechanism

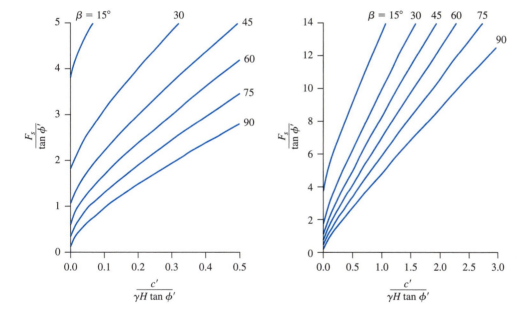

Figure 15.27 Michalowski's analysis for stability of simple slopes

Example 15.6

A slope with $\beta = 45°$ is to be constructed with a soil that has $\phi' = 20°$ and $c' = 24$ kN/m^2. The unit weight of the compacted soil will be 18.9 kN/m^3.

    a. Find the critical height of the slope.

    b. If the height of the slope is 10 m, determine the factor of safety with respect to strength.

Solution
Part a
We have

$$m = \frac{c'}{\gamma H_{cr}}$$

From Figure 15.23, for $\beta = 45°$ and $\phi' = 20°$, $m = 0.06$. So

$$H_{cr} = \frac{c'}{\gamma m} = \frac{24}{(18.9)(0.06)} = \mathbf{21.1\,m}$$

Part b
If we assume that full friction is mobilized, then, referring to Figure 15.23 (for $\beta = 45°$ and $\phi'_d = \phi' = 20°$), we have

$$m = 0.06 = \frac{c'_d}{\gamma H}$$

or

$$c'_d = (0.06)(18.9)(10) = 11.34 \text{ kN/m}^2$$

Thus,

$$F_{\phi'} = \frac{\tan \phi'}{\tan \phi'_d} = \frac{\tan 20}{\tan 20} = 1$$

and

$$F_{c'} = \frac{c'}{c'_d} = \frac{24}{11.34} = 2.12$$

Since $FS_{c'} \neq FS_{\phi'}$, this is not the factor of safety with respect to strength.
Now we can make another trial. Let the developed angle of friction, ϕ'_d, be equal to 15°. For $\beta = 45°$ and the friction angle equal to 15°, we find from Figure 15.23.

$$m = 0.083 = \frac{c'_d}{\gamma H}$$

or

$$c'_d = (0.083)(18.9)(10) = 15.69 \text{ kN/m}^2$$

For this trial,

$$F_{\phi'} = \frac{\tan \phi'}{\tan \phi'_d} = \frac{\tan 20}{\tan 15} = 1.36$$

and

$$F_{c'} = \frac{c'}{c'_d} = \frac{24}{15.69} = 1.53$$

Similar calculations of $F_{\phi'}$ and $F_{c'}$ for various assumed values of ϕ'_d are given in the following table.

ϕ'_d	tan ϕ'_d	$F_{\phi'}$	m	c'_d (kN/m²)	$F_{c'}$
20	0.364	1.0	0.06	11.34	2.12
15	0.268	1.36	0.083	15.69	1.53
10	0.176	2.07	0.105	19.85	1.21
5	0.0875	4.16	0.136	25.70	0.93

The values of $F_{\phi'}$ are plotted against their corresponding values of $F_{c'}$ in Figure 15.28, from which we find

$$FS_{c'} = FS_{\phi'} = FS_s = \mathbf{1.42}$$

Note: We could have found the value of FS_s from Figure 15.25c. Since $\beta = 45°$, it is a slope of 1V:1H. For this slope

$$\frac{c'}{\gamma H} = \frac{24}{(18.9)(10)} = 0.127$$

From Figure 15.25c, for $c'/\gamma H = 0.127$, the value of $FS_s \approx \mathbf{1.4}$

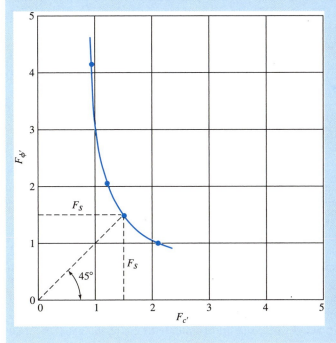

Figure 15.28

Example 15.7

Solve Example 15.6 using Michalowski's solution.

Solution

Part a

For critical height (H_{cr}), $FS_s = 1$. Thus,

$$\frac{c'}{\gamma H \tan \phi'} = \frac{24}{(18.9)(H_{cr})(\tan 20)} = \frac{3.49}{H_{cr}}$$

$$\frac{FS_s}{\tan \phi'} = \frac{1}{\tan 20} = 2.747$$

$$\beta = 45°$$

From Figure 15.27, for $\beta = 45°$ and $F_s/\tan \phi' = 2.747$, the value of $c'/\gamma H \tan \phi' \approx 0.17$. So

$$\frac{3.49}{H_{cr}} = 0.17; \quad H_{cr} = \mathbf{20.5\,m}$$

Part b

$$\frac{c'}{\gamma H \tan \phi'} = \frac{24}{(18.9)(10)(\tan 20)} = 0.349$$

$$\beta = 45°$$

From Figure 15.27, $F_s/\tan \phi' = 4$.

$$F_s = 4 \tan \phi' = (4)(\tan 20) = \mathbf{1.46}$$ ∎

15.10 Ordinary Method of Slices

Stability analysis by using the method of slices can be explained with the use of Figure 15.29a, in which *AC* is an arc of a circle representing the trial failure surface. The soil above the trial failure surface is divided into several vertical slices. The width of each slice need not be the same. Considering a unit length perpendicular to the cross section shown, the forces that act on a typical slice (*n*th slice) are shown in Figure 15.29b. W_n is the weight of the slice. The forces N_r and T_r, respectively, are the normal and tangential components of the reaction R. P_n and P_{n+1} are the normal forces that act on the sides of the slice. Similarly, the shearing forces that act on the sides of the slice are T_n and T_{n+1}. For simplicity, the pore water pressure is assumed to be zero. The forces P_n, P_{n+1}, T_n, and T_{n+1} are difficult to determine. However, we can make an approximate assumption that the resultants of P_n and T_n are equal in magnitude to the resultants of P_{n+1} and T_{n+1} and that their lines of action coincide.

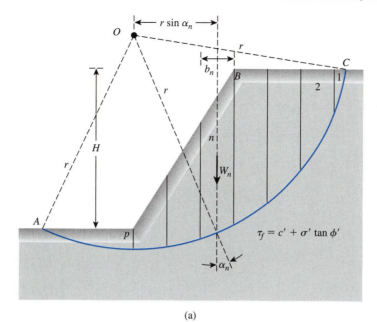

(a)

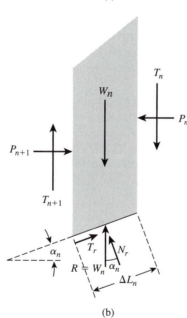

(b)

Figure 15.29
Stability analysis by ordinary
method of slices: (a) trial failure surface;
(b) forces acting on nth slice

For equilibrium consideration,

$$N_r = W_n \cos \alpha_n$$

The resisting shear force can be expressed as

$$T_r = \tau_d(\Delta L_n) = \frac{\tau_f(\Delta L_n)}{F_s} = \frac{1}{F_s}[c' + \sigma' \tan \phi']\Delta L_n \qquad (15.46)$$

The normal stress, σ', in Eq. (15.46) is equal to

$$\frac{N_r}{\Delta L_n} = \frac{W_n \cos \alpha_n}{\Delta L_n}$$

For equilibrium of the trial wedge ABC, the moment of the driving force about O equals the moment of the resisting force about O, or

$$\sum_{n=1}^{n=p} W_n r \sin \alpha_n = \sum_{n=1}^{n=p} \frac{1}{F_s} \left(c' + \frac{W_n \cos \alpha_n}{\Delta L_n} \tan \phi' \right) (\Delta L_n)(r)$$

or

$$F_s = \frac{\displaystyle\sum_{n=1}^{n=p} (c' \Delta L_n + W_n \cos \alpha_n \tan \phi')}{\displaystyle\sum_{n=1}^{n=p} W_n \sin \alpha_n} \tag{15.47}$$

[*Note:* ΔL_n in Eq. (15.47) is approximately equal to $(b_n)/(\cos \alpha_n)$, where b_n = the width of the nth slice.]

Note that the value of α_n may be either positive or negative. The value of α_n is positive when the slope of the arc is in the same quadrant as the ground slope. To find the minimum factor of safety—that is, the factor of safety for the critical circle—one must make several trials by changing the center of the trial circle. This method generally is referred to as the *ordinary method of slices.*

For convenience, a slope in a homogeneous soil is shown in Figure 15.29. However, the method of slices can be extended to slopes with layered soil, as shown in Figure 15.30. The general procedure of stability analysis is the same. However, some minor points should be kept in mind. When Eq. (15.47) is used for the factor of safety calculation, the values of ϕ' and c' will not be the same for all slices. For example, for slice No. 3 (see Figure 15.30), we have to use a friction angle of $\phi' = \phi'_3$ and cohesion $c' = c'_3$; similarly, for slice No. 2, $\phi' = \phi'_2$ and $c' = c'_2$.

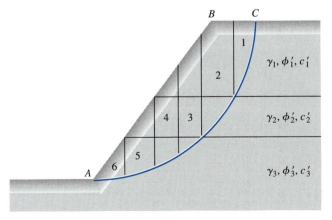

Figure 15.30
Stability analysis, by ordinary method of slices, for slope in layered soils

It is of interest to note that if total shear strength parameters (that is, $\tau_f = c + \tan \phi$) were used, Eq. (15.47) would take the form

$$F_s = \frac{\sum\limits_{n=1}^{n=p}(c\Delta L_n + W_n \cos \alpha_n \tan \phi)}{\sum\limits_{n=1}^{n=p} W_n \sin \alpha_n} \tag{15.48}$$

Example 15.8

For the slope shown in Figure 15.31, find the factor of safety against sliding for the trial slip surface AC. Use the ordinary method of slices.

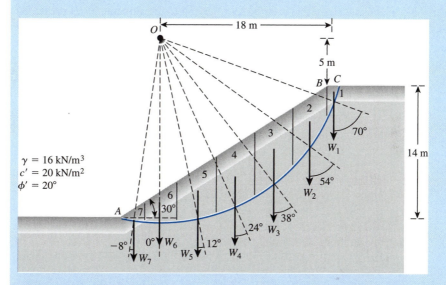

Figure 15.31 Stability analysis of a slope by ordinary method of slices

Solution
The sliding wedge is divided into seven slices. Now the following table can be prepared:

Slice no. (1)	W (kN/m) (2)	α_n (deg) (3)	$\sin \alpha_n$ (4)	$\cos \alpha_n$ (5)	ΔL_n (m) (6)	$W_n \sin \alpha_n$ (kN/m) (7)	$W_n \cos \alpha_n$ (kN/m) (8)
1	22.4	70	0.94	0.342	2.924	21.1	7.66
2	294.4	54	0.81	0.588	6.803	238.5	173.1
3	435.2	38	0.616	0.788	5.076	268.1	342.94
4	435.2	24	0.407	0.914	4.376	177.1	397.8
5	390.4	12	0.208	0.978	4.09	81.2	381.8
6	268.8	0	0	1	4	0	268.8
7	66.58	−8	−0.139	0.990	3.232	−9.25	65.9
					Σ Col. 6 = 30.501 m	Σ Col. 7 = 776.75 kN/m	Σ Col. 8 = 1638 kN/m

$$F_s = \frac{(\Sigma \text{ Col. 6})(c') + (\Sigma \text{ Col. 8})\tan \phi'}{\Sigma \text{ Col. 7}}$$

$$= \frac{(30.501)(20) + (1638)(\tan 20)}{776.75} = \mathbf{1.55}$$ ∎

15.11 Bishop's Simplified Method of Slices

In 1955, Bishop proposed a more refined solution to the ordinary method of slices. In this method, the effect of forces on the sides of each slice are accounted for to some degree. We can study this method by referring to the slope analysis presented in Figure 15.29. The forces that act on the nth slice shown in Figure 15.29b have been redrawn in Figure 15.32a. Now, let $P_n - P_{n+1} = \Delta P$ and $T_n - T_{n+1} = \Delta T$. Also, we can write

$$T_r = N_r(\tan \phi_d') + c_d' \Delta L_n = N_r\left(\frac{\tan \phi'}{F_s}\right) + \frac{c' \Delta L_n}{F_s} \qquad (15.49)$$

Figure 15.32b shows the force polygon for equilibrium of the nth slice. Summing the forces in the vertical direction gives

$$W_n + \Delta T = N_r \cos \alpha_n + \left[\frac{N_r \tan \phi'}{F_s} + \frac{c' \Delta L_n}{F_s}\right]\sin \alpha_n$$

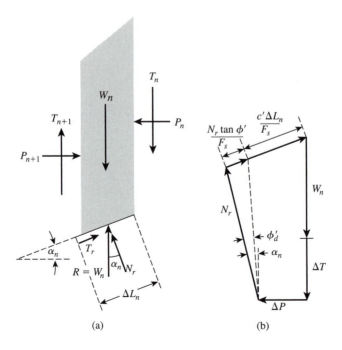

Figure 15.32
Bishop's simplified method of slices: (a) forces acting on the nth slice; (b) force polygon for equilibrium

or

$$N_r = \frac{W_n + \Delta T - \dfrac{c'\Delta L_n}{F_s}\sin \alpha_n}{\cos \alpha_n + \dfrac{\tan \phi' \sin \alpha_n}{F_s}} \tag{15.50}$$

For equilibrium of the wedge ABC (Figure 15.29a), taking the moment about O gives

$$\sum_{n=1}^{n=p} W_n r \sin \alpha_n = \sum_{n=1}^{n=p} T_r r \tag{15.51}$$

where

$$T_r = \frac{1}{F_s}(c' + \sigma' \tan \phi')\Delta L_n$$

$$= \frac{1}{F_s}(c'\Delta L_n + N_r \tan \phi') \tag{15.52}$$

Substitution of Eqs. (15.50) and (15.52) into Eq. (15.51) gives

$$F_s = \frac{\displaystyle\sum_{n=1}^{n=p}(c'b_n + W_n \tan \phi' + \Delta T \tan \phi')\dfrac{1}{m_{\alpha(n)}}}{\displaystyle\sum_{n=1}^{n=p} W_n \sin \alpha_n} \tag{15.53}$$

where

$$m_{\alpha(n)} = \cos \alpha_n + \frac{\tan \phi' \sin \alpha_n}{F_s} \tag{15.54}$$

Figure 15.33 shows the variation of $m_{\alpha(n)}$ with α_n and $\tan \phi'/F_s$.

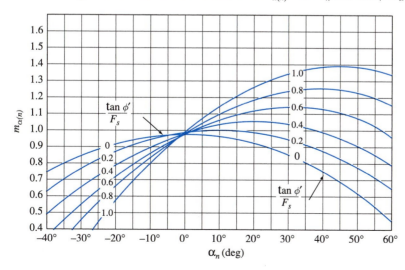

Figure 15.33 Variation of $m_{\alpha(n)}$ with α_n and $\tan \phi'/F_s$ [Eq. (15.54)]

For simplicity, if we let $\Delta T = 0$, Eq. (15.53) becomes

$$F_s = \frac{\sum\limits_{n=1}^{n=p}(c'b_n + W_n \tan \phi')\dfrac{1}{m_{\alpha(n)}}}{\sum\limits_{n=1}^{n=p} W_n \sin \alpha_n} \qquad (15.55)$$

Note that the term F_s is present on both sides of Eq. (15.55). Hence, we must adopt a trial-and-error procedure to find the value of F_s. As in the method of ordinary slices, a number of failure surfaces must be investigated so that we can find the critical surface that provides the minimum factor of safety.

Bishop's simplified method is probably the most widely used. When incorporated into computer programs, it yields satisfactory results in most cases. The ordinary method of slices is presented in this chapter as a learning tool only. It is used rarely now because it is too conservative.

15.12 *Stability Analysis by Method of Slices for Steady-State Seepage*

The fundamentals of the ordinary method of slices and Bishop's simplified method of slices were presented in Sections 15.10 and 15.11, respectively, and we assumed the pore water pressure to be zero. However, for steady-state seepage through slopes, as is the situation in many practical cases, the pore water pressure must be considered when effective shear strength parameters are used. So we need to modify Eqs. (15.47) and (15.55) slightly.

Figure 15.34 shows a slope through which there is steady-state seepage. For the nth slice, the average pore water pressure at the bottom of the slice is equal to $u_n = h_n \gamma_w$. The total force caused by the pore water pressure at the bottom of the nth slice is equal to $u_n \Delta L_n$.

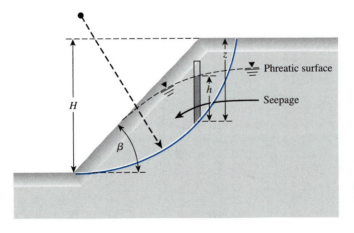

Figure 15.34
Stability analysis of slope with steady-state seepage

Thus, Eq. (15.47) for the ordinary method of slices will be modified to read as follows.

$$F_s = \frac{\sum\limits_{n=1}^{n=p}(c'\Delta L_n + (W_n \cos \alpha_n - u_n \Delta L_n)] \tan \phi'}{\sum\limits_{n=1}^{n=p} W_n \sin \alpha_n} \tag{15.56}$$

Similarly, Eq. (15.55) for Bishop's simplified method of slices will be modified to the form

$$F_s = \frac{\sum\limits_{n=1}^{n=p}[c'b_n + (W_n - u_n b_n)\tan \phi'] \dfrac{1}{m_{(\alpha)n}}}{\sum\limits_{n=1}^{n=p} W_n \sin \alpha_n} \tag{15.57}$$

Note that W_n in Eqs. (15.56) and (15.57) is the *total weight* of the slice.

Using Eq. (15.57), Bishop and Morgenstern (1960) developed tables for the calculation of F_s for simple slopes. The principles of these developments can be explained as follows. In Eq. (15.50),

$$W_n = \text{total weight of the } n\text{th slice} = \gamma b_n z_n \tag{15.58}$$

where z_n = average height of the nth slice. Also in Eq. (15.57),

$$u_n = h_n \gamma_w$$

So, we can let

$$r_{u(n)} = \frac{u_n}{\gamma z_n} = \frac{h_n \gamma_w}{\gamma z_n} \tag{15.59}$$

Note that $r_{u(n)}$ is a nondimensional quantity. Substituting Eqs. (15.58) and (15.59) into Eq. (15.57) and simplifying, we obtain

$$F_s = \left[\frac{1}{\sum\limits_{n=1}^{n=p} \dfrac{b_n}{H}\dfrac{z_n}{H} \sin \alpha_n}\right] \times \sum\limits_{n=1}^{n=p}\left\{ \frac{\dfrac{c'}{\gamma H}\dfrac{b_n}{H} + \dfrac{b_n}{H}\dfrac{z_n}{H}[1 - r_{u(n)}]\tan \phi'}{m_{\alpha(n)}}\right\} \tag{15.60}$$

For a steady-state seepage condition, a weighted average value of $r_{u(n)}$ can be taken, which is a constant. Let the weighted averaged value of $r_{u(n)}$ be r_u. For most practical cases, the value of r_u may range up to 0.5. Thus,

$$F_s = \left[\frac{1}{\sum\limits_{n=1}^{n=p} \dfrac{b_n}{H}\dfrac{z_n}{H} \sin \alpha_n}\right] \times \sum\limits_{n=1}^{n=p}\left\{ \frac{\left[\dfrac{c'}{\gamma H}\dfrac{b_n}{H} + \dfrac{b_n}{H}\dfrac{z_n}{H}(1 - r_u)\tan \phi'\right]}{m_{\alpha(n)}}\right\} \tag{15.61}$$

The factor of safety based on the preceding equation can be solved and expressed in the form

$$F_s = m' - n' r_u \tag{15.62}$$

where m' and n' = stability coefficients. Table 15.3 gives the values of m' and n' for various combinations of $c'/\gamma H$, D, ϕ', and β.

Table 15.3 Values of m' and n' [Eq. (15.62)]

a. Stability coefficients m' and n' for $c'/\gamma H = 0$

	Stability coefficients for earth slopes							
	Slope 2:1		Slope 3:1		Slope 4:1		Slope 5:1	
ϕ'	m'	n'	m'	n'	m'	n'	m'	n'
10.0	0.353	0.441	0.529	0.588	0.705	0.749	0.882	0.917
12.5	0.443	0.554	0.665	0.739	0.887	0.943	1.109	1.153
15.0	0.536	0.670	0.804	0.893	1.072	1.139	1.340	1.393
17.5	0.631	0.789	0.946	1.051	1.261	1.340	1.577	1.639
20.0	0.728	0.910	1.092	1.213	1.456	1.547	1.820	1.892
22.5	0.828	1.035	1.243	1.381	1.657	1.761	2.071	2.153
25.0	0.933	1.166	1.399	1.554	1.865	1.982	2.332	2.424
27.5	1.041	1.301	1.562	1.736	2.082	2.213	2.603	2.706
30.0	1.155	1.444	1.732	1.924	2.309	2.454	2.887	3.001
32.5	1.274	1.593	1.911	2.123	2.548	2.708	3.185	3.311
35.0	1.400	1.750	2.101	2.334	2.801	2.977	3.501	3.639
37.5	1.535	1.919	2.302	2.558	3.069	3.261	3.837	3.989
40.0	1.678	2.098	2.517	2.797	3.356	3.566	4.196	4.362

b. Stability coefficients m' and n' for $c'/\gamma H = 0.025$ and $D = 1.00$

	Stability coefficients for earth slopes							
	Slope 2:1		Slope 3:1		Slope 4:1		Slope 5:1	
ϕ'	m'	n'	m'	n'	m'	n'	m'	n'
10.0	0.678	0.534	0.906	0.683	1.130	0.846	1.367	1.031
12.5	0.790	0.655	1.066	0.849	1.337	1.061	1.620	1.282
15.0	0.901	0.776	1.224	1.014	1.544	1.273	1.868	1.534
17.5	1.012	0.898	1.380	1.179	1.751	1.485	2.121	1.789
20.0	1.124	1.022	1.542	1.347	1.962	1.698	2.380	2.050
22.5	1.239	1.150	1.705	1.518	2.177	1.916	2.646	2.317
25.0	1.356	1.282	1.875	1.696	2.400	2.141	2.921	2.596
27.5	1.478	1.421	2.050	1.882	2.631	2.375	3.207	2.886
30.0	1.606	1.567	2.235	2.078	2.873	2.622	3.508	3.191
32.5	1.739	1.721	2.431	2.285	3.127	2.883	3.823	3.511
35.0	1.880	1.885	2.635	2.505	3.396	3.160	4.156	3.849
37.5	2.030	2.060	2.855	2.741	3.681	3.458	4.510	4.209
40.0	2.190	2.247	3.090	2.993	3.984	3.778	4.885	4.592

(*continued*)

Table 15.3 (*continued*)

c. Stability coefficients m' and n' for $c'/\gamma H = 0.025$ and $D = 1.25$

	Stability coefficients for earth slopes							
	Slope 2:1		Slope 3:1		Slope 4:1		Slope 5:1	
ϕ'	m'	n'	m'	n'	m'	n'	m'	n'
10.0	0.737	0.614	0.901	0.726	1.085	0.867	1.285	1.014
12.5	0.878	0.759	1.076	0.908	1.299	1.098	1.543	1.278
15.0	1.019	0.907	1.253	1.093	1.515	1.311	1.803	1.545
17.5	1.162	1.059	1.433	1.282	1.736	1.541	2.065	1.814
20.0	1.309	1.216	1.618	1.478	1.961	1.775	2.334	2.090
22.5	1.461	1.379	1.808	1.680	2.194	2.017	2.610	2.373
25.0	1.619	1.547	2.007	1.891	2.437	2.269	2.879	2.669
27.5	1.783	1.728	2.213	2.111	2.689	2.531	3.196	2.976
30.0	1.956	1.915	2.431	2.342	2.953	2.806	3.511	3.299
32.5	2.139	2.112	2.659	2.686	3.231	3.095	3.841	3.638
35.0	2.331	2.321	2.901	2.841	3.524	3.400	4.191	3.998
37.5	2.536	2.541	3.158	3.112	3.835	3.723	4.563	4.379
40.0	2.753	2.775	3.431	3.399	4.164	4.064	4.958	4.784

d. Stability coefficients m' and n' for $c'/\gamma H = 0.05$ and $D = 1.00$

	Stability coefficients for earth slopes							
	Slope 2:1		Slope 3:1		Slope 4:1		Slope 5:1	
ϕ'	m'	n'	m'	n'	m'	n'	m'	n'
10.0	0.913	0.563	1.181	0.717	1.469	0.910	1.733	1.069
12.5	1.030	0.690	1.343	0.878	1.688	1.136	1.995	1.316
15.0	1.145	0.816	1.506	1.043	1.904	1.353	2.256	1.567
17.5	1.262	0.942	1.671	1.212	2.117	1.565	2.517	1.825
20.0	1.380	1.071	1.840	1.387	2.333	1.776	2.783	2.091
22.5	1.500	1.202	2.014	1.568	2.551	1.989	3.055	2.365
25.0	1.624	1.338	2.193	1.757	2.778	2.211	3.336	2.651
27.5	1.753	1.480	1.380	1.952	3.013	2.444	3.628	2.948
30.0	1.888	1.630	2.574	2.157	3.261	2.693	3.934	3.259
32.5	2.029	1.789	2.777	2.370	3.523	2.961	4.256	3.585
35.0	2.178	1.958	2.990	2.592	3.803	3.253	4.597	3.927
37.5	2.336	2.138	3.215	2.826	4.103	3.574	4.959	4.288
40.0	2.505	2.332	3.451	3.071	4.425	3.926	5.344	4.668

(*continued*)

Table 15.3 (continued)

e. Stability coefficients m' and n' for $c'/\gamma H = 0.05$ and $D = 1.25$

	Stability coefficients for earth slopes							
	Slope 2:1		Slope 3:1		Slope 4:1		Slope 5:1	
ϕ'	m'	n'	m'	n'	m'	n'	m'	n'
10.0	0.919	0.633	1.119	0.766	1.344	0.886	1.594	1.042
12.5	1.065	0.792	1.294	0.941	1.563	1.112	1.850	1.300
15.0	1.211	0.950	1.471	1.119	1.782	1.338	2.109	1.562
17.5	1.359	1.108	1.650	1.303	2.004	1.567	2.373	1.831
20.0	1.509	1.266	1.834	1.493	2.230	1.799	2.643	2.107
22.5	1.663	1.428	2.024	1.690	2.463	2.038	2.921	2.392
25.0	1.822	1.595	2.222	1.897	2.705	2.287	3.211	2.690
27.5	1.988	1.769	2.428	2.113	2.957	2.546	3.513	2.999
30.0	2.161	1.950	2.645	2.342	3.221	2.819	3.829	3.324
32.5	2.343	2.141	2.873	2.583	3.500	3.107	4.161	3.665
35.0	2.535	2.344	3.114	2.839	3.795	3.413	4.511	4.025
37.5	2.738	2.560	3.370	3.111	4.109	3.740	4.881	4.405
40.0	2.953	2.791	3.642	3.400	4.442	4.090	5.273	4.806

f. Stability coefficients m' and n' for $c'/\gamma H = 0.05$ and $D = 1.50$

	Stability coefficients for earth slopes							
	Slope 2:1		Slope 3:1		Slope 4:1		Slope 5:1	
ϕ'	m'	n'	m'	n'	m'	n'	m'	n'
10.0	1.022	0.751	1.170	0.828	1.343	0.974	1.547	1.108
12.5	1.202	0.936	1.376	1.043	1.589	1.227	1.829	1.399
15.0	1.383	1.122	1.583	1.260	1.835	1.480	2.112	1.690
17.5	1.565	1.309	1.795	1.480	2.084	1.734	2.398	1.983
20.0	1.752	1.501	2.011	1.705	2.337	1.993	2.690	2.280
22.5	1.943	1.698	2.234	1.937	2.597	2.258	2.990	2.585
25.0	2.143	1.903	2.467	2.179	2.867	2.534	3.302	2.902
27.5	2.350	2.117	2.709	2.431	3.148	2.820	3.626	3.231
30.0	2.568	2.342	2.964	2.696	3.443	3.120	3.967	3.577
32.5	2.798	2.580	3.232	2.975	3.753	3.436	4.326	3.940
35.0	3.041	2.832	3.515	3.269	4.082	3.771	4.707	4.325
37.5	3.299	3.102	3.817	3.583	4.431	4.128	5.112	4.735
40.0	3.574	3.389	4.136	3.915	4.803	4.507	5.543	5.171

(continued)

Table 15.3 (*continued*)

g. **Stability coefficients m' and n' for $c'/\gamma H = 0.075$ and toe circles**

	Stability coefficients for earth slopes							
	Slope 2:1		Slope 3:1		Slope 4:1		Slope 5:1	
ϕ'	m'	n'	m'	n'	m'	n'	m'	n'
20	1.593	1.158	2.055	1.516	2.498	1.903	2.934	2.301
25	1.853	1.430	2.426	1.888	2.980	2.361	3.520	2.861
30	2.133	1.730	2.826	2.288	3.496	2.888	4.150	3.461
35	2.433	2.058	3.253	2.730	4.055	3.445	4.846	4.159
40	2.773	2.430	3.737	3.231	4.680	4.061	5.609	4.918

h. **Stability coefficients m' and n' for $c'/\gamma H = 0.075$ and $D = 1.00$**

	Stability coefficients for earth slopes							
	Slope 2:1		Slope 3:1		Slope 4:1		Slope 5:1	
ϕ'	m'	n'	m'	n'	m'	n'	m'	n'
20	1.610	1.100	2.141	1.443	2.664	1.801	3.173	2.130
25	1.872	1.386	2.502	1.815	3.126	2.259	3.742	2.715
30	2.142	1.686	2.884	2.201	3.623	2.758	4.357	3.331
35	2.443	2.030	3.306	2.659	4.177	3.331	5.024	4.001
40	2.772	2.386	3.775	3.145	4.785	3.945	5.776	4.759

i. **Stability coefficients m' and n' for $c'/\gamma H = 0.075$ and $D = 1.25$**

	Stability coefficients for earth slopes							
	Slope 2:1		Slope 3:1		Slope 4:1		Slope 5:1	
ϕ'	m'	n'	m'	n'	m'	n'	m'	n'
20	1.688	1.285	2.071	1.543	2.492	1.815	2.954	2.173
25	2.004	1.641	2.469	1.957	2.972	2.315	3.523	2.730
30	2.352	2.015	2.888	2.385	3.499	2.857	4.149	3.357
35	2.728	2.385	3.357	2.870	4.079	3.457	4.831	4.043
40	3.154	2.841	3.889	3.428	4.729	4.128	5.603	4.830

j. **Stability coefficients m' and n' for $c'/\gamma H = 0.075$ and $D = 1.50$**

	Stability coefficients for earth slopes							
	Slope 2:1		Slope 3:1		Slope 4:1		Slope 5:1	
ϕ'	m'	n'	m'	n'	m'	n'	m'	n'
20	1.918	1.514	2.199	1.728	2.548	1.985	2.931	2.272
25	2.308	1.914	2.660	2.200	3.083	2.530	3.552	2.915
30	2.735	2.355	3.158	2.714	3.659	3.128	4.128	3.585
35	3.211	2.854	3.708	3.285	4.302	3.786	4.961	4.343
40	3.742	3.397	4.332	3.926	5.026	4.527	5.788	5.185

(*continued*)

Table 15.3 (*continued*)

k. Stability coefficients m' and n' for $c'/\gamma H = 0.100$ and toe circles

	Stability coefficients for earth slopes							
	Slope 2:1		Slope 3:1		Slope 4:1		Slope 5:1	
ϕ'	m'	n'	m'	n'	m'	n'	m'	n'
20	1.804	2.101	2.286	1.588	2.748	1.974	3.190	2.361
25	2.076	1.488	2.665	1.945	3.246	2.459	3.796	2.959
30	2.362	1.786	3.076	2.359	3.770	2.961	4.442	3.576
35	2.673	2.130	3.518	2.803	4.339	3.518	5.146	4.249
40	3.012	2.486	4.008	3.303	4.984	4.173	5.923	5.019

l. Stability coefficients m' and n' for $c'/\gamma H = 0.100$ and $D = 1.00$

	Stability coefficients for earth slopes							
	Slope 2:1		Slope 3:1		Slope 4:1		Slope 5:1	
ϕ'	m'	n'	m'	n'	m'	n'	m'	n'
20	1.841	1.143	2.421	1.472	2.982	1.815	3.549	2.157
25	2.102	1.430	2.785	1.845	3.458	2.303	4.131	2.743
30	2.378	1.714	3.183	2.258	3.973	2.830	4.751	3.372
35	2.692	2.086	3.612	2.715	4.516	3.359	5.426	4.059
40	3.025	2.445	4.103	3.230	5.144	4.001	6.187	4.831

m. Stability coefficients m' and n' for $c'/\gamma H = 0.100$ and $D = 1.25$

	Stability coefficients for earth slopes							
	Slope 2:1		Slope 3:1		Slope 4:1		Slope 5:1	
ϕ'	m'	n'	m'	n'	m'	n'	m'	n'
20	1.874	1.301	2.283	1.558	2.751	1.843	3.253	2.158
25	2.197	1.642	2.681	1.972	3.233	2.330	3.833	2.758
30	2.540	2.000	3.112	2.415	3.753	2.858	4.451	3.372
35	2.922	2.415	3.588	2.914	4.333	3.458	5.141	4.072
40	3.345	2.855	4.119	3.457	4.987	4.142	5.921	4.872

n. Stability coefficients m' and n' for $c'/\gamma H = 0.100$ and $D = 1.50$

	Stability coefficients for earth slopes							
	Slope 2:1		Slope 3:1		Slope 4:1		Slope 5:1	
ϕ'	m'	n'	m'	n'	m'	n'	m'	n'
20	2.079	1.528	2.387	1.742	2.768	2.014	3.158	2.285
25	2.477	1.942	2.852	2.215	3.297	2.542	3.796	2.927
30	2.908	2.385	3.349	2.728	3.881	3.143	4.468	3.614
35	3.385	2.884	3.900	3.300	4.520	3.800	5.211	4.372
40	3.924	3.441	4.524	3.941	5.247	4.542	6.040	5.200

To determine F_s from Table 15.3, we must use the following step-by-step procedure:

Step 1: Obtain ϕ', β, and $c'/\gamma H$.
Step 2: Obtain r_u (weighted average value).
Step 3: From Table 15.3, obtain the values of m' and n' for $D = 1$, 1.25, and 1.5 (for the required parameters ϕ', β, r_u, and $c'/\gamma H$).
Step 4: Determine F_s, using the values of m' and n' for each value of D.
Step 5: The required value of F_s is the smallest one obtained in Step 4.

15.13 *Other Solutions for Steady-State Seepage Condition*

Spencer's Solution

Bishop's simplified method of slices described in Sections 15.11 and 15.12 satisfies the equations of equilibrium with respect to the moment but not with respect to the forces. Spencer (1967) has provided a method to determine the factor of safety (F_s) by taking into account the interslice forces (P_n, T_n, P_{n+1}, T_{n+1}, as shown in Figure 15.32), which does satisfy the equations of equilibrium with respect to moment and forces. The details of this method of analysis are beyond the scope of this text; however, the final results of Spencer's work are summarized in this section in Figure 15.35. Note that r_u, as shown in Figure 15.35, is the same as that defined by Eq. (15.61).

In order to use the charts given in Figure 15.35 and to determine the required value of F_s, the following step-by-step procedure needs to be used.

Step 1: Determine c', γ, H, β, ϕ', and r_u for the given slope.
Step 2: Assume a value of F_s.
Step 3: Calculate $c'/[F_{s\,\text{(assumed)}}\,\gamma H]$.
$$\uparrow$$
$$\text{Step 2}$$
Step 4: With the value of $c'/F_s\gamma H$ calculated in Step 3 and the slope angle β, enter the proper chart in Figure 15.35 to obtain ϕ'_d. Note that Figures 15.35 a, b, and c, are, respectively, for r_u of 0, 0.25, and 0.5, respectively.
Step 5: Calculate $F_s = \tan \phi'/\tan \phi'_d$.
$$\uparrow$$
$$\text{Step 4}$$
Step 6: If the values of F_s as assumed in Step 2 are not the same as those calculated in Step 5, repeat Steps 2, 3, 4, and 5 until they are the same.

Michalowski's Solution

Michalowski (2002) used the kinematic approach of limit analysis similar to that shown in Figures 15.26 and 15.27 to analyze slopes with steady-state seepage. The results of this analysis are summarized in Figure 15.36 for $r_u = 0.25$ and $r_u = 0.5$. Note that Figure 15.27 is applicable for the $r_u = 0$ condition.

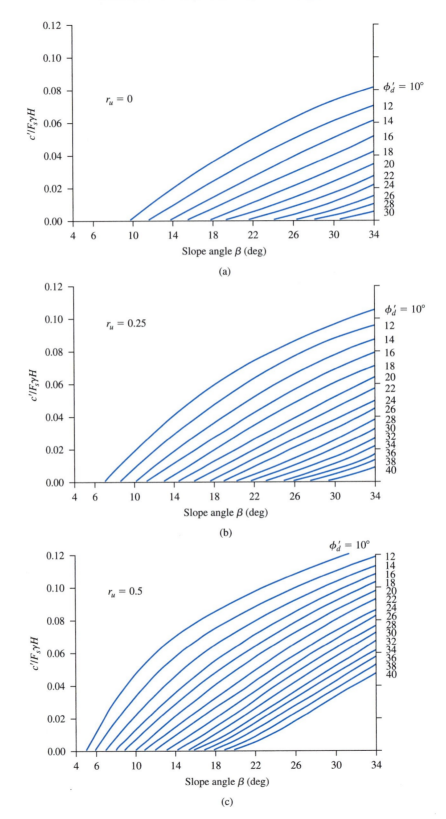

Figure 15.35 Spencer's solution—plot of $c'/F_s\gamma H$ versus β

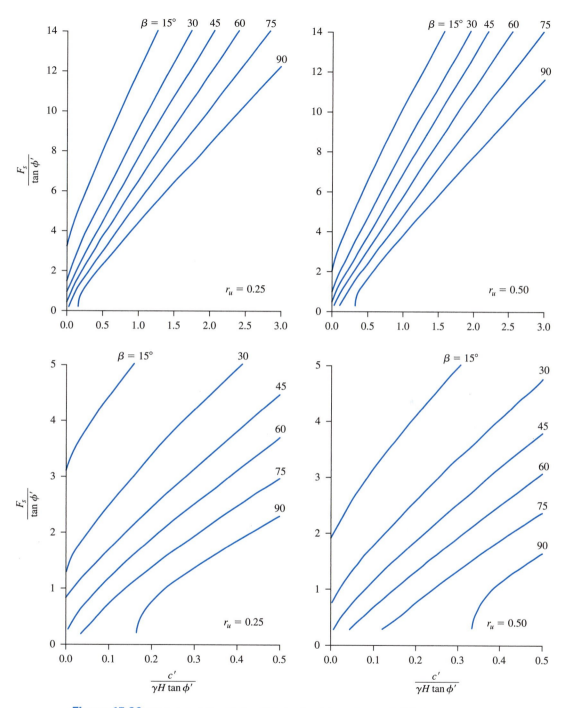

Figure 15.36 Michalowski's solution for steady-state seepage condition

Example 15.9

A given slope under steady-state seepage has the following: $H = 21.62$ m, $\phi' = 25°$, slope: 2H:1V, $c' = 20$ kN/m^2, $\gamma = 18.5$ kN/m^3, $r_u = 0.25$. Determine the factor of safety, F_s. Use Table 15.3.

Solution

$$\beta = \tan^{-1}\left(\frac{1}{2}\right) = 26.57°$$

$$\frac{c'}{\gamma H} = \frac{20}{(18.5)(21.62)} = 0.05$$

Now the following table can be prepared.

β (deg)	ϕ' (deg)	$c'/\gamma H$	D	m' [a]	n' [b]	$F_s = \dfrac{}{m' - n'r_u}$ [c]
26.57	25	0.05	1.00	1.624[d]	1.338[d]	1.29
26.57	25	0.05	1.25	1.822[e]	1.595[e]	1.423
26.57	25	0.05	1.5	2.143[f]	1.903[f]	1.667

[a] From Table 15.3
[b] From Table 15.3
[c] Eq. (15.59); $r_u = 0.25$
[d] Table 15.3
[e] Table 15.3
[f] Table 15.3

So,

$$F_s \simeq 1.29 \qquad \blacksquare$$

Example 15.10

Solve Example 15.9 using Spencer's solution (Figure 15.35).

Solution

Given: $H = 21.62$ m, $\beta = 26.57°$, $c' = 20$ kN/m^2, $\gamma = 18.5$ kN/m^3, $\phi' = 25°$, and $r_u = 0.25$. Now the following table can be prepared.

β (deg)	$F_{s(\text{assumed})}$	$\dfrac{c'}{F_{s(\text{assumed})}\gamma H}$	ϕ'_d [a] (deg)	$F_{s(\text{calculated})} = \dfrac{\tan \phi'}{\tan \phi'_d}$
26.57	1.1	0.0455	18	1.435
26.57	1.2	0.0417	19	1.354
26.57	1.3	0.0385	20	1.281
26.57	1.4	0.0357	21	1.215

[a] From Figure 15.35b

Figure 15.37 shows a plot of $F_{s(\text{assumed})}$ against $F_{s(\text{calculated})}$, from which $F_s \simeq 1.3$.

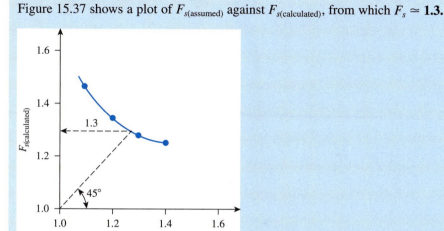

Figure 15.37

Example 15.11

Solve Example 15.9 using Michalowski's solution (Figure 15.36).

Solution

$$\frac{c'}{\gamma H \tan \phi'} = \frac{20}{(18.5)(21.62)(\tan 25)} = 0.107$$

For $r_u = 0.25$, from Figure 15.36, $\dfrac{F_s}{\tan \phi'} \approx 3.1$ So,

$$F_s = (3.1)(\tan 25) = 1.45$$

15.14 *A Case History of Slope Failure*

Ladd (1972) reported the results of a study of the failure of a slope that had been constructed over a sensitive clay. The study was conducted in relation to a major improvement program of Interstate Route 95 in Portsmouth, New Hampshire, which is located 80 kilometer north of Boston on the coast. To study the stability of the slope, a test embankment was built to failure during the spring of 1968. The test embankment was heavily instrumented. The general subsoil condition at the test site, the section of the test embankment, and the instruments placed to monitor the performance of the test section are shown in Figure 15.38.

The ground water level at the test section was at an elevation of +6 m (mean sea level). The general physical properties of the soft to very soft gray silty clay layer as shown in Figure 15.38 are as follows:

Natural moisture content = $50 \pm 5\%$
Undrained shear strength as obtained from field vane shear tests = 12 ± 2.4 kN/m²

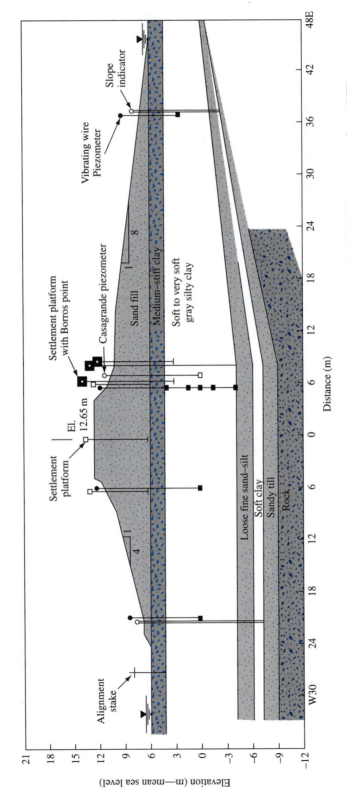

Figure 15.38 Cross-section through the centerline of the experimental test section looking north (*After Ladd, 1972. With permission from ASCE.*)

Remolded shear strength $= 1.2 \pm 0.24 \text{ kN/m}^2$
Liquid limit $= 35 \pm 5$
Plastic limit $= 20 \pm 2$

During construction of the test embankment, fill was placed at a fairly uniform rate within a period of about one month. Failure of the slope (1 vertical: 4 horizontal) occurred on June 6, 1968, at night. The height of the embankment at failure was 6.55 m Figure 15.39 shows the actual failure surface of the slope. The rotated section shown in Figure 15.39 is the "before failure" section rotated through an angle of 13 degrees about a point 13.72, El. 15.55 m.

Ladd (1972) reported the total stress ($\phi = 0$ concept) stability analysis of the slope that failed by using Bishop's simplified method (Section 15.11). The variation of the undrained shear strengths (c_u) used for the stability analysis is given below. Note that these values have not been corrected via Eq. (12.46).

Elevation (m—mean sea level)	c_u as obtained from vane shear strength tests (kN/m²)
6.0 to 4.6	47.9
4.6 to 3.0	19.2
3.0 to 1.5	11.5
1.5 to 0	12.0
0 to −0.75	14.4
−0.75 to −1.5	11.3
−1.5 to −3.0	12.7
−3.0 to −4.0	14.4

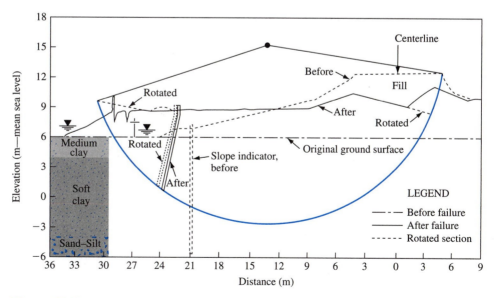

Figure 15.39 Cross-section of the experimental test section before and after failure (*After Ladd, 1972. With permission from ASCE.*)

The factor of safety (F_s) as obtained from the stability analysis for the critical circle of sliding was 0.88. The critical circle of sliding is shown in Figure 15.40. The factor of safety for *actual surface of sliding* as obtained by using Bishop's simplified method was 0.92. For comparison purposes, the actual surface of sliding is also shown in Figure 15.40. Note that the bottom of the actual failure surface is about 0.91 m above the theoretically determined critical failure surface.

Ladd (1972) also reported the stability analysis of the slope based on the average undrained shear strength variation of the clay layer as determined by using the Stress History And Normalized Soil Engineering Properties (SHANSEP). The details of obtaining c_u by this procedure are beyond the scope of this text. However, the final results are given in the following table.

Elevation (m—mean sea level)	c_u as obtained from vane shear strength tests (kN/m²)
6.0 to 4.6	47.9
4.6 to 3.0	16.1
3.0 to 1.5	11.0
1.5 to 0	12.5
0 to −0.75	14.4
−0.75 to −1.5	15.3
−1.5 to −3.0	17.0
−3.0 to −4.0	19.2

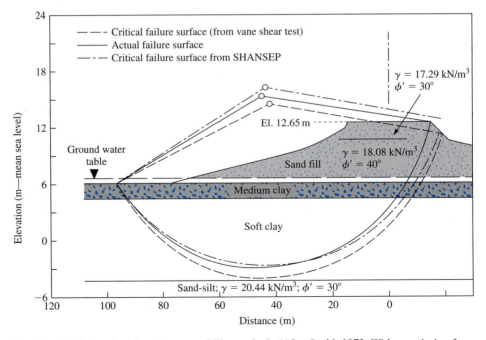

Figure 15.40 Results of total stress stability analysis (*After Ladd, 1972. With permission from ASCE.*) (*Note:* SHANSEP = Stress History And Normalized Soil Engineering Properties)

Using the preceding average value of c_u, Bishop's simplified method of stability analysis yields the following results:

Failure surface	Factor of safety, F_s
actual failure surface	1.02
critical failure surface	1.01

Figure 15.40 also shows the critical failure surface as determined by using the values of c_u obtained from SHANSEP.

  Based on the preceding results, we can draw the following conclusions:

a. The actual failure surface of a slope with limited height is an arc of a circle.
b. The disagreement between the predicted critical failure surface and the actual failure surface is primarily due to the shear strength assumptions. The c_u values obtained from SHANSEP give an $F_s \simeq 1$, and the critical failure surface is practically the same as the actual failure surface.

  This case study is another example that demonstrates the importance of proper evaluation of soil parameters for prediction of the stability of various structures.

15.15 Morgenstern's Method of Slices for Rapid Drawdown Condition

Morgenstern (1963) used Bishop's method of slices (Section 15.11) to determine the factor of safety, F_s, during rapid drawdown. In preparing the solution, Morgenstern used the following notation (Figure 15.41):

- L = height of drawdown
- H = height of embankment
- β = angle that the slope makes with the horizontal

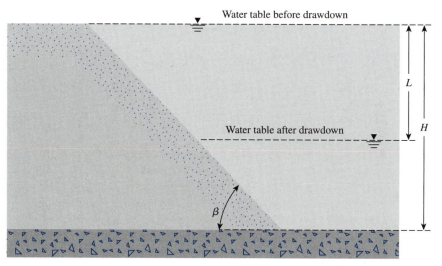

Figure 15.41 Stability analysis for rapid drawdown condition

Morgenstern also assumed that

1. The embankment is made of homogeneous material and rests on an impervious base.
2. Initially, the water level coincides with the top of the embankment.
3. During drawdown, pore water pressure does not dissipate.
4. The unit weight of saturated soil $(\gamma_{sat}) = 2\gamma_w$ (γ_w = unit weight of water).

Figures 15.42 through 15.44 provide the drawdown stability charts developed by Morgenstern.

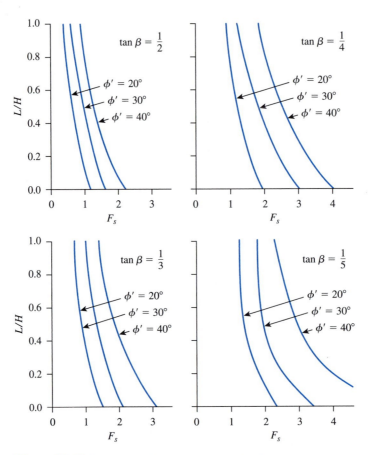

Figure 15.42 Morgenstern's drawdown stability chart for $c'/\gamma H = 0.0125$

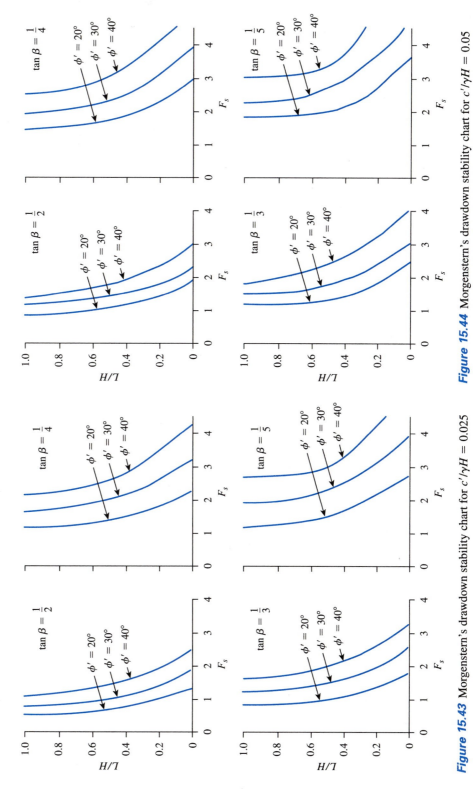

Figure 15.43 Morgenstern's drawdown stability chart for $c'/\gamma H = 0.025$

Figure 15.44 Morgenstern's drawdown stability chart for $c'/\gamma H = 0.05$

15.16	**Fluctuation of Factor of Safety of Slopes in Clay Embankment on Saturated Clay**

Figure 15.45a shows a clay embankment constructed on a *saturated soft clay*. Let P be a point on a potential failure surface APB that is an arc of a circle. Before construction of the embankment, the pore water pressure at P can be expressed as

$$u = h\gamma_w \qquad (15.63)$$

Under ideal conditions, let us assume that the height of the fill needed for the construction of the embankment is placed uniformly, as shown in Figure 15.45b. At time $t = t_1$, the embankment height is equal to H, and it remains constant thereafter (that is, $t > t_1$). The average shear stress increase, τ, on the potential failure surface caused by the construction of the embankment also is shown in Figure 15.45b. The value of τ will increase linearly with time up to time $t = t_1$ and remain constant thereafter.

The pore water pressure at point P (Figure 15.45a) will continue to increase as construction of the embankment progresses, as shown in Figure 15.45c. At time $t = t_1$, $u = u_1 > h\gamma_w$. This is because of the slow rate of drainage from the clay layer. However, after construction of the embankment is completed (that is, $t > t_1$), the pore water pressure gradually will decrease with time as the drainage (thus consolidation) progresses. At time $t \simeq t_2$,

$$u = h\gamma_w$$

For simplicity, if we assume that the embankment construction is rapid and that practically no drainage occurs during the construction period, the average *shear strength* of the clay will remain constant from $t = 0$ to $t = t_1$, or $\tau_f = c_u$ (undrained shear strength). This is shown in Figure 15.45d. For time $t > t_1$, as consolidation progresses, the magnitude of the shear strength, τ_f, will gradually increase. At time $t \geq t_2$—that is, after consolidation is completed—the average shear strength of the clay will be equal to $\tau_f = c' + \sigma' \tan \phi'$ (drained shear strength) (Figure 15.45d). The factor of safety of the embankment along the potential surface of sliding can be given as

$$F_s = \frac{\text{Average shear strength of clay, } \tau_f, \text{ along sliding surface (Figure 15.45d)}}{\text{Average shear stress, } \tau, \text{ along sliding surface (Figure 15.45b)}}$$

$$(15.64)$$

The general nature of the variation of the factor of safety, F_s, with time is shown in Figure 15.45e. As we can see from this figure, the magnitude of F_s initially decreases with time. At the end of construction (time $t = t_1$), the value of the factor of safety is a minimum. Beyond this point, the value of F_s continues to increase with drainage up to time $t = t_2$.

Cuts in Saturated Clay

Figure 15.46a shows a cut slope in a saturated soft clay in which APB is a circular potential failure surface. During advancement of the cut, the average shear stress, τ, on the potential failure surface passing through P will increase. The maximum value of the average shear stress, τ, will be attained at the end of construction—that is, at time $t = t_1$. This property is shown in Figure 15.46b.

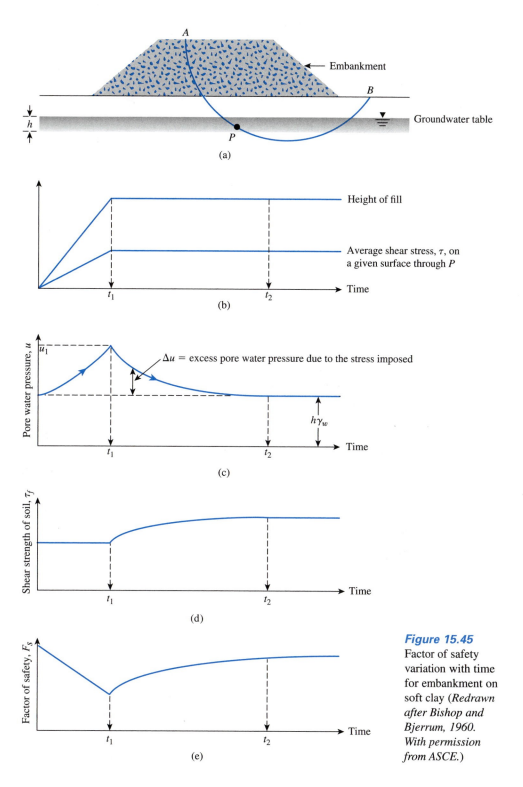

Figure 15.45
Factor of safety
variation with time
for embankment on
soft clay (*Redrawn
after Bishop and
Bjerrum, 1960.
With permission
from ASCE.*)

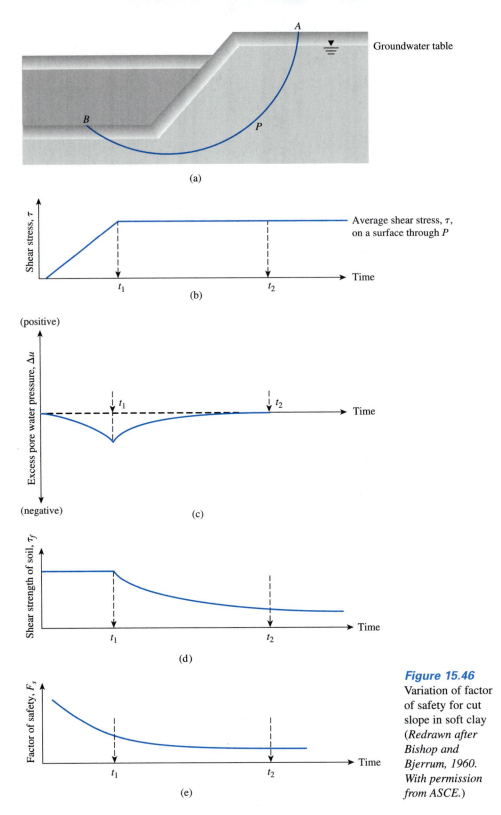

(a)

(b)

(c)

(d)

(e)

Figure 15.46
Variation of factor of safety for cut slope in soft clay (*Redrawn after Bishop and Bjerrum, 1960. With permission from ASCE.*)

Because of excavation of the soil, the effective overburden pressure at point P will decrease, which will induce a reduction in the pore water pressure. The variation of the net change of pore water pressure, Δu, is shown in Figure 15.46c. After excavation is complete (time $t > t_1$), the net negative excess pore water pressure will gradually dissipate. At time $t \geq t_2$, the magnitude of Δu will be equal to 0.

The variation of the average shear strength, τ_f, of the clay with time is shown in Figure 15.46d. Note that the shear strength of the soil after excavation gradually decreases. This decrease occurs because of dissipation of the negative excess pore water pressure.

If the factor of safety of the cut slope, F_s, along the potential failure surface is defined by Eq. (15.64), its variation will be as shown in Figure 15.46e. Note that the magnitude of F_s decreases with time, and its minimum value is obtained at time $t \geq t_2$.

Problems

15.1 Refer to Figure 15.47. Given: $\beta = 30°$, $\gamma = 15.5$ kN/m³, $\phi' = 20°$, and $c' = 15$ kN/m². Find the height, H, that will have a factor of safety, F_s, of 2 against sliding along the soil-rock interface.

15.2 For the slope shown in Figure 15.47, find the height, H, for critical equilibrium. Given: $\beta = 22°$, $\gamma = 15.72$ kN/m³, $\phi' = 15°$, and $c' = 12$ kN/m².

15.3 Refer to Figure 15.48. If there were seepage through the soil and the groundwater table coincided with the ground surface, what would be the value of F_s? Use $H = 8$ m, ρ_{sat} (saturated density of soil) $= 1900$ kg/m³, $\beta = 20°$, $c' = 18$ kN/m², and $\phi' = 25°$.

15.4 For the infinite slope shown in Figure 15.48, find the factor of safety against sliding along the plane AB, given that $H = 7.62$ m, $G_s = 2.6$, $e = 0.5$, $\phi' = 22°$, and $c' = 28.75$ kN/m². Note that there is seepage through the soil and that the groundwater table coincides with the ground surface.

15.5 For a finite slope such as that shown in Figure 15.10, assume that the slope failure would occur along a plane (Culmann's assumption). Find the height of the slope for critical equilibrium. Given: $\phi' = 25°$, $c' = 19.2$ kN/m², $\gamma = 18.08$ kN/m³, and $\beta = 50°$.

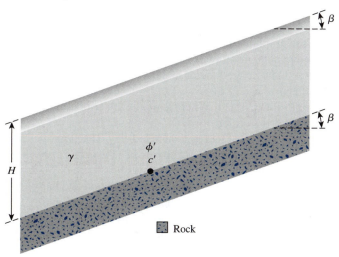

Figure 15.47

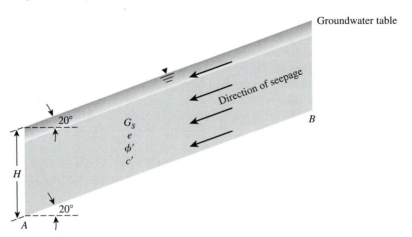

Groundwater table

Direction of seepage

20°

G_s

e

ϕ'

c'

B

H

20°

A

Figure 15.48

15.6 Refer to Figure 15.10. Using the soil parameters given in Problem 15.5, find the height of the slope, H, that will have a factor of safety of 2 against sliding. Assume that the critical surface for sliding is a plane.

15.7 Refer to Figure 15.10. Given that $\phi' = 15°$, $c' = 9.6$ kN/m², $\gamma = 18.08$ kN/m³, $\beta = 50°$, and $H = 3.66$ m, determine the factor of safety with respect to sliding. Assume that the critical surface for sliding is a plane.

15.8 Determine the height of a finite slope (1 vertical to 2 horizontal) that should have a factor of safety of 2 against sliding. For the soil, the following values are given: $c' = 18$ kN/m², $\phi' = 20°$, and $\rho = 1700$ kg/m³. Assume the critical surface for sliding to be a plane.

15.9 A cut slope is to be made in a saturated clay. Given: $c_u = 30$ kN/m² ($\phi = 0$ condition) and $\gamma = 17$ kN/m³. The slope makes an angle, β, of 60° with the horizontal. Determine the maximum depth up to which the cut could be made. Assume that the critical surface for sliding is circular. What is the nature of the critical circle (that is, toe, slope, or midpoint)?

15.10 For the cut slope described in Problem 15.9, if we need a factor of safety of 2.0 against sliding, how deep should the cut be made?

15.11 Using the graph given in Figure 15.13, determine the height of a slope (1 vertical to 1 horizontal) in saturated clay having an undrained shear strength of 25 kN/m². The desired factor of safety against sliding is 2.5 Given: $\gamma = 18$ kN/m³ and $D = 1.2$.

15.12 Refer to Problem 15.11. What should be the critical height of the slope? What is the nature of the critical circle?

15.13 A cut slope was excavated in a saturated clay. The slope angle, β, is equal to 40° with the horizontal. Slope failure occurred when the cut reached a depth of 8.5 m. Previous soil explorations showed that a rock layer was located at a depth of 12 m below the ground surface. Assuming an undrained condition and that $\gamma = 18.5$ kN/m³:

a. Determine the undrained cohesion of the clay (Figure 15.13).

b. What was the nature of the critical circle?

c. With reference to the top of the slope, at what distance did the surface of sliding intersect the bottom of the excavation?

15.14 A clay slope is built over a layer of rock. Determine the factor of safety if $k_h = 0.4$ for the slope with values
- Height, $H = 16$ m
- Slope angle, $\beta = 30°$
- Saturated unit weight of soil, $\gamma_{sat} = 17$ kN/m³
- Undrained shear strength, $c_u = 50$ kN/m²

15.15 For a slope in clay where $H = 15.2$ m, $\gamma = 18.08$ kN/m³, $\beta = 60°$, and $c_u = 47.9$ kN/m², determine the factor of safety for $k_h = 0.3$.

15.16 Refer to Figure 15.49. Use Figure 15.27 ($\phi' > 0$) to solve the following.
- **a.** If $n' = 2$, $\phi' = 20°$, $c' = 20$ kN/m², and $\gamma = 16$ kN/m³, find the critical height of the slope.
- **b.** If $n' = 1.5$, $\phi' = 25°$, $c' = 35.9$ kN/m², and $\gamma = 17.29$ kN/m³, find the critical height of the slope.

15.17 Refer to Figure 15.49. Using Figure 15.23, find the factor of safety, F_s, with respect to sliding for a slope with the following.
- Slope: 2H:1V
- $\phi' = 10°$
- $c' = 33.5$ kN/m²
- $\gamma = 17.29$ kN/m³
- $H = 15.2$ m

15.18 Repeat Problem 15.17 with the following.
- Slope: 1H:1V
- $\phi' = 20°$
- $c' = 19.2$ kN/m²
- $\gamma = 18.08$ kN/m³
- $H = 9.15$ m

15.19 Repeat Problem 15.17 with the following.
- Slope: 2.5H:1V
- $\phi' = 12°$
- $c' = 24$ kN/m²
- $\gamma = 16.5$ kN/m³
- $H = 12$ m

15.20 Referring to Figure 15.50 and using the ordinary method of slices, find the factor of safety with respect to sliding for the following trial cases.
- **a.** $\beta = 45°$, $\phi' = 20°$, $c' = 19.2$ kN/m², $\gamma = 18.08$ kN/m³, $H = 12.2$ m, $\alpha = 30°$, and $\theta = 70°$
- **b.** $\beta = 45°$, $\phi' = 15°$, $c' = 18$ kN/m², $\gamma = 17.1$ kN/m³, $H = 5$ m, $\alpha = 30°$, and $\theta = 80°$

15.21 Determine the minimum factor of safety of a slope with the following parameters: $H = 7.62$ m, $\beta = 26.57°$, $\phi' = 20°$, $c' = 14.4$ kN/m², $\gamma = 18.86$ kN/m³, and $r_u = 0.5$. Use Bishop and Morgenstern's method.

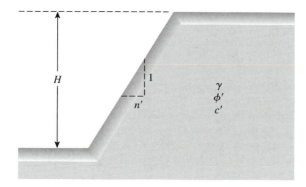

Figure 15.49

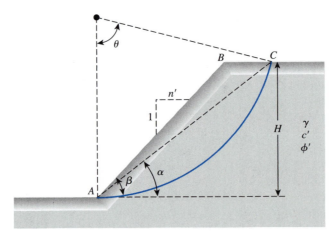

Figure 15.50

15.22 Determine the minimum factor of safety of a slope with the following parameters: $H = 6$ m, $\beta = 18.43°$, $\phi' = 20°$, $c' = 6$ kN/m^2, $\gamma = 20$ kN/m^3, and $r_u = 0.5$. Use Bishop and Morgenstern's method.

15.23 Use Spencer's chart to determine the value of F_s for the given slope: $\beta = 20°$, $H = 15$ m, $\phi' = 15°$, $c' = 20$ kN/m^2, $\gamma = 17.5$ kN/m^3, and $r_u = 0.5$.

15.24 For a slope, given:
- Slope: 3H:1V
- $H = 12.63$ m
- $\phi' = 25°$
- $c' = 12$ kN/m^2
- $\gamma = 19$ kN/m^3
- $r_u = 0.25$

Using Spencer's chart, determine the factor of safety, F_s.

References

BISHOP, A. W. (1955). "The Use of Slip Circle in the Stability Analysis of Earth Slopes," *Geotechnique,* Vol. 5, No. 1, 7–17.

BISHOP, A. W., and BJERRUM, L. (1960). "The Relevance of the Triaxial Test to the Solution of Stability Problems," *Proceedings,* Research Conference on Shear Strength of Cohesive Soils, ASCE, 437–501.

BISHOP, A. W., and MORGENSTERN, N. R. (1960). "Stability Coefficients for Earth Slopes," *Geotechnique,* Vol. 10, No. 4, 129–147.

CRUDEN, D. M., and VARNES, D. J. (1966). "Landslide Types and Processes," *Special Report 247,* Transportation Research Board, 36–75.

CULMANN, C. (1875). *Die Graphische Statik,* Meyer and Zeller, Zurich.

FELLENIUS, W. (1927). *Erdstatische Berechnungen,* rev. ed., W. Ernst u. Sons, Berlin.

KOPPULA, S. D. (1984). "On Stability of Slopes on Clays with Linearly Increasing Strength," *Canadian Geotechnical Journal,* Vol. 21, No. 3, 557–581.

LADD, C. C. (1972). "Test Embankment on Sensitive Clay," *Proceedings,* Conference on Performance of Earth and Earth-Supported Structures, ASCE, Vol. 1, Part 1, 101–128.

MICHALOWSKI, R. L. (2002). "Stability Charts for Uniform Slopes," *Journal of Geotechnical and Geoenvironmental Engineering,* ASCE, Vol. 128, No. 4, 351–355.

MORGENSTERN, N. R. (1963). "Stability Charts for Earth Slopes During Rapid Drawdown," *Geotechnique,* Vol. 13, No. 2, 121–133.

SINGH, A. (1970). "Shear Strength and Stability of Man-Made Slopes," *Journal of the Soil Mechanics and Foundations Division*, ASCE, Vol. 96, No. SM6, 1879–1892.

SPENCER, E. (1967). "A Method of Analysis of the Stability of Embankments Assuming Parallel Inter-Slice Forces," *Geotechnique,* Vol. 17, No. 1, 11–26.

TAYLOR, D. W. (1937). "Stability of Earth Slopes," *Journal of the Boston Society of Civil Engineers,* Vol. 24, 197–246.

TERZAGHI, K., and PECK, R. B. (1967). *Soil Mechanics in Engineering Practice,* 2nd ed., Wiley, New York.

16 Soil-Bearing Capacity for Shallow Foundations

The lowest part of a structure generally is referred to as the *foundation.* Its function is to transfer the load of the structure to the soil on which it is resting. A properly designed foundation transfers the load throughout the soil without overstressing the soil. Overstressing the soil can result in either excessive settlement or shear failure of the soil, both of which cause damage to the structure. Thus, geotechnical and structural engineers who design foundations must evaluate the bearing capacity of soils.

Depending on the structure and soil encountered, various types of foundations are used. Figure 16.1 shows the most common types of foundations. A *spread footing* is simply an enlargement of a load-bearing wall or column that makes it possible to spread the load of the structure over a larger area of the soil. In soil with low load-bearing capacity, the size of the spread footings required is impracticably large. In that case, it is more economical to construct the entire structure over a concrete pad. This is called a *mat foundation.*

Pile and *drilled shaft foundations* are used for heavier structures when great depth is required for supporting the load. Piles are structural members made of timber, concrete, or steel that transmit the load of the superstructure to the lower layers of the soil. According to how they transmit their load into the subsoil, piles can be divided into two categories: friction piles and end-bearing piles. In the case of friction piles, the superstructure load is resisted by the shear stresses generated along the surface of the pile. In the end-bearing pile, the load carried by the pile is transmitted at its tip to a firm stratum.

In the case of drilled shafts, a shaft is drilled into the subsoil and then is filled with concrete. A metal casing may be used while the shaft is being drilled. The casing may be left in place or may be withdrawn during the placing of concrete. Generally, the diameter of a drilled shaft is much larger than that of a pile. The distinction between piles and drilled shafts becomes hazy at an approximate diameter of 1 m, and the definitions and nomenclature are inaccurate.

Spread footings and mat foundations generally are referred to as *shallow foundations,* whereas pile and drilled-shaft foundations are classified as *deep foundations.* In a more general sense, shallow foundations are foundations that have a depth-of-embedment-to-width ratio of approximately less than four. When the depth-of-embedment-to-width ratio of a foundation is greater than four, it may be classified as a deep foundation.

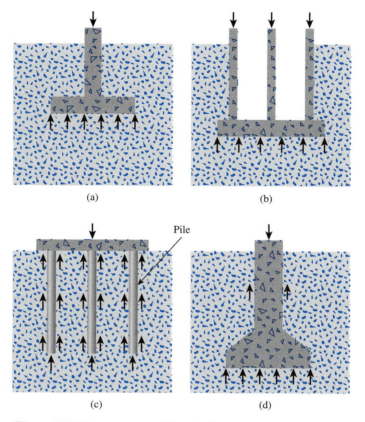

Figure 16.1 Common types of foundations: (a) spread footing; (b) mat foundation; (c) pile foundation; (d) drilled shaft foundation

In this chapter, we discuss the soil-bearing capacity for shallow foundations. As mentioned before, for a foundation to function properly, (1) the settlement of soil caused by the load must be within the tolerable limit, and (2) shear failure of the soil supporting the foundation must not occur. Compressibility of soil—consolidation and elasticity theory—was introduced in Chapter 10. This chapter introduces the load-carrying capacity of shallow foundations based on the criteria of shear failure in soil.

16.1   *Ultimate Soil-Bearing Capacity for Shallow Foundations*

To understand the concept of the ultimate soil-bearing capacity and the mode of shear failure in soil, let us consider the case of a long rectangular footing of width B located at the surface of a dense sand layer (or stiff soil) shown in Figure 16.2a. When a uniformly distributed load of q per unit area is applied to the footing, it settles. If the uniformly distributed load (q) is increased, the settlement of the footing gradually increases. When the value of $q = q_u$ is reached (Figure 16.2b), bearing capacity failure occurs; the footing undergoes a very large settlement without any further increase of q. The soil on one or both sides of

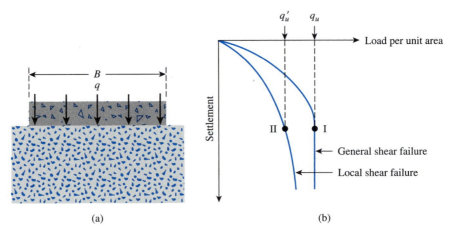

Figure 16.2 Ultimate soil-bearing capacity for shallow foundation: (a) model footing; (b) load settlement relationship

the foundation bulges, and the slip surface extends to the ground surface. The load-settlement relationship is like Curve I shown in Figure 16.2b. In this case, q_u is defined as the ultimate bearing capacity of soil.

The bearing capacity failure just described is called a *general shear failure* and can be explained with reference to Figure 16.3a. When the foundation settles under the application of a load, a triangular wedge-shaped zone of soil (marked I) is pushed down,

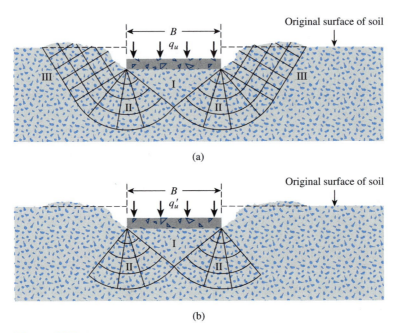

Figure 16.3 Modes of bearing capacity failure in soil: (a) general shear failure of soil; (b) local shear failure of soil

and, in turn, it presses the zones marked II and III sideways and then upward. At the ultimate pressure, q_u, the soil passes into a state of plastic equilibrium and failure occurs by sliding.

If the footing test is conducted instead in a loose-to-medium dense sand, the load-settlement relationship is like Curve II in Figure 16.2b. Beyond a certain value of $q = q'_u$, the load-settlement relationship becomes a steep, inclined straight line. In this case, q'_u is defined as the ultimate bearing capacity of soil. This type of soil failure is referred to as *local shear failure* and is shown in Figure 16.3b. The triangular wedge-shaped zone (marked I) below the footing moves downward, but unlike general shear failure, the slip surfaces end somewhere inside the soil. Some signs of soil bulging are seen, however.

16.2 *Terzaghi's Ultimate Bearing Capacity Equation*

In 1921, Prandtl published the results of his study on the penetration of hard bodies (such as metal punches) into a softer material. Terzaghi (1943) extended the plastic failure theory of Prandtl to evaluate the bearing capacity of soils for shallow strip footings. For practical considerations, a long wall footing (length-to-width ratio more than about five) may be called a *strip footing*. According to Terzaghi, a foundation may be defined as a shallow foundation if the depth D_f is less than or equal to its width B (Figure 16.4). He also assumed that, for ultimate soil-bearing capacity calculations, the weight of soil above the base of the footing may be replaced by a uniform surcharge, $q = \gamma D_f$.

The failure mechanism assumed by Terzaghi for determining the ultimate soil-bearing capacity (general shear failure) for a rough strip footing located at a depth D_f measured from the ground surface is shown in Figure 16.5a. The soil wedge *ABJ* (Zone I) is an elastic zone. Both *AJ* and *BJ* make an angle ϕ' with the horizontal. Zones marked II (*AJE* and *BJD*) are the radial shear zones, and zones marked III are the Rankine passive zones. The rupture lines *JD* and *JE* are arcs of a logarithmic spiral, and *DF* and *EG* are straight lines. *AE, BD, EG,* and *DF* make angles of $45 - \phi'/2$

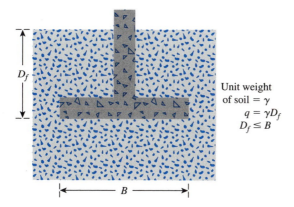

Figure 16.4 Shallow strip footing

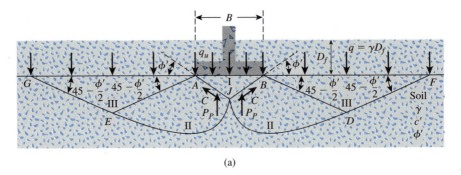

(a)

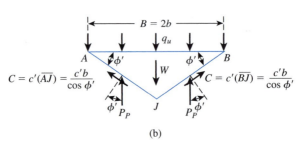

(b)

Figure 16.5 Terzaghi's bearing capacity analysis

degrees with the horizontal. The equation of the arcs of the logarithmic spirals *JD* and *JE* may be given as

$$r = r_o e^{\theta \tan \phi'}$$

If the load per unit area, q_u, is applied to the footing and general shear failure occurs, the passive force P_p is acting on each of the faces of the soil wedge *ABJ*. This concept is easy to conceive of if we imagine that *AJ* and *BJ* are two walls that are pushing the soil wedges *AJEG* and *BJDF,* respectively, to cause passive failure. P_p should be inclined at an angle δ' (which is the angle of wall friction) to the perpendicular drawn to the wedge faces (that is, *AJ* and *BJ*). In this case, δ' should be equal to the angle of friction of soil, ϕ'. Because *AJ* and *BJ* are inclined at an angle ϕ' to the horizontal, the direction of P_p should be vertical.

Now let us consider the free-body diagram of the wedge *ABJ* as shown in Figure 16.5b. Considering the unit length of the footing, we have, for equilibrium,

$$(q_u)(2b)(1) = -W + 2C \sin \phi' + 2P_p \tag{16.1}$$

where $b = B/2$

W = weight of soil wedge *ABJ* = $\gamma b^2 \tan \phi'$

C = cohesive force acting along each face , *AJ* and *BJ*, that is equal to the unit cohesion times the length of each face = $c'b/(\cos \phi')$

Thus,

$$2bq_u = 2P_p + 2bc' \tan \phi' - \gamma b^2 \tan \phi' \tag{16.2}$$

or

$$q_u = \frac{P_p}{b} + c' \tan \phi' - \frac{\gamma b}{2} \tan \phi' \qquad (16.3)$$

The passive pressure in Eq. (16.2) is the sum of the contribution of the weight of soil γ, cohesion c', and surcharge q and can be expressed as

$$P_p = \tfrac{1}{2}\gamma(b \tan \phi')^2 K_\gamma + c'(b \tan \phi')K_c + q(b \tan \phi')K_q \qquad (16.4)$$

where K_γ, K_c, and K_q are earth-pressure coefficients that are functions of the soil friction angle, ϕ'.

Combining Eqs. (16.3) and (16.4), we obtain

$$q_u = c'N_c + qN_q + \frac{1}{2}\gamma BN_\gamma$$

where

$$N_c = \tan \phi'(K_c + 1) \qquad (16.5)$$

$$N_q = K_q \tan \phi' \qquad (16.6)$$

$$N_\gamma = \tfrac{1}{2} \tan \phi'(K_\gamma \tan \phi' - 1) \qquad (16.7)$$

The terms N_c, N_q, and N_γ are, respectively, the contributions of cohesion, surcharge, and unit weight of soil to the ultimate load-bearing capacity. It is extremely tedious to evaluate K_c, K_q, and K_γ. For this reason, Terzaghi used an approximate method to determine the ultimate bearing capacity, q_u. The principles of this approximation are the following.

1. If $c' = 0$ and surcharge $(q) = 0$ (that is, $D_f = 0$), then
$$q_u = q_\gamma = \tfrac{1}{2}\gamma BN_\gamma \qquad (16.8)$$

2. If $\gamma = 0$ (that is, weightless soil) and $q = 0$, then
$$q_u = q_c = c'N_c \qquad (16.9)$$

3. If $\gamma = 0$ (weightless soil) and $c' = 0$, then
$$q_u = q_q = qN_q \qquad (16.10)$$

By the method of superimposition, when the effects of the unit weight of soil, cohesion, and surcharge are considered, we have

$$q_u = q_c + q_q + q_\gamma = c'N_c + qN_q + \tfrac{1}{2}\gamma BN_\gamma \qquad (16.11)$$

Equation (16.11) is referred to as *Terzaghi's bearing capacity equation*. The terms N_c, N_q, and N_γ are called the *bearing capacity factors*. The values of these factors are given in Table 16.1.

Table 16.1 Terzaghi's Bearing Capacity Factors—N_c, N_q and N_γ—Eqs. (16.11), (16.12), and (16.13), respectively

ϕ' (deg)	N_c	N_q	N_γ^a	ϕ' (deg)	N_c	N_q	N_γ^a
0	5.70	1.00	0.00	26	27.09	14.21	9.84
1	6.00	1.10	0.01	27	29.24	16.90	11.60
2	6.30	1.22	0.04	28	31.61	17.81	13.70
3	6.62	1.35	0.06	29	34.24	19.98	16.18
4	6.97	1.49	0.10	30	37.16	22.46	19.13
5	7.34	1.64	0.14	31	40.41	25.28	22.65
6	7.73	1.81	0.20	32	44.04	28.52	26.87
7	8.15	2.00	0.27	33	48.09	32.23	31.94
8	8.60	2.21	0.35	34	52.64	36.50	38.04
9	9.09	2.44	0.44	35	57.75	41.44	45.41
10	9.61	2.69	0.56	36	63.53	47.16	54.36
11	10.16	2.98	0.69	37	70.01	53.80	65.27
12	10.76	3.29	0.85	38	77.50	61.55	78.61
13	11.41	3.63	1.04	39	85.97	70.61	95.03
14	12.11	4.02	1.26	40	95.66	81.27	116.31
16	12.86	4.45	1.52	41	106.81	93.85	140.51
16	13.68	4.92	1.82	42	119.67	108.75	171.99
17	14.60	5.45	2.18	43	134.58	126.50	211.56
18	15.12	6.04	2.59	44	161.95	147.74	261.60
19	16.56	6.70	3.07	45	172.28	173.28	325.34
20	17.69	7.44	3.64	46	196.22	204.19	407.11
21	18.92	8.26	4.31	47	224.55	241.80	512.84
22	20.27	9.19	5.09	48	258.28	287.85	650.67
23	21.75	10.23	6.00	49	298.71	344.63	831.99
24	23.36	11.40	7.08	50	347.50	416.14	1072.80
25	25.13	12.72	8.34				

[a]From Kumbhojkar (1993)

For square and circular footings, Terzaghi suggested the following equations for ultimate soil-bearing capacity:

The square footing is

$$q_u = 1.3c'N_c + qN_q + 0.4\gamma BN_\gamma \qquad (16.12)$$

The circular footing is

$$q_u = 1.3c'N_c + qN_q + 0.3\gamma BN_\gamma \qquad (16.13)$$

where B = diameter of the footing.

Krizek (1965) gave the following approximations for the values of N_c, N_q, and N_γ with a maximum deviation of 15%.

$$N_c = \frac{228 + 4.3\phi'}{40 - \phi'} \tag{16.14}$$

$$N_q = \frac{40 + 5\phi'}{40 - \phi'} \tag{16.15}$$

and

$$N_\gamma = \frac{6\phi'}{40 - \phi'} \tag{16.16}$$

where ϕ' = soil friction angle, in degrees. Equations (16.14) through (16.16) are valid for $\phi' = 0°$ to $35°$.

Equation (16.11) was derived on the assumption that the bearing capacity failure of soil takes place by general shear failure. In the case of local shear failure, we may assume that

$$\bar{c}' = \tfrac{2}{3}c' \tag{16.17}$$

and

$$\tan \bar{\phi}' = \tfrac{2}{3} \tan \phi' \tag{16.18}$$

The ultimate bearing capacity of soil for a strip footing may be given by

$$q_u' = \bar{c}'N_c' + qN_q' + \tfrac{1}{2}\gamma BN_\gamma' \tag{16.19}$$

The modified bearing capacity factors N_c', N_q', and N_γ' are calculated by using the same general equation as that for N_c, N_q, and N_γ, but by substituting $\bar{\phi}' = \tan^{-1}(\tfrac{2}{3} \tan \phi')$ for ϕ'. The values of the bearing capacity factors for a local shear failure are given in Table 16.2. The ultimate soil-bearing capacity for square and circular footings for the local shear failure case now may be given as follows [similar to Eqs. (16.12) and (16.13)]:
The square footing is

$$q_u' = 1.3\bar{c}'N_c' + qN_q' + 0.4\gamma BN_\gamma' \tag{16.20}$$

The circular footing is

$$q_u' = 1.3\bar{c}'N_c' + qN_q' + 0.3\gamma BN_\gamma' \tag{16.21}$$

For an undrained condition with $\phi = 0$ and $\tau_f = c_u$, the bearing capacity factors are $N_\gamma = 0$ and $N_q = 1$. Also, $N_c = 5.7$. In that case, Eqs. (16.11), (16.12), and (16.13) (which are the cases for general shear failure) take the forms

Table 16.2 Terzaghi's Modified Bearing Capacity Factors—N'_c, N'_q, and N'_γ—Eqs. (16.19), (16.20), and (16.21), respectively

ϕ' (deg)	N'_c	N'_q	N'_γ	ϕ' (deg)	N'_c	N'_q	N'_γ
0	5.70	1.00	0.00	26	16.53	6.05	2.59
1	5.90	1.07	0.005	27	16.30	6.54	2.88
2	6.10	1.14	0.02	28	17.13	7.07	3.29
3	6.30	1.22	0.04	29	18.03	7.66	3.76
4	6.51	1.30	0.055	30	18.99	8.31	4.39
5	6.74	1.39	0.074	31	20.03	9.03	4.83
6	6.97	1.49	0.10	32	21.16	9.82	5.51
7	7.22	1.59	0.128	33	22.39	10.69	6.32
8	7.47	1.70	0.16	34	23.72	11.67	7.22
9	7.74	1.82	0.20	35	25.18	12.75	8.35
10	8.02	1.94	0.24	36	26.77	13.97	9.41
11	8.32	2.08	0.30	37	28.51	16.32	10.90
12	8.63	2.22	0.35	38	30.43	16.85	12.75
13	8.96	2.38	0.42	39	32.53	18.56	14.71
14	9.31	2.55	0.48	40	34.87	20.50	17.22
16	9.67	2.73	0.57	41	37.45	22.70	19.75
16	10.06	2.92	0.67	42	40.33	25.21	22.50
17	10.47	3.13	0.76	43	43.54	28.06	26.25
18	10.90	3.36	0.88	44	47.13	31.34	30.40
19	11.36	3.61	1.03	45	51.17	35.11	36.00
20	11.85	3.88	1.12	46	55.73	39.48	41.70
21	12.37	4.17	1.35	47	60.91	44.54	49.30
22	12.92	4.48	1.55	48	66.80	50.46	59.25
23	13.51	4.82	1.74	49	73.55	57.41	71.45
24	14.14	5.20	1.97	50	81.31	65.60	85.75
25	14.80	5.60	2.25				

$$q_u = 5.7c_u + q \qquad \text{(strip footing)} \tag{16.22}$$

and

$$q_u = (1.3)(5.7)c_u + q = 7.41c_u + q \qquad \text{(square and circular footing)}$$
$$\tag{16.23}$$

16.3   *Effect of Groundwater Table*

In developing the bearing capacity equations given in the preceding section we assumed that the groundwater table is located at a depth much greater than the width, B of the footing. However, if the groundwater table is close to the footing, some changes are required in the second and third terms of Eqs. (16.11) to (16.13), and Eqs. (16.19) to (16.21). Three different conditions can arise regarding the location of the groundwater table with respect to the bottom of the foundation. They are shown in Figure 16.6. Each of these conditions is briefly described next.

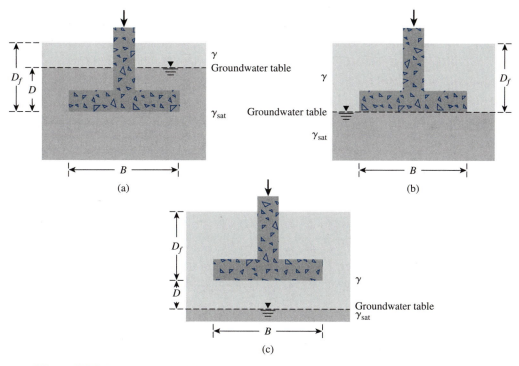

Figure 16.6 Effect of the location of groundwater table on the bearing capacity of shallow foundations: (a) Case I; (b) Case II; (c) Case III

Case I (Figure 16.6a) If the groundwater table is located at a distance D above the bottom of the foundation, the magnitude of q in the second term of the bearing capacity equation should be calculated as

$$q = \gamma(D_f - D) + \gamma'D \qquad (16.24)$$

where $\gamma' = \gamma_{sat} - \gamma_w$ = effective unit weight of soil. Also, the unit weight of soil, γ, that appears in the third term of the bearing capacity equations should be replaced by γ'.

Case II (Figure 16.6b) If the groundwater table coincides with the bottom of the foundation, the magnitude of q is equal to γD_f. However, the unit weight, γ, in the third term of the bearing capacity equations should be replaced by γ'.

Case III (Figure 16.6c) When the groundwater table is at a depth D below the bottom of the foundation, $q = \gamma D_f$. The magnitude of γ in the third term of the bearing capacity equations should be replaced by γ_{av}.

$$\gamma_{av} = \frac{1}{B}[\gamma D + \gamma'(B - D)] \qquad \text{(for } D \leq B\text{)} \qquad (16.25a)$$

$$\gamma_{av} = \gamma \qquad \text{(for } D > B\text{)} \qquad (16.25b)$$

16.4 *Factor of Safety*

Generally, a factor of safety, F_s, of about 3 or more is applied to the ultimate soil-bearing capacity to arrive at the value of the allowable bearing capacity. An F_s of 3 or more is not considered too conservative. In nature, soils are neither homogeneous nor isotropic. Much uncertainty is involved in evaluating the basic shear strength parameters of soil.

There are two basic definitions of the allowable bearing capacity of shallow foundations. They are gross allowable bearing capacity, and net allowable bearing capacity.

The *gross allowable bearing capacity* can be calculated as

$$q_{all} = \frac{q_u}{F_s} \tag{16.26}$$

As defined by Eq. (16.26) q_{all} is the allowable load per unit area to which the soil under the foundation should be subjected to avoid any chance of bearing capacity failure. It includes the contribution (Figure 16.7) of (a) the dead and live loads above the ground surface, $W_{(D+L)}$; (b) the self-weight of the foundation, W_F; and (c) the weight of the soil located immediately above foundation, W_S. Thus,

$$q_{all} = \frac{q_u}{F_s} = \left[\frac{W_{(D+L)} + W_F + W_S}{A} \right] \frac{1}{F_s} \tag{16.27}$$

where A = area of the foundation.

The *net allowable bearing capacity* is the allowable load per unit area of the foundation in excess of the existing vertical effective stress at the level of the foundation. The vertical effective stress at the foundation level is equal to $q = \gamma D_f$. So, the net ultimate load is

$$q_{u(net)} = q_u - q \tag{16.28}$$

Hence,

$$q_{all(net)} = \frac{q_{u(net)}}{F_s} = \frac{q_u - q}{F_s} \tag{16.29}$$

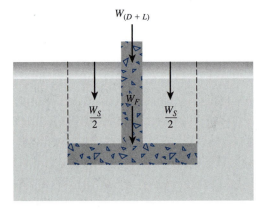

$W_{(D+L)}$

W_F

$\dfrac{W_S}{2}$ $\dfrac{W_S}{2}$

Figure 16.7 Contributions to q_{all}

If we assume that the weight of the soil and the weight of the concrete from which the foundation is made are approximately the same, then

$$q = \gamma D_f \simeq \frac{W_S + W_F}{A}$$

Hence,

$$q_{\text{all(net)}} = \frac{W_{(D+L)}}{A} = \frac{q_u - q}{F_s} \tag{16.30}$$

Example 16.1

A continuous foundation is shown in Figure 16.8. Using Terzaghi's bearing capacity factors, determine the gross allowable load per unit area (q_{all}) that the foundation can carry. Given:

- $\gamma = 17.29$ kN/m^3
- $c' = 9.6$ kN/m^2
- $\phi' = 20°$
- $D_f = 0.9$ m
- $B = 1.2$ m
- Factor of safety = 3

Assume general shear failure.

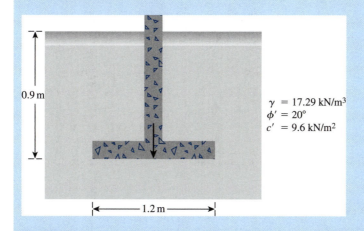

0.9 m

$\gamma = 17.29$ kN/m^3
$\phi' = 20°$
$c' = 9.6$ kN/m^2

1.2 m

Figure 16.8

Solution
From Eq. (16.11),

$$q_u = c'N_c + qN_q + \tfrac{1}{2}\gamma BN_\gamma$$

From Table 16.1, for $\phi' = 20°$,

$$N_c = 17.69 \qquad N_q = 7.44 \qquad N_\gamma = 3.64$$

Also,

$$q = \gamma D_f = (17.29)(0.9) = 15.56 \text{ kN/m}^2$$

So,

$$q_u = (9.6)(17.69) + (15.56)(7.44) + (\tfrac{1}{2})(17.29)(1.2)(3.64)$$

$$= 323.35 \text{ kN/m}^2$$

From Eq. (16.26),

$$q_{all} = \frac{q_u}{F_s} = \frac{323.35}{3} \approx \mathbf{108 \ kN/m^2} \qquad \blacksquare$$

Example 16.2

A square foundation is shown in Figure 16.9. The footing will carry a gross mass of 30,000 kg. Using a factor of safety of 3, determine the size of the footing—that is, the size of B. Use Eq. (16.12).

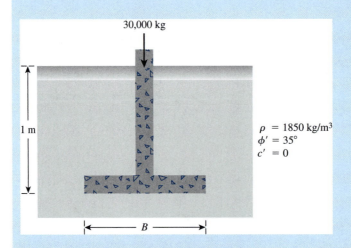

30,000 kg

1 m

$\rho = 1850 \text{ kg/m}^3$
$\phi' = 35°$
$c' = 0$

B

Figure 16.9

Solution

It is given that soil density = 1850 kg/m³. So

$$\gamma = \frac{1850 \times 9.81}{1000} = 18.15 \text{ kN/m}^3$$

Total gross load to be supported by the footing is

$$\frac{(30,000)9.81}{1000} = 294.3 \text{ kN} = Q_{all}$$

From Eq. (16.12)

$$q_u = 1.3c'N_c + qN_q + 0.4\gamma BN_\gamma$$

With a factor of safe+ty of 3,

$$q_{all} = \frac{q_u}{3} = \frac{1}{3}(1.3c'N_c + qN_q + 0.4\gamma BN_\gamma) \qquad \text{(a)}$$

Also,

$$q_{all} = \frac{Q_{all}}{B^2} = \frac{294.3}{B^2} \qquad \text{(b)}$$

From Eqs. (a) and (b),

$$\frac{294.3}{B^2} = \frac{1}{3}(1.3c'N_c + qN_q + 0.4\gamma BN_\gamma) \qquad \text{(c)}$$

From Table 16.1, for $\phi' = 35°$, $N_c = 57.75$, $N_q = 41.44$, and $N_\gamma = 45.41$. Substituting these values into Eq. (c) yields

$$\frac{294.3}{B^2} = \frac{1}{3}[(1.3)(0)(57.75) + (18.15 \times 1)(41.44) + 0.4(18.15)(B)(45.41)]$$

or

$$\frac{294.3}{B^2} = 250.7 + 109.9$$

The preceding equation may now be solved by trial and error, and from that we get

$$B \simeq \mathbf{0.95\,m}$$

$\blacksquare$

16.5 *General Bearing Capacity Equation*

After the development of Terzaghi's bearing capacity equation, several investigators worked in this area and refined the solution (that is, Meyerhof, 1951 and 1963; Lundgren and Mortensen, 1953; Balla, 1962; Vesic, 1973; and Hansen, 1970). Different solutions show that the bearing capacity factors N_c and N_q do not change much. However, for a given value of ϕ', the values of N_γ obtained by different investigators vary widely. This difference is because of the variation of the assumption of the wedge shape of soil located directly below the footing, as explained in the following paragraph.

While deriving the bearing capacity equation for a strip footing, Terzaghi used the case of a rough footing and assumed that the sides AJ and BJ of the soil wedge ABJ (see Figure 16.5a) make an angle ϕ' with the horizontal. Later model tests (for example, DeBeer and Vesic, 1958) showed that Terzaghi's assumption of the general nature of the rupture surface in soil for bearing capacity failure is correct. However, tests have shown that the sides AJ and BJ of the soil wedge ABJ make angles of about $45 + \phi'/2$ degrees (instead of ϕ') with the horizontal. This type of failure mechanism is shown in Figure 16.10. It consists of a Rankine active zone ABJ (Zone I), two radial shear zones (Zones II), and two Rankine passive zones (Zones III). The curves JD and JE are arcs of a logarithmic spiral.

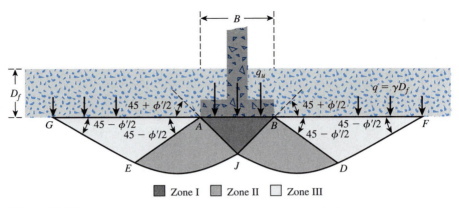

Figure 16.10 Soil-bearing capacity calculation—general shear failure

On the basis of this type of failure mechanism, the ultimate bearing capacity of a strip footing may be evaluated by the approximate method of superimposition described in Section 16.2 as

$$q_u = q_c + q_q + q_\gamma \tag{16.31}$$

where q_c, q_q, and q_γ are the contributions of cohesion, surcharge, and unit weight of soil, respectively.

Reissner (1924) expressed q_q as

$$q_q = qN_q \tag{16.32}$$

where

$$N_q = e^{\pi \tan \phi'} \tan^2\left(45 + \frac{\phi'}{2}\right) \tag{16.33}$$

Prandtl (1921) showed that

$$q_c = c'N_c \tag{16.34}$$

where

$$N_c = (N_q - 1)\cot \phi'$$
$$\uparrow$$
$$\text{Eq. (16.33)} \tag{16.35}$$

Meyerhof (1963) expressed q_γ as

$$q_\gamma = \tfrac{1}{2}B\gamma N_\gamma \tag{16.36}$$

where

$$N_\gamma = (N_q - 1)\tan(1.4\phi')$$
$$\uparrow$$
$$\text{Eq. (16.33)} \tag{16.37}$$

Combining Eqs. (16.31), (16.32), (16.34), and (16.36), we obtain

$$q_u = c'N_c + qN_q + \tfrac{1}{2}\gamma BN_\gamma \qquad (16.38)$$

This equation is in the same general form as that given by Terzaghi [Eq. (16.11)]; however, the values of the bearing capacity factors are not the same. The values of N_q, N_c, and N_γ, defined by Eqs. (16.33), (16.35), and (16.37), are given in Table 16.3. But for all practical purposes, Terzaghi's bearing capacity factors will yield good results. Differences in bearing capacity factors are usually minor compared with the unknown soil parameters.

The soil-bearing capacity equation for a strip footing given by Eq. (16.38) can be modified for general use by incorporating the following factors:

Depth factor:	To account for the shearing resistance developed along the failure surface in soil above the base of the footing.
Shape factor:	To determine the bearing capacity of rectangular and circular footings.

Table 16.3 Bearing Capacity Factors N_c, N_q, and N_γ [Eqs. (16.33), (16.35) and (16.37)]

ϕ' (deg)	N_c	N_q	N_γ	ϕ' (deg)	N_c	N_q	N_γ
0	5.14	1.00	0	26	22.25	11.85	8.002
1	5.38	1.09	0.002	27	23.94	13.20	9.463
2	5.63	1.20	0.010	28	25.80	14.72	11.190
3	5.90	1.31	0.023	29	27.86	16.44	13.236
4	6.19	1.43	0.042	30	30.14	18.40	15.668
5	6.49	1.57	0.070	31	32.67	20.63	18.564
6	6.81	1.72	0.106	32	35.49	23.18	22.022
7	7.16	1.88	0.152	33	38.64	26.09	26.166
8	7.53	2.06	0.209	34	42.16	29.44	31.145
9	7.92	2.25	0.280	35	46.12	33.30	37.152
10	8.35	2.47	0.367	36	50.59	37.75	44.426
11	8.80	2.71	0.471	37	55.63	42.92	53.270
12	9.28	2.97	0.596	38	61.35	48.93	64.073
13	9.81	3.26	0.744	39	67.87	55.96	77.332
14	10.37	3.59	0.921	40	75.31	64.20	93.690
15	10.98	3.94	1.129	41	83.86	73.90	113.985
16	11.63	4.34	1.375	42	93.71	85.38	139.316
17	12.34	4.77	1.664	43	105.11	99.02	171.141
18	13.10	5.26	2.009	44	118.37	115.31	211.406
19	13.93	5.80	2.403	45	133.88	134.88	262.739
20	14.83	6.40	2.871	46	152.10	158.51	328.728
21	15.82	7.07	3.421	47	173.64	187.21	414.322
22	16.88	7.82	4.066	48	199.26	222.31	526.444
23	18.05	8.66	4.824	49	229.93	265.51	674.908
24	19.32	9.60	5.716	50	266.89	319.07	873.843
25	20.72	10.66	6.765				

Inclination factor: To determine the bearing capacity of a footing on which the direction of load application is inclined at a certain angle to the vertical.

Thus, the modified general ultimate bearing capacity equation can be written as

$$q_u = c' \lambda_{cs} \lambda_{cd} \lambda_{ci} N_c + q \lambda_{qs} \lambda_{qd} \lambda_{qi} N_q + \tfrac{1}{2} \lambda_{\gamma s} \lambda_{\gamma d} \lambda_{\gamma i} \gamma B N_\gamma \qquad (16.39)$$

where λ_{cs}, λ_{qs}, and $\lambda_{\gamma s}$ = shape factors
$\quad \lambda_{cd}$, λ_{qd}, and $\lambda_{\gamma d}$ = depth factors
$\quad \lambda_{ci}$, λ_{qi}, and $\lambda_{\gamma i}$ = inclination factors

The approximate values of these shape, depth, and inclination factors recommended by Meyerhof are given in Table 16.4.

Table 16.4 Meyerhof's Shape, Depth, and Inclination Factors for a Rectangular Footing[a]

Shape factors

For $\phi = 0°$:

$$\lambda_{cs} = 1 + 0.2 \left(\frac{B}{L} \right)$$

$$\lambda_{qs} = 1$$
$$\lambda_{\gamma s} = 1$$

For $\phi' \geq 10°$:

$$\lambda_{cs} = 1 + 0.2 \left(\frac{B}{L} \right) \tan^2 \left(45 + \frac{\phi'}{2} \right)$$

$$\lambda_{qs} = \lambda_{\gamma s} = 1 + 0.1 \left(\frac{B}{L} \right) \tan^2 \left(45 + \frac{\phi'}{2} \right)$$

Depth factors

For $\phi = 0°$:

$$\lambda_{cd} = 1 + 0.2 \left(\frac{D_f}{B} \right)$$

$$\lambda_{qd} = \lambda_{\gamma d} = 1$$

For $\phi' \geq 10°$:

$$\lambda_{cd} = 1 + 0.2 \left(\frac{D_f}{B} \right) \tan \left(45 + \frac{\phi'}{2} \right)$$

$$\lambda_{qd} = \lambda_{\gamma d} = 1 + 0.1 \left(\frac{D_f}{B} \right) \tan \left(45 + \frac{\phi'}{2} \right)$$

Inclination factors

$$\lambda_{ci} = \left(1 - \frac{\alpha°}{90°} \right)^2$$

$$\lambda_{qi} = \left(1 - \frac{\alpha°}{90°} \right)^2$$

$$\lambda_{\gamma i} = \left(1 - \frac{\alpha°}{\phi'°} \right)^2$$

[a] B = width of footing; L = length of footing

For undrained condition, if the footing is subjected to vertical loading (that is, $\alpha = 0°$), then

$$\phi = 0$$

$$c = c_u$$

$$N_\gamma = 0$$

$$N_q = 1$$

$$N_c = 1$$

$$\lambda_{ci} = \lambda_{qi} = \lambda_{\gamma i} = 1$$

So Eq. (16.39) transforms to

$$q_u = 5.14c_u\left[1 + 0.2\left(\frac{B}{L}\right)\right]\left[1 + 0.2\left(\frac{D_f}{B}\right)\right] + q \qquad (16.40)$$

16.6 A Case History for Evaluation of the Ultimate Bearing Capacity

Several documented cases of large-scale field-load tests used to determine the ultimate bearing capacity of shallow foundations are presently available. One of these field-load tests is discussed in this section. The results of this test are compared with the theories presented in this chapter.

Skempton (1942) reported a field-load test in clay for a large foundation with $B = 2.44$ m and $L = 2.74$ m. This test also was reported by Bishop and Bjerrum (1960). Figure 16.11 shows a diagram of the foundation and the soil profile. The variation of the undrained cohesion (c_u) of the soil profile also is shown in Figure 16.11. The average moisture content, liquid limit, and plastic limit of the clay underlying the foundation were 50, 70, and 28%, respectively. The foundation was loaded to failure immediately after contruction. The net ultimate bearing capacity was determined to be 119.79 kN/m².

The net ultimate bearing capacity was defined in Eq. (16.28) as

$$q_{u(net)} = q_u - q$$

From Eq. (16.39), for the vertical loading condition and $\phi = 0°$ (*note: $N_q = 1$, $N_\gamma = 0$, $\lambda_{qs} = 1$, and $\lambda_{qd} = 1$*),

$$q_u = c_u\lambda_{cs}\lambda_{cd}N_c + q$$

So,

$$q_{u(net)} = (c_u\lambda_{cs}\lambda_{cd}N_c + q) - q = c_u\lambda_{cs}\lambda_{cd}N_c$$

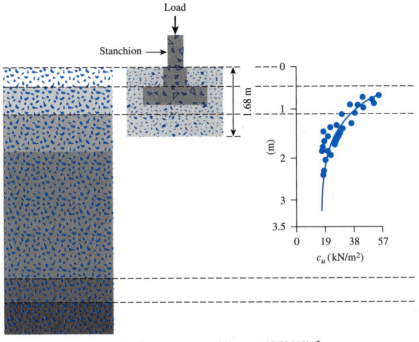

Net foundation pressure at failure = 119.79 kN/m²

Hard core   Top soil   Firm brown clay   Mottled clay   Soft blue clay
Peaty clay   Firm sandy clay

Figure 16.11 Skempton's field-load test on a foundation supported by a saturated clay *(After Bishop and Bjerrum, 1960. With permission from ASCE.)*

For the case under consideration, $c_u \approx 16.8$ kN/m² (see Figure 16.11) and $N_c = 5.14$ (see Table 16.3). From Table 16.4,

$$\lambda_{cs} = 1 + 0.2\left(\frac{B}{L}\right) = 1 + 0.2\left(\frac{2.44}{2.74}\right) = 1.2$$

$$\lambda_{cd} = 1 + 0.2\left(\frac{D_f}{B}\right) = 1 + 0.2\left(\frac{1.68}{2.44}\right) = 1.138$$

Hence,

$$q_{u(net)} = (16.8)(1.2)(1.138)(5.14) = 117.9 \text{ kN/m}^2$$

So, for this field-load test,

$$\frac{q_{u(net\text{-}theory)}}{q_{u(net\text{-}actual)}} = \frac{117.9}{119.79} \approx 0.98$$

Thus, the agreement between the theoretical estimate and the field-load test result is fairly good. The slight variation between them may be because of the estimation of the average value of c_u.

Bishop and Bjerrum (1960) cited several end-of-construction failures of footings on saturated clay. The data for these failures are given in Table 16.5. We can see from this table that in all cases $q_{u(net-theory)}/q_{u(net-actual)}$ is approximately 1. This finding confirms that the design of shallow foundations based on the net ultimate bearing capacity is a reliable technique.

Table 16.5 End-of-Construction Failures of Footings—Saturated Clay Foundation: $\phi = 0$ Condition[*]

| Locality | Data of clay | | | | | $\dfrac{q_{u(net-theory)}}{q_{u(net-actual)}}$ |
	w(%)	LL	PL	PI	$\dfrac{w - PL}{PI}$	
Loading test, Marmorerá	10	35	25	20	−0.25	0.92
Kensal Green	—	—	—	—	—	1.02
Silo, Transcona	50	110	30	80	0.25	1.09
Kippen	50	70	28	42	0.52	0.95
Screw pile, Lock Ryan	—	—	—	—	—	1.05
Screw pile, Newport	—	—	—	—	—	1.07
Oil tank, Fredrikstad	45	55	25	30	0.67	1.08
Oil tank A, Shellhaven	70	87	25	62	0.73	1.03
Oil tank B, Shellhaven	—	—	—	—	—	1.05
Silo, U.S.A.	40	—	—	—	—	0.98
Loading test, Moss	9	—	—	—	—	1.10
Loading test, Hagalund	68	55	20	35	1.37	0.93
Loading test, Torp	27	24	16	8	1.39	0.96
Loading test, Rygge	45	37	19	18	1.44	0.95

[*]*After Bishop and Bjerrum (1960). With permission from ASCE.*
Note: w = moisture content, LL = liquid limit; PL = plastic limit; PI = plasticity index

Example 16.3

A square footing is shown in Figure 16.12. Determine the safe gross load (factor of safety of 3) that the footing can carry. Use Eq. (16.39).

Solution
From Eq. (16.39),

$$q_u = c'\lambda_{cs}\lambda_{cd}N_c + q\lambda_{qs}\lambda_{qd}N_q + \tfrac{1}{2}\gamma'\lambda_{\gamma s}\lambda_{\gamma d}BN_\gamma$$

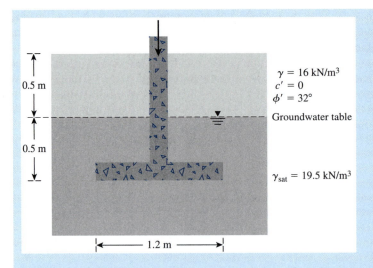

$\gamma = 16$ kN/m³
$c' = 0$
$\phi' = 32°$

Groundwater table

$\gamma_{sat} = 19.5$ kN/m³

0.5 m

0.5 m

1.2 m

Figure 16.12

(*Note*: λ_{ci}, λ_{qi}, and $\lambda_{\gamma i}$ are all equal to 1 because the load is vertical.)
Because $c' = 0$,

$$q_u = q\lambda_{qs}\lambda_{qd}N_q + \tfrac{1}{2}\gamma'\lambda_{\gamma s}\lambda_{\gamma d}BN_\gamma$$

From Table 16.3 for $\phi' = 32°$, $N_q = 23.18$ and $N_\gamma = 22.02$. From Table 16.4,

$$\lambda_{qs} = \lambda_{\gamma s} = 1 + 0.1\left(\frac{B}{L}\right)\tan^2\left(45 + \frac{\phi'}{2}\right)$$

$$= 1 + 0.1\left(\frac{1.2}{1.2}\right)\tan^2\left(45 + \frac{32}{2}\right) = 1.325$$

$$\lambda_{qd} = \lambda_{\gamma d} = 1 + 0.1\left(\frac{D_f}{B}\right)\tan\left(45 + \frac{\phi'}{2}\right)$$

$$= 1 + 0.1\left(\frac{1}{1.2}\right)\tan\left(45 + \frac{32}{2}\right) = 1.15$$

The groundwater table is located above the bottom of the foundation, so, from Eq. (16.23),

$$q = (0.5)(16) + (0.5)(19.5 - 9.81) = 12.845 \text{ kN/m}^2$$

Thus,

$$q_u = (12.845)(1.325)(1.15)(23.18) + (\tfrac{1}{2})(19.5 - 9.81)(1.325)(1.15)(1.2)(22.02)$$

$$= 453.7 + 195.1 = 648.8 \text{ kN/m}^2$$

$$q_{all} = \frac{q_u}{3} = \frac{648.8}{3} = 216.3 \text{ kN/m}^2$$

Hence, the gross load is as follows:

$$Q = q_{all}(B^2) = 216.3(1.2)^2 = \textbf{311.5 kN}$$

Ultimate Load for Shallow Foundations Under Eccentric Load

One-Way Eccentricity

To calculate the bearing capacity of shallow foundations with eccentric loading, Meyerhof (1953) introduced the concept of *effective area*. This concept can be explained with reference to Figure 16.13, in which a footing of length L and width B is subjected to an eccentric load, Q_u. If Q_u is the ultimate load on the footing, it may be approximated as follows:

1. Referring to Figures 16.13b and 16.13c, calculate the effective dimensions of the foundation. If the eccentricity (e) is in the x direction (Figure 16.13b), the *effective dimensions* are

$$X = B - 2e$$

and

$$Y = L$$

However, if the eccentricity is in the y direction (Figure 16.13c), the effective dimensions are

$$Y = L - 2e$$

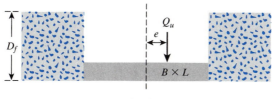

(a) Section

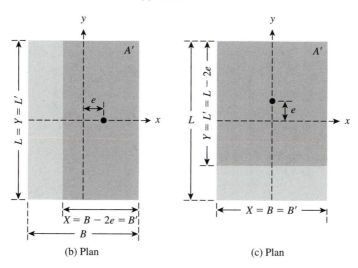

(b) Plan   (c) Plan

Figure 16.13 Ultimate load for shallow foundation under eccentric load

and

$$X = B$$

2. The lower of the two effective dimensions calculated in Step 1 is the *effective width* (*B'*) and the other is the *effective length* (*L'*). Thus,

$$B' = X \text{ or } Y, \text{ whichever is smaller}$$

$$L' = X \text{ or } Y, \text{ whichever is larger}$$

3. So the effective area is equal to *B'* times *L'*. Now, using the effective width, we can rewrite Eq. (16.39) as

$$q_u = c' \lambda_{cs} \lambda_{cd} N_c + q \lambda_{qs} \lambda_{qd} N_q + \tfrac{1}{2} \lambda_{\gamma s} \lambda_{\gamma d} \gamma B' N_\gamma \qquad (16.41)$$

Note that the preceding equation is obtained by substituting *B'* for *B* in Eq. (16.39). While computing the shape and depth factors, one should use *B'* for *B* and *L'* for *L*.

4. Once the value of q_u is calculated from Eq. (16.41), we can obtain the total gross ultimate load as follows:

$$Q_u = q_u(B'L') = q_u A' \qquad (16.42)$$

where A' = effective area.

Purkayastha and Char (1977) carried out stability analysis of eccentrically loaded *continuous foundations on granular soil* (i.e., $c' = 0$) using the method of slices. Based on that analysis, they proposed that

$$R_k = 1 - \frac{q_{u(\text{eccentric})}}{q_{u(\text{centric})}} \qquad (16.43)$$

where R_k = reduction factor

$q_{u(\text{eccentric})}$ = ultimate bearing capacity of eccentrically loaded continuous foundations

$$= \frac{Q_{u(\text{eccentric})}}{B}$$

$q_{u(\text{centric})}$ = ultimate bearing capacity of centrally loaded continuous foundations

$$= \frac{Q_{u(\text{centric})}}{B}$$

The magnitude of R_k can be expressed as

$$R_k = a\left(\frac{e}{B}\right)^k \tag{16.44}$$

where a and k are functions of the embedment ratio D_f/B (Table 16.6).
    Hence, combining Eqs. (16.43) and (16.44) gives

$$Q_{u(\text{eccentric})} = Q_{u(\text{centric})}\left[1 - a\left(\frac{e}{B}\right)^k\right] \tag{16.45}$$

where $Q_{u(\text{eccentric})}$ and $Q_{u(\text{centric})}$ = ultimate load per unit length, respectively, for eccentrically and centrically loaded foundations.

Two-Way Eccentricity

When foundations are subjected to loads with two-way eccentricity, as shown in Figure 16.14, the effective area is determined such that the centroid coincides with the load. The procedure

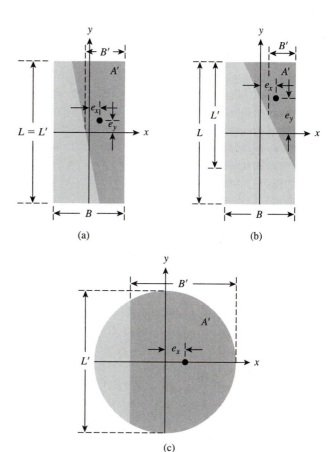

Table 16.6 Variations of a and k [Eq. (16.44)]

D_f/B	a	k
0	1.862	0.73
0.25	1.811	0.785
0.5	1.754	0.80
1.0	1.820	0.888

Figure 16.14 Foundation subjected to two-way eccentricity

for finding the effective dimensions, B' and L', are beyond the scope of this text and readers may refer to Das (2007). Once B' and L' are determined, Eqs. (16.41) and (16.42) may be used to determine the ultimate load.

Example 16.4

A rectangular footing 1.5m $\times$ 1m is shown in Figure 16.15. Determine the magnitude of the gross ultimate load applied eccentrically for bearing capacity failure in soil.

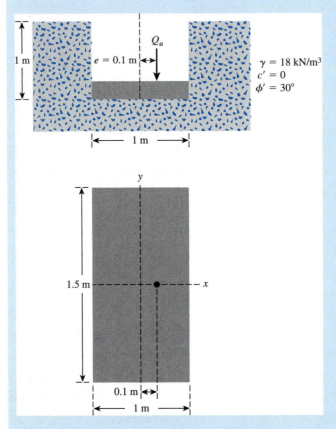

Figure 16.15

Solution
From Figure 16.13b and 16.15,

$$X = B - 2e = 1 - 2e = 1 - (2)(0.1) = 0.8 \text{ m}$$
$$Y = L = 1.5 \text{ m}$$

So, effective width $(B') = 0.8$ m and effective length $(L') = 1.5$ m. From Eq. (16.39),

$$q_u = q\lambda_{qs}\lambda_{qd}N_q + \tfrac{1}{2}\lambda_{\gamma s}\lambda_{\gamma d}\gamma B'N_\gamma$$

From Table 16.3 for $\phi' = 30°$, $N_q = 18.4$ and $N_\gamma = 15.668$. From Table 16.4,

$$\lambda_{qs} = \lambda_{\gamma s} = 1 + 0.1\left(\frac{B'}{L'}\right)\tan^2\left(45 + \frac{\phi'}{2}\right)$$

$$= 1 + 0.1\left(\frac{0.8}{1.5}\right)\tan^2\left(45 + \frac{30}{2}\right) = 1.16$$

$$\lambda_{qd} = \lambda_{\gamma d} = 1 + 0.1\left(\frac{D_f}{B'}\right)\tan\left(45 + \frac{\phi'}{2}\right)$$

$$= 1 + 0.1\left(\frac{1}{0.8}\right)\tan\left(45 + \frac{30}{2}\right) = 1.217$$

So

$$q_u = (1 \times 18)(1.16)(1.217)(18.4)$$

$$+ \left(\tfrac{1}{2}\right)(1.16)(1.217)(18)(0.8)(15.668) = 627 \text{ kN/m}^2$$

Hence, from Eq. (16.42),

$$Q_u = q_u(B'L') = (627)(0.8)(1.5) = \textbf{752 kN} \qquad \blacksquare$$

Example 16.5

Consider an eccentrically loaded continuous foundation supported by a granular soil. Given: $B = 1.5$ m, $D_f = 0.75$ m, load eccentricity $e/B = 0.1$, $\gamma = 17.5$ kN/m^3, $\phi' = 30°$, and $c' = 0$. Use the reduction factor method [Eq. (16.45)] and determine the gross ultimate load per unit length that the foundation can carry.

Solution
From Eq. (16.45),

$$Q_{u(\text{eccentric})} = Q_{u(\text{centric})}\left[1 - a\left(\frac{e}{B}\right)^k\right]$$

Also, $D_f/B = 0.75/1.5 = 0.5$. From Table 16.6, $a = 1.754$ and $k = 0.8$. Thus,

$$Q_{u(\text{centric})} = qN_q\lambda_{qd} + \frac{1}{2}\lambda_{\gamma d}BN_\gamma$$

(*Note*: The shape factors are all equal to one, since it is a continuous foundation.)

$$q = \gamma D_f = (17.5)(0.75) = 13.125 \text{ kN/m}^2$$

From Table 16.3 for $\phi = 30°$, $N_q = 18.4$ and $N_\gamma = 15.668$. Therefore,

$$\lambda_{qd} = \lambda_{\gamma d} = 1 + 0.1\left(\frac{D_f}{B}\right)\tan\left(45 + \frac{\phi'}{2}\right)$$

$$= 1 + 0.1\left(\frac{0.75}{1.5}\right)\tan\left(45 + \frac{30}{2}\right) = 1.087$$

$$Q_{u(\text{centric})} = (13.125)(18.4)(1.087) + (0.5)(1.087)(1.5)(15.668) = 275.28 \text{ kN/m}^2$$

Hence,

$$Q_{u(\text{eccentric})} = (275.28)[1 - (1.754)(0.1)^{0.8}] = 198.76 \text{ kN} \approx \textbf{199 kN} \quad \blacksquare$$

16.8   *Bearing Capacity of Sand Based on Settlement*

Obtaining undisturbed specimens of cohesionless sand during a soil exploration program is usually difficult. For this reason, the results of standard penetration tests (SPTs) performed during subsurface exploration are commonly used to predict the allowable soil-bearing capacity of foundations on sand. (The procedure for conducting SPTs is discussed in detail in Chapter 18.)

Meyerhof (1956) proposed correlations for the *net allowable bearing capacity* (q_{net}) based on settlement (elastic). It was further revised by Meyerhof (1965) based on the field performance of foundations. The correlations can be expressed as follows.

$$q_{net}(\text{kN/m}^2) = \frac{N_{60}}{0.05}F_d\left[\frac{S_e(\text{mm})}{25}\right] \quad (\text{for } B \le 1.22\,\text{m}) \quad (16.46)$$

$$q_{net}(kN/m^2) = \frac{N_{60}}{0.08}\left(\frac{B + 0.3}{B}\right)^2 F_d\left[\frac{S_e(mm)}{25}\right] \quad \text{(for } B > 1.22\,m) \quad (16.47)$$

where B = foundation width (m)
$\quad S_e$ = settlement

In Eqs. (16.46) through (16.47),

$\quad N_{60}$ = field standard penetration number based on 60% average energy ratio
$\quad S_e$ = allowable settlement (elastic)

$$F_d = \text{depth factor} = 1 + 0.33\left(\frac{D_f}{B}\right) \le 1.33 \quad (16.48)$$

The N_{60} values referred to in Eqs. (16.46) and (16.47) are the average values between the bottom of the foundation and $2B$ below the bottom.

COMPARISON WITH FIELD SETTLEMENT OBSERVATION

Meyerhof (1965) compiled the observed maximum settlement (S_e) for several mat foundations constructed on sand and gravel. These are shown in Table 16.7 (Column 5) along with the values of B, q_{net}, and N_{60}.
From Eq. (16.47), we can write

$$S_e(mm) = \frac{25q_{net}}{\left(\dfrac{N_{60}}{0.08}\right)\left(\dfrac{B + 0.3}{B}\right)^2 F_d} \quad (16.49)$$

As can be seen from Table 16.7, the widths B for the mats are large. Hence

$$\left(\frac{B + 0.3}{B}\right)^2 \approx 1$$

$$F_d = 1 + 0.33\frac{D_f}{B} \approx 1$$

So

$$S_e(mm) \approx \frac{25q_{net}}{12.5N_{60}} \approx \frac{2q_{net}(kN/m^2)}{N_{60}} \quad (16.50)$$

Using the actual values of q_{net} and N_{60} given in Table 16.7, the magnitudes of S_e have been calculated via Eq. (16.50). These are shown in Column 6 of Table 16.7 as $S_{e(predicted)}$. The ratio of $S_{e(predicted)}/S_{e(observed)}$ is shown in Column 7. This ratio varies from 0.84 to 3.6. Hence, it can be concluded that the allowable net bearing capacity for a given allowable settlement calculated using the empirical relation is safe and conservative.

Table 16.7 Observed and Calculated Maximum Settlement of Mat Foundations on Sand and Gravel

Structure (1)	B (m) (2)	N_{60} (3)	q_{net} (kN/m²) (4)	$S_{e(observed)}$ (mm) (5)	$S_{e(predicted)}$ (mm) (6)	$\dfrac{S_{e(predicted)}}{S_{e(observed)}}$ (7)
T. Edison Sao Paulo, Brazil	18.3	15	229.8	15.2	30.6	2.0
Banco de Brasil Sao Paulo, Brazil	22.9	18	239.4	27.9	26.6	0.95
Iparanga Sao Paulo, Brazil	9.1	9	306.4	35.6	68.1	1.91
C.B.I., Esplanada Sao Paulo, Brazil	14.6	22	383.0	27.9	34.8	1.25
Riscala Sao Paulo, Brazil	4.0	20	229.8	12.7	23	1.81
Thyssen Dusseldorf, Germany	22.6	25	239.4	24.1	19.2	0.8
Ministry Dusseldorf, Germany	15.9	20	220.2	21.6	22.0	1.02
Chimney Cologne, Germany	20.4	10	172.4	10.2	34.5	3.38

Columns 2, 3, 4, 5—Compiled from Meyerhof (1965)
Column 6—from Eq. (16.50)

16.9   *Plate-Load Test*

In some cases, conducting field-load tests to determine the soil-bearing capacity of foundations is desirable. The standard method for a field-load test is given by the American Society for Testing and Materials (ASTM) under Designation D-1194 (ASTM, 1997). Circular steel bearing plates 162 to 760 mm in diameter and 305 mm × 305 mm square plates are used for this type of test.

A diagram of the load test is shown in Figure 16.16. To conduct the test, one must have a pit of depth D_f excavated. The width of the test pit should be at least four times the width of the bearing plate to be used for the test. The bearing plate is placed on the soil at the bottom of the pit, and an incremental load on the bearing plate is applied. After the application of an incremental load, enough time is allowed for settlement to occur. When the settlement of the bearing plate becomes negligible, another incremental load is applied. In this manner, a load-settlement plot can be obtained, as shown in Figure 16.17.

From the results of field load tests, the ultimate soil-bearing capacity of actual footings can be approximated as follows:

For clays,

$$q_{u(footing)} = q_{u(plate)} \tag{16.51}$$

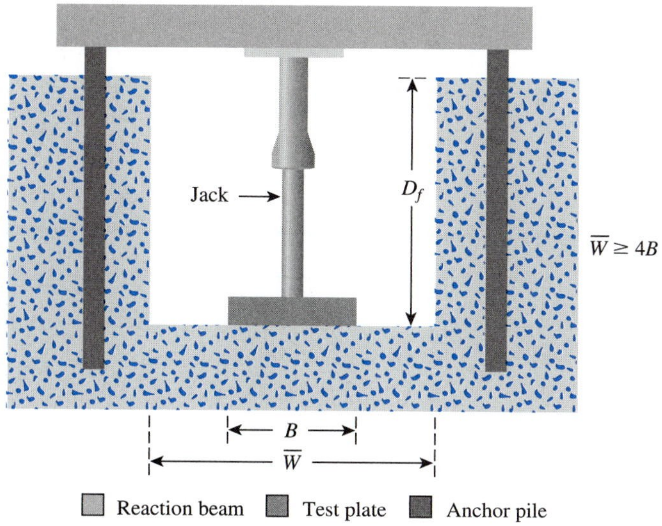

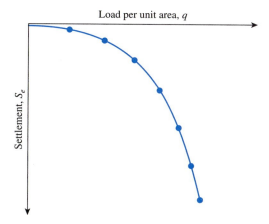

Jack

D_f

$\overline{W} \geq 4B$

B

$\overline{W}$

Reaction beam   Test plate   Anchor pile

Figure 16.16 Diagram of plate load test

Load per unit area, q

Settlement, S_e

Figure 16.17 Typical load-settlement curve obtained from plate load test

For sandy soils,

$$q_{u\,(\text{footing})} = q_{u\,(\text{plate})} \frac{B_{(\text{footing})}}{B_{(\text{plate})}} \qquad (16.52)$$

For a given intensity of load q, the settlement of the actual footing also can be approximated from the following equations:

In clay,

$$S_{e\,(\text{footing})} = S_{e\,(\text{plate})} \frac{B_{(\text{footing})}}{B_{(\text{plate})}} \qquad (16.53)$$

In sandy soil,

$$S_{e(\text{footing})} = S_{e(\text{plate})}\left[\frac{2B_{(\text{footing})}}{B_{(\text{footing})} + B_{(\text{plate})}}\right]^2 \tag{16.54}$$

Example 16.6

The ultimate bearing capacity of a 700-mm diameter plate as determined from field load tests is 280 kN/m². Estimate the ultimate bearing capacity of a circular footing with a diameter of 1.5 m. The soil is sandy.

Solution
From Eq. (16.52),

$$q_{u(\text{footing})} = q_{u(\text{plate})}\frac{B_{(\text{footing})}}{B_{(\text{plate})}} = 280\left(\frac{1.5}{0.7}\right)$$

$$= \mathbf{680 \ kN/m^2}$$ ∎

Example 16.7

The results of a plate-load test in a sandy soil are shown in Figure 16.18. The size of the plate is 0.305 m × 0.305 m. Determine the size of a square column foundation that should carry a load of 2500 kN with a maximum settlement of 25 mm.

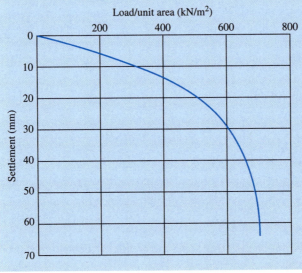

Figure 16.18

Solution

The problem has to be solved by trial and error using the following table and Eq. (16.54).

Q (kN) (1)	Assumed width, B_F (m) (2)	$q = \dfrac{Q}{B_F^2}$ (kN/m^2) (3)	$S_{e(plate)}$ corresponding to q in Column 3 from Fig. 16.18 (mm) (4)	$S_{e(footing)}$ using Eq. (16.56) (mm) (5)
2500	4.0	156.25	4.0	13.81
2500	3.0	277.80	8.0	26.37
2500	3.2	244.10	6.8	22.67

So a column footing with dimensions of **3.2 m $\times$ 3.2 m** will be appropriate. ■

16.10 *Summary and General Comments*

In this chapter, theories for estimating the ultimate and allowable bearing capacities of shallow foundations were presented. Procedures for field-load tests and estimation of the allowable bearing capacity of granular soil based on limited settlement criteria were discussed briefly.

Several building codes now used in the United States and elsewhere provide presumptive bearing capacities for various types of soil. It is extremely important to realize that they are *approximate values only*. The bearing capacity of foundations depends on several factors:

1. Subsoil stratification
2. Shear strength parameters of the subsoil
3. Location of the groundwater table
4. Environmental factors
5. Building size and weight
6. Depth of excavation
7. Type of structure

Hence, it is important that the allowable bearing capacity at a given site be determined based on the findings of soil exploration at that site, past experience of foundation construction, and fundamentals of the geotechnical engineering theories for bearing capacity.

The allowable bearing capacity relationships based on settlement considerations such as those given in Section 16.9 do not take into account the settlement caused by consolidation of the clay layers. Excessive settlement usually causes the building to crack, which ultimately may lead to structural failure. Uniform settlement of a structure does not produce cracking; on the other hand, differential settlement may produce cracks and damage to a building.

Problems

16.1 A continuous footing is shown in Figure 16.19. Using Terzaghi's bearing capacity factors, determine the gross allowable load per unit area (q_{all}) that the footing can carry. Assume general shear failure. Given: $\gamma = 18.08$ kN/m³, $c' = 28.75$ kN/m², $\phi' = 25°$, $D_f = 1.07$ m, $B = 1.22$ m, and factor of safety $= 3$.

16.2 Repeat Problem 16.1 with the following: $\gamma = 17.5$ kN/m³, $c' = 14$ kN/m², $\phi' = 20°$, $D_f = 1.0$ m, $B = 1.2$ m, and factor of safety $= 3$.

16.3 Repeat Problem 16.1 with the following: $\gamma = 17.7$ kN/m³, $c_u = 48$ kN/m², $\phi = 0°$, $D_f = 0.6$ m, $B = 0.8$ m, and factor of safety $= 4$.

16.4 Repeat Problem 16.1 using Eq. (16.39).

16.5 Repeat Problem 16.2 using Eq. (16.39).

16.6 Repeat Problem 16.3 using Eq. (16.39).

16.7 A square footing is shown in Figure 16.20. Determine the gross allowable load, Q_{all}, that the footing can carry. Use Terzaghi's equation for general shear failure ($F_s = 3$). Given: $\gamma = 16.5$ kN/m³, $\gamma_{sat} = 18.55$ kN/m³, $c' = 0$, $\phi' = 35°$, $B = 1.5$ m, $D_f = 1.22$ m, and $h = 0.61$ m.

16.8 Repeat Problem 16.7 with the following: density of soil above the groundwater table, $\rho = 1800$ kg/m³; saturated soil density below the groundwater table, $\rho_{sat} = 1980$ kg/m³; $c' = 23.94$ kN/m²; $\phi' = 25°$; $B = 1.8$ m; $D_f = 1.2$ m; and $h = 2$ m.

16.9 A square footing ($B \times B$) must carry a gross allowable load of 188 kN. The base of the footing is to be located at a depth of 0.91 m below the ground surface. For the soil, we are given that $\gamma = 17.29$ kN/m³, $c' = 9.6$ kN/m², and $\phi' = 20°$. If the required factor of safety is 3, determine the size of the footing. Use Terzaghi's bearing capacity factors and general shear failure of soil.

16.10 Repeat Problem 16.9 with the following: gross allowable load $= 411$ kN; $D_f = 0.61$ m; $\gamma = 18.08$ kN/m³; $c' = 0$; $\phi' = 35°$; and required factor of safety $= 3$.

16.11 Repeat Problem 16.7 using Eq. (16.39).

16.12 A square footing is shown in Figure 16.21. The footing is subjected to an eccentric load. For the following cases, determine the gross allowable load that the footing could carry. Use Meyerhof's procedure and $F_s = 4$.

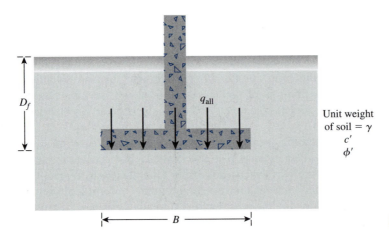

Figure 16.19

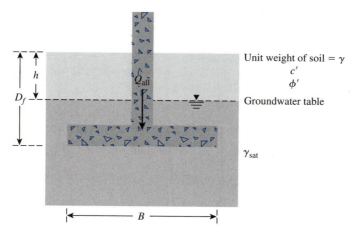

Figure 16.20

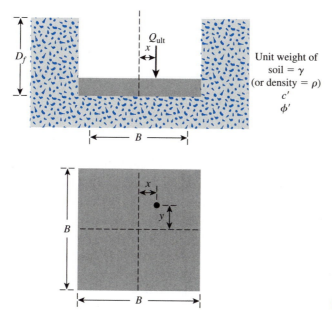

Figure 16.21

a. $\gamma = 17.29$ kN/m³, $c' = 0$, $\phi' = 35°$, $B = 1.52$ m, $D_f = 1.07$ m, $x = 0.18$ m, $y = 0$

b. $\gamma = 18.87$ kN/m³, $c' = 19.17$ kN/m², $\phi' = 25°$, $B = 1.83$ m, $D_f = 1.37$ m, $x = 0$, $y = 0.15$ m

c. $\rho = 1950$ kg/m³, $c' = 0$, $\phi' = 40°$, $B = 3$ m, $D_f = 1.4$ m, $x = 0.3$ m, $y = 0$

16.13 A plate-load test was conducted in a sandy soil in which the size of the bearing plate was 0.3 m × 0.3 m. The ultimate load per unit area (q_u) for the test was found to be 184.5 kN/m². Estimate the total allowable load (Q_{all}) for a footing of size 1.8 m × 1.8 m. Use a factor of safety of 4.

16.14 A plate load test (bearing plate of 762 mm diameter) was conducted in clay. The ultimate load per unit area (q_u) for the test was found to be 248.9 kN/m². What should be the total allowable load (Q_{all}) for a column footing with a diameter of 2 m? Use a factor of safety of 3.

References

AMERICAN SOCIETY FOR TESTING AND MATERIALS (1997). *Annual Book of Standards,* ASTM, Vol. 04.08, West Conshohocken, Pa.

BALLA, A. (1962). "Bearing Capacity of Foundations," *Journal of the Soil Mechanics and Foundations Division*, ASCE, Vol. 89, No. SM5, 12–34.

BISHOP, A. W., and BJERRUM, L. (1960). "The Relevance of the Triaxial Test to the Solution of Stability Problems," *Proceedings,* Research Conference on Shear Strength of Cohesive Soils, ASCE, 437–501.

DAS, B. M. (2007), *Principles of Foundation Engineering,* 6th ed., Thomson, Toronto, Canada.

HANSEN, J. B. (1970). "A Revised and Extended Formula for Bearing Capacity," Danish Geotechnical Institute, *Bulletin No. 28,* Copenhagen.

KUMBHOJKAR, A. S. (1993). "Numerical Evaluation of Terzaghi's N_γ," *Journal of Geotechnical Engineering,* ASCE, Vol. 119, No. GT3, 598–607.

LUNDGREN, H., and MORTENSEN, K. (1953). "Determination by the Theory of Elasticity of the Bearing Capacity of Continuous Footing on Sand," *Proceedings*, 3rd International Conference on Soil Mechanics and Foundation Engineering, Vol. I, 409–412.

MEYERHOF, G. G. (1951). "The Ultimate Bearing Capacity of Foundations," *Geotechnique,* Vol. 2, No. 4, 301–331.

MEYERHOF, G. G. (1953). "The Bearing Capacity of Foundations Under Eccentric and Inclined Loads," *Proceedings,* 3rd International Conference on Soil Mechanics and Foundation Engineering, Vol. 1, 440–445.

MEYERHOF, G. G. (1956). "Penetration Tests and Bearing Capacity of Cohesionless Soils," *Journal of the Soil Mechanics and Foundations Division,* American Society of Civil Engineers, Vol. 82, No. SM1, pp. 1–19.

MEYERHOF, G. G. (1963). "Some Recent Research on the Bearing Capacity of Foundations," *Canadian Geotechnical Journal,* Vol. 1, 16–26.

MEYERHOF, G. G. (1965). "Shallow Foundations," *Journal of the Soil Mechanics and Foundations Division,* ASCE, Vol. 91, No. SM2, pp. 21–31.

PRANDTL, L. (1921). "Über die Eindringungsfestigkeit (Harte) plastischer Baustoffe und die Festigkeit von Schneiden," *Zeitschrift für Angewandte Mathematik und Mechanik,* Basel, Switzerland, Vol. 1, No. 1, 15–20.

PURKAYASTHA, R. D., and CHAR, R. A. N. (1977). "Stability Analysis for Eccentrically Loaded Footings," *Journal of the Geotechnical Engineering Division*, ASCE, Vol. 103, No. 6, 647–651.

REISSNER, H. (1924). "Zum Erddruckproblem," *Proceedings,* 1st International Congress of Applied Mechanics, 295–311.

SKEMPTON, A. W. (1942). "An Investigation of the Bearing Capacity of a Soft Clay Soil," *Journal of the Institute of Civil Engineers,* London, Vol. 18, 307–321.

TERZAGHI, K. (1943). *Theoretical Soil Mechanics,* Wiley, New York.

VESIC, A. S. (1973). "Analysis of Ultimate Loads of Shallow Foundations," *Journal of the Soil Mechanics and Foundations Division*, ASCE, Vol. 99, No. SM1, 45–73.

17 Landfill Liners and Geosynthetics

Enormous amounts of solid waste are generated every year in the United States and other industrialized countries. These waste materials, in general, can be classified into four major categories: (1) municipal waste, (2) industrial waste, (3) hazardous waste, and (4) low-level radioactive waste. Table 17.1 lists the waste material generated in 1984 in the United States in these four categories (Koerner, 1994).

The waste materials generally are placed in landfills. The landfill materials interact with moisture received from rainfall and snow to form a liquid called *leachate*. The chemical composition of leachates varies widely, depending on the waste material involved. Leachates are a main source of groundwater pollution; therefore, they must be contained properly in all landfills, surface impoundments, and waste piles within some type of liner system. In the following sections of this chapter, various types of liner systems and the materials used in them are discussed.

17.1 Landfill Liners—Overview

Until about 1982, the predominant liner material used in landfills was clay. Proper clay liners have a hydraulic conductivity of about 10^{-7} cm/sec or less. In 1984, the U.S. Environmental Protection Agency's minimum technological requirements for hazardous-waste landfill design and construction were introduced by the U.S. Congress in Hazardous and Solid Waste amendments. In these amendments, Congress stipulated that all new landfills should have double liners and systems for leachate collection and removal.

Table 17.1 Waste Material Generation in the United States

Waste type	Approximate quantity in 1984 (millions of metric tons)
Municipal	300
Industrial (building debris, degradable waste, nondegradable waste, and near hazardous)	600
Hazardous	150
Low-level radioactive	15

To understand the construction and functioning of the double-liner system, we must review the general properties of the component materials involved in the system—that is, clay soil and geosynthetics (such as geotextiles, geomembranes, and geonets).

Section 17.2 gives the details for the compaction of clay soils in the field for liner construction. A brief review of the essential properties of geosynthetics is given in Sections 17.3 through 17.6.

<table>
<tr><td>**17.2**</td></tr>
</table>

Compaction of Clay Soil for Clay Liner Construction

It was shown in Chapter 6 (Section 6.5) that, when a clay is compacted at a lower moisture content, it possesses a flocculent structure. Approximately at the optimum moisture content of compaction, the clay particles have a lower degree of flocculation. A further increase in the moisture content at compaction provides a greater degree of particle orientation; however, the dry unit weight decreases, because the added water dilutes the concentration of soil solids per unit volume.

Figure 17.1 shows the results of laboratory compaction tests on a clay soil as well as the variation of hydraulic conductivity on the compacted clay specimens. From the laboratory test results shown, the following observations can be made:

1. For a given compaction effort, the hydraulic conductivity, k, decreases with the increase in molding moisture content, reaching a minimum value at about the optimum moisture content (that is, approximately where the soil has a higher unit weight with the clay particles having a lower degree of flocculation). Beyond the optimum moisture content, the hydraulic conductivity increases slightly.
2. For similar compaction effort and dry unit weight, a soil will have a lower hydraulic conductivity when it is compacted on the wet side of the optimum moisture content.

Benson and Daniel (1990) conducted laboratory compaction tests by varying the size of clods of moist clayey soil. These tests show that, for similar compaction effort and molding moisture content, the magnitude of k decreases with the decrease in clod size.

In some compaction work in clayey soils, the compaction must be done in a manner so that a certain specified upper level of hydraulic conductivity of the soil is achieved. Examples of such works are compaction of the core of an earth dam and installation of clay liners in solid-waste disposal sites.

To prevent groundwater pollution from leachates generated from solid-waste disposal sites, the U.S. Environmental Protection Agency (EPA) requires that clay liners have a hydraulic conductivity of 10^{-7} cm/sec or less. To achieve this value, the contractor must ensure that the soil meets the following criteria (Environmental Protection Agency, 1989):

1. The soil should have at least 20% fines (fine silt and clay-sized particles).
2. The plasticity index *(PI)* should be greater than 10. Soils that have a *PI* greater than about 30 are difficult to work with in the field.
3. The soil should not include more than 10% gravel-sized particles.
4. The soil should not contain any particles or chunks of rock that are larger than 25 to 50 mm.

In many instances, the soil found at the construction site may be somewhat nonplastic. Such soil may be blended with imported clay minerals (like sodium bentonite) to

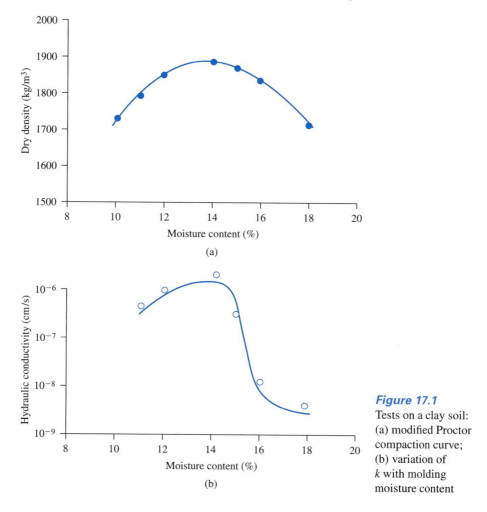

Figure 17.1
Tests on a clay soil:
(a) modified Proctor
compaction curve;
(b) variation of
k with molding
moisture content

achieve the desired range of hydraulic conductivity. In addition, during field compaction, a heavy sheepsfoot roller can introduce larger shear strains during compaction that create a more dispersed structure in the soil. This type of compacted soil will have an even lower hydraulic conductivity. Small lifts should be used during compaction so that the feet of the compactor can penetrate the full depth of the lift.

There are a few published studies on the variation of hydraulic conductivity of mixtures of nonplastic soils and Bentonite. Sivapullaiah, *et al.* (2000) evaluated the hydraulic conductivity of Bentonite-sand and Bentonite-silt mixtures in the laboratory. Based on this study the following correlation was developed

$$\log k = \frac{e - 0.0535(LL) - 5.286}{0.0063(LL) + 0.2516} \tag{17.1}$$

where k = hydraulic conductivity of Bentonite-nonplastic soil mixture (m/sec)
e = void ratio of compacted mixture
LL = liquid limit of mixture (%)

Equation (17.1) is valid for mixtures having a liquid limit greater than 50%.

The size of the clay clods has a strong influence on the hydraulic conductivity of a compacted clay. Hence, during compaction, the clods must be broken down mechanically to as small as possible. A very heavy roller used for compaction helps to break them down.

Bonding between successive lifts is also an important factor; otherwise, permeant can move through a vertical crack in the compacted clay and then travel along the interface between two lifts until it finds another crack, as is shown schematically in Figure 17.2. Poor bonding can increase substantially the overall hydraulic conductivity of a compacted clay. An example of poor bonding was seen in a trial pad construction in Houston in 1986. The trial pad was 0.91 m thick and built in six, 15.2 mm lifts. The results of the hydraulic conductivity tests for the compact soil from the trial pad are given in Table 17.2. Note that, although the laboratory-determined values of k for various lifts are on the order of 10^{-7} to 10^{-9} cm/sec, the actual overall value of k increased to the order of 10^{-4}. For this reason, scarification and control of the moisture content after compaction of each lift are extremely important in achieving the desired hydraulic conductivity.

In the construction of clay liners for solid-waste disposal sites where it is required that $k \leq 10^{-7}$ cm/sec, it is important to establish the moisture content–unit weight criteria in the laboratory for the soil to be used in field construction. This helps in the development of proper specifications.

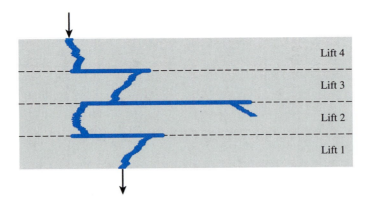

Figure 17.2 Pattern of flow through a compacted clay with improper bonding between lifts (*After U.S. Environmental Protection Agency, 1989*)

Table 17.2 Hydraulic Conductivity from Houston Liner Tests[*]

Location	Sample	Laboratory k (cm/sec)
Lower lift	76 mm tube	4×10^{-9}
Upper lift	76 mm tube	1×10^{-9}
Lift interface	76 mm tube	1×10^{-7}
Lower lift	Block	8×10^{-5}
Upper lift	Block	1×10^{-8}

Actual overall $k = 1 \times 10^{-4}$ cm/sec

[*]After Environmental Protection Agency (1989)

Daniel and Benson (1990) developed a procedure to establish the moisture content–unit weight criteria for clayey soils to meet the hydraulic conductivity requirement. The following is a step-by-step procedure to develop the criteria.

Step 1: Conduct Proctor tests to establish the dry unit weight versus molding moisture content relationships (Figure 17.3a).

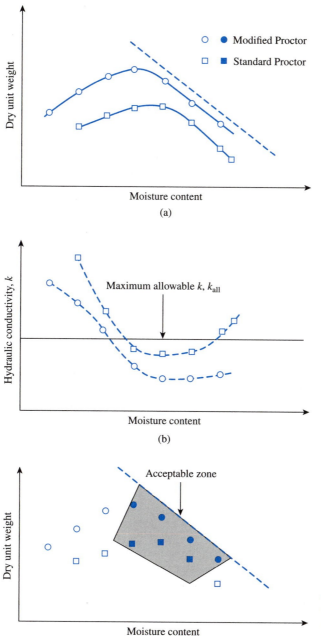

Figure 17.3

(a) Proctor curves; (b) variation of hydraulic conductivity of compacted specimens; (c) determination of acceptable zone

Step 2:   Conduct permeability tests on the compacted soil specimens (from Step 1), and plot the results, as shown in Figure 17.3b. In this figure, also plot the maximum allowable value of k (that is, k_{all}).

Step 3:   Replot the dry unit weight–moisture content points (Figure 17.3c) with different symbols to represent the compacted specimens with $k > k_{all}$ and $k \leq k_{all}$.

Step 4:   Plot the acceptable zone for which k is less than or equal to k_{all} (Figure 17.3c).

17.3   *Geosynthetics*

In general, geosynthetics are fabric-like material made from polymers such as polyester, polyethylene, polypropylene, polyvinyl chloride (PVC), nylon, chlorinated polyethylene, and others. The term *geosynthetics* includes the following:

- Geotextiles
- Geomembranes
- Geogrids
- Geonets
- Geocomposites

Each type of geosynthetic performs one or more of the following five major functions:

1. Separation
2. Reinforcement
3. Filtration
4. Drainage
5. Moisture barrier

Geosynthetics have been used in civil engineering construction since the late 1970s, and their use currently is growing rapidly. In this chapter, it is not possible to provide detailed descriptions of manufacturing procedures, properties, and uses of all types of geosynthetics. However, an overview of geotextiles, geomembranes, and geonets is given. For further information, refer to a geosynthetics text, such as that by Koerner (1994).

17.4   *Geotextiles*

Geotextiles are textiles in the traditional sense; however, the fabrics usually are made from petroleum products such as polyester, polyethylene, and polypropylene. They also may be made from fiberglass. Geotextiles are not prepared from natural fabrics, which decay too quickly. They may be *woven, knitted, or nonwoven.*

Woven geotextiles are made of two sets of parallel filaments or strands of yarn systematically interlaced to form a planar structure. *Knitted geotextiles* are formed by interlocking a series of loops of one or more filaments or strands of yarn to form a planar structure. *Nonwoven geotextiles* are formed from filaments or short fibers arranged in an oriented or a random pattern in a planar structure. These filaments or short fibers first are arranged into a loose web. They then are bonded by using one or a combination of the following processes:

- *Chemical bonding*—by glue, rubber, latex, cellulose derivative, and so forth
- *Thermal bonding*—by heat for partial melting of filaments
- *Mechanical bonding*—by needle punching

The *needle-punched nonwoven* geotextiles are thick and have high in-plane hydraulic conductivity.

Geotextiles have four major uses:

1. *Drainage:* The fabrics can channel water rapidly from soil to various outlets.
2. *Filtration:* When placed between two soil layers, one coarse grained and the other fine grained, the fabric allows free seepage of water from one layer to the other. At the same time, it protects the fine-grained soil from being washed into the coarse-grained soil.
3. *Separation:* Geotextiles help keep various soil layers separate after construction. For example, in the construction of highways, a clayey subgrade can be kept separate from a granular base course.
4. *Reinforcement:* The tensile strength of geotextiles increases the load-bearing capacity of the soil.

Geotextiles currently available commercially have thicknesses that vary from about 0.25 to 7.6 mm. The mass per unit area of these geotextiles ranges from about 150 to 700 g/cm^2.

One of the major functions of geotextiles is filtration. For this purpose, water must be able to flow freely through the fabric of the geotextile (Figure 17.4). Hence, the *cross-plane hydraulic conductivity* is an important parameter for design purposes. It should be realized that geotextile fabrics are compressible, however, and their thickness may change depending on the effective normal stress to which they are being subjected. The change in thickness under normal stress also changes the cross-plane hydraulic

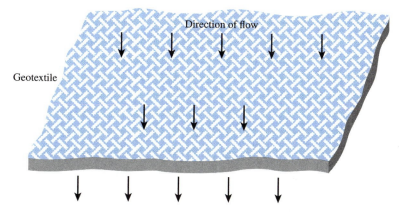

Figure 17.4 Cross-plane flow through geotextile

conductivity of a geotextile. Thus, the cross-plane capability is generally expressed in terms of a quantity called *permittivity,* or

$$P = \frac{k_n}{t} \tag{17.2}$$

where P = permittivity

k_n = hydraulic conductivity for cross-plane flow

t = thickness of the geotextile

In a similar manner, to perform the function of drainage satisfactorily, geotextiles must possess excellent in-plane permeability. For reasons stated previously, the in-plane hydraulic conductivity also depends on the compressibility, and, hence, the thickness of the geotextile. The in-plane drainage capability can thus be expressed in terms of a quantity called *transmissivity,* or

$$T = k_p t \tag{17.3}$$

where T = transmissivity

k_p = hydraulic conductivity for in-plane flow (Figure 17.5)

The units of k_n and k_p are cm/sec, however, the unit of permittivity, P, is sec^{-1} or min^{-1}. In a similar manner, the unit of transmissivity, T, is $\text{m}^3/\text{sec} \cdot \text{m}$. Depending on the type of geotextile, k_n and P, and k_p and T can vary widely. The following are some typical values for k_n, P, k_p, and T.

- *Hydraulic conductivity, k_n:* 1×10^{-3} to 2.5×10^{-1} cm/sec
- *Permittivity, P:* 2×10^{-2} to $2.0 \ \text{sec}^{-1}$
- *Hydraulic conductivity, k_p:*
  Nonwoven: 1×10^{-3} to 5×10^{-2} cm/sec
  Woven: 2×10^{-3} to 4×10^{-3} cm/sec
- *Transmissivity T:*
  Nonwoven: 2×10^{-6} to 2×10^{-9} $\text{m}^3/\text{sec} \cdot \text{m}$
  Woven: 1.5×10^{-8} to 2×10^{-8} $\text{m}^3/\text{sec} \cdot \text{m}$

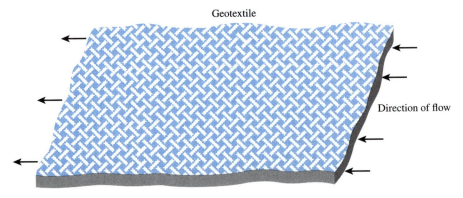

Figure 17.5 In-plane flow in geotextile

When a geotextile is being considered for use in the design and construction of landfill liners, certain properties must be measured by tests on the geotextile to determine its applicability. A partial list of these tests follows.

1. Mass per unit area
2. Percentage of open area
3. Equivalent opening size
4. Thickness
5. Ultraviolet resistivity
6. Permittivity
7. Transmissivity
8. Puncture resistance
9. Resistance to abrasion
10. Compressibility
11. Tensile strength and elongation properties
12. Chemical resistance

17.5 *Geomembranes*

Geomembranes are impermeable liquid or vapor barriers made primarily from continuous polymeric sheets that are flexible. The type of polymeric material used for geomembranes may be *thermoplastic* or *thermoset*. The thermoplastic polymers include PVC, polyethylene, chlorinated polyethylene, and polyamide. The thermoset polymers include ethylene vinyl acetate, polychloroprene, and isoprene-isobutylene. Although geomembranes are thought to be impermeable, they are not. Water vapor transmission tests show that the hydraulic conductivity of geomembranes is in the range of 10^{-10} to 10^{-13} cm/sec; hence, they are only "essentially impermeable."

Many scrim-reinforced geomembranes manufactured in single piles have thicknesses that range from 0.25 to about 0.4 mm. These single piles of geomembranes can be laminated together to make thicker geomembranes. Some geomembranes made from PVC and polyethylene may be as thick as 4.5 to 5 mm.

The following is a partial list of tests that should be conducted on geomembranes when they are to be used as landfill liners.

1. Density
2. Mass per unit area
3. Water vapor transmission capacity
4. Tensile behavior
5. Tear resistance
6. Resistance to impact
7. Puncture resistance
8. Stress cracking
9. Chemical resistance
10. Ultraviolet light resistance
11. Thermal properties
12. Behavior of seams

The most important aspect of construction with geomembranes is the preparation of seams. Otherwise, the basic reason for using geomembranes as a liquid or vapor barrier will be defeated. Geomembrane sheets generally are seamed together in the factory to prepare larger sheets. These larger sheets are field seamed into their final position. There are several types of seams, some of which are described briefly.

Lap Seam with Adhesive
- A solvent adhesive is used for this type of seam (Figure 17.6a). After application of the solvent, the two sheets of geomembrane are overlapped, then roller pressure is applied.

Lap Seam with Gum Tape
- This type of seam (Figure 17.6b) is used mostly in dense thermoset material, such as isoprene-isobutylene.

Tongue-and-Groove Splice
- A schematic diagram of the tongue-and-groove splice is shown in Figure 17.6c. The tapes used for the splice are double sided.

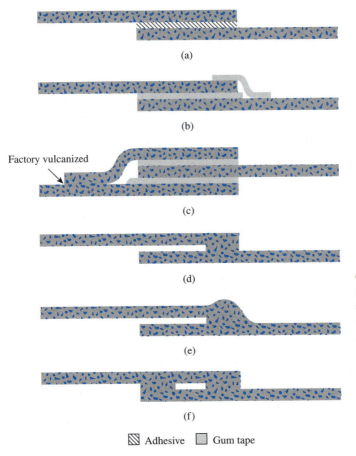

(a)

(b)

Factory vulcanized

(c)

(d)

(e)

(f)

⬛ Adhesive ▢ Gum tape

Figure 17.6

Configurations of field geomembrane seams: (a) lap seam; (b) lap seam with gum tape; (c) tongue-and-groove splice; (d) extrusion weld lap seam; (e) fillet weld lap seam; (f) double hot air or wedge seam (*After U.S. Environmental Protection Agency, 1989*)

Extrusion Weld Lap Seam

- Extrusion or fusion welding is done on geomembranes made from polyethylene. A ribbon of molten polymer is extruded between the two surfaces to be joined (Figure 17.6d).

Fillet Weld Lap Seam

- This seam is similar to an extrusion weld lap seam; however, for fillet welding, the extrudate is placed over the edge of the seam (Figure 17.6e).

Double Hot Air or Wedge Seam

- In the hot air seam, hot air is blown to melt the two opposing surfaces. For melting, the temperatures should rise to about 260°C or more. After the opposite surfaces are melted, pressure is applied to form the seam (Figure 17.6f). For hot wedge seams, an electrically heated element like a blade is passed between the two opposing surfaces of the geomembrane. The heated element helps to melt the geomembrane, after which pressure is applied by a roller to form the seam.

17.6   *Geonets*

Geonets are formed by the continuous extrusion of polymeric ribs at acute angles to each other. They have large openings in a netlike configuration. The primary function of geonets is drainage. Figure 17.7 is a photograph of a typical piece of geonet. Most geonets currently available are made of medium-density and high-density polyethylene. They are available in rolls with widths of 1.8 to 2.1 m and lengths of 30 to 90 m. The approximate aperture sizes vary from 30 mm $\times$ 30 mm to about 6 mm $\times$ 6 mm. The thickness of geonets available commercially can vary from 3.8 to 7.6 mm.

Seaming of geonets is somewhat more difficult. For this purpose, staples, threaded loops, and wire sometimes are used.

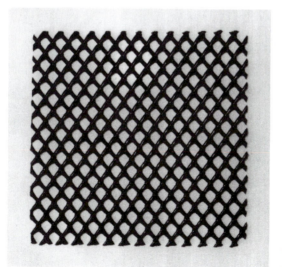

Figure 17.7
Geonet (*Courtesy of Braja M. Das, Henderson, Nevada*)

17.7 Single Clay Liner and Single Geomembrane Liner Systems

Until about 1982—that is, before the guidelines for the minimum technological requirements for hazardous-waste landfill design and construction were mandated by the U.S. Environmental Protection Agency—most landfill liners were *single clay liners*. Figure 17.8 shows the cross section of a single clay liner system for a landfill. It consists primarily of a compacted clay liner over the native foundation soil. The thickness of the compacted clay liner varies between 0.9 and 1.8 m. The *maximum* required hydraulic conductivity, k, is 10^{-7} cm/sec. Over the clay liner is a layer of gravel with perforated pipes for leachate collection and removal. Over the gravel layer is a layer of filter soil. The filter is used to protect the holes in the perforated pipes against the movement of fine soil particles. In most cases, the filter is medium coarse to fine sandy soil. It is important to note that this system does not have any leak-detection capability.

Around 1982, single layers of geomembranes also were used as a liner material for landfill sites. As shown in Figure 17.9, the geomembrane is laid over native foundation soil. Over the geomembrane is a layer of gravel with perforated pipes for leachate collection and removal. A layer of filter soil is placed between the solid waste material and the gravel. As in the single clay liner system, no provision is made for leak detection.

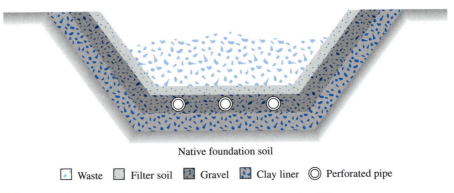

Native foundation soil

▢ Waste   ▢ Filter soil   ▢ Gravel   ▢ Clay liner   ◯ Perforated pipe

Figure 17.8 Cross section of single clay liner system for a landfill

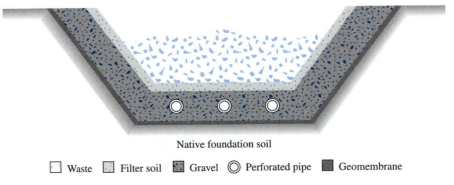

Native foundation soil

▢ Waste   ▢ Filter soil   ▢ Gravel   ◯ Perforated pipe   ▣ Geomembrane

Figure 17.9 Cross section of single geomembrane liner system for a landfill

17.8 *Recent Advances in the Liner Systems for Landfills*

Since 1984, most landfills developed for solid and hazardous wastes have double liners. The two liners are an upper primary liner and a lower secondary liner. Above the top liner is a *primary* leachate collection and removal system. In general, the primary leachate collection system must be able to maintain a leachate head of 0.3 m or less. Between the primary and secondary liners is a system for leak detection, collection, and removal (LDCR) of leachates. The general guidelines for the primary leachate collection system and the LDCR system are as follows:

1. It can be a granular drainage layer or a geosynthetic drainage material, such as a geonet.
2. If a granular drainage layer is used, it should have a minimum thickness of 0.3 m.
3. The granular drainage layer (or the geosynthetic) should have a hydraulic conductivity, k, greater than 10^{-2} cm/sec.
4. If a granular drainage layer is used, it should have a granular filter or a layer of geotextile over it to prevent clogging. A layer of geotextile also is required over the geonet when it is used as the drainage layer.
5. The granular drainage layer, when used, must be chemically resistant to the waste material and the leachate that are produced. It also should have a network of perforated pipes to collect the leachate effectively and efficiently.

In the design of the liner systems, the compacted clay layers should be at least 1 m thick, with $k \leq 10^{-7}$ cm/sec. Figures 17.10 and 17.11 show schematic diagrams of two double-liner systems. In Figure 17.10, the primary leachate collection system is made of a granular material with perforated pipes and a filter system above it. The primary liner is a geomembrane. The LDCR system is made of a geonet. The secondary liner is a *composite*

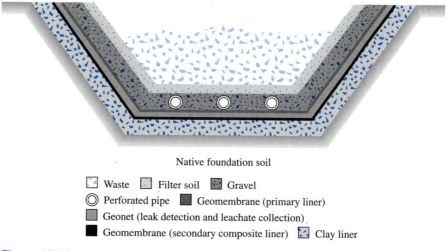

Native foundation soil

☐ Waste ☐ Filter soil ▨ Gravel
◎ Perforated pipe ■ Geomembrane (primary liner)
▨ Geonet (leak detection and leachate collection)
■ Geomembrane (secondary composite liner) ▨ Clay liner

Figure 17.10 Cross section of double-liner system (note the secondary composite liner)

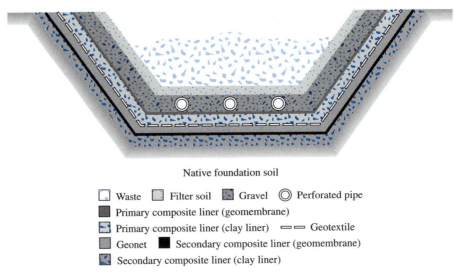

Native foundation soil

☐ Waste   ▨ Filter soil   ▨ Gravel   ◎ Perforated pipe

■ Primary composite liner (geomembrane)

▨ Primary composite liner (clay liner)   ══ Geotextile

▨ Geonet   ■ Secondary composite liner (geomembrane)

▨ Secondary composite liner (clay liner)

Figure 17.11 Cross section of double-liner system (note the primary and secondary composite liners)

liner made of a geomembrane with a compacted clay layer below it. In Figure 17.11, the primary leachate collection system is similar to that shown in Figure 17.10; however, the primary and secondary liners are both composite liners (geomembrane-clay). The LDCR system is a geonet with a layer of geotextile over it. The layer of geotextile acts as a filter and separator.

The geomembranes used for landfill lining must have a minimum thickness of 0.76 mm, however, all geomembranes that have a thickness of 0.76 mm may not be suitable in all situations. In practice, most geomembranes used as liners have thicknesses ranging from 1.8 to 2.54 mm.

17.9   *Leachate Removal Systems*

The bottom of a landfill must be graded properly so that the leachate collected from the primary collection system and the LDCR system will flow to a low point by gravity. Usually, a grade of 2% or more is provided for large landfill sites. The low point of the leachate collection system ends at a sump. For a primary leachate collection, a manhole is located at the sump, which rises through the waste material. Figure 17.12 shows a schematic diagram of the leachate removal system with a low-volume sump. A typical leachate removal system for high-volume sumps (for primary collection) is shown in Figure 17.13.

Leachate can be removed from the LDCR system by means of pumping, as shown in Figure 17.14, or by gravity monitoring, as shown in Figure 17.15. When leachate is removed by pumping, the plastic pipe used for removal must penetrate the primary liner. On the other hand, if gravity monitoring is used, the pipe will penetrate the secondary liner.

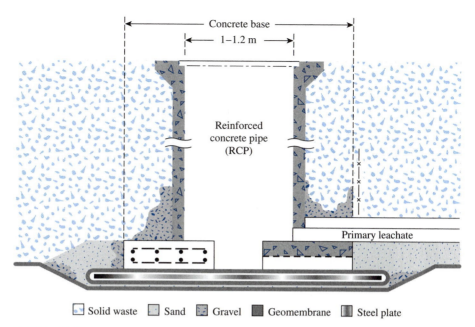

Concrete base

1–1.2 m

Reinforced
concrete pipe
(RCP)

Primary leachate

☐ Solid waste   ☐ Sand   ☐ Gravel   ■ Geomembrane   ☐ Steel plate

Figure 17.12 Primary leachate removal system with a low-volume sump (*After U.S. Environmental Protection Agency, 1989*).

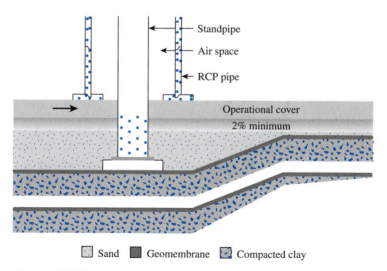

Standpipe

Air space

RCP pipe

Operational cover
2% minimum

☐ Sand   ■ Geomembrane   ☐ Compacted clay

Figure 17.13 Primary leachate removal system with a high-volume sump (*After U.S. Environmental Protection Agency, 1989*)

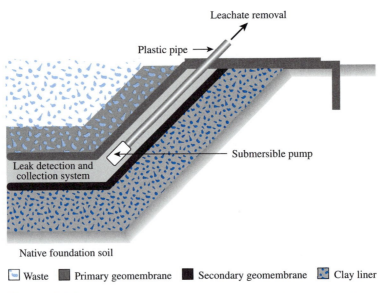

Waste Primary geomembrane Secondary geomembrane Clay liner

Figure 17.14 Secondary leak detection, collection, and removal (LDCR) system—by means of pumping. (*Note:* The plastic pipe penetrates the primary geomembrane)

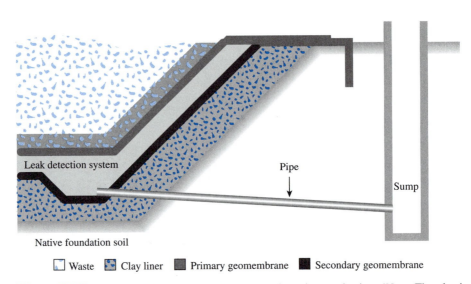

Waste Clay liner Primary geomembrane Secondary geomembrane

Figure 17.15 Secondary LDCR system, by means of gravity monitoring. (*Note:* The plastic pipe penetrates the secondary geomembrane)

Closure of Landfills

When the landfill is complete and no more waste can be placed into it, a cap must be put on it (Figure 17.16). This cap will reduce and ultimately eliminate leachate generation. A schematic diagram of the layering system recommended by the U.S. Environmental Protection Agency (1979, 1986) and Koerner (1994) for hazardous-waste landfills is shown in Figure 17.17. Essentially, it consists of a compacted clay cap over the solid waste, a geomembrane liner, a drainage layer, and a cover of topsoil. The manhole used for leachate collection penetrates the landfill cover. Leachate removal continues until its generation is stopped. For hazardous-waste landfill sites, the EPA (1989) recommends this period to be about 30 years.

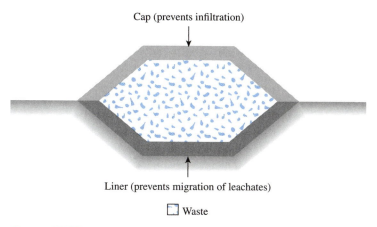

Figure 17.16 Landfill with liner and cap

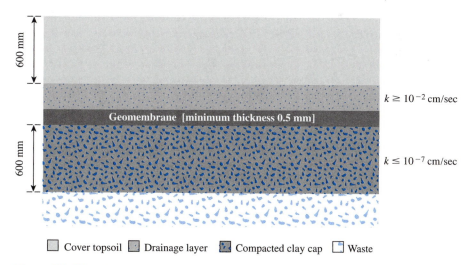

Figure 17.17 Schematic diagram of the layering system for landfill cap

17.11   *Summary and General Comments*

This chapter provided a brief overview of the problems associated with solid- and hazardous-waste landfills. The general concepts for the construction of landfill liners using compacted clayey soil and geosynthetics (that is, geotextiles, geomembranes, and geonets) were discussed. Several areas were not addressed, however, because they are beyond the scope of the text. Areas not discussed include the following:

1. *Selection of material.* The chemicals contained in leachates generated from hazardous and nonhazardous waste may interact with the liner materials. For this reason, it is essential that representative leachates are used to test the chemical compatibility so that the liner material remains intact during the periods of landfill operation and closure, and possibly longer. Selection of the proper leacheates becomes difficult because of the extreme variations encountered in the field. The mechanical properties of geomembranes also are important. Properties such as workability, creep, stress cracking, and the thermal coefficient of expansion should be investigated thoroughly.

2. *Stability of side-slope liner.* The stability and slippage checks of the side-slope liners of a landfill site are important and complicated because of the variation of the frictional characters of the composite materials involved in liner construction. For a detailed treatment of this topic, refer to any book on geosynthetics (e.g., Koerner, 1994).

3. *Leak-response action plan.* It is extremely important that any leaks or clogging of the drainage layer(s) in a given waste-disposal site be detected as quickly as possible. Leaks or cloggings are likelihoods at a site even with good construction quality control. Each waste-disposal facility should have a *leak-response action plan.*

References

BENSON, C. H. and DANIEL, D. E. (1990). "Influence of Clods on Hydraulic Conductivity of Compacted Clay," *Journal of Geotechnical Engineering,* ASCE, Vol. 116, No. 8, 1231–1248.

DANIEL, D. E. and BENSON, C. H. (1990). "Water Content-Density Criteria for Compacted Soil Liners," *Journal of Geotechnical Engineering,* ASCE, Vol. 116, No. 12, 1811–1830.

KOERNER, R. M. (1994). *Designing with Geosynthetics,* 3rd ed., Prentice-Hall, Englewood Cliffs, N.J.

SIVAPULLAIAH, P. V., SRIDHARAN, A., and STALIN, V. K. (2000). "Hydraulic Conductivity of Bentonite-Sand Mixtures," *Canadian Geotechnical Journal,* Vol. 37, No. 2, 406–413.

U.S. ENVIRONMENTAL PROTECTION AGENCY (1979). *Design and Construction for Solid Waste Landfills,* Publication No. EPA-600/2-79-165, Cincinnati, Ohio.

U.S. ENVIRONMENTAL PROTECTION AGENCY (1986). *Cover for Uncontrolled Hazardous Waste Sites,* Publication No. EPA-540/2-85-002, Cincinnati, Ohio.

U.S. ENVIRONMENTAL PROTECTION AGENCY (1989). *Requirements for Hazardous Waste Landfill Design, Construction, and Closure,* Publication No. EPA-625/4-89-022, Cincinnati, Ohio.

18 Subsoil Exploration

The preceding chapters reviewed the fundamental properties of soils and their behavior under stress and strain in idealized conditions. In practice, natural soil deposits are not homogeneous, elastic, or isotropic. In some places, the stratification of soil deposits even may change greatly within a horizontal distance of 15 to 30 m. For foundation design and construction work, one must know the actual soil stratification at a given site, the laboratory test results of the soil samples obtained from various depths, and the observations made during the construction of other structures built under similar conditions. For most major structures, adequate subsoil exploration at the construction site must be conducted. The purposes of subsoil exploration include the following:

1. Determining the nature of soil at the site and its stratification
2. Obtaining disturbed and undisturbed soil samples for visual identification and appropriate laboratory tests
3. Determining the depth and nature of bedrock, if and when encountered
4. Performing some *in situ* field tests, such as permeability tests (Chapter 7), vane shear tests (Chapter 12), and standard penetration tests
5. Observing drainage conditions from and into the site
6. Assessing any special construction problems with respect to the existing structure(s) nearby
7. Determining the position of the water table

This chapter briefly summarizes subsoil exploration techniques. For additional information, refer to the *Manual of Foundation Investigations* of the American Association of State Highway and Transportation Officials (1967).

18.1 Planning for Soil Exploration

A soil exploration program for a given structure can be divided broadly into four phases:

1. *Compilation of the existing information regarding the structure:* This phase includes gathering information such as the type of structure to be constructed and its future use, the requirements of local building codes, and the column and load-bearing wall

loads. If the exploration is for the construction of a bridge foundation, one must have an idea of the length of the span and the anticipated loads to which the piers and abutments will be subjected.

2. *Collection of existing information for the subsoil condition:* Considerable savings in the exploration program sometimes can be realized if the geotechnical engineer in charge of the project thoroughly reviews the existing information regarding the subsoil conditions at the site under consideration. Useful information can be obtained from the following sources:
   a. Geologic survey maps
   b. County soil survey maps prepared by the U.S. Department of Agriculture and the Soil Conservation Service
   c. Soil manuals published by the state highway departments
   d. Existing soil exploration reports prepared for the construction of nearby structures

   Information gathered from the preceding sources provides insight into the type of soil and problems that might be encountered during actual drilling operations.

3. *Reconnaissance of the proposed construction site:* The engineer visually should inspect the site and the surrounding area. In many cases, the information gathered from such a trip is invaluable for future planning. The type of vegetation at a site, in some instances, may indicate the type of subsoil that will be encountered. The accessibility of a site and the nature of drainage into and from it also can be determined. Open cuts near the site provide an indication about the subsoil stratification. Cracks in the walls of nearby structure(s) may indicate settlement from the possible existence of soft clay layers or the presence of expansive clay soils.

4. *Detailed site investigation:* This phase consists of making several test borings at the site and collecting disturbed and undisturbed soil samples from various depths for visual observation and for laboratory tests. No hard-and-fast rule exists for determining the number of borings or the depth to which the test borings are to be advanced. For most buildings, at least one boring at each corner and one at the center should provide a start. Depending on the uniformity of the subsoil, additional test borings may be made. Table 18.1 gives guidelines for initial planning of borehole spacing.

The test borings should extend through unsuitable foundation materials to firm soil layers. Sowers and Sowers (1970) provided a rough estimate of the minimum depth

Table 18.1 Spacing of Borings

Project	Boring spacings (m)
One-story buildings	25–30
Multistory buildings	15–25
Highways	250–300
Earth dams	25–50
Residential subdivision planning	60–100

of borings (unless bedrock is encountered) for multistory buildings. They can be given by the following equations, applicable to light steel or narrow concrete buildings:

$$z_b\,(\text{m}) = 3S^{0.7} \tag{18.1}$$

or to heavy steel or wide concrete buildings:

$$z_b\,(\text{m}) = 6S^{0.7} \tag{18.2}$$

In Eqs. (18.1) and (18.2), z_b is the approximate depth of boring and S is the number of stories.

The American Society of Civil Engineers (1972) recommended the following rules of thumb for estimating the boring depths for buildings.

1. Estimate the variation of the net effective stress increase, $\Delta\sigma'$, that will result from the construction of the proposed structure with depth. This variation can be estimated by using the principles outlined in Chapter 10. Determine the depth D_1 at which the value of $\Delta\sigma'$ is equal to 10% of the average load per unit area of the structure.
2. Plot the variation of the effective vertical stress, σ'_o, in the soil layer with depth. Compare this with the net stress increase variation, $\Delta\sigma'$, with depth as determined in Step 1. Determine the depth D_2 at which $\Delta\sigma' = 0.05\sigma'_o$.
3. The smaller of the two depths, D_1 and D_2, is the approximate minimum depth of the boring.

When the soil exploration is for the construction of dams and embankments, the depth of boring may range from one-half to two times the embankment height.

The general techniques used for advancing test borings in the field and the procedure for the collection of soil samples are described in the following sections.

18.2 Boring Methods

The test boring can be advanced in the field by several methods. The simplest is the *use of augers*. Figure 18.1 shows two types of hand augers that can be used for making boreholes up to a depth of about 3 to 5 m. They can be used for soil exploration work for highways and small structures. Information regarding the types of soil present at various depths is obtained by noting the soil that holds to the auger. The soil samples collected in this manner are disturbed, but they can be used to conduct laboratory tests such as grain-size determination and Atterberg limits.

When the boreholes are to be advanced to greater depths, the most common method is to use continuous-flight augers, which are power operated. The power for drilling is delivered by truck- or tractor-mounted drilling rigs. Continuous-flight augers are available commercially in 1 to 1.5 m sections. During the drilling operation, section after section of auger can be added and the hole extended downward. Continuous-flight augers can be

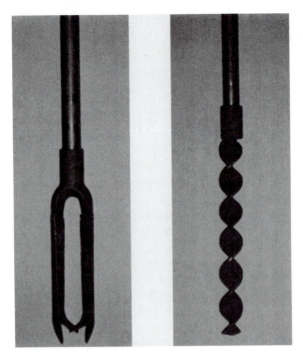

Figure 18.1 Hand augers: (a) Iwan auger; (b) slip auger
(*Courtesy of Braja M. Das, Henderson, Nevada*)

solid stem or hollow stem. Some of the commonly used solid-stem augers have outside diameters of 67 mm, 83 mm, 102 mm and 114 mm, The inside and outside diameters of some hollow-stem augers are given in Table 18.2.

Flight augers bring the loose soil from the bottom of the hole to the surface. The driller can detect the change in soil type encountered by the change of speed and the sound of drilling. Figure 18.2 shows a drilling operation with flight augers. When solid-stem augers are used, the auger must be withdrawn at regular intervals to obtain soil samples and to conduct other operations such as standard penetration tests. Hollow-stem augers have a distinct advantage in this respect—they do not have to be removed at frequent intervals for sampling or other tests. As shown in Figure 18.3, the outside of the auger acts like a casing. A removable plug is attached to the bottom of the auger by means of a center rod.

Table 18.2 Dimensions of Commonly Used Hollow-Stem Augers

Inside diameter (mm)	Outside diameter (mm)
63.5	158.75
69.85	187.8
76.2	203.2
88.9	228.6
101.6	254.0

Figure 18.2 Drilling with flight augers (*Courtesy of Danny R. Anderson, PE, of Professional Service Industries, Inc, El Paso, Texas.*)

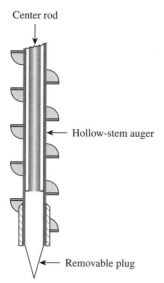

Figure 18.3
Schematic diagram of hollow-stem auger with removable plug

During the drilling, the plug can be pulled out with the auger in place, and soil sampling and standard penetration tests can be performed. When hollow-stem augers are used in sandy soils below the groundwater table, the sand might be pushed several feet into the stem of the auger by excess hydrostatic pressure immediately after the removal of the plug. In such conditions, the plug should not be used. Instead, water inside the hollow stem should be maintained at a higher level than the groundwater table.

Rotary drilling is a procedure by which rapidly rotating drilling bits attached to the bottom of drilling rods cut and grind the soil and advance the borehole down. Several types of drilling bits are available for such work. Rotary drilling can be used in sand, clay, and rock (unless badly fissured). Water or drilling mud is forced down the drilling rods to the bits, and the return flow forces the cuttings to the surface. Drilling mud is a slurry prepared by mixing bentonite and water (bentonite is a montmorillonite clay formed by the weathering of volcanic ash). Boreholes with diameters ranging from 50 to 200 mm can be made easily by using this technique.

Wash boring is another method of advancing boreholes. In this method, a casing about 2 to 3 m long is driven into the ground. The soil inside the casing then is removed by means of a *chopping bit* that is attached to a drilling rod. Water is forced through the drilling rod, and it goes out at a very high velocity through the holes at the bottom of the chopping bit (Figure 18.4). The water and the chopped soil particles rise upward in the drill hole and overflow at the top of the casing through a T-connection. The wash water then is collected in a container. The casing can be extended with additional pieces as the borehole progresses; however, such extension is not necessary if the borehole will stay open without caving in.

Percussion drilling is an alternative method of advancing a borehole, particularly through hard soil and rock. In this technique, a heavy drilling bit is raised and lowered to chop the hard soil. Casing for this type of drilling may be required. The chopped soil particles are brought up by the circulation of water.

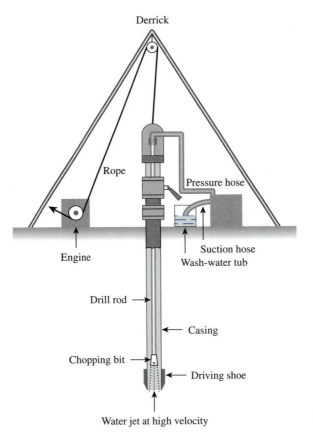

Figure 18.4 Wash boring

Common Sampling Methods

During the advancement of the boreholes, soil samples are collected at various depths for further analysis. This section briefly discusses some of the methods of sample collection.

Sampling by Standard Split Spoon

Figure 18.5 shows a diagram of a split-spoon sampler. It consists of a tool-steel driving shoe at the bottom, a steel tube (that is split longitudinally into halves) in the middle, and a coupling at the top. The steel tube in the middle has inside and outside diameters of 34.9 mm and 50.8 mm respectively. Figure 18.6 shows a photograph of an unassembled split-spoon sampler.

When the borehole is advanced to a desired depth, the drilling tools are removed. The split-spoon sampler is attached to the drilling rod and then lowered to the bottom of the borehole (Figure 18.7). The sampler is driven into the soil at the bottom of the borehole by means of hammer blows. The hammer blows occur at the top of the drilling rod. The hammer weighs 623 N. For each blow, the hammer drops a distance of 0.762 m. The number of blows required for driving the sampler through three 152.4 mm intervals is recorded. The sum of the number of blows required for driving the last two 152.4 mm

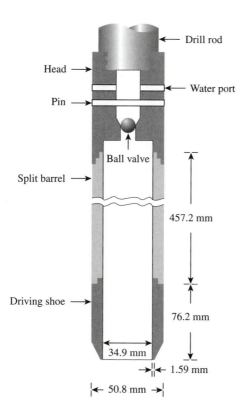

Figure 18.5 Diagram of standard split-spoon sampler

Figure 18.6 Split-spoon sampler, unassembled (*Courtesy of ELE International*)

Figure 18.7 Drilling rod with split-spoon sampler lowered to the bottom of the borehole (*Courtesy of Braja M. Das, Henderson, Nevada*)

intervals is referred to as the *standard penetration number, N.* It also commonly is called the *blow count.* The interpretation of the standard penetration number is given in Section 18.5. After driving is completed, the sampler is withdrawn and the shoe and coupling are removed. The soil sample collected inside the split tube then is removed and transported to the laboratory in small glass jars. Determination of the standard penetration number and collection of split-spoon samples usually are done at 1.5 m intervals.

At this point, it is important to point out that there are several factors that will contribute to the variation of the standard penetration number, *N,* at a given depth for similar soil profiles. These factors include SPT hammer efficiency, borehole diameter, sampling method, and rod length factor (Skempton, 1986; Seed, *et al.,* 1985). The two most common types of SPT hammers used in the field are the *safety hammer* and *donut hammer.* They commonly are dropped by a rope with *two wraps around a pulley.*

The SPT hammer energy efficiency can be expressed as

$$E_r(\%) = \frac{\text{actual hammer energy to the sampler}}{\text{input energy}} \times 100 \qquad (18.3)$$

$$\text{Theoretical input energy} = Wh \qquad (18.4)$$

where W = weight of the hammer ≈ 0.623 kN
   h = height of drop ≈ 0.76 mm

So,

$$Wh = (0.623)(0.76) = 0.474 \text{ kN-m}$$

In the field, the magnitude of E_r can vary from 30 to 90%. The standard practice now in the U.S. is to express the *N*-value to an average energy ratio of 60% ($\approx N_{60}$). Thus, correcting for field procedures and on the basis of field observations, it appears reasonable to standardize the field penetration number as a function of the input driving energy and its dissipation around the sampler into the surrounding soil, or

$$N_{60} = \frac{N\eta_H\eta_B\eta_S\eta_R}{60}$$ (18.5)

where N_{60} = standard penetration number corrected for field conditions
N = measured penetration number
η_H = hammer efficiency (%)
η_B = correction for borehole diameter
η_S = sampler correction
η_R = correction for rod length

Based on the recommendations of Seed, *et al.* (1985) and Skempton (1986), the variations of η_H, η_B, η_S, and η_R are summarized in Table 18.3.

Table 18.3 Variations of η_H, η_B, η_S, and η_R [Eq. (18.5)]

1. Variation of η_H

Country	Hammer type	Hammer release	η_H (%)
Japan	Donut	Free fall	78
	Donut	Rope and pulley	67
United States	Safety	Rope and pulley	60
	Donut	Rope and pulley	45
Argentina	Donut	Rope and pulley	45
China	Donut	Free fall	60
	Donut	Rope and pulley	50

2. Variation of η_B

Diameter (mm)	η_B
60–120	1
150	1.05
200	1.15

3. Variation of η_S

Variable	η_S
Standard sampler	1.0
With liner for dense sand and clay	0.8
With liner for loose sand	0.9

4. Variation of η_R

Rod length (m)	η_R
>10	1.0
6–10	0.95
4–6	0.85
0–4	0.75

Sampling by Thin-Wall Tube

Sampling by thin-wall tube is used for obtaining fairly undisturbed soil samples. The thin-wall tubes are made of seamless, thin tubes and commonly are referred to as *Shelby tubes* (Figure 18.8). To collect samples at a given depth in a borehole, one first must remove the drilling tools. The sampler is attached to a drilling rod and lowered to the bottom of the borehole. After this, it is pushed hydraulically into the soil. It then is spun to shear off the base and is pulled out. The sampler with the soil inside is sealed and taken to the laboratory for testing. Most commonly used thin-wall tube samplers have outside diameters of 76.2 mm.

Sampling by Piston Sampler

Piston samplers are particularly useful when highly undisturbed samples are required. The cost of recovering such samples is, of course, higher. Several types of piston samplers can be used; however, the sampler proposed by Osterberg (1952) is the most advantageous (Figure 18.9). It consists of a thin-wall tube with a piston. Initially, the piston closes the end of the thin-wall tube. The sampler first is lowered to the bottom of the borehole (Figure 18.9a), then the thin-wall tube is pushed into the soil hydraulically—past the piston. After this, the pressure is released through a hole in the piston rod (Figure 18.9b). The presence of the piston prevents distortion in the sample by neither letting the soil squeeze into the sampling tube very fast nor admitting excess soil. Samples obtained in this manner consequently are disturbed less than those obtained by Shelby tubes.

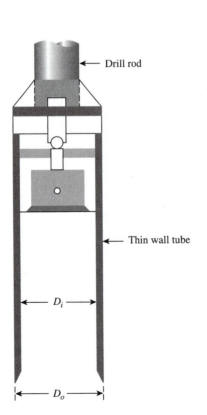

Figure 18.8 Thin-wall tube sampler

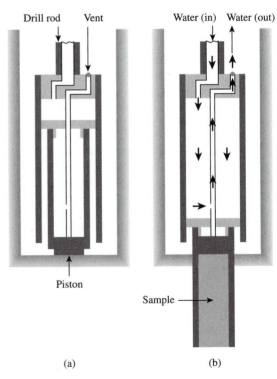

Figure 18.9 Piston sampler: (a) sampler lowered to bottom of borehole; (b) pressure released through hole in piston rod

18.4 Sample Disturbance

The degree of disturbance of the sample collected by various methods can be expressed by a term called the *area ratio,* which is given by

$$A_r(\%) = \frac{D_o^2 - D_i^2}{D_i^2} \times 100 \tag{18.6}$$

where D_o = outside diameter of the sampler
D_i = inside diameter of the sampler

A soil sample generally can be considered undisturbed if the area ratio is less than or equal to 10%. The following is a calculation of A_r for a standard split-spoon sampler and a 50.8 mm Shelby tube:

For the standard spit-spoon sampler, $D_i = 35.05$ mm and $D_o = 50.8$ mm. Hence,

$$A_r(\%) = \frac{(50.8)^2 - (35.05)^2}{(35.05)^2} \times 100 = 110\%$$

For the Shelby-tube sampler (50.8 mm) diameter, $D_i = 47.63$ mm and $D_o = 50.8$ mm. Hence,

$$A_r(\%) = \frac{(50.8)^2 - (47.63)^2}{(47.63)^2} \times 100 = 13.8\%$$

The preceding calculation indicates that the sample collected by split spoons is highly disturbed. The area ratio (A_r) of the 50.8 mm diameter Shelby tube samples is slightly higher than the 10% limit stated previously. For practical purpose, however, it can be treated as an undisturbed sample.

The disturbed but representative soil samples recovered by split-spoon samplers can be used for laboratory tests, such as grain-size distribution, liquid limit, plastic limit, and shrinkage limit. However, undisturbed soil samples are necessary for performing tests such as consolidation, triaxial compression, and unconfined compression.

18.5 Correlations for Standard Penetration Test

The procedure for conducting standard penetration tests was outlined in Section 18.3. The standard penetration number, N_{60}, is commonly used to correlate several useful physical parameters of soil. Some of these are briefly described next.

Cohesive Soil

The consistency of clay soils can be estimated from the standard penetration number N_{60}. In order to achieve that, Szechy and Vargi (1978) calculated the *consistency index (CI)* as

$$CI = \frac{LL - w}{LL - PL} \tag{18.7}$$

where w = natural moisture content

$\quad LL$ = liquid limit

$\quad PL$ = plastic limit

The approximate correlation between CI, N_{60}, and the unconfined compression strength (q_u) is given in Table 18.4.

It is important to point out that the correlation between N_{60} and q_u given in Table 18.4 is approximate. The sensitivity, S_t, of clay soil also plays an important role in the actual N_{60} value obtained in the field. In any case, for clays of a given geology, a reasonable correlation between N_{60} and q_u can be obtained, as shown in Figure 18.10. In this figure, the notation p_a is the atmospheric pressure (in the same unit as q_u). For the data shown in Figure 18.10, the reported regression is given by (Kulhawy and Mayne, 1990).

$$\frac{q_u}{p_a} = 0.58N_{60}^{0.72} \tag{18.8}$$

Table 18.4 Approximate Correlation between CI, N_{60}, and q_u

Standard penetration number, N_{60}	Consistency	CI	Unconfined compression strength, q_u (kN/m²)
<2	Very soft	<0.5	<25
2 to 8	Soft to medium	0.5 to 0.75	25 to 80
8 to 15	Stiff	0.75 to 1.0	80 to 150
15 to 30	Very stiff	1.0 to 1.5	150 to 400
>30	Hard	>1.5	>400

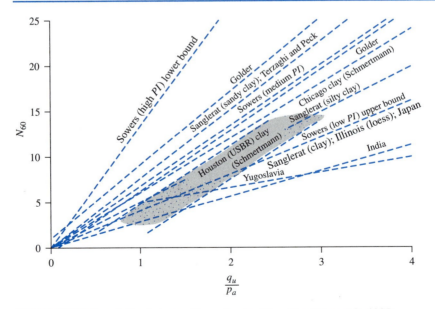

Figure 18.10 Correlation between N_{60} and q_u/p_a (*Based on Djoenaidi, 1985*)

Granular Soil

In granular soils, the standard penetration number is highly dependent on the effective overburden pressure, σ'_o.

A number of empirical relationships have been proposed to convert the field-standard penetration number N_{60} to a *standard effective overburden pressure, σ_o',* of 96 kN/m². The general form is

$$(N_1)_{60} = C_N N_{60} \tag{18.9}$$

Several correlations have been developed over the years for the correction factor, C_N. In this text, two of these are recommended for use.

Correlation of Liao and Whitman (1986)

$$C_N = 9.78 \sqrt{\frac{1}{\sigma_o'}} \tag{18.10}$$

where σ_o' = effective overburden pressure in kN/m²
   or,

$$C_N = \left[\frac{1}{\left(\dfrac{\sigma_o'}{p_a} \right)} \right]^{0.5} \tag{18.11}$$

where p_a = atmosphere pressure (≈ 100 kN/m²)

Correlation of Skempton (1986)

$$C_N = \frac{2}{1 + 0.01\sigma_o'} \tag{18.12}$$

where σ_o' = effective overburden pressure in kN/m²
   or,

$$C_N = \frac{2}{1 + \left(\dfrac{\sigma_o'}{p_a} \right)} \tag{18.13}$$

Table 18.5 shows approximate correlations for the standard penetration number, $(N_1)_{60}$, and relative density, D_r.

Cubrinovski and Ishihara (1999) proposed a correlation between N_{60} and the relative density of granular soils, D_r, in the form

$$D_r (\%) = \left[\frac{N_{60} \left(0.23 + \dfrac{0.06}{D_{50}} \right)^{1.7}}{9} \left(\dfrac{98}{\sigma_o'} \right) \right]^{0.5} (100) \tag{18.14}$$

Table 18.5 Approximate Relationship Between Corrected Standard Penetration Number and Relative Density of Sand

Corrected standard penetration number, $(N_1)_{60}$	Relative density, D_r (%)
0–5	0–5
5–10	5–30
10–30	30–60
30–50	60–95

where σ'_o = effective overburden pressure in kN/m^2
$\quad\quad\; D_{50}$ = sieve size through which 50% of soil will pass (mm)

The drained angle of friction of *granular soils, ϕ',* also has been correlated to the standard penetration number. Peck, Hanson, and Thornburn (1974) gave a correlation between $(N_1)_{60}$ and ϕ' in a graphic form, which can be approximated as (Wolff, 1989)

$$\phi' \text{ (deg)} = 27.1 + 0.3(N_1)_{60} - 0.00054(N_1)_{60}^2 \tag{18.15}$$

Schmertmann (1975) also provided a correlation for N_{60} versus σ'_o. After Kulhawy and Mayne (1990), this correlation can be approximated as

$$\phi' = \tan^{-1}\left[\frac{N_{60}}{12.2 + 20.3\left(\dfrac{\sigma'_o}{p_a}\right)}\right]^{0.34} \tag{18.16}$$

where p_a = atmospheric pressure (same unit as σ'_o).

The standard penetration number is a useful guideline in soil exploration and the assessment of subsoil conditions, provided that the results are interpreted correctly. Note that all equations and correlations relating to the standard penetration numbers are approximate. Because soil is not homogeneous, a wide variation in the N_{60} value may be obtained in the field. For soil deposits that contain large boulders and gravel, the standard penetration numbers may be erratic.

Example 18.1

Following are the results of a standard penetration test in sand. Determine the corrected standard penetration numbers, $(N_1)_{60}$, at various depths. Note that the water table was not observed within a depth of 10.5 m below the ground surface. Assume that the average unit weight of sand is 17.3 kN/m^3. Use Eq. (18.11).

Depth, z (m)	N_{60}
1.5	8
3.0	7
4.5	12
6.0	14
7.5	13

Solution
From Eq. (18.11)

$$C_N = \left[\frac{1}{\left(\dfrac{\sigma'_o}{p_a}\right)}\right]^{0.5}$$

$$p_a \approx 100 \text{ kN/m}^2.$$

Now the following table can be prepared.

Depth, z (m)	σ'_o (kN/m²)	C_N	N_{60}	$(N_1)_{60}$
1.5	25.95	1.96	8	≈ 16
3.0	51.90	1.39	7	≈ 10
4.5	77.85	1.13	12	≈ 14
6.0	103.80	0.98	14	≈ 14
7.5	129.75	0.87	13	≈ 11

Example 18.2

Solve Example 18.1 using Eq. (18.13).

Solution
From Eq. (18.13),

$$C_N = \frac{2}{1 + \left(\dfrac{\sigma'_o}{p_a}\right)} \quad (Note: p_a \approx 100 \text{ kN/m}^2)$$

Now we can prepare the following table.

Depth, z (m)	σ'_o (kN/m²)	C_N	N_{60}	$(N_1)_{60}$
1.5	25.95	1.59	8	≈ 13
3.0	51.90	1.32	7	≈ 9
4.5	77.85	1.2	12	≈ 13
6.0	103.80	0.98	14	≈ 14
7.5	129.75	0.87	13	≈ 11

Example 18.3

Refer to Example 18.1. Using Eq. (18.16), estimate the average soil friction angle, ϕ', from $z = 0$ to $z = 7.5$ m.

Solution
From Eq. (18.16),

$$\phi' = \tan^{-1}\left[\frac{N_{60}}{12.2 + 20.3\left(\dfrac{\sigma'_o}{p_a}\right)}\right]^{0.34}$$

$$p_a = 100 \text{ kN/m}^2$$

Now the following table can be prepared.

Depth, z (m)	σ'_o (kN/m²)	N_{60}	ϕ' (deg) [Eq. (18.16)]
1.5	25.95	8	37.5
3.0	51.9	7	33.8
4.5	77.85	12	36.9
6.0	103.8	14	36.7
7.5	129.75	13	34.6

Average $\phi' \approx 36°$

■

18.6   *Other* In Situ *Tests*

Depending on the type of project and the complexity of the subsoil, several types of *in situ* tests can be conducted during the exploration period. In many cases, the soil properties evaluated from the *in situ* tests yield more representative values. This better accuracy results primarily because the sample disturbance during soil exploration is eliminated. Following are some of the common tests that can be conducted in the field.

Vane Shear Test

The principles and the application of the vane shear test were discussed in Chapter 12. When soft clay is encountered during the advancement of a borehole, the undrained shear strength of clay, c_u, can be determined by conducting a vane shear test in the borehole. This test provides valuable information about the strength in undisturbed clay.

Borehole Pressuremeter Test

The pressuremeter is a device that originally was developed by Menard in 1965 for *in situ* measurement of the stress–strain modulus. This device basically consists of a pressure cell and two guard cells (Figure 18.11). The test involves expanding the pressure cell inside a borehole and measuring the expansion of its volume. The test data are interpreted on the basis of the theory of expansion of an infinitely thick cylinder of soil. Figure 18.12 shows the variation of the pressure-cell volume with changes in the cell pressure. In this figure, Zone I represents the reloading portion, during which the soil around the borehole is pushed back to its initial state—that is, the state it was in before drilling. Zone II represents a pseudoelastic zone, in which the cell volume versus cell pressure is practically linear. The zone marked III is the plastic zone. For the pseudoelastic zone,

$$E_s = 2(1 + \mu_s)V_o \frac{\Delta p}{\Delta V} \tag{18.17}$$

where E_s = modulus of elasticity of soil
$\quad \mu_s$ = Poisson's ratio of soil
$\quad V_o$ = cell volume corresponding to pressure p_o (that is, the cell pressure corresponding to the beginning of Zone II)
$\Delta p/\Delta V$ = Slope of straight-line plot of Zone II

Water pressure (for expansion of main cell)

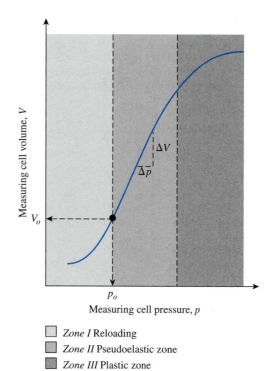

Zone I Reloading
Zone II Pseudoelastic zone
Zone III Plastic zone

Gas pressure (for expansion of guard cell)

Guard cell   Measuring cell

Figure 18.11 Schematic diagram for pressuremeter test

Figure 18.12 Relationship between measuring pressure and measuring volume for Menard pressuremeter

Menard recommended a value of $\mu_s = 0.33$ for use in Eq. 18.17, but other values can be used. With $\mu_s = 0.33$,

$$E_s = 2.66 \, V_o \frac{\Delta p}{\Delta V} \tag{18.18}$$

From the theory of elasticity, the relationship between the modulus of elasticity and the shear modulus can be given as

$$E_s = 2(1 + \mu_s)G_s \tag{18.19}$$

where G_s = shear modulus of soil. Hence, combining Eqs. (18.17) and (18.19) gives

$$G = V_o \frac{\Delta p}{\Delta V} \tag{18.20}$$

Pressuremeter test results can be used to determine the at-rest earth-pressure coefficient, K_o (Chapter 13). This coefficient can be obtained from the ratio of p_o and σ'_o (σ'_o = effective vertical stress at the depth of the test), or

$$K_o = \frac{p_o}{\sigma'_o} \tag{18.21}$$

Note that p_o (see Figure 18.12) represents the *in situ* lateral pressure.

The pressuremeter tests are very sensitive to the conditions of a borehole before the test.

Cone Penetration Test

The Dutch cone penetrometer is a device by which a 60° cone with a base area of 10 cm^2 (Figure 18.13) is pushed into the soil, and the cone end resistance, q_c, to penetration is measured. Most cone penetrometers that are used commonly have friction sleeves that follow the point. This allows independent determination of the cone resistance (q_c) and the frictional resistance (f_c) of the soil above it. The friction sleeves have an exposed surface area of about 150 cm^2.

The penetrometer shown in Figure 18.13 is a *mechanical-friction cone penetrometer*. At the present time, *electrical-friction cone penetrometers* also are used for field investigation.

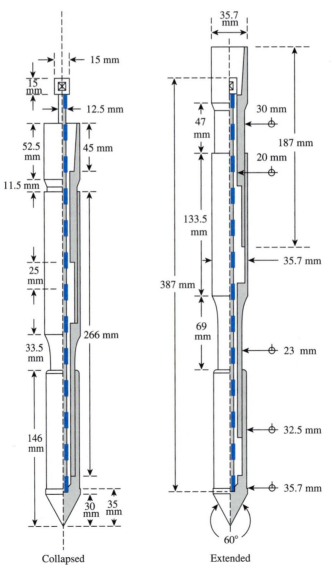

Figure 18.13
Dutch cone penetrometer with friction sleeve (From *Annual Book of ASTM Standards,* 04.08. Copyright © 1991 American Society for Testing and Materials. Reprinted with permission.)

Collapsed   Extended

One of the major advantages of the cone penetration test is that boreholes are not necessary to conduct the test. Unlike the standard penetration test, however, soil samples cannot be recovered for visual observation and laboratory tests.

Robertson and Campanella (1983) provided correlations between the vertical effective stress (σ'_o), drained soil friction angle (ϕ'), and q_c for sand. The relationship between σ'_o, ϕ', and q'_c can be approximated (after Kulhawy and Mayne, 1990) as

$$\phi' = \tan^{-1}\left[0.1 + 0.38 \log\left(\frac{q_c}{\sigma'_o}\right) \right] \tag{18.22}$$

The cone penetration resistance also has been correlated with the equivalent modulus of elasticity, E_s, of soils by various investigators. Schmertmann (1970) gave a simple correlation for sand as

$$E_s = 2q_c \tag{18.23}$$

Trofimenkov (1974) also gave the following correlations for the modulus of elasticity in sand and clay:

$$E_s = 3q_c \quad \text{(for sands)} \tag{18.24}$$

$$E_s = 7q_c \quad \text{(for clays)} \tag{18.25}$$

Correlations such as Eqs. (18.23) through (18.25) can be used in the calculation of the elastic settlement of foundations (Chapter 11).

Anagnostopoulos, *et al.* (2003) have provided several correlations based on a large number of field-test results conducted on a wide variety of soils. The correlations obtained from this study are summarized below.

Case 1. Correlation between Undrained Shear Strength (c_u) and q_c.

$$c_u = \frac{q_c - \sigma_o}{N_k} \tag{18.26}$$

where σ_o = total vertical stress

$\quad N_K$ = bearing capacity factor ($\approx$ 18.3 for all cones)

Consistent units need to be used in Eq. (18.26). The values of c_u in the field tests were equal to or less than about 250 kN/m².

Case 2. Correlation between Sleeve-Frictional Resistance (f_c) and c_u.
For mechanical cones,

$$f_c = 1.26c_u \tag{18.27}$$

For electric cones,

$$f_c = c_u \tag{18.28}$$

Average for all cones

$$f_c = 1.21c_u \tag{18.29}$$

Case 3. Correlation Between q_c and N_{60}.
Based on a large number of field test results on various types of soil with mean grain size (D_{50}) varying from 0.001 mm to about 7 to 8 mm, Anagnostopoulos, *et al.* (2003) provided the following correlation between q_c, N_{60}, and D_{50}.

$$\frac{\left(\dfrac{q_c}{p_a}\right)}{N_{60}} = 7.64 D_{50}^{0.26} \tag{18.30}$$

where p_a = atmospheric pressure (≈ 100 kN/m²)
D_{50} = mean grain size, in mm

Case 4. Correlation Between Friction Ratio (R_f) and D_{50}.
The friction ratio can be defined as

$$R_f = \frac{f_c}{q_c} \tag{18.31}$$

Based on the soils investigated (with D_{50} ranging from 0.001 mm to about 7 to 8 mm), the correlation between R_f and D_{50} can be given as.

$$R_f(\%) = 1.45 - 1.36 \log(D_{50}) \quad \text{(electric cone)} \tag{18.32}$$

and

$$R_f(\%) = 0.7811 - 1.611 \log(D_{50}) \quad \text{(mechanical cone)} \tag{18.33}$$

where D_{50} is in mm.

18.7 Rock Coring

It may be necessary to core rock if bedrock is encountered at a certain depth during drilling. It is always desirable that coring be done for at least 3 m. If the bedrock is weathered or irregular, the coring may need to be extended to a greater depth. For coring, a core barrel is attached to the drilling rod. A coring bit is attached to the bottom of the core barrel. The cutting element in the bit may be diamond, tungsten, or carbide. The coring is advanced by rotary drilling. Water is circulated through the drilling rod during coring, and the cuttings are washed out. Figure 18.14a shows a diagram of rock coring by the use of a single-tube core barrel. Rock cores obtained by such barrels can be fractured because of torsion. To avoid this problem, one can use double-tube core barrels (Figure 18.14b). Table 18.6 gives the details of various types of casings and core barrels, diameters of core barrel bits, and diameters of core samples obtained. The core samples smaller than the BX size tend to break away during coring.

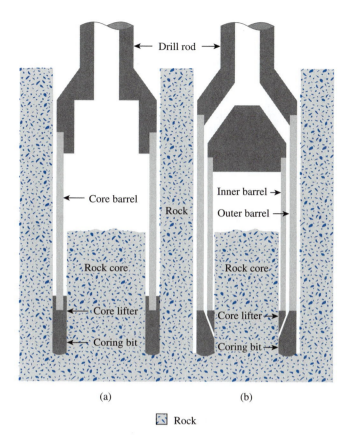

Drill rod

Core barrel

Rock

Rock core

Core lifter

Coring bit

Inner barrel →

Outer barrel →

Rock core

Core lifter ↔

Coring bit ↔

(a)

(b)

⬚ Rock

Figure 18.14
Rock coring: (a) single-tube
core barrel; (b) double-tube
core barrel

Table 18.6 Details of Core Barrel Designations, Bits, and Core Samples

Casing and core barrel designation	Outside diameter of core barrel bit, mm	Diameter of core sample, mm
EX	36.5	22.2
AX	47.6	28.6
BX	58.7	41.3
NX	74.6	54.0

On the basis of the length of the rock core obtained from each run, the following quantities can be obtained for evaluation of the quality of rock.

$$\text{Recovery ratio} = \frac{\text{Length of rock core recovered}}{\text{Length of coring}} \qquad (18.34)$$

$$\text{Rock quality designation } (RQD) = \frac{\begin{array}{c}\Sigma \text{ Length of rock pieces} \\ \text{recovered having lengths of} \\ \text{101.6 mm or more}\end{array}}{\text{Length of coring}} \qquad (18.35)$$

Table 18.7 Qualitative Description of Rocks Based on *RQD*

RQD	Rock quality
1–0.9	Excellent
0.9–0.75	Good
0.75–0.5	Fair
0.5–0.25	Poor
0.25–0	Very poor

A recovery ratio equal to 1 indicates intact rock. However, highly fractured rocks have a recovery ratio of 0.5 or less. Deere (1963) proposed the classification system in Table 18.7 for *in situ* rocks on the basis of their *RQD*.

18.8   *Soil Exploration Report*

At the end of the soil exploration program, the soil and rock samples collected from the field are subjected to visual observation and laboratory tests. Then, a soil exploration report is prepared for use by the planning and design office. Any soil exploration report should contain the following information:

1. Scope of investigation
2. General description of the proposed structure for which the exploration has been conducted
3. Geologic conditions of the site
4. Drainage facilities at the site
5. Details of boring
6. Description of subsoil conditions as determined from the soil and rock samples collected
7. Groundwater table as observed from the boreholes
8. Details of foundation recommendations and alternatives
9. Any anticipated construction problems
10. Limitations of the investigation

The following graphic presentations also need to be attached to the soil exploration report:

1. Site location map
2. Location of borings with respect to the proposed structure
3. Boring logs (Figure 18.15)
4. Laboratory test results
5. Other special presentations

The boring log is the graphic presentation of the details gathered from each borehole. Figure 18.15 shows a typical boring log.

BORING LOG

PROJECT TITLE ____Shopping center_____

LOCATION __Intersection Hill Street and Miner Street_____ DATE __June 7, 1997__

BORING NUMBER __4__ TYPE OF BORING __Hollow-stem auger__ GROUND ELEVATION __40.3 m__

DESCRIPTION OF SOIL	DEPTH (m) AND SAMPLE NUMBER		STANDARD PENETRATION NUMBER, N_{60}	MOISTURE CONTENT, w (%)	COMMENTS
Tan sandy silt		0.3			
- - - - - - - -		0.6			
		0.9			
		1.2			
Light brown silty clay (CL)	SS-1	1.5 / 1.8	13	11	Liquid limit = 32 $PI = 9$
		2.1			
		2.4			
		2.7			
Groundwater table June 14, 1997 ▼	SS-2	3.0 / 3.3	5	24	
		3.6			
		3.9			
		4.0			
Soft clay (CL)	ST-1	4.3 / 4.6	6	28	Liquid limit = 44 $PI = 26$ q_u = unconfined compression strength = 40.7 kN/m²
		4.9			
		5.2			
		5.5			
Compact sand and gravel End of boring @ 6.4 m	SS-3	5.8 / 6.1 / 6.4	32		

Figure 18.15 Typical boring log (*Note:* SS = split-spoon sample; ST = Shelby tube sample)

Problems

18.1 Determine the area ratio of a Shelby tube sampler that has outside and inside diameters of 88.9 mm and 85.7 mm, respectively.

18.2 Repeat Problem 18.1 with outside diameter = 114 mm. and inside diameter = 111 mm.

18.3 The following are the results of a standard penetration test in sand. Determine the corrected standard penetration numbers, $(N_1)_{60}$, at the various depths given. Note that the water table was not observed within a depth of 12 m below the ground surface. Assume that the average unit weight of sand is 15.5 kN/m^3. Use Eq. (18.10).

Depth (m)	N_{60}
1.5	8
3	7
4.5	12
6	14
7.5	13

18.4 Using the values of N_{60} given in Problem 18.3 and Eq. (18.16), estimate the average soil friction angle, ϕ'. Use $p_a \approx 100$ kN/m^2.

18.5 The following are the results of a standard penetration test in dry sand.

Depth (m)	N_{60}
1.5	6
3	12
4.5	17
6	21
7.5	23

For the sand deposit, assume the mean grain size, D_{50}, to be 0.26 mm and the unit weight of sand to be 15.5 kN/m^3. Estimate the variation of relative density with depth. Use Eq. (18.14).

18.6 A boring log in a sandy soil is shown in Figure 18.16. Determine the corrected standard penetration numbers from Eq. (18.13).

18.7 **a.** From the results of Problem 18.3, estimate a design value of $(N_1)_{60}$ (corrected standard penetration number) for the construction of a shallow foundation.

  b. Refer to Eq. (16.47). For a 2-m-square column foundation in plan, what allowable load could the column carry? The bottom of the foundation is to be located at a depth of 1.0 m below the ground surface. The maximum tolerable settlement is 25 mm.

18.8 For the soil profile described in Problem 18.5, estimate the variation of the cone penetration resistance, q_c, with depth. Use Eq. (18.30).

18.9 In a layer of saturated clay, a cone penetration was conducted. At a depth of 10 m below the ground surface, q_c was determined to be 1400 kN/m^2. Estimate the undrained shear strength of the clay. Given: unit weight of saturated clay = 18 kN/m^2.

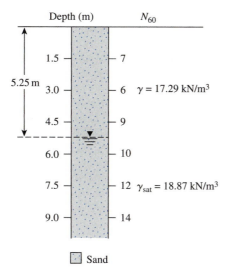

Figure 18.16

18.10 The average cone penetration resistance at a certain depth in a sandy soil is 205 kN/m². Estimate the modulus of elasticity of the soil at that depth.

18.11 During a field exploration program, rock was cored for a length of 1.2 m and the length of the rock core recovered was 0.9 m. Determine the discovery ratio.

References

AMERICAN ASSOCIATION OF STATE HIGHWAY AND TRANSPORTATION OFFICIALS (1967). *Manual of Foundation Investigations,* National Press Building, Washington, D.C.

AMERICAN SOCIETY OF CIVIL ENGINEERS (1972). "Subsurface Investigation for Design and Construction of Foundations of Buildings, Part I," *Journal of the Soil Mechanics and Foundations Division,* ASCE, Vol. 98, No. SM5, 481–490.

AMERICAN SOCIETY FOR TESTING AND MATERIALS (1991). *Annual Book of ASTM Standards,* Vol. 04.08. Philadelphia, Pa.

ANAGNOSTOPOULOS, A., KOUKIS, G., SABATAKAKIS, N., and TSIAMBAOS, G. (2003). "Empirical Correlation of Soil Parameters Based on Cone Penetration Tests (CPT) for Greek Soils," *Geotechnical and Geological Engineering,* Vol. 21, No. 4, 377–387.

CUBRINOVSKI, M., and ISHIHARA, K. (1999). "Empirical Correlation Between SPT N–Value and Relative Density for Sandy Soils," *Soils and Foundations,* Vol. 39, No. 5, 61–92.

DEERE, D. U. (1963). "Technical Description of Rock Cores for Engineering Purposes," *Felsmechanik und Ingenieurgeologie,* Vol. 1, No. 1, 16–22.

DJOENAIDI, W. J. (1985). "A Compendium of Soil Properties and Correlations." Master's thesis, University of Sydney, Australia.

KULHAWY, F. H., and MAYNE, P. W. (1990). *Manual on Estimating Soil Properties for Foundation Design,* Final Report (EL-6800) submitted to Electric Power Research Institute (EPRI), Palo Alto, Calif.

LIAO, S., and WHITMAN, R. V. (1986). "Overburden Correction Factor for SPT in Sand," *Journal of Geotechnical Engineering,* ASCE, Vol. 112, No. 3, 373–377.

MENARD, L. (1965). "Rules for Calculation of Bearing Capacity and Foundation Settlement Based on Pressuremeter Tests," *Proceedings,* 6th International Conference on Soil Mechanics and Foundation Engineering, Montreal, Canada, Vol. 2, 295–299.

OSTERBERG, J. O. (1952). "New Piston-Type Sampler," *Engineering News Record,* April 24.

PECK, R. B., HANSON, W. E., and Thornburn, T. H. (1974). *Foundation Engineering,* 2nd ed., Wiley, New York.

ROBERTSON, P. K., and CAMPANELLA, R. G. (1983). "Interpretation of Cone Penetration Tests. Part I: Sand," *Canadian Geotechnical Journal,* Vol. 20, No. 4, 718–733.

SCHMERTMANN, J. H. (1970). "Static Cone to Compute Static Settlement Over Sand," *Journal of the Soil Mechanics and Foundations Division,* ASCE, Vol. 96, No. SM3, 1011–1043.

SCHMERTMANN, J. H. (1975). "Measurement of *in situ* Shear Strength," *Proceedings,* Specialty Conference on *in situ* Measurement of Soil Properties, ASCE, Vol. 2, 57–138.

SEED, H. B., TOKIMATSU, K., HARDER, L. F., and CHUNG, R. M. (1985). "Influence of SPT Procedures in Soil Liquefaction Resistance Evaluations," *Journal of Geotechnical Engineering,* ASCE, Vol. 111, No. 12, 1426–1445.

SKEMPTON, A. W. (1986). "Standard Penetration Test Procedures and the Effect in Sands of Overburden Pressure, Relative Density, Particle Size, Aging and Overconsolidation," *Geotechnique,* Vol. 36, No. 3, 425–447.

SOWERS, G. B., and SOWERS, G. F. (1970). *Introductory Soil Mechanics and Foundations,* Macmillan, New York.

SZECHY, K., and VARGA, L., (1978). *Foundation Engineering—Soil Exploration and Spread Foundation,* Akademiai Kiado, Budapest Hungary.

TROFIMENKOV, J. G. (1974). "General Reports: Eastern Europe," *Proceedings,* European Symposium of Penetration Testing, Stockholm, Sweden, Vol. 2.1, 24–39.

WOLFF, T. F. (1989). "Pile Capacity Prediction Using Parameter Functions," in *Predicted and Observed Axial Behavior of Piles, Results of a Pile Prediction Symposium,* sponsored by Geotechnical Engineering Division, ASCE, Evanston, Ill., June 1989, ASCE Geotechnical Special Publication No. 23, 96–106.

Answers to Selected Problems

Chapter 2

2.1 $C_u = 5.13$; $C_c = 1.48$

2.3 $C_u = 4.33$; $C_c = 0.73$

2.5   **b.** $D_{10} = 0.11$ mm; $D_{30} = 0.17$ mm; $D_{60} = 0.3$ mm

      c.   2.73

      d.   0.88

2.7   **b.** $D_{10} = 0.23$ mm; $D_{30} = 0.33$ mm; $D_{60} = 0.48$ mm

      c.   2.09

      d.   0.99

2.9   Gravel = 0%; Sand = 6%; Silt = 52%; Clay = 42%

2.11   0.0052 mm

Chapter 3

3.5 $w = 15.6\%$

      $\gamma = 18.99$ kN/m^3

      $\gamma_d = 16.43$ kN/m^3

      $e = 0.59$

      $n = 0.37$

      $S = 70.6\%$

3.7   **a.**   16.91 kN/m^3

      b.   2.44

      c.   0.417

3.9   **a.**   15.89 kN/m^3

      b.   0.648

      c.   0.39

      d.   44.5%

3.11   **a.**   1423.7 kg/m^3

      b.   0.479

      c.   53.5%

      d.   222 kg/m^3

3.13   **a.**   20.59 kN/m^3

      b.   1.99%

3.15 **a.** 18.46 kN/m^3
      b. 15.19 kN/m^3
      c. 77.7%
3.17 **a.** 0.65
      b. 19.92 kN/m^3
3.19 0.311 kN
3.21 18.8 kN/m^3
3.23 **a.** 74.6%
      b. 17.47 kN/m^3

Chapter 4

4.1 **a.** 28.5
     b. 16.3
4.3 **a.** 39.7
     b. 21
4.5 $SL = 19.4\%$
     $SR = 1.85$

Chapter 5

5.1

Soil	Classification
A	Clay
B	Sandy clay
C	Loam
D	Sandy clay and sandy clay loam (borderline)
E	Sandy Loam

5.3

Soil	Symbol	Group name
1	SP	Poorly graded sand
2	MH	Elastic silt with sand
3	CH	Fat clay
4	SC	Clayey sand
5	SC	Clayey sand
6	GM-GC	Silty clayey gravel with sand
7	CH	Fat clay with sand

Chapter 6

6.1

w (%)	γ_{zav} (kN/m^3)
5	23.72
8	22.11
10	21.16
12	20.29
15	19.10

6.3

w (%)	ρ_d @ S (%) (kg/m³)		
	80	90	100
10	2002	2059	2107
20	1601	1676	1741

6.5 $e = 0.358$
$S = 94\%$

6.7   89.9%

6.9   Pit B

6.11 $\rho_d = 1626.4 \text{ kg/m}^3$
$R = 96.7\%$

6.13   15.94 kN/m³

6.15   20.65; Rating: Fair

Chapter 7

7.1   0.056 cm/sec

7.3 $h = 43.03$ cm
$v = 0.0212$ cm/sec

7.5   0.31 cm²

7.7   0.207 m³/min

7.9 7.18×10^{-5} m³/sec/m

7.11   17.38 cm/min

7.13   0.0354 cm/sec

7.15 1.52×10^{-6} cm/sec

7.17   36.8

Chapter 8

8.1   0.009 cm/sec

8.3   0.518 m³/m/day

8.5   7.2 m³/day/m

8.7   2.1 m³/m/day

8.9   0.271 m³/m/day

Chapter 9

9.1

Point	kN/m²		
	σ	u	σ'
A	0	0	0
B	26.4	0	26.4
C	60.93	17.95	42.98
D	108.88	41.89	66.99

9.3

Point	σ	u	σ'
	kN/m²		
A	0	0	0
B	45	0	45
C	109	39.24	69.76
D	199	88.29	110.71

9.5

Point	σ	u	σ'
	kN/m²		
A	0	0	0
B	65	0	65
C	126.95	29.43	97.52
D	153.94	44.158	109.79

9.7 4.15 m

9.9 0.019 m³/min

9.11 4.88

9.13

Depth (m)	σ	u	σ'
	kN/m²		
0	0	0	0
5	87	0	87
		−19.13	106.13
8	140.76	0	140.76
11.5	205.37	34.34	171.03

Chapter 10

10.1 $\sigma_1 = 129.24$ kN/m²
$\sigma_3 = 30.76$ kN/m²
$\sigma_n = 51.03$ kN/m²
$\tau_n = 39.82$ kN/m²

10.3 $\sigma_1 = 68.9$ kN/m²
$\sigma_3 = 161.1$ kN/m²
$\sigma_n = 138.5$ kN/m²
$\tau_n = 39.7$ kN/m²

10.5 $\sigma_1 = 95$ kN/m²
$\sigma_3 = 30$ kN/m²
$\sigma_n = 94.2$ kN/m²
$\tau_n = 7.1$ kN/m²

10.7 1 kN/m²

10.9 15.32 kN/m²

10.11 674 kN/m

10.13 24.54 kN/m^2
10.15 @ *A*: 211.95 kN/m^2
@ *B*: 203.8 kN/m^2
@ *C*: 18.81 kN/m^2

10.17

γ (m)	$\Delta\sigma_z$ (kN/m^2)
0	87.4
0.6	86.4
1.2	83.5
2.4	62.04
3.6	19.9

10.19 **a.** 20 kN/m^2
b. 46.7 kN/m^2
c. 6.59 kN/m^2

Chapter 11

11.1 10.9 mm
11.3 **b.** 310 kN/m^2
c. 0.53
11.5 **b.** 47 kN/m^2
c. 0.136
11.7 229 mm
11.9 0.596
11.11 159.6 days
11.13 **a.** 0.00034 m^2/kN
b. 6.67 × 10^{-8} cm/sec
11.15 **a.** 230.1 days
b. 186 mm
11.17 66.7 days
11.19 **a.** 31.25%
b. 102.6 days
c. 25.65 days

Chapter 12

12.1 **a.** 41.2°
b. 0.739 kN
12.3 37.5°
12.5 138 kN/m^2
12.7 473.5 kN/m^2
12.9 127.7 kN/m^2
12.11 **a.** 20.8°
b. 55.4°
c. $\sigma' = 338.7$ kN/m^2
$\tau = 128.7$ kN/m^2

12.13 496.8 kN/m^2

12.15 $\phi = 15.6°$

$\phi' = 24.4°$

12.17 **a.** 18°

b. 64.9 kN/m^2

12.19 $\Delta\sigma_{d(f)} = 124$ kN/m^2

$\Delta u_{d(f)} = 48.37$ kN/m^2

12.21 −93.5 kN/m^2

12.23 16.24 kN/m^2

Chapter 13

13.1 $P_o = 281.55$ kN/m; $\bar{z} = 2.33$ m

13.3 $P_o = 205.7$ kN/m; $\bar{z} = 2$ m

13.5 $P_a = 57.41$ kN/m; $\bar{z} = 152$ m

13.7 $P_a = 37.44$ kN/m; $\bar{z} = 1.33$ m

13.9 $P_a = 182.05$ kN/m; $\bar{z} = 0.81$ m

13.11 $P_p = 645.8$ kN/m; $\bar{z} = 1.67$ m

13.13 $P_a = 34.31$ kN/m; $\bar{z} = 0.89$ m

13.15 $P_a = 141.1$ kN/m; $\bar{z} = 2.04$ m

13.17 402.6 kN/m. P_p acts at 1.33 m from the bottom of the wall inclined at an angle $\beta = -6.64°$.

13.19 **b.** 1.75 m

c. 47.01 kN/m

d. $P_a = 76.34$ kN/m; $\bar{z} = 0.94$ m

13.21 10.02 kN/m

13.23 Part 1: 85.14 kN/m

Part 2: 84.4 kN/m

13.25 68.3 kN/m

Chapter 14

14.1 1533 kN/m

14.3 861.8 kN/m

14.5 710.2 kN/m

14.7 267.2 kN/m

Chapter 15

15.1 1.63 m

15.3 0.98

15.5 31.53 m

15.7 1.76

15.9 9.19 m; toe failure

15.11 3.86 m

15.13 **a.** 27.5 kN/m^2

b. Mid-point circle

c. 5.95 m

15.15 0.58
15.17 1.5
15.19 1.75
15.21 0.75
15.23 1.09

Chapter 16

16.1 353.5 kN/m^2
16.3 71 kN/m^2
16.5 138.7 kN/m^2
16.7 657.2 kN
16.9 1.25 m
16.11 897.7 kN
16.13 896.67 kN

Chapter 18

18.1 7.6%

18.3

Depth (m)	$(N_1)_{60}$
1.5	16
3.0	10
4.5	14
6.0	14
7.5	12

18.5

Depth (m)	D_r (%)
1.5	87
3.0	87
4.5	84
6.0	81
7.5	76

18.7 211.8 kN/m^2
18.9 66.7 kN/m^2
18.11 75%

Index

CONVERSION FACTORS FROM SI TO ENGLISH UNITS

Length:		
1 m	=	3.281 ft
1 cm	=	3.281×10^{-2} ft
1 mm	=	3.281×10^{-3} ft
1 m	=	39.37 in.
1 cm	=	0.3937 in.
1 mm	=	0.03937 in.

Area:		
1 m^2	=	10.764 ft^2
1 cm^2	=	10.764×10^{-4} ft^2
1 mm^2	=	10.764×10^{-6} ft^2
1 m^2	=	1550 in.^{2}
1 cm^2	=	0.155 in.^{2}
1 mm^2	=	0.155×10^{-2} in.^{2}

Volume:		
1 m^3	=	35.32 ft^3
1 cm^3	=	35.32×10^{-4} ft^3
1 m^3	=	61,023.4 in.^{3}
1 cm^3	=	0.061023 in.^{3}

Force:		
1 N	=	0.2248 lb
1 kN	=	224.8 lb
1 kgf	=	2.2046 lb
1 kN	=	0.2248 kip
1 kN	=	0.1124 U.S. ton
1 metric ton	=	2204.6 lb
1 N/m	=	0.0685 lb/ft

Stress:		
1 N/m^2	=	20.885×10^{-3} lb/ft^2
1 kN/m^2	=	20.885 lb/ft^2
1 kN/m^2	=	0.01044 U.S. ton/ft^2
1 kN/m^2	=	20.885×10^{-3} kip/ft^2
1 kN/m^2	=	0.145 lb/in.^{2}

Unit weight:		
1 kN/m^3	=	6.361 lb/ft^3
1 kN/m^3	=	0.003682 lb/in.^{3}

Moment:		
1 N $\cdot$ m	=	0.7375 lb-ft
1 N $\cdot$ m	=	8.851 lb-in.

Energy:		
1 J	=	0.7375 ft-lb

Moment of inertia:		
1 mm^4	=	2.402×10^{-6} in.^{4}
1 m^4	=	2.402×10^{6} in.^{4}

Section modulus:		
1 mm^3	=	6.102×10^{-5} in.^{3}
1 m^3	=	6.102×10^{4} in.^{3}

Hydraulic conductivity:		
1 m/min	=	3.281 ft/min
1 cm/min	=	0.03281 ft/min
1 mm/min	=	0.003281 ft/min
1 m/sec	=	3.281 ft/sec
1 mm/sec	=	0.03281 ft/sec
1 m/min	=	39.37 in./min
1 cm/sec	=	0.3937 in./sec
1 mm/sec	=	0.03937 in./sec

Coefficient of consolidation:		
1 cm^2/sec	=	0.155 in.^{2}/sec
1 m^2/yr	=	4.915×10^{-5} in.^{2}/sec
1 cm^2/sec	=	1.0764×10^{-3} ft^2/sec